Proceedings
of the 5th International Yellow River Forum on Ensuring Water Right of the River's Demand and Healthy River Basin Maintenance

Volume Ⅰ

Yellow River Conservancy Press

图书在版编目(CIP)数据

第五届黄河国际论坛论文集/尚宏琦,骆向新主编. —郑州：
黄河水利出版社,2015.9
ISBN 978 – 7 – 5509 – 0399 – 9

Ⅰ.①第⋯　Ⅱ.①尚⋯　②骆⋯　Ⅲ.①黄河 – 河道整治 –
国际学术会议 – 文集　Ⅳ.①TV882.1 – 53

中国版本图书馆 CIP 数据核字(2012)第 314288 号

出　版　社:黄河水利出版社
　　　　地址:河南省郑州市顺河路黄委会综合楼 14 层　　　　邮政编码:450003
发行单位:黄河水利出版社
　　　　发行部电话:0371 – 66026940、66020550、66028024、66022620(传真)
　　　　E-mail:hhslcbs@126.com
承印单位:河南省瑞光印务股份有限公司
开本:787 mm × 1 092 mm　1/16
印张:149.75
印数:1—1 000
版次:2015 年 9 月第 1 版　　　　　　印次:2015 年 9 月第 1 次印刷

定价(全五册):960.00 元(US $155.00)

Under the Auspices of

Ministry of Water Resources, People's Republic of China

Sponsored & Hosted by

Yellow River Conservancy Commission(YRCC), Ministry of Water Resources, P. R. China

China Yellow River Foundation(CYRF)

Editing Committee of Proceedings of the 5th International Yellow River Forum on Ensuring Water Right of the River's Demand and Healthy River Basin Maintenance

Welcome

(preface)

The 5th International Yellow River Forum (IYRF) is sponsored by Yellow River Conservancy Commission (YRCC) and China Yellow River Foundation (CYRF). On behalf of the Organizing Committee of the conference, I warmly welcome you from over the world to Zhengzhou to attend the 5th IYRF. I sincerely appreciate the valuable contributions of all the delegates.

As an international academic conference, IYRF aims to set up a platform of wide exchange and cooperation for global experts, scholars, managers and stakeholders in water and related fields. Since the initiation in 2003, IYRF has been hosted for four times successfully, which shows new concepts and achievements of the Yellow River management and water management in China, demonstrates the new scientific results in nowadays world water and related fields, and promotes water knowledge sharing and cooperation in the world.

The central theme of the 5th IYRF is "Ensuring water right of the river's demand and healthy river basin maintenance". The Organizing Committee of the 5th IYRF has received near one thousand paper abstracts. Reviewed by the Technical Committee, part of the abstracts are finally collected into the Technical Paper Abstracts of the 5th IYRF.

An ambience of collaboration, respect, and innovation will once again define the forum environment, as experts researchers, representatives from national and local governments, international organizations, universities, research institutions and civil communities gather to discuss, express and listen to the opportunities, challenges and solutions to ensure the sustainable water resources management.

We appreciate the generous supports from the co – sponsors, including domestic and abroad governments and organizations. We also would like to thank the members of the Organizing Committee and the Technical Committee for their great supports and the hard work of the secretariat, as well as all the experts and authors for their outstanding contributions to the 5th IYRF.

Finally, I would like to present my best wishes to the success of the 5th IYRF, and hope every participant to have a good memory about the forum!

Chen Xiaojiang
Chairman of the Organizing Committee, IYRF
Commissioner of YRCC, MWR, China
Zhengzhou, September 2012

Contents

Main Themes

A. Advanced River Basin and Water Resources Management with Social and Economic Development

B. Ensuring Water Right of the River's Demand with Strategy and Measures of Keeping Healthy River

C. Strict and Efficient Management of Water Resources in the River Basin

4

Main Themes

Sustainable Yellow River Water Resources Management

Chen Xiaojiang

Yellow River Conservancy Commission of Ministry of Water Resources, Zhengzhou,450003, China

The Yellow River is the mother river of China, and has cradled the glorious Chinese civilization for five thousand years. Since the founding of P. R. China, under the leadership of the Chinese Government, the Yellow River management and development has made remarkable achievements. In the process of building a well – off society and national modernization, the global and strategic position of the Yellow River management work become even more prominent.

After the turn of this century, based on the scientific judge of the complicated situation of water resources as well as the social and economic development trend in China, the Chinese Government came up with a new concept of water management, that is ' People – oriented, harmonious coexistence between human beings and water, the sustainable utilization of water resources for supporting sustainable development of economic and society '. The Yellow River Conservancy Commission took active explores and practices about integrated water resources management. As a consequence, we successfully coped with issues such as drought, water shortage, eco – system deterioration and so forth, and achieved no zero – flow for continuous 13 years. All these significantly contributed to the social and economic development of the basin and related regions.

The sustainable development of river basin and water rights of rivers are the topics received worldwide concern. Thus, I would like to introduce some practice and experiences of integrated water resources management in the Yellow River basin.

1   To implement available water supply allocation and guarantee the water rights of multiple parties

The Yellow River is the first river in China that conducts water allocation in the basin scale. In 1987, the Chinese Government approved the available water supply allocation scheme of the Yellow River. According to this scheme, in the normal year, the annual runoff of the Yellow River is 5.8×10^{10} m^3. Of all the water resources, 3.7×10^{10} m^3 is allocated for water demand of the 11 provinces (regions, cities), while the remaining 2.1×10^{10} m^3 is allocated for sediment transport and ecological base flow within the river channel. The scheme considers the water demand of both economic development and the river itself, and it provides fundamental support for water resources exploration and management of the whole river basin.

2   To implement integrated water resources management and regulation, and maximally satisfy the water demands of different stakeholders

In 1999, the Yellow River Conservancy Commission was authorized by the national government to implement the integrated water resources management and regulation in the Yellow River basin. In 2002, the water resources management system with the incorporation of river basin and region was clearly defined in the newly issued Water Law. In 2006, Regulations of the Yellow River Water Regulation was issued by the national government. This is the first national administrative regulations to regulate water dispatching of large rivers, clears the responsibility and authority of basin and regional management and scheduling, and establishes a new mode for water resources management of the Yellow River, namely, 'the national authority is responsible for water allocation; river basin authority is responsible for implementation of the water allocation scheme; and the provinces (autonomous regions) are responsible for water utilization and distribution, double control of the total water use and section flow, integrated regulation of important water intakes and reservoirs'. In terms of the benefit of the ten years water resources regulation from 1999 ~ 2000, in the river basin and related regions, the accumulative increment of GDP is 3.504×10^{11} Yuan and that of grain production is 3.72×10^7 t.

3 To probe into water rights system and promote water – saving society construction

Strictly implement water extraction license system and water resources argumentation system for construction project, take control on the total water consumption and enhance water use efficiency. In 2003, pilot projects about water right transfer were carried out in Ningxia and Inner Mongolia. In these projects, the new industrial projects invested in improving the irrigation constructions. And the water saved in the irrigation was bought by the new industrial projects. The benefits of water transfer are remarkable. Up to now, 39 water right transfer projects has been approved in Ningxia and Inner Mongolia, the total transfer of water is 3.37×10^8 m^3, and the investment of water – saving projects is 2.5×10^9 Yuan, the average investment for per cubic meter of water is 7.46 Yuan. Through water right transfer, water resources are moved effectively towards industries with high water use efficiency. What is more, under the premise of no increment in total water use, it can satisfy the newly increased water demand of social – economical development, and promote the adjustment of industrial structure and transformation of economic development mode.

4 To strengthen water resources protection and obviously improve the water ecological environment

We have built the water conservation planning system of the basin, defined 346 important water function zones which have been approved by the State Council, checked and ratified the pollutant carrying capacity, and made the control requirements of the total pollutants discharged into rivers in different level years. We focused the 56 buffer zones at the provincial boundaries, strengthened the management of water function zones in accordance with the law, and had completed the verification of nearly 2,000 sewage outlets along the river. We also improved the emergency response mechanism for sudden water pollution incident, and have successfully dealt with a number of water pollution incidents. We sought to improve the overall capability of water quality monitoring and the construction of monitoring network system. As a result, the monitoring coverage of the important water function zone has been gradually increased, and 75 provincial sections have been achieved full coverage of water quality monitoring.

Currently, the water quality of the Yellow River main stream was significantly improved, and the deterioration trend of estuary ecosystem has been effectively curbed. In main stream, the reach length with water quality of Ⅰ ~ Ⅲ category accounts for 85% of the total length, increased by 15% in comparison with that in 2000. In addition, the area of estuarine delta wetlands increased over 250 km^2, biodiversity is significantly improved, and compared with 1990, bird species in the Yellow River Delta National Nature Reserve increased by 109 kinds.

After all these efforts, we have gained considerable achievements for mitigating the contradiction between supply and demand of the Yellow River water resources, as well as providing powerful support for sustainable social – economic development. However, the essential characteristic of the Yellow River, namely the water resources shortage, has remained unchanged. In the future, with the increase of water demand, the contradiction between water supply and demand is likely to be more serious, and the mission of water resources management and protection would be more complicated and arduous.

The Chinese Government attaches great importance to the issues of water. In 2011, a water conservancy conference with the highest level was held by the central government. In this conference, significant strategic deployment about speeding up the reform in water conservancy section was put forward. Meanwhile, the water conservancy was raised to the strategic height and it is related to security of flood control, water supply, provisions supply, economy, ecology and the whole nation. In addition, the strict water resources management was taken as the strategic measure to accelerate the transformation of economic development mode in the conference. In this January, the state council issued *the opinions about the-most-strict implementation of the water resources management system.*

The nature of the – most – strict water resources management system is to promote sustainable use of the water resources and economic and social sustainable development in the basin, while the core of which is the implementation of the "three red lines", namely: the water resources utilization total quantity control, water resources utilization efficiency control and the effluent sewage volume control of water function zone. Meanwhile, we should make full use of the water resources functions such as basic, constraint, controlling and guiding; establish the ' bottom – up mechanism' of water resources saving and protection. These are beneficial for the basin economic structure adjustment and development mode transformation, as well as for the harmony among social – economic development, water resources and water environment carrying capacity, so as to achieve the harmony, sustainability, co – development and mutual promotion between the water resources utilization and social – economic development.

Therefore, we will take the following measures.

4. 1　Strengthen the top – level design

To achieve constantly improving the planning system of management and development of the Yellow River, we should carry out review and formulation of the comprehensive and specialized planning, for instance, river basin comprehensive planning, river basin (regions) planning of Long-term water supply and demand, basin irrigation development planning, river water resources protection planning. Meanwhile, it is essential to strengthen guidance and restraint function of the planning, so that the overall layout of the Yellow River management and development, as well as the optimization of water resources allocation, conservation and protection of water resources, are consistent with the national major function oriented zoning, and are also in agreement with population, resources, environment, with the economic – social development in the basin.

4. 2　Strengthen the system construction.

The systems to be constructed or reinforced include the following aspects.

Firstly, further establish the – most – strict water resource management system, such as resources argumentation for all the construction projects, water extraction license and pollutant discharge outlet approval. It is also necessary to carry out the planning water resources argumentation to impose strict constraints on the construction of high water consumption and heavy pollution projects in water shortage areas and ecological fragile areas, comprehensively promote the construction of water – saving and anti – pollution social construction.

What is more, for the regions where water resources utilization and total amount of pollutant discharge has exceeded the red line, the further approval for more water and pollutant discharge will be limited. In these areas, water right transfer is recommended to get the water licenses for the new projects.

4. 3　Strengthen supervision and coordination.

To form a powerful force to implement the most strict water resources management, we will seek to build up a water resources management system and mechanism that engaging the basin and regional organizations, and also establish the water resources consultation mechanism that involving all the stakeholders. As a constraint index, the "three red lines" will be included in the economic and social development comprehensive evaluation index system of the provinces (autonomous regions) along the Yellow River. In addition, it is of great importance to perfect the public participation mechanism, and regularly publish the implementation progress of "three red lines", so as to develop a supervision mechanism with wide social participation.

4. 4　Strengthen ability construction

It is required to enforce the fundamental capability building for the most strict water resources management and constantly enhance the water quality monitoring network of water use sections and provincial sections. Within the next three years, we will strive to equip all the provincial sections and 90% of all water use amount in the main stream with the water quality and quantity monitoring facilities. Based on the good job of water allocation on the tributaries, the scope of water monitoring should be gradually expanded. Moreover, we should further improve our capability for water resources monitoring and management, through improving the water dispatching and management system and the decision support system for water resource conservation.

Water is an origin of life, an essential element for production, and a basis for ecology. It is an inevitable choice for sustainable development and river basin water rights protection to implement of the most strict water resources management system, and it needs the co – participation and Long-term efforts of the whole society. We would like to cooperate with the colleagues from all over the world, to share some successful experiences, to crack the tough issues we are facing, and to contribute to the sustainable economic – social development and improvement of water ecological environment.

A World Pact for Better Basin Management

Jean-François DONZIER
The International Network of Basin Organizations (INBO)

The International Office for Water
21, rue de Madrid – 75008 Paris (France)

Abstract: Initiated by the International Network of Basin Organizations (INBO), its Regional Networks in Africa, America, Asia, Europe and the Mediterranean, and 12 French metropolitan and overseas Basin Committees, this "World Pact" was signed on Friday, 16 March 2012 in Marseilles by 69 organizations from all continents, the basins of which are concerning 33 countries.

The signatories commit themselves to apply in their respective basins the management principles recognized as the most relevant and most effective using the field experience acquired by the INBO Member Organizations for over 18 years.

Today, more than 100 organizations worldwide have still signed the "Pact", in particular during the last RIO + 20 UN Conference!

Key words: river basin management, world pact, river basin organizations, good governance

1 Introduction

Initiated by the International Network of Basin Organizations (INBO), its Regional Networks in Africa, America, Asia, Europe and the Mediterranean, and 12 French metropolitan and overseas Basin Committees, this "World Pact" was signed on Friday, 16 March 2012 in Marseilles by 69 organizations, participating in the 6th WWF6, coming from all continents, the basins of which are concerning 33 countries.

The signatories commit themselves to apply in their respective basins the management principles recognized as the most relevant and most effective using the field experience acquired by the INBO Member Organizations for over 18 years.

They affirm as a prerequisite that the basins of rivers, lakes and aquifers, whether local, national or transboundary, are the suited areas in which organize the joint management of water resources, aquatic ecosystems and all water-related activities in order to cope with the global changes related to rapid world population growth, migration, increasing urbanization, climate change ....

They express their will to commit themselves alongside their national governments and international institutions for:

(1) Improving water governance, facilitating the creation of Basin Organizations where they do not exist yet, strengthening existing organizations;

(2) Organizing dialogue with the stakeholders recognized in their basins;

(3) Based on prior assessment, facilitating the agreement of the various stakeholders on a "shared vision of the future of their basin" and developing, through dialogue and transparency, Management Plans for setting out the goals to be achieved in the medium and long term;

(4) Developing successive action and investment plans that meet the economic, social and environmental priorities of the basins;

(5) Making better use of water and ensuring low consumption of this scarce resource by better control of the demand, while encouraging more efficient uses, and according to the case, the use of non-conventional resources;

(6) Better taking into account the significance of ecosystems and of the services they provide in making decisions for the development and management of river basins;

(7) Mobilizing the financial resources needed for carrying out governance reforms, ensuring a long-term good basin management and implementing the action and investment plans needed and

ensuring their lasting operation;

(8) Organizing in each basin a harmonized data collection as part of Integrated Information Systems;

(9) Supporting initiatives of Regional Cooperation Institutions for harmonizing policies and legislation in the field of water and for developing joint action plans at the basin level;

(10) Strengthening institutional and technical cooperation with counterpart Basin Organizations in their region or other parts of the world;

(11) Organizing better liaison with Research Organizations to better focus their work on the priority aspects of basin management and rapidly disseminate their findings in the field.

They also committed themselves to promote the "Pact" to other Basin Organizations that were not able to come to Marseilles for inviting them to join quickly and also becoming signatories.

Among the 69 signatories, there were African Transboundary Basin Organizations, and, for example, the Basin Committees of Quebec and the Brazilian Basin Committees represented by their national Associations, as well as the pilot Basin Organizations of Cambodia, Laos and Vietnam in the Mekong River Basin, etc.

The commitment also provides for the establishment of a symbolic basin passport to reinforce the feeling of citizenship in their river basin.

Today, and in particular during the last RIO + 20 UN Conference, more than 100 organizations worldwide have still signed the "Pact"!

Every interested organization around the Wold is invited to join us by signing the "Pact".

2 The world pact for better basin management

We, representatives of river, lake or aquifer basin organizations, from different parts of the world, subscribe to the present "World Pact for better basin management" in view of the development of integrated and joint water resources management at national, regional and transboundary level to meet the challenges facing our planet.

Indeed, we must achieve the Millennium Development Goals and ensure "green growth" and face the global changes associated with the rapid world population growth, migration, increasing urbanization, climate change, etc.

Our efforts should indeed allow fighting against natural disasters, reliably meeting the drinking water needs of urban and rural populations to improve hygiene and health and prevent epidemics, securing food sufficiency, developing industry, energy production, waterways transport, tourism and recreational activities, preventing and controlling pollution of all kinds to preserve aquatic ecosystems, support fish production and more generally preserve the biodiversity of water-related environments.

All these stakes cannot be tackled on a sectoral or local basis, or separately from each other. The search for solutions must instead involve all stakeholders in an integrated and joint approach, organized in cooperation with the river basin units and for the sustainable use of water resources.

2.1 Part 1: Prior Declaration

Through our commitment to this Pact, we recognize that:

(1) It is becoming imperative to introduce and/or permanently strengthen new forms of governance of water resources, such as those already recommended at Dublin (1991), Rio (1992), Paris (1998), The Hague (2000), Johannesburg (2002), Kyoto (2003), Mexico (2006) and Istanbul (2009) in particular;

(2) River, lake and aquifer basins are the relevant territory for the organization of joint management of water resources, aquatic ecosystems, and all water-related activities;

(3) The different ecosystems found in river basins are very important both for biodiversity and environmental services, including for the regulation of the hydrological cycles and risk prevention, as well as for the treatment of pollution;

(4) The basins of transboundary rivers, lakes and aquifers are to be paid special attention and

be jointly managed by the riparian Countries;

(5) The establishment and strengthening of basin organizations in best suited forms, especially international commissions, authorities or other transboundary basin organizations, facilitate dialogue, cooperation, information exchange and implementation of joint projects and actions, for sharing benefits, anticipating the future and preventing potential conflicts between the stakeholders concerned;

(6) It is necessary to increase regional integration by harmonizing policies and laws and by implementing the regional programmes of common interest needed to improve surface and groundwater resources management at basin level;

(7) It is necessary to create or strengthen the funding dedicated to the management of water resources and aquatic environments and generally of the "great water cycle";

(8) It is useful to develop or strengthen federating frameworks for facilitating bilateral or multilateral initiatives in this strategic field of basin management;

(9) The stakeholders of the civil society and the local authorities should be better associated and involved in the management of the basins where they live;

(10) It is necessary to increase cooperation among basin organizations around the world and in each region to facilitate the sharing of experience and know-how on best practices in river basin management and their adaptation to different contexts.

2.2　Part 2: Commitments of Basin Organizations

Recognizing the need of urgent actions, We, representatives of the organizations, signatories of the "World Pact for better basin management", express our will, in what comes under our statutory powers and within the limits of our own resources, to commit ourselves alongside our national governments and international institutions for:

(1) Acting to improve water governance, facilitating the creation of basin organizations where they do not exist, strengthening existing organizations, helping the concerned authorities develop useful reforms and policies for sustainable water management and the programmes needed for their implementation in the field;

(2) Supporting processes of sustainable, integrated, joint and participative management of water resources and environments organized on the appropriate scale of local, national or transboundary basins according to the case;

(3) Organizing dialogue with the stakeholders recognized in our basins and their effective participation, to achieve a truly shared vision of the future, to identify the necessary agreements on priorities and the resources to mobilize, coordinate initiatives and projects, analyze the results;

(4) Based on a prior assessment, facilitating the agreement of the various stakeholders on a "shared vision" of the future of their basin and developing, through dialogue and transparency, management plans or basin master plans for setting out the goals to be achieved in the medium and long term;

(5) Developing successive action and investment plans that meet the economic, social and environmental priorities of the basins, set out in the management plans, and establishing mechanisms for evaluating their results while using suited performance indicators;

(6) Making better use of water and ensure low consumption of this scarce resource by better control of the demand, encouraging more efficient uses, and according to the case, the use of unconventional resources, the reuse of treated wastewater or artificial recharge of aquifers for sustainable development in particular;

(7) Better taking into account the significance of ecosystems and of their services in planning decisions for the development and management of our river basins;

(8) Implementing priority actions especially needed in drinking water supply, sanitation, health, energy, agriculture and fishing, waterways transport, protection against risks and biodiversity conservation, thus contributing to sustainable development and poverty alleviation;

(9) Mobilizing the financial resources in an adequate manner, using mechanisms for guaranteeing their sustainability, for carrying out these governance reforms, ensuring a long-term good ba-

sin governance and implementing the stakeholders' action and investment plans needed and ensuring their lasting operation;

(10) Organizing in each basin, in cooperation with the major data producers and managers, harmonized data collection as part of Integrated Information Systems, which are permanent, reliable, representative, inter-operable and easily accessible, allowing a precise vision of the encountered situations and their evolution;

(11) Supporting initiatives of regional cooperation institutions for harmonizing policies and legislation in the field of water and for developing and implementing joint action plans particularly at the basin level, and monitoring legal changes.

We wish that donors would recognize the essential role of basin organizations in sustainable water resources management, by financially supporting projects related to the above commitments and principles.

In order to fulfil our commitment, we will also take the necessary actions to:

(1) Strengthen institutional and technical cooperation with counterpart basin organizations in our region or other parts of the world, particularly within the existing INBO networks to quickly disseminate best practices, jointly develop innovative solutions and ensure the required training of the various stakeholders involved;

(2) Organize better liaison with research organizations to better focus their work on the priority aspects of basin management and rapidly disseminate their practical findings.

We are fully committed to report, on the occasion of the next World Water Forum in 2015, our actions and to share the progress made by our basin organizations.

We are also committed to promote this Pact to the other basins organizations and extend them the invitation to join us and therefore becoming signatories to the same as well.

Come and sign the "Pact" with us!!

Yellow River Basin Water Resources Protection and Rivers & Lakes Health Security

Liao Yiwei[1] and *Si Yiming*[2]

1. Yellow River Conservancy Commission, Zhengzhou, 450003, China
2. Yellow River Basin Water Resources Protection Bureau, Zhengzhou, 450004, China

Abstract: It is one of the four goals raised by "The CPC (Communist Party of China) Central Committee and State Council decision 2011 on speeding up the development of water conservancy reform" that to set up water resources protection and rivers & lakes health security system in 2020 basically. Construction of the water resources protection and rivers & lakes health security system with Yellow River characteristics is an important goal of the task of the Yellow River harnessing under the new situation. Four water problems of "Flood is still a serious problem for the development of some districts, water supply and water demand are unbalanced, water environment is polluted and there are a series of ecological environmental problems" are the epitome of water problems in China. In the Yellow River Basin, natural conditions are complex and river basin and water regimes are special; present situation of the Yellow River Basin of "low carrying capacity, high environmental stress and fragile ecology" is determined by natural endowment and human impact. Realization of the goals of protection of water quality, water quantity and aquatic ecosystem in the Yellow River Basin matters to the sustainable economic and social development of the river basin and will be very important to solve water problems and guarantee water security in China.

Key words: sustainable development, rivers & lakes health, water resources protection, system construction

1 Understanding of water resources protection and rivers & lakes health security

1.1 Basic concept and scope of water resources protection and rivers & lakes health security

Connotation and denotation of water resources protection is constantly enriched and expanded with the development of productivity, technological progress and deepening of people's understanding. The following definition of water resources protection is widely used in the academic circles presently: all kinds of measures in terms of law, administration, economy, technology and education etc. that are adopted to prevent depletion and pollution of water source and other problems of unsustainable use of water resources due to inappropriate development and utilization. Statically, water resources protection includes protection of water quantity, control of water quality, protection of occurrence conditions of water resources, sustaining of carrying capacity of water resources and restriction of threshold value allowed by resilience of aquatic ecosystem; dynamically, water resources protection is to ensure activity of water cycle and hydrodynamic force that complies with natural ecological characteristics of water resources.

Currently, the domestic water resources protection function is implemented by water resources administration department through water resources and ecological protection activities as stipulated the laws, and also includes responsibilities of the water conservancy department in prevention and control of water pollution. At present, work of water resources protection in the river basin mainly includes the following contents: ①Construct a water resources protection management system taking the river basin as the unit, combination of the river basin with the district, joint cooperation between water conservancy department and environmental protection department and other departments and extensive participation of the public in accordance with system regulations of water resources management in Water Law of the People's Republic of China. ②Implement the most strict water resources management system. Promote the optimization and adjustment of layout, structure and function of economic and social development to increase utilization efficiency of water resources taking the carrying capacity of water resources as the constraint; conduct comprehensive execution and treatment and overall supervision to promote water conservation and pollution reduction and prevention and control of pollution taking the pollutants carrying capacity of river as the constraint and management of water function zones as the emphasis. ③Guarantee the safety of water quality and quantity of drinking water. ④Scientifically regulate water resources, plan domestic water, production water and ecological water as a whole to maintain ecological basic flow and water level of the watercourse and lake & reservoir and promote protection and restoration of aquatic ecosystem protection etc.

Concept of water resources protection in China was initially formed in the early 1970s and it develops and extends gradually and tends to be perfect in the present stage; while concept of rivers & lakes health was proposed later than concept of water resources protection, and it is a derivative concept under the idea of China's sustainable development and ecological civilization in the 21st century and has a larger denotation and inclusiveness compared with water resources protection.

In terms of the natural system, core connotation of rivers & lakes health is to maintain ecosystem diversity of rivers & lakes and its basic condition is compliance with water quality standard and satisfaction of water quantity for each function of rivers, namely environmental flow, one of topics of this sub-forum.

The scope involved in rivers & lakes health and its overall evaluation factor shall mainly include: runoff continuity, i. e. long and stable water flow conditions; water and sediment passage, i. e. stable watercourse and balanced water and sediment relationship; good water quality, i. e. realization of water quality goal required by satisfaction of water body function; river ecology, i. e. diversified biological structures and dynamically balanced aquatic ecosystem; and social service function, i. e. mainly includes generation, irrigation, shipping and water use etc.

Based on the above explanation, concept and scope of rivers & lakes health security includes water resources protection and the latter is an important supporting component of the former.

1. 2 Outstanding contradictions and problems encountered by the Yellow River Basin Water Resources Protection and Rivers & Lakes Health Security

1. 2. 1 Serious water pollution

The Yellow River Basin is rich in energy and mineral resources, so it has a very important strategic position in domestic economic and social development. For rate of exploitation of resource and energy is higher, pollutants emission intensity of point pollution source is larger. With continuous acceleration of urbanization of the river basin and relative lagging of wastewater treatment in the city, domestic sewage in the city is increasingly becoming one of the important factors of serious water pollution in the Yellow River basin.

Once, the Yellow River Basin (as 2% water resources) received about 6% waste water and 7% COD emission of the nation and gross receiving pollutants had exceeded its water environment carrying capacity, the waste water discharged to the Yellow River Basin is doubled in the 21st century compared to the 1980s, and maintained at 40×10^6 t for a long time, the year-round compliance rate of water quality in the water function zones of the river basin was only 43. 8% that is lower than national average level; the Yellow River Basin is facing more severe water pollution situation compared with other river basins.

1. 2. 2 Scarce water resources

Years of average natural runoff of the river in the Yellow River Basin is only 2% of that of the entire country, and annual runoff per capita is only 1/4 of that of the entire country. With the acceleration of industrialization and urbanization, the demands for water resources by economic and social development and improvement of ecological environment in the Yellow River Basin will still increase rigidly in the future and contradiction between supply and demand of water resources of the Yellow River Basin will become increasingly prominent before inter-basin water transfer project takes effect.

1. 2. 3 Fragile ecological environment

Since the middle of the 20th century, because of large adverse changes of climate, precipitation and other natural factors worldwide, more importantly, destruction and interference of human factor, serious problems appear in the Yellow River Basin, especially upper and middle reaches, such as vegetation degradation, water loss and soil erosion and wetland shrinkage. Although these problems have received great attention, have been governed with a larger effort and restrained and relieved to a certain extent in recent years, the problem of fragile ecological environment in the Yellow River Basin is still very outstanding in general.

2 Main practices and effects of the Yellow River Basin water resources protection and rivers & lakes health security

2. 1 Proposal of idea "maintaining the healthy life of the Yellow River"

As known to all, the Yellow River has a prominent sticking point of "less water but much sediment and uncoordinated water and sediment relationship". The Yellow River lacks of water resources, but it undertakes the task of supplying water to this river basin and the district irrigated by the Yellow River downstream which accounts for 15% of the nation's agricultural acreage and 12% of the nation's population as well as more than 50 large and medium-sized cities. For a long time, main channel of the Yellow River have silted up and shrunk severely, the situation of "secondary suspended river" has been severe, contradiction between supply and demand of water resources and water pollution has worsened, all of these problems reflect survival crisis of the Yellow River. Therefore, the Yellow River Conservancy Commission (YRCC) formally put forward a new river management idea, "maintaining the healthy life of the Yellow River" and identified it as the ultimate goal of the Yellow River management in 2004 from the strategic height of sustainable development.

Afterwards, "maintaining the healthy life of the Yellow River" is gradually enriched and evolved to a complete river management system, i. e. "1493 river management system" from a kind of river management idea, Fig. 1.

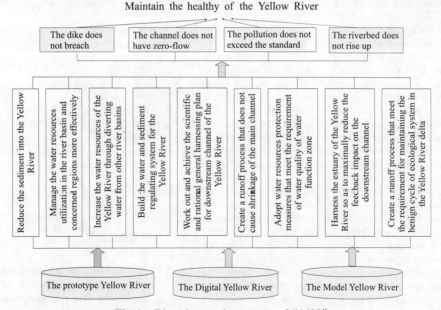

Fig. 1   River harnessing system of "1493"

From the diagram, we can see that there is a very strong correlation between "maintaining the healthy life of the Yellow River" and Yellow River Basin water resources protection and rivers & lakes health security system and work practices of water quality, water quantity and aquatic ecosystem carried out under the guidance of this river management system has a very important exploration meaning and guiding role for construction of the Yellow River Basin water resources protection and rivers & lakes health security system.

2.2 Important practices and remarkable effects

2.2.1 Continuous strengthen management of water function zones and strictly control the gross of pollutants discharged into rivers, which make water quality of the Yellow River mainstream get better obviously

Water Law of the People's Republic of China revised and issued in 2002 specifies management system of water function zones. New management system of water resources protection is established according to this law, the most important management practice of water resources protection of the river basin is started there from and this law has also laid the foundation for "management of red line limited within pollutants carrying capacity for water function zones".

Water function zoning in the river basin is the basis of water resources protection and supervision. Through the efforts during the Tenth Five-year Plan and the Eleventh Five-year Plan, water function zoning of the Yellow River Basin has been approved by the State Council and People's Government of each province (district); more than 720 various water function zones are designated in the whole river basin, 346 water function zones are listed into national important water function zones and the river length of the zoning is near 17,000 km.

Management methods of water function zone, the view of supervision and management of sewage outlet to rivers formulated and many supporting laws and regulations and rules issued by the Yellow River Conservancy Commission and each province (autonomous region) of the river basin in succession constitute a relatively perfect system of management laws and regulations of water function zones in the river basin; pollutants carrying capacity of the river basin is verified basically, control index of red line limited within pollutants carrying capacity for water function zones has been confirmed by most of provinces (autonomous regions) and compliance evaluation system of water function zones is established initially.

In specific management, the Yellow River Basin Water Resources Protection Bureau performs its obligations of management of buffer zone at the provincial boundary according to the law and strictly control the gross of pollutants discharged into rivers by firmly grasping the important starting point, namely management of sewage outlet to rivers and actively promotes the implementation of energy conservation and emission reduction by each province (autonomous region) of the river basin and increase efforts to prevent and control of water pollution.

Practices show that continuous strengthening of management of water function zone plays an active and important role in effectively improving water environment quality of the Yellow River Basin and improvement of water quality of the Yellow River mainstream is especially remarkable. According to analysis result for water quality change trend of the Yellow River mainstream in the recent decade provided by the Yellow River Basin Water Environment Monitoring Center: for water quality change trend in the recent decade, the period of 2002 ~ 2003 is taken as the inflection point, water quality tended to worsen before 2002, and was

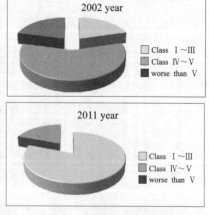

Fig. 2　The comparison diagram for water quality proportion of the Yellow River mainstream in 2002 and 2011)

restrained initially after 2003 and has been improved greatly as a whole from 2006 to present. In 2002, the river length of classes I ~ III water accounted for 14.7% and the river length of classes IV ~ V water and class V bad water accounted for 51.7% and 33.6% respectively in the evaluate length of 3,613 km of the Yellow River mainstream. In 2011, the proportion of the river length of classes I ~ III water was increased to about 70%, the proportion of the river length of classes IV ~ V water was reduced to about 29% and class V bad water quality hardly appeared, see Fig. 2.

2.2.2 Give play to unique advantages of river basin organizations and dispose various sudden water pollution accidents quickly to guarantee water supply quality safety of the Yellow River

In 2003, the YRCC established the first emergency mechanism of sudden water pollution incident. Since the Eleventh Five-year Plan, YRCC and water conservancy department and environmental protection department of the provinces (autonomous regions) along the Yellow River have disposed more than 40 major water pollution accidents properly, which effectively guarantees water supply quality safety of the Yellow River. Successful disposal of Weihe River oil pollution accident not only gives play to the advantages of the river basin organizations but also reflects the importance and necessity of joint disposal between the river basin and the district and between water conservancy department and environmental protection department.

In 2011, the YRCC and water conservancy department and environmental protection department of 8 provinces (autonomous regions) in the river basin jointly signed Opinion on Framework of Information Notification and Communication Coordination Mechanism of Sudden Water Pollution Accident in the Yellow River Basin; this achievement is not only the prolongation of emergency mechanism of sudden water pollution accident on the aspect of river basin but also an important practice for establishing and perfecting the water resources protection management system that combines the river basin with the district with joint cooperation between water conservancy department and environmental protection department adopted. This mechanism plays an important part in the emergency disposal of sudden water pollution accident in the future.

2.2.3 Deepen the study of environment flow and actively explore protection and restoration on the Yellow River downstream and aquatic ecosystem protection and rehabilitation in the Yellow River estuary delta, which produce a remarkable effect

The ecological environment water demand of the Yellow River mainstream is one of the important evaluation indicators in the Yellow River health indicators system. Several research topics have been carried out, such as "Study on the Ecological Environment Water Demand of the Yellow River Mainstream" which emphasize on the Yellow River harnessing, "Study on the Key Technology for Health & Restoration of the Yellow River" which is one of the key projects scientifically supported during the national "Eleventh Five-year Plan" period, "Study on the Ecological Environment Water Demand of the Yellow River Estuary" jointly made by China and Holland, and "Study on the Functional Non-stream Cutting Off Indicators of Important Reaches of the Yellow River Mainstream & Branches" which is a nonprofit project supported by the Ministry of Water Resources, and important achievements have been made. Compared with the studies on other rivers in China, the relevant studies on the environmental stream of the Yellow River introduced new concepts, thoughts, technologies and methods of foreign countries and combined ecology, hydrology and hydraulics. Also, some interested parties participated in the studies. Great breakthroughs and innovations have been made in many aspects.

Based on the above achievements, some exploratory practices to restore and regulate the environmental base flow and ecology of the Yellow River were carried out. Three ecological water compensations in the estuary delta before flood period to regulate water and sediment were conducted respectively in 2008, 2009 and 2010 in a row, and significant ecological benefits have been achieved.

2.2.4 Tamp the basic supporting capacity of water quality monitoring and informatization and strengthen international cooperation and exchange to provide technical support for the Yellow River Basin water resources protection and rivers & lakes health security

At present, water quality monitoring network system of the Yellow River Basin has begun to take shape, and 269 water quality monitoring stations have been established in the entire river basin by river basin organization and water conservancy department of each province (autonomous region) , which can grasp the situation of water quality change of the important reaches of the river basin basically. In the period of "ten-fifth" and "eleven-fifth" plan, it has been basically formed the modern water quality monitoring system of " combination of routine monitoring and automatic monitoring, sentinel surveillance combined with motor patrol, combination of regular monitoring and real-time monitoring, strengthen the emergency supervisory monitoring and achieving water quality monitoring information management" , by accelerating water quality monitoring and capacity building. During the Eleven-Fifth Plan, water quality monitoring capability of the Yellow River basin was improved significantly. Monitoring coverage of provincial section and water quality of water function zones in the river basin have been improved continuously, in which, the monitoring coverage of provincial section has reached 100% and the monitoring coverage of important water function zones in the river basin has reached 57% presently.

Under the framework of engineering system "Digital Yellow River" , the first digital monitoring center, namely monitoring system of the Yellow River basin water resources protection that is applied to water resources protection of the river basin is built in China. Major achievements include research & development and construction of three application systems such as consultation environment, analog display and supervision management, monitoring management and emergency management. This center provides strong information technology support for emergent consultation and emergent disposal of sudden water pollution accident for many times and plays an important role in the supervision and management of the Yellow River basin water resources protection.

For a long time, the Yellow River Basin Water Resources Protection Bureau has carried out cooperation and exchange in many aspects with America, Holland, European Union and other countries and international organizations in terms of protection of water resources and aquatic ecosystem, in which, the cooperation achievements including hydrochemical characteristics of sediment-laden river, study of ecological water demand of freshwater wetland in the Yellow River estuary, development of early warning system of water pollution in the middle reach of the Yellow River, calculation method of pollutants carrying capacity and ecological and biological monitoring methods have been applied in the protection and management of the Yellow River basin water resources and provide powerful technical support.

3 Basic conception and work plan of the Yellow River Basin water resources protection and rivers & lakes health security

3.1 Basic thought

Guided by the scientific outlook, based on the most strict water resources regulation and the national water function zones on key rivers & lakes, to improve the management system of water resources protection "taking river basin as unit, combination of the river basin and the district" , construct the Yellow River Basin water resources protection and rivers & lakes health security system, deepen the study of basic laws and key technologies, improve basic supporting capacity and realize the goals of protection of water resources and aquatic ecosystem in each stage to provide powerful safeguard for economic and social sustainable development and ecological civilization construction of the river basin.

Focus on the following 5 aspects in the course of building the Yellow River Basin water resources protection and rivers & lakes health security system.

3.1.1 Do well in top-level design and strive to build a water resources protection system combining the river basin with the district and adopting management at different levels and categories

Taking a new round of water resources protection planning of the river basin and implementation of approval opinion on administration authority of sewage outlet to rivers of the Yellow River Basin issued by the Ministry of Water Resources as an opportunity, build a water resources protection system combining the river basin with the district and adopting management at different levels and categories. It is necessary to ensure this system can not only give full play to 4 roles "comprehensive coordination, technical guidance, administrative enforcement and supervision and check" of the organizations in the river basin but also give full play to function advantage and supporting role of local department of water administration.

3.1.2 Tamp the work in the early period of the base and strive to build a scientific and reasonable assessment indicator system with strong operability

Define assessment and evaluation indicators necessary for management of red line carrying pollutants and corresponding overall and staged control goals, and further strengthen local assessment to make the basis of red line management more sufficient and the standard more specific.

3.1.3 Emphasize capacity building and strive to build a research system of water quality monitoring and application technology supporting supervision and management and compliance assessment

Emphasize the collection, sorting, application and other basic work of the important information of the river basin and strengthen the administrative enforcement, monitoring management, informatization and research of application technology and talent team building.

3.1.4 Strengthen basic construction and strive to build a water resources protection and aquatic ecosystem restoration engineering system taking rivers & lakes health as the central task

Attach great importance to connect water systems of rivers and lakes, improve natural purification capacity of the rivers, pay attention to construction and protection of wetland, plan the layout and regulation of sewage outlet to rivers, strengthen the protection of drinking water source and take ecological water compensation measures if permitted by the conditions to improve the water environment and restore the aquatic ecosystem.

3.1.5 Emphasize the public participation and strive to improve socialized management level of water resources protection and rivers & lakes health security

In the course of building the system, continuously perfect and strengthen socialized management to fully reflect the idea and connotation of socialized management of water resources protection as a social public welfare undertaking. We shall emphasize good link with environmental protection department and relevant departments and industries and shall particularly emphasize wide participation and social supervision of all sectors of the society, especially all interested parties.

3.2 Staged goal

3.2.1 Water resources protection goal

In 2015, compliance rate of water quality in the important water function zones of the Yellow River basin will reach 60%;

In 2020, compliance rate of water quality in the important water function zones of the Yellow River basin will reach 79%;

In 2030, compliance rate of water quality in the important water function zones of the Yellow River basin will reach above 95%.

3.2.2 Aquatic ecosystem protection goal

In 2020, ecological environment water demand of important control section of the Yellow River mainstream will be guaranteed basically, ecology of key protected zones including river source and river estuary will be restored moderately and worsening trend of aquatic ecosystem of the river basin will be controlled initially.

In 2030, lower limit water use of ecological environment will be guaranteed basically and the aquatic ecosystem will be improved.

3.3 System framework

Basic framework of the Yellow River Basin water resources protection and rivers & lakes health security system to be built in 2020 basically is as shown in Fig. 3.

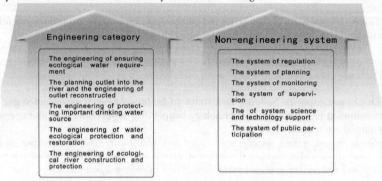

Fig. 3 Framework of the Yellow River Basin Water Resources Protection and Rivers & Lakes Health Security System

3.4 Emphases in the near future

3.4.1 Steadily promote "management of red line limited within pollutants carrying capacity for water function zones" of the Yellow River Basin

Complete the check of pollutants carrying capacity of important water function zones and formulation of discharge limitation plan in stages. Recheck pollutants carrying capacity of important water function zones and formulate limitation plans for total discharge capacity of water function zones of important rivers & lakes in 2015 and 2020. Design management systems at different levels and categories of important water function zones of the river basin and water quality compliance assessment system. Further strengthen basic management of important water function zones including basic information survey and basic database construction.

3.4.2 Compile the planning of new round of the Yellow River basic water resources protection with great effort

This round of planning will endeavor to make breakthroughs in the construction of the following three aspects: the top-level design of water resources protection, overall arrangement of water quality, water quantity and aquatic ecosystem and the engineering system of water resources and aquatic ecosystem protection.

3.4.3 Continuously strengthen supervision and management of sewage outlet to rivers

Based on the opinion on division of administration authority of sewage outlet to rivers of the river basin approved by the Ministry of Water Resources, normalize and sort out the relationship between agreement of set of sewage outlet to rivers and assessment and approval of environmental effect, water-drawing permit and review of construction project in the watercourse according to the law. Strictly approve set of sewage outlet to rivers according to the law and restrict the approval of sewage outlet to rivers and new water drawing for the district whose discharge capacity has exceeded limited total discharge capacity of water function zones.

3.4.4 Solidly carry out safety control of drinking water source

Regularly carry out safety assessment of important drinking water source of the river basin and comprehensively carry out compliance construction of safety control of important drinking water source.

3.4.5 Further explore the protection and restoration of aquatic ecosystem

Establish reasonable rivers & lakes health evaluation indicator system and a scientific evaluation standard system and steadily propel the evaluation of rivers & lakes health. Deepen the study of ecological water demand of rivers & lakes and support pilot demonstration of joint regulation and control of water quality and water quantity of typical river basin. Study key technologies including biological habitat monitoring and aquatic life monitoring.

3.4.6 Quickly improve basic supporting capacity of water resources protection and rivers & lakes health security

Perfect water quality monitoring network of the river basin and realize goals of water quality monitoring coverage of important water function zones in each stage. Build an automatic monitoring system of water quality at the provincial boundaries of the Yellow River mainstream and accelerate construction of aquatic ecosystem laboratory. Further improve informatization level and administrative enforcement ability.

4 Conclusions

Construction of the water resources protection and rivers & lakes health security system with the Yellow River characteristics is a quite challenging major topic and system engineering. As one of main bearers of this important task, The Yellow River Basin Water Resources Protection Bureau will learn and absorb domestic and foreign advanced management concepts and research achievements and continuously explore and innovate on the basis of existing practice achievements and experience to strive to make greater breakthroughs and contributions.

Strategic Analysis for Water Ecology Security in China—from EU Perspective

Philippe Bergeron[1] , *Wen Huina*[2] , *Xia Peng*[3] and *Shi Yang*[2]

1. EU – China River Basin Management Programme
2. Yellow River Water Resources Protection Bureau, Zhengzhou, 450004, China
3. Development and Research Center, Ministry of Water Resources, Beijing, 100038, China

Abstract: China No. 1 document in 2011 and the three red lines of water resources policy objectives defined for Ministry of Water Recourses (MWR) is a formidable and unique water resources challenge unprecedented in the world. The EU on the other hand has accumulated an unmatched wealth of good policy practices aiming at controlling water use and pollution release and securing a high level of water resources protection across the union. The objective of this Strategic Analysis of Water Ecology Security is to provide decision makers of an insight on the rationale for policy action desirable to successfully implement the "three red lines" defined in the No 1 document of MWR based on EU experience with good policy practices.

Based on an assessment of the current Chinese practices for water resources protection and water ecology security and a comparison with EU and EU Member States policy experience in the field, the success of the three red lines policy under MWR is expected to critically hinges on the capacity of the Chinese government at all level (central, provincial & local) to thoroughly and effectively monitor and enforce the policy.

Currently the Chinese regulations for the permitting of water abstraction as well as wastewater discharge into water bodies do not allow adequate monitoring for compliance and enforcement. The permits needed are not integrated (quantity, efficiency and quality) and critically, the permits are too simply worded and do not cover in adequate unambiguous details and precision the obligations of the permit with regard to Emission Limit Values (ELV), Mixing zone, BAT requirements, monitoring and inspection, reporting requirement, sources and data to be reported, specific operation condition changes which would require a reassessment of the permit, penalties to be incurred in case of breach of permit, and more.

In the EU an integrated water permit will be written by experience inspectors which specify in very great details and great unambiguous precision what are the obligations of the permittee. It is this attention to details and the unambiguous precision in the written permit drafted by an experienced inspector, which makes all the difference when it comes to compliance inspection and enforcement procedures.

The key message is that to enable effective compliance and monitoring of the achievement of the three red line policy, a new water resources regulatory process having at its core a new integrated water resources impact permit for water users and pollution discharger merging together the three dimensions of the red lines (quantity, efficiency, quality) is an absolute necessity.

Key words: permit, three red lines

1 Main dimensions of water ecology security.

Water Ecology Security in China is of particular importance considering the huge water resources challenges faced by the country. China faces 4 main water resources challenges:

(1) Shortage of water in the Northern and North – Western regions causing at time acute water scarcity in large urban areas such as Beijing, Tianjin and more.

(2) Excess of water in the South which often leads to flooding.

(3) Polluted water all over China due to limited pollution control and heavy effluent release

from human settlement, industry and agriculture.

(4) Muddy water due to uncontrolled erosion in upstream catchments of most river basins.

All these four challenges are dramatically affecting the ecology of the water bodies in the country and therefore impact on the national ecology security. Climate change is expected to exacerbate the difficulties with more extreme rainfall events mostly between June to September which sees 60% ~ 80% of the annual total. This leads to a growing mismatch between water availability and water demand especially in the North and the coastal areas where the engines of economic development are located.

Water ecology security in China can be rationalized into four dimensions which include:

(1) Pollution prevention and control.

(2) Maintenance of environmental flow.

(3) Maintenance of groundwater level.

(4) Restore and sustain river morphology and ecology.

These four dimensions are briefly outlined below.

1.1 Pollution prevention and control

Pollution prevention and control may be the most critical issue because water pollution from human settlements, industry and agriculture impact negatively and may destroy the ecology of water bodies, sometime in irreversible ways. In the presence of severe water pollution no meaningful water ecology revival or restoration can succeed.

Pollution prevention and control aims at avoiding discharge of pollutants into water bodies to enable water ecology security. Important strategy elements include among others:

(1) Permitting the discharge of any pollution into receiving water to better control it.

(2) Defining emission limit value for human settlement, industry and agriculture for the discharge of pollutant into river.

(3) Defining environmental quality standards in receiving water and water biota compatible to thriving natural ecological life.

(4) Prevent spills and accidents involving hazardous substances and limit the damage when occurring though emergency preparedness.

(5) Eliminate or restrict use of products or processes that are causes of water pollution.

(6) Promote a life cycle approach to industrial sites forcing operators to return sites to their original state when the activity is over.

(7) Encourage the use of clean process technology which avoid or minimize the need for discharge of wastewater into water bodies.

(8) Promote recover, recycling and reuse policies that avoid the release of pollutants into receiving waters.

(9) Impose pollution charges to dischargers of pollution into water bodies with charges preferably corresponding to the economic cost of the discharge.

(10) Establish environmental liability and compensation regulatory requirement for polluters forcing them to reassess their water use policies.

(11) Provide market based incentives to good performers.

1.2 Maintenance of environmental flow

Environmental flows can be described as the quality, quantity, and timing of water flows required to maintain the components, functions, processes, and resilience of aquatic ecosystems which provide goods and services to people. The flows of many rivers are increasingly being modified in China when water is withdrawn massively for agriculture and urban use some times over very large distances like in the 3 South North Transfer Canals and not returned to the river through drainage or groundwater flows, or retained for hydropower. Thus the flow of many rivers in China is being reduced or seasonally altered changing the size and frequency of floods, the length and severity of droughts, and adversely affecting ecosystems.

Important strategy elements to secure environmental flows in rivers include among others:

(1) Clarify and strengthen the legal standing of environmental water allocations.

(2) Recognize environmental flows in water resources policies and legislation.

(3) Include environmental water provisions in basin water resources planning.

(4) Develop studies to demonstrate the benefits from environmental water allocation.

(5) Define guidance methodologies for setting environmental objectives in basin plans.

(6) Specify clear requirements for stakeholder involvement in the decision making mechanisms.

(7) Appoint an independent authority to audit implementation.

1.3 Maintenance of groundwater level

Groundwater plays a crucial role in complex natural systems by: ①providing base flow to rivers; ②supporting aquatic ecosystems as well as riparian and terrestrial vegetation; ③maintaining a geochemical balance; and ④preventing earth subsidence. In many part of China ground water is overexploited (ground water extraction in the Hai river is estimated by the World bank to be about 50% percent greater than the sustainable yield) and is also subject to chemical pollution.

To assist in minimizing abstraction – induced stress on groundwater, the following strategy elements can be deployed among others:

(1) Strong permitting and enforcement of water abstraction and pollution control.

(2) Water demand management by heavy users to reduce total consumption by introducing various incentives to save water.

(3) Innovative technical solutions to reduce loss and leakage and improve the usefulness of the same volume of water.

(4) Cost covering tariffs, to put water in line with other commodities and encourage consumers to find ways to reduce water consumption.

(5) Education and involvement of all stakeholders in decision – making, especially at local level through user associations.

(6) Aquifer classification to provide a framework for implementing differentiated protection.

(7) Reserve determinations to allow for the role of groundwater in sustaining aquatic ecosystems to be understood and promoted within the context of a balance between use and protection.

(8) Land – use zoning that restricts potentially polluting developments on important or sensitive aquifer systems.

(9) Establishment of intensive groundwater monitoring programmes to track pollution and excessive abstraction trends.

(10) Programme to use groundwater in instances where comparison to surface water resources shows it to be economically and environmentally superior.

(11) Promote relevant and applied groundwater – related research, so that practicing hydro – geologists have both knowledge and the appropriate tools to manage the groundwater resources in an integrated manner.

1.4 Restore and sustain river morphology and ecology

Existing, relatively intact ecosystems in aquatic environment are the keystone for conserving biodiversity, and provide the biota and other natural materials needed for water ecology security. When the ecology of water bodies is impaired, the prevention of further degradation should be the prime objective. Restoration of aquatic ecosystem is often a complementary activity to be combined with protection and preservation of river eco – systems.

To assist in restoring and sustaining rivers ecology and morphology the following strategy elements may be considered among others:

(1) Identify the root cause of degradation that may include the cumulative effects of numerous impacts upstream and upslope as well as downstream modification such as dams and channelization.

(2) Develop clear achievable and measurable goals.

(3) Focus on feasibility taking into account scientific, hydrological, financial, social and other considerations.

(4) Design for self – sustainability by minimizing the need for continuous maintenance of the site, such as supplying artificial sources of water, vegetation management, or frequent repairing of damage done by high water events. This also involves favouring ecological integrity, as an ecosystem in good condition is more likely to have the ability to adapt to changes.

(5) Restore native species and avoid non – native species that may out – compete natives because they are expert colonizers of disturbed areas and lack natural controls.

(6) Favour natural fixes and bioengineering techniques that combine live plants with dead plants or inorganic materials, to produce living, functioning systems to prevent erosion, control sediment and other pollutants, and provide habitat.

(7) Provide adaptive management that monitor to help determine whether additional actions or adjustments are needed and adapt where changes are necessary.

2 Possible strategy in the priority area—an integrated water resources impact permit

Considering the high pollution still affecting many water bodies (rivers and lakes) in the country, pollution prevention and control is certainly the dimension that needs the most urgent attention and action. Human, industrial or agricultural water pollution can destroys the ecology of water bodies. Experience in the EU has shown that the reduction and elimination of water pollution in water bodies can often be reversible. Efficient water pollution control and prevention can lay the necessary foundation upon which the ecological restoration and revival of water bodies' biota, fauna and flora can be developed.

The proposed strategy below is aimed at effectively addressing the priority area of pollution control and prevention identified earlier as the essential precursor of water ecology security targeted by MWR under the No. 1 Document.

The strategy intends to create a new, innovative and national level "Water Resources Impact Permit" without which the enforcement of the three red lines based would not be possible.

The proposed new permit is expected to tackle together the three redlines targets (water quantity, water efficiency, water quality). As such it is partially integrated and can benefit from the experience of a number of features of the EU IPPC permit.

Based on the EU experience the new permit policy could be formulated along the following 3 mutually reinforcing principles.

1 site: every operating site using water and causing significant abstraction and/or discharge into a receiving water (also beyond WFZ major rivers boundaries) is allocated "three red lines" targets (quantity, efficiency, quality).

1 permit: every operating site allocated "three red lines targets" is subject to a new "three red lines" water resources impact permitting process and a new permit issued at the state level but implemented at the provincial level with oversight from the central level allowing strictest reporting, monitoring and enforcement as required by the No 1 document.

1 competent authority: every operating site allocated "three red lines targets" is subject to mandatory periodic inspections coordinated by a single competent authority to control compliance with the requirements of the "three red lines" permit.

The paragraphs below summarizes the main aspects of the proposed policy

2. 1 Sites subjected to the permitting process

Sites subjected to the 3 red lines permit would be limited to large plants handling major quantity of pollutant and/or hazardous substances. In an initial phase this could mean largest operated sites responsible for the generation of 40% of the untreated waste water emissions potentially released into a WFZ or 40% of the water abstracted within in a WFZ. In a second phase, the coverage could be extended to the sites covering perhaps 80% of the pollution generated by operators. Smaller remaining operators would remain excluded as there would not altogether amount to more than 20% of the generated pollution or 20% of the abstracted water.

Sites subjected to the 3 redlines permitting process would be specified in detail and documented in an annex to the policy and could include all plants consuming significant quantity of water. Sites handling hazardous substances would be defined precisely in other annexes of the policy through the specification of minimal quantities of either generic hazardous substances in line with international classification (explosive, oxidizing, flammable, toxic, dangerous et al.) as well as well identified particularly toxic or dangerous substances that would need to be handled by a site to require a permit.

Sites falling under the categories of operation requiring a permit would have to notify the regulating competent authority within a defined period to ensure the authority has a complete list of operators without having to chase them. Failure to notify could mean prohibition to operate.

The new permit wouldn't be an additional permit. It would rather consolidate all the existing authorizations and permits linked to water abstraction and wastewater discharge of operators into a new single permit that would replace existing ones after it has been issued. The new single integrated permit would provide the tools necessary for forceful inspections and for unambiguous compliance control and enforcement.

2.2 Scope of the "three red lines permit"

The "three red lines permit" would be issued as a new state level permit that could with time replace existing provincial or local level permit.

To acquire a permit, an operator site will have to submit to the regulating competent authority application requiring extensive information about abstraction of water, processes involving water and discharge of wastewater. The competent authority will check the application validity and to which extend the information submitted is complete. Issues of commercial confidentiality would be considered and when justified and approved by the regulator respected but complete information would have to be submitted by the operating site so that a thorough and complete permit can be assessed and issued.

Part of the information required in a permit application would be a water resources impact study which should document: ①the extent of the potential impact (including geographical area and size of the affected population; ②any effects on specifically protected areas, species or other assets of particular significance; ③the trans – boundary nature of the impact; ④the magnitude and complexity of the impact; ⑤the probability of the impact; and ⑥the duration, frequency and reversibility of the impact.

Sites handling hazardous substances would be requested to submit additional information linked to the risk of accident such as safety measures, a safety plan and an emergency response plan worked out by independent experts certified by the relevant competent authority.

The permit issued by the competent authority would require the cooperation of a group of experienced permit writers well versed in inspection and enforcement covering the three red lines criteria (water quantity, efficiency and quality). Through the extensive use of check – lists, guidance documentation and negotiation between experts working for the competent authority and the applicant, agreement would be reached on technologies upgrade or change that would be implemented by the applicant operator to ensure emission reduction targets are met on certain deadlines. These necessary changes and their timing would be written down in the permit.

The permit document would end up being a substantial document specifying in detail among others:

(1) The receiver of the permit (operator site).

(2) The scope of the permit (all water handling processes of the site).

(3) Clearly defined and specified list of obligation of the permit holder in terms of water abstraction quantity, water use efficiency and treated wastewater discharge.

(4) Emission limit values to be fixed to every relevant pollutant and how it will be monitored and reported.

(5) Expected receiving water quality and specified mixing zone, location of receiving water

functional zone monitoring point, carrying capacity of receiving water, with assumptions used in its calculation (flow, velocity, decay coefficient, temperature et al.).

(6) The flows and loads in the discharge and receiving waters to be used as the basis of calculation of impact and likely compliance with receiving water quality and the downstream control point.

(7) Full water resources impact calculation will most likely utilize river water quality modelling systems to resolve the calculations and put into the context of other abstractions and discharges in the area. Extensive guidance and training on the application of these models will be required.

(8) Statistical means of assessing compliance or failure with abstraction, discharge and WFZ standards for flow volume and quality.

(9) Efficient technologies to be progressively harnessed and the timing of their introduction.

(10) List of requirements executable by the operator as well as by the inspection authority for monitoring and reporting.

(11) Special and detailed obligations concerning all important inspection procedures especially regarding discontinuous measurements and continuous measurement.

(12) Rights of access for regulatory inspectors to access the site and take samples without advance notice.

(13) Temporal limitation of the permit.

(14) Costs of inspection.

(15) Penalties to be incurred in case of breaching the permits conditionalities.

2.3　Functions of the competent authority

To ensure the water resources impact permit satisfies its objective of enforcing the three red lines targets, the permit should be coordinated and supervised by a single competent authority at the national level.

The main functions of the competent authority will include the following:

(1) Issue the policy and publish nationwide the request to notify by relevant operators.

(2) Process the notification received by the operators and ensure validity.

(3) Submit to notifying operators application forms with time line for submitting an application for a "three red lines permit".

(4) Check of completeness of application received.

(5) Coordinate the participation of other special authorities and experts in the permit assessment process to cover the three red lines elements (water quantity, efficiency and quality) .

(6) Coordinate the consultation and participation of the public in relation to the EIA review.

(7) Evaluate whether the preconditions to issue a permit are fulfilled.

(8) Determine and coordinate the obligations to be fixed in the written permit.

(9) Coordinate monitoring and inspection.

In many developed countries such a permit would be managed by the water directorate of the environmental agency or ministry. China is but an exceptional case. Due to the primordial importance of water quantity over water quality in the country (even if there were no pollution in the rivers, China would still face a water resources challenge due to the serious water scarcity in the Northern part of the country and around the major the engines of growth in the country), the case can certainly be made for entrusting the coordination of this new water resource impact permit to the MWR.

3　Proposed 3 phases strategy for implementation

The establishment of a new national "Water Resources Impact" permit at the scale and depth necessary to make the No. 1 document succeed and the targets of the three red lines achieved, will take time to develop in China. Consequently a phased approach allowing the regulatory framework to progressively converge to an optimal and efficient status compatible with the three targets of the red lines is desirable.

Building up on the Elbe river restoration highlighted earlier, it is recommended that the implementation of the No 1 document start soonest by addressing first the largest and most heavy polluters in each WFZ. At that time, the capacity of MWR and other relevant authorities to issue a formal permit won't be available. It is therefore recommended that during the first phase of the implementation the requirements imposed on largest polluters of each WFZ to achieve three red lines is embedded in a protocol or proto – permit to still allow strict enforcement. This will allow the progressive build – up of the expertise and capacity by relevant authorities across the country to handle the full – fledged procedures required in a second phase to operate smoothly the issuance of "three red lines permits" without chocking the capacity to operate of the operator sites that will need the new permit to operate.

The proposed implementation plan foresees three phases as follow.

3.1 1st phase: urgent initial improvement during 12th FYP (2012 ~ 2015)

The focus of this phase will be to significantly reduce large water abstraction and wastewater discharge at points sources by targeting a first tiers of largest operators representing the abstractors and emitters with highest "three red lines" improvement potential. The target will be to achieve an overall cumulated improvement of around 30% at the macro level in terms of pollution reduction and efficiency of water use in the WFZs. Water quantity usage during the period may not be reduced significantly due to the rapid growth of development which will call for additional water usage for new industry. As a whole, the quantity of water use should shows a trend toward a stabilization of overall usage compatible with the overall quantitative targets of the No 1 document.

During this phase the targeted companies will be selected by the competent authority (MWR) based on WFZ pollution load scenarios allowing to identify highest pollution load reduction opportunities in each WFZ.

Operators corresponding to identified most polluting and most promising sectors with production capacity above defined thresholds would be requested to "notify" the competent authority of their existence and their key production and processes capacity. This will be then used by the managers of the WFZs with the help of modelling tools and scenarios to identify largest abstractors or polluters to be targeted for rapid pollution load reduction under this 1st phase.

During this phase finally targeted companies will be then asked based on a "three red lines" application documentation, to agree on a "protocol" for three red lines improvement allowing efficient and unambiguous monitoring, reporting and enforcement that will be negotiated between operating sites and competent authority. The development of these protocols which will be represent proto – permits, will allow the competent authorities to test and prepare the development and launch of formal permits starting with the second phase and build the capacity of competent authorities for permit development coordination, issuance and related compliance monitoring, inspection and enforcement.

3.2 2nd phase: establishment of the three red lines permitting during the 13th FYP (2016 ~ 2020)

The attention will here be turned to a second tiers of most significant abstraction and discharge points sources to target the bulk of the Chinese pollution (around 80%) reaching water bodies or excessive abstraction depleting groundwater levels. The "three red lines" water resources impact permit will be officially launched and addresses in a staged approach for each WFZ, successive groups of operators starting always with the most polluting and those with the highest risk of accident hazards in each WFZ. During this period non – point sources will also be targeted through improved agricultural practices. Finally during this phase "ecological zones" will be identified and documented based on the morphology of the river and the degree of artificial modification incurred so far. For these zones an assessment of main pressures and impacts using as basis the WFZ scenarios but adapted now to these newly defined ecological zones will be developed and documented as basis for defining "good ecological status" objectives for all river bodies to be implemented in the third phase.

As the water pollution will become less severe, this phase will also be an important one to start the ecological restoration works. That is then the time when the experience and expertise accumulated in the EU about the implementation of the WFD will start to become useful. See chapter 6 presenting several aspects of promoting good ecology in water and water ecology restoration in water bodies.

3.3 3rd phase: fulfillment of the No. 1 objectives during the 14th and 15th FYP (2021 ~ 2030)

At the beginning of this phase, the pollution load in river is expected to have been significantly reduced to levels compatible with the establishment of genuine "good ecological status". In addition MWR will at that stage, be equipped with an efficient and enforceable water resources impact permitting tool and legal enforcement capability. This will open the door for the fine tuning of pollution control and prevention in water bodies for any individual chemical substance considered a priority substance and deserving attention for reduction as part of the EQS improvement process. This third phase will focus then on incrementally improve and restore the ecology of the water bodies using instruments and tools similar to the one defined and applied in the WFD in the EU.

4 Inspections, incentives and penalties

Inspections are the most important element of enforcement and compliance efforts. Under the new permit, much stronger and thorough inspections would need to be deployed to all permit holding sites. Inspection should be conducted by government inspectors, or by independent parties hired by and reporting back to the responsible enforcement agency. There are a number of different types of inspection activity that need to be deployed for effective compliance monitoring and enforcement of the "three red lines" permit.

Among others the following inspection types may be considered:

(1) Walk – through inspection: this provides a quick survey of a process, where an inspector checks on general issues, e. g. control equipment and working practices. This type of inspections helps to determine whether more extensive inspection is needed. These inspections can be announced or can also be unexpected.

(2) Compliance evaluation inspection: this involves an intensive examination of a particular technological process or a whole facility, but does not include sampling. It would consider records, interview staff, review self – monitoring data, examine control equipment, et al.

(3) Sampling inspection: this type of inspection includes the visual and record examination described above, but also includes collecting and analysing physical samples. This is the most resource intensive type of inspection.

Criteria to decide on frequency and type of inspection would include:

(1) The potential hazard of the site to water bodies.

(2) The complexity of the inspection needed to evaluate compliance.

(3) The history of the site in relation to compliance.

(4) The availability of self – monitoring information.

(5) Specific inspections resulting from accidents or request from the local population.

Inspection of a site should systematically lead to inspection report specifying: ①does the operator have an accurate permit? ② Is the correct water consumption and pollution release monitoring equipment installed? ③Is all monitoring equipment properly maintained and accurately operated? ④Are all records properly maintained? ⑤Does the plant comply with all emission limits and other operating conditions? ⑥ Is the plant implementing agreed upgrading requirements? ⑦ Do the process's management plants include compliance requirements? ⑧ Are there any signs of deliberate falsification of records, equipment, et al. ?

Incentives under the new permit is essentially the access to innovative and efficient technologies that allow an operator to lower its water consumption, minimize pollution release and improve productivity, quality and efficiency, leading to lower cost and higher profitability. Through the integrated nature of the permitting process, regulatory experts can assist the operator win access to best available technology adapted to its specific process and production situation.

The penalty system for the enforcement of the three red lines permits needs to be effective, proportionate and dissuasive. Effectiveness means that penalties are capable of ensuring compliance with the policy and achieving the desired objective. Proportionality implies that penalties adequately reflect the gravity of the violation and do not go beyond what is necessary to achieve the desired objective. Dissuasiveness requires that penalties have a deterrent effect on the offender which should be prevented from repeating the offence and on the other potential offenders to commit the said offence.

Penalties should include a broad "toolkit" of civil sanctions for regulators to promote and enforce regulatory compliance. This may include administrative monetary sanctions and the strengthening of statutory notices to work alongside criminal law for worse and repeating offenders to combat non – compliance.

Reference

Chris Chubb, Martin Griffiths, Simon Spooner. Regulation for Water Quality Management – Handbook on EU Principles and Practice [M]. Zhengzhou: Yellow River Conservancy Press, 2012.

Yellow River Action on Climate Changes

Su Maolin

Yellow River Conservancy Commission, Zhengzhou, 450004, China

Abstract:In recent years, extreme weather events such as high temperatures, extreme cold, drought, rainstorm and floods occurred frequently, the effects of climate changes become increasingly apparent. It has impacted on the world in varying degrees and caused general concern of the international community. Affected by the climate changes and the interference of human activities, the temporal and spatial distribution of Water Resources in the Yellow River Basin has made a great change. As a response to climate changes on the impact of economic and social development, Yellow River Conservancy Commission (YRCC) adopted a series of policies and countermeasures to address climate changes in recent years, which include: ①strengthening hydrological monitoring and forecasting work to support powerfully water and sediment regulation of the Yellow River; ②forming initially water and sediment regulation engineering system; ③enhancing water resources effective operation with most strict water resources management; ④ reinforcing water resources allocation management to establish water – saving society.
Key words: climate changes, extreme climate events, water resources, Yellow River Basin, countermeasures

The Yellow River Basin is located in the warm temperature monsoon climate zone, having both humid and semi – humid climate area, and both arid and semi – arid climate area as well, the geographical location and climatic zone determines the complexity of the climate in the basin, the difference of weather and climate factors such as air temperature and precipitation is obvious between different regions and different years in the basin, so the Yellow River is a river with frequent flood and drought disasters and shortage of water resources.

1 Characteristics of climate changes in the Yellow River Basin

1.1 Climate changes in the Yellow River Basin

In the period from 1961 to 2005, the annual average air temperature rose about 1.3 ℃ in the Yellow River Basin, about 0.3 ℃/10 a. The Ningxia and Inner Mongolia reach is the most remarkable reach of warming, about 0.42 ℃/10 a. And winter is the most remarkable season of warming, about 0.39 ℃/10 a.

The average annual precipitation is about 446 mm in the period from 1956 to 2010 in the Yellow River Basin. The precipitation has a declining trend since the 1960s to the 1990s, and it declined most dramatically in the 1990s, the average annual precipitation declined about 10% in the 1990s compared with the 1960s. The precipitation increased since 2000, which close to the historical average.

1.2 Extreme weather and climate events

1.2.1 Rainstorm

On July 4 – 5, 2002, the maximum rainfall in 24 h reached to 274.4 mm at Zichang hydrological station, Qingjianhe River, it is a heavy rainstorm regarded as 500 years to meet.

In mid – August to mid – October, 2003, it appeared six continuous rainstorm processes in Jinghe River and Weihe River, both rainfall range and last time have not happened in the recent 50 years.

On September 19, 2010, the rainfall in 6 h reached to 185 mm at Linjiaping hydrological station, Qiushuihe River, which is more than the rainfall occurring once every 200 years.

2 – 14 July 27, 2012, the rainfall in 12 h reached to 227 mm at Shenjiawan hydrological station, Jialuhe River, which is the largest precipitation in 12 h on record.

1.2.2 Draught

The precipitation in most areas of the Yellow River Basin was significantly less in the period from July 2002 to June 2003, and it appeared rare drought in the lower Yellow River.

It appeared severe drought in most areas of the middle and lower reaches of the Yellow River in the winter of 2008, with no precipitation falling for more than 100 d in some local areas.

It appeared severe drought in the areas downstream of Sanmenxia, the Yellow River Basin, in the period from October 2010 to January 2011, with no effective precipitation falling for more than 80 d, and over 100 d in some local areas.

1.2.3 Air temperature in ice – flood period

Most years since 1990, the air temperature in winter is higher in the Yellow River Basin, and the changing range of 10 – day and month average air temperature is in a large scope. Take Baotou meteorological station as example, the extreme event, which 10 – day average temperature ranked amongst the top three in the same period in historical record, appeared 28 times since 2000, among which the extreme highest/lowest 10 – day average temperature appeared 7 times. In the winter of 2009/2010, the average air temperature of the second 10 – day of November reached to – 8.4 ℃, which is the extreme lowest on record in the same period, while the average air temperature of the last 10 – day of February raised to 3 ℃, which is the extreme highest on record in the same period.

1.3 Response of Yellow River hydrology and water resources to climate changes

On July 4, 2002, the flood peek reached up to 4,670 m^3/s at Zichang hydrological station, Qingjianhe River, which is regarded as the flood of 100 years to meet. In September, 2011, it appeared the largest flood since 1981 and 1982 respectively, at Weihe River and Yiluohe River. In July, 2012, it appeared the largest flood since 1986 at Lanzhou hydrological station, a major controlling station at the upper Yellow River, and appeared the largest flood since 1989 at Wubao hydrological station, a major controlling station at the middle Yellow River.

The total runoff of a hydrological year, July 2002 to June 2003, was just 9.6×10^9 m^3 at Tangnaihai hydrological station, a major controlling station at the upper Yellow River, which is the smallest annual runoff of the station on record. While the total runoff reached to 13×10^9 m^3 just in the two months of July and August, 2012, at Tangnaihai hydrological station, which is more than doubled the average runoff of July and August, and is the largest runoff on record.

In the ice – flood period of 2009/2010, it occurred repeatedly three times of freeze – up and breakup in Ningxia reach, with the temperature rising and falling dramatically. It occurred three severe ice flood disasters in Ningxia and Inner Mongolia reach since 2000, and it happened the most severe ice flood in 2008 within 40 years, resulting in dyke breach.

2 Action on climate changes in the Yellow River Basin

2.1 The building of prefect flood management and water resources allocation engineering system

The flood management engineering system has been built initially, which is consisted of reservoirs, embankments, flood detention districts and shelter engineering projects on beach, and the water resources allocation engineering system has also been built, which is mainly consisted of large – scale reservoirs. The embankment of the Yellow River reaches up to 2,846 km. The joint operation of Longyangxia and Liujiaxia Reservoir on the upper Yellow River reduces the magnitude of flood occurring once every 100 years, in Qinghai, Gansu, Ningxia and Inner Mongolia reaches, also controls and mitigates the ice flood disaster in the Ningxia and Inner Mongolia reach. The joint operation of Sanmenxia and Xiaolangdi Reservoir on the mainstream of the middle Yellow River, as well as the reservoirs on the tributary of the Yellow River, improves the flood control standard of the river dikes in the lower Yellow River from once every 60 years to once every 1,000 years. The total capacity of the reservoirs of various types amount to $7.8 \times 10^9 \ m^3$, by the operation of all the reservoirs, the annual and inter – annual distribution of runoff is adjusted, ensuring the watershed and relevant areas demand for water.

2.2 Strengthening construction of non – project measures to improve the ability of flood and water resources management

The ability of hydrological monitoring automation and informatization is improved, by strengthening the construction of hydrology monitoring infrastructure, and increasing the covering density of monitoring network, to monitor and control the rain and water regime in the whole Yellow River Basin effectively; a rainstorm and flood numerical warning and forecasting system is built integrated multi – model and multi – method, according to the feature of rainstorm and flood in different areas of the basin, improving the ability of flood forecasting and warning, and extending the lead time for flood forecasting; the building of middle and long term runoff forecasting system, enhance the ability of flood and drought trend forecasting and water resources regime analysis; the building of flood and water regulation scheme and model of the Yellow River, which is coupled with the flood and runoff forecasting system, make the management and controlling of flood and flow is more effective.

In September 2011, the flood peak at Huayuankou hydrological station would reach to 7,600 m^3/s if the flood forming in the middle Yellow River routing according to the natural flood process. In fact, the flood peak at Huayuankou reduced to 3,120 m^3/s after reservoir operation and water regulation. In 2012, for the flood forming in the upper Yellow River, the flood peak at Lanzhou hydrological station would reach to 5,920 m^3/s after restoring calculation, however the actually flood peak was just 3,860 m^3/s by the jointed operation of Longyangxia and Liujiaxia Reservoir and water regulation.

2.3 Enhancing water resources allocation and implementing the most strict water resources

The State Council approved the "Water Allocation Plan of the Yellow River" in 1987 and promulgated the" Regulations on Water Regulation of the Yellow River" in July 2006. According to these plan and regulation, YRCC established a working mechanism for water regulation. Combination of the Yellow River middle and long – term runoff forecast, including year, quarter, month, and demand plan for water in watershed and relevant areas, the annual water allocation plan among provinces in the Yellow River Basin is developed, with implementation of annual allocation, months planning, and 10 – day regulation.

In 2010, YRCC issued the "Yellow River Water Drawing Permit Management Implementation Details 'and' Yellow River Water Rights Transfer Management Implementation Measures", and initially established the most strict water resources management system framework of the Yellow River, as a result, the Yellow River water resources management turned from mere water supply management to both water supply management and water demand management.

2.4 reinforcing water resources allocation management to establish water – saving society

In recent years, the Yellow River vigorously promotes the establishment of water – saving society, focusing on the construction of water – saving irrigation area, promoting the canal lining transformation and irrigation mode reform, implicating sprinkler irrigation and micro – irrigation technology, to improve the efficiency of irrigation water use. In addition to irrigation district water – saving innovation, the industrial and urban water – saving is also strengthened.

Actively promotes the construction of the Yellow River water rights system, and carries out the transformation of the Yellow River water rights. For Qinghai, Ningxia, Henan, Shandong and other provinces, the water allocation indicators are refined to cities, and counties, to determine the water rights. The pilot program of water rights transfer is developed in Ningxia and Inner Mongolia, by the implementation of water rights transfer, the traditional agriculture is driven to transform to modern agriculture, which is market – oriented, industrialization, large – scale and intensive, improving the efficiency and effectiveness of water resources use.

3　Capacity building to address climate changes should be further strengthened in urgent need

3.1　To improve water and sediment regulation engineering system

The feature of the Yellow River is "Less content of water than sediment, different source of water and sediment and unbalance between water and sand", which cause the complex ecological characteristics. To solve the water problem of the Yellow River, it need to consider the coordination of flood, sediment and water resources, satisfying the demand of water resources, to maintain the ecological system of river. So the YRCC proposes the plan of water and sediment regulation engineering system.

The water and sediment regulation engineering system mainly includes: building and improving the flood and water resources impoundment and regulation structures, which are made up of key water control project in main stream and controlling reservoir in tributaries, shaping river main channel to meet moderate flood and sediment transport, establishing flood and water regulation mechanism, to realize scientific, intelligent and daily water regulation.

3.2　To strengthen water resources forecasting and monitoring capability

With the deepening of the Yellow River management and development, the existing water and sediment monitoring and forecasting ability is difficult to adapt to the new requirements, it is mainly reflected in the following: hydrological network layout is unreasonable, with low density and lack of function, hydrological infrastructure is weak and forecasting ability is low, the number of cross – sections for deposit measurement is insufficient in Ningxia and Inner Mongolia reach, the rainstorm and flood warning and forecasting system of Toudaoguai to Sanmenxia region has not yet been established, sediment forecasting still can not meet the requirements of water and sediment jointed regulation.

In recent, it is urgently need to further strengthen the study on rainfall-runoff law to build water and sediment monitoring and forecasting system, which is mainly including: the hydrological network deployment and hydrological monitoring infrastructure construction, the improvement of water regime forecast communication networks, the enhancement of forecasting ability, strengthening the forecasting ability of hydrology and water resources, and the construction of information management of hydrology and water resources.

3.3　To enhance flood management and research on water resources allocation fighting the shortage of water resources in future

According to forecasts, the demand for water resources will continue to increase with the socio – economic development of the Yellow River Basin, while the amount of water resources in the basin will further reduce by the impact of climate changes and human activities, so the YRCC organizes and develops the study on further optimization allocation of water resources in the basin and inter – basin water diversion. In the same time, the YRCC actively promotes the building of Yellow River water rights system, to carry out the transfer of the Yellow River water rights. Continue to carry out the construction of water – saving society. According to the flood management ideas and concepts of the new era, coordinating flood control and disaster mitigation and promoting the beneficial, to realize effective management of the flood and rational utilization of flood resources. Efforts to increase the effective supply of water resources, while ensuring the safety of flood control to maintain good aquatic ecosystems of the Yellow River Basin.

3.4 To deepen the impact study on climate changes

Recently, the Yellow River Basin has carried out a series of research on climate changes and impact, and has made some progress, but there is a great gap comparing with the grand goal of the Yellow River. It needs further studies in the aspects of identification and assessment of impact of climate changes on the Yellow River water resources, and forecast of climate changes trends in the Yellow River Basin. It needs to deepen the impact study on climate changes, to explore climate change law and identify the impact of climate changes, and develop proper policies and measures to address climate changes.

4 Conclusions

The Yellow River Basin is in the stage of rapid socio – economic development, the task of economic development and ecological environment improvement is very difficult, YRCC has recognized that it is facing a significant challenge in terms of addressing climate change. Guided by the scientific concept of development, and combined with the reality of the Yellow River Basin, YRCC will strengthen the capacity building to address climate change. The issue of climate changes is a common challenge facing the international community, it needs the cooperation of countries in the world and the international community to solve the issue. YRCC will continue to develop international cooperation, implementing technical cooperation projects on climate changes, to improve the ability of responding to climate change.

Developments in Turkey in the Context of Participatory Approach Based on River Basin Management

Nermin ÇİÇEK and *Özge Hande SAHTİYANCI*

The Ministry of Forestry and Water Affairs, General Directorate of Water Management, Ankara, 06560, Turkey

Abstract: In this study, the developments in the area of participatory approach based on river basin management in Turkey have been indicated. This study serves a guide for showing what has been done in Turkey in the area of basin management and participation of related parts to this process. The need of water management in Turkey as a result of semi – arid climate, increasing population, increase of industry etc. has stated. Water pollution which is another important pressure of water resources make the water management crucial has stressed. In the first step of the study, the institutional structure of Turkey in water sector has been showed. Water is an intersection of many sectors for this reason it has many stakeholders such as public bodies, private sector, non-governmental organizations and local people. It needs to manage in a holistic way which includes the participation of all stakeholders in a harmonious way. In addition, the structure of "Water Management Coordination Committee" which established for the coordination of related parts in water sector has been stressed. One of the importance of "Water Management coordination Committee" is that the approval of programme of measures of river basin management plans is done by this Committee and that makes the implementation of programme of measures very esaier. The progress in this area as a result of adaptation of EU Directives has been explained. Turkey has strived various efforts in this issue. As a result of adaptation to EU acquis one of the legislative improvements "Draft By – Law on Protection of Basins and Preparation of Management Plans" has been explained. By – Law includes the procedures and principles of planning and protection of quantity and quality of groundwater and surface water by a holistic approach and principles of preparation of river basin management plans. Furthermore, regional management structures which will be formed during the preparation of "River Basin Management Plans" have been set forth. In the following stage the concept of "River Basin Protection Action Plans" which will be converted to "River Basin Management Plans" has been examined and the efforts for participation of all stakeholders to decision and implementation phases have been defined. The projects, studies and improvements in the area of basin management have explained.

Key words: water resources, participatory approach, management, basin, Turkey

1 Introduction

In the recent years, throughout the world the water resources have gained more significance than ever. In the Middle East, where Turkey is also located, the scarcity of water is being experienced severely. This region is in a semi – arid climate; therefore, the potential of water resources is low. On the other hand, rapidly increasing population of the countries in the region has also influenced the demand for water.

One of the important problems of water resources is water pollution in Turkey. Water pollution is determined in some of the present surface and groundwater in our country due to reasons such as rapid population growth, increase of industry, increase of fertilizer and herbicide use in agriculture and not to be aware of environmental conscious.

The success of water resources management is achieved only by accurately determining the whole relationships among processes effecting the hydrologic system. In this stage, the definition of the system as a basin within its natural boundaries and determination of its available yield will enable a rational and effective water resources management. The sustainability approach, which meets the long-term water demands without creating undesirable effects on the system at the end of active water usage period, must be used.

Within this line, the ongoing works have accelerated through the EU Adaptation Process. The projects conducted in the framework of alignment to the EU Environment Acquis are realized through the participation of all stakeholders. Establishing effective mechanisms for active public participation in planning and decision – making process must be highly prioritized. These mechanisms must be adapted to the appropriate scale, target groups and issues, and they must ensure transparency and accessibility.

2 Institutional structure of Turkey

In Turkey, water management is defined as the aim and responsibilities of the Ministry of Forestry and Water Affairs. The Ministry of Forestry and Water Affairs which is competent authority on subjects related with water, has a general coordination task in terms of development and implementation of Turkey's water policy including coordination of adaptation of Turkish Water Legislation with the EU acquis as well as subjects such as water management and protection of water. The Ministry is in cooperation with other Ministries, public bodies and other stakeholders related with water management issues. One of the responsibilities of the Ministry of Forestry and Water Affairs is conducting necessary coordination within related parts in the river basin management plans.

Other than Ministry of Forestry and Water Affairs, the Ministries have responsibilities about water. Tab. 1 is Turkish governmental organization and their yasks in water management.

Tab. 1 Turkish governmental organization and their tasks in water management

Organization		Main tasks and responsibilities
Ministry of forestry and water affairs	General directorate of water management	1. Define policies in order to preserve, enhance and consumption of water resources. 2. Provide coordination of water management on national and international basis. 3. In order to protect and improve the ecological and chemical quality of water resources considering the protection – consumption balance, including coastal waters, to prepare basin management plans on river basis, to conduct complementary river basin management legislation process. 4. Define precautions with related institutions in addition to evaluate, update and supervise the implementations about pollution prevention on basin basis. 5. Define the aim, principles and receiving environment standards together with the related institutions to monitor water quality and to make it monitored. 6. Define strategy policies about floods and to prepare flood management plans. 7. Conduct necessary coordination related with sectoral basis appropriation of water in compliance with the river basin management plans. 8. Follow processes originated from international convents and other legislation related with protection and management of water resources, to steer the aims related with trans – boundary and verge boundary waters interoperate with other public bodies. 9. Build up international water database information system.
	State hydraulic works	1. Planning, construction and financing of water and wastewater treatment plants. 2. Construct dams and hydroelectric power plants. 3. Build irrigation and drainage systems. 4. Perform all studies for surveys, investigation, conservation and utilization of ground water. 5. Responsible for allocation and registration of ground water. 6. Responsible for flood control.
Ministry of environment and urbanization		1. Responsible from the control, inspection and sanction of discharges. 2. Preparation of environmental impact assessment and environment plans.
Ministry of health		Responsible for water intended for human consumption.
Ministry of food, agriculture and livestock		Responsible from nitrate bylaw

3 Establishment of water management coordination committee

Our country is in a semi-arid region for this reason enhancement of water quality and quantity and achievement of sustainable water usage gains more importance. For this purpose all related institutions need to act in coordination and cooperation.

Coordination and cooperation of institutions is necessary for preparation of river basin management plans, making water related investments, carrying out of other legal, administrative and technical issues.

Water Management Coordination Committee is established for the determination of measures to the protection of water resources in a holistic way, achievement of coordination and corporation of different sectors, enhancement of water investments, attainment of goals stated in national and international documents and implementation of institution's responsibilities stated in the river basin management plans.

The Committee is formed from Ministry of Forestry and Water Affairs, Ministry of Environment and Urbanization, Ministry of Internal Affairs, Ministry of Foreign Affairs, Ministry of Health, Ministry of Food, Agriculture and Livestock, Ministry of Science, Industry and Technology, Ministry of Energy and Natural Resources, Ministry of Culture and Tourism, Ministry of Development, Ministry of European Union and Turkish Water Institute.

Related institutions and organizations, universities, non – governmental organizations, employee associations and private sector representatives can be called and they can participate to sub – committee and committee works.

One of the responsibilities of Committee is that; programme of measures formed for each basin during the studies of river basin management plans will be submitted to Water Management Coordination Committee for approval. By this approval programme of measures will be implemented in the basins.

4 Draft by – law on protection of basins and preparation of management plans

General Directorate of Water Management is conducting the studies about preparation of Draft By – Law on Protection of Basins and Preparation of Management Plans. The aim of this By – Law is determination of procedures and principles of planning and protection of quantity and quality of groundwater and surface water by a holistic approach. This By – Law also includes principles of preparation of river basin management plans.

In the process of preparation of River Basin Management Plans, the Ministry of Forestry and Water Affairs enables the active participation of all related agencies and institutes by receiving opinions about plans. Raising awareness in the public and participation of stakeholders is one of the important points of successful implementation of river basin management plans. For the coordination and participation of stakeholders, committees will be established in the basin scale.

Basin management committees will be established in each basin for planning and protection of quantity and quality of surface water and groundwater by preparing, implementing, auditing and assessing of river basin management plans. Basin management committees involve the provinces in the basin. The head of the committee will be the Governor of the province which has more severe water management problems. The committees comprised from provincial directorates of Ministry of Forestry and Water Affairs, Ministry of Environment and Urbanization, Ministry of Internal Affairs, Ministry of Foreign Affairs, Ministry of Health, Ministry of Food, Agriculture and Livestock, Ministry of Science, Industry and Technology, Ministry of Energy and Natural Resources, Ministry of Culture and Tourism, Ministry of Development, Ministry of European Union and local authorities, universities and NGOs.

5 River basin protection action plans and river basin management plans

In Turkey 25 river basins are identified. General Directorate of Water Management has prepared 11 River Basin Protection Action Plans and the Project "Preparing River Basin Protection Action Plans for 14 Basins" is going on. In the year 2013, River Basin Protection Action Plans for 25 basins will be completed. By the year 2023, these 25 River Basin Protection Action plans will have been converted to River Basin Management Plans. "Draft Büyük Menderes River Basin Management Plan" is prepared as an output of the Twinning Project "Capacity Building Support to the Water Sector in Turkey". "Conversion of River Basin Protection Action Plans to River Bain Management Plans Project" which includes 5 basins (B. Menderes, Meriç – Ergene, Susurluk, Gediz, Akarçay Basins) will start in 2013.

In the scope of River Basin Protection Action Plans; quantity, properties and pollution of surface and groundwater, pressure and impacts as a result of industrial, agricultural and economical activities are determined, water quality maps are formed, for the protection of river basins and reducing of pollution; programme of measures in the short, medium and long terms is prepared. The content of River Basin Protection Action Plans is as below:

(1) General description of the river basin district.
(2) Field surveys and determination of environmental background.
(3) Water quality classification.
(4) Calculation of pollution loads.
(5) Prominent environmental problems of the basin and their solutions.
(6) Planning of urban wastewater treatment plans.
(7) Preparation of River Basin Protection Action plans.
(8) Submission of data to GIS.

Preparation of River Basin Protection Action plans for 11 basins (Marmara, Susurluk, Kuzey Ege, Küçük Menderes, Büyük Menderes, Burdur, Konya Kapalı, Ceyhan, Seyhan, Kızılırmak, Yeilşrmak) has finished in 2010. Preparation of River Basin Protection Action Plans for 14 Basins (Antalya, Doğu Akdeniz, Batı Karadeniz, Fırat – Dicle, Doğu Karadeniz, Batı Akdeniz, Çoruh, Aras, Asi, Meriç – Ergene, Van, Akarçay, Gediz, Sakarya) started in December 2011 and will be ended in December 2013.

Fig. 1 is river basins of Turkey.

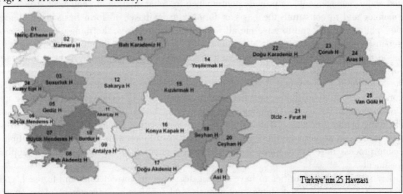

Fig. 1   River basins of Turkey

6   Participatory approach in river basin protection action plans

During the Preparation of River Basin Protection Action Plans of 14 Basins Project (2011 ~ 2013) there will be 3 meetings with stakeholders for every basin. The schedule of the meetings is as below(see Tab. 2).

Until now, Opening Meetings are conducted for 14 basins. For each meeting all related institutions, public bodies, industry sectors, non – governmental organizations and local people are invited to the meetings. In these meetings, firstly river basin protection action plans are explained to

38

people attained meetings. In the second part of the meetings contribution of participants is asked. The participants asked the questions arised in their heads about the project and they added more information about the site. Participants of the opening meetings is Ministry of forestry and water affairs, TUBITAK, Provincial Directorate of Food, Agriculture and Livestock , Provincial Directorate of Science, Industry and Technology, Provincial Directorate of Health, Provincial Directorate of Environment and Urbanization, Regional Directorate of State Hydraulic Works, General Directorate of Environmental Impact Assessment, Permission and Audit, Governorships, District Governorships, Municipalities, Regional Directorate of Bank of Provinces, Provincial Special Administration, Universities, Development Agencies, Research Institutes.

Tab. 2　Conducted meetings in the project

Opening meetings	February ~ March 2012
1. Stakeholder meeting	February ~ March 2013
2. Stakeholder meeting	September ~ October 2013

7　Conclusions

The main reasons of water related problems are rapid and unplanned urbanization depending on population growth, industrialization, intense agricultural activities, misuse of the lands and global warming. But the real problem is related on "Integrated Management" which is developed by the targets such as planned and economical usage of water which is a natural resource and has no alternative. It includes determination and prevention of the problems which threaten water resources, protection of water and water related ecosystems and sustainable economical growth.

For accomplishment of the implementation, preparation of appropriate legislation is important to form corporate infrastructure to facilitate the applications within the frame of obligations in the EU membership process and to remove the legal overlapping, to remove corporate conflicts, to realize integrated water management, to provide continuance by rehabilitating of ecological structure of water, to monitor and to constitute the basis of forming a database, to form basis to implement the principle "polluter pays" and to integrate all water quality directives. Forming a mechanism to support inter-institutional coordination is also important to implement the directives. Especially forming basin based management models for structuring of studies in basins by the owners of the basins will carry out better and effective solutions.

The Ministry of Forestry and Water Affairs process his development and vision in such a way, he is working compatible with the benefits of the country in EU membership process. Especially studies on harmonization of the Directives and studies on reorganization are the indicators of truly comprehending of the process. "Water Management Coordination Committee" is formed to put our target and strategies, to activate the organizations and institutions with the aim of evaluation of EU membership process in terms of water sector and implementation of our national based applications and investments.

Turkey specified his priorities to have correct decisions on correct time and to reach good status targets and to preserve good status continuously and forming his investment plans considering defined time frames. Time has become an important concept for Turkey who wants to be an EU member. For this reason all the prepared plans will be realized and put into practice.

References

The Ministry of Environment and Forestry, Capacity Building for Water Sector in Turkey.
Sarıkaya H Z, Çiçek N. "Management of Water Sources, EU Membership Process and the Applications of the Ministry of Environment and Forestry.
Çiçek N. Water Framework Directive and EU and Turkey Approach within the Example of Büyük Menderes River Basin Management Plan[R]. Selçuk University, 2010.

Practice and Pondering on Water Governance and Stewardship in Yellow River Basin

Guo Guoshun[1] and *Niu Yuguo*[2]

1. Yellow River Conservancy Commission, Zhengzhou, 450003, China
2. Yellow River Henan Bureau, Zhengzhou, 450003, China

Abstract: Water is the source of life, key of production and base of ecology. Water is so important that it has been under the management of the government. Water resources in drainage basin are an important part of national water resources and under the integral management and regulation of basin administrative organization—national agency. When shortage of water resources becomes more and more serious and water crisis gets severe, each country is seeking solution and changing the intrinsic administrative mode, one of which is to encourage public and stakeholders to participate water management in order to improve the performance of water management and meet the requirement of different stakeholders. The special report begins with notion, role, necessity and development tendency. Based on the analysis of water management and public participation of Henan Yellow River, relative experience is summered up about public participation in water management in Henan Yellow River, that is, attracting the public to pay attention to and take part in water management through publicizing, comprehensively promoting to make governmental information public, ensuring civil right to know, declare, defend and hearing in accordance with provisions. Water – taking hearing system has been put into practice. Linkage mechanism of Yellow River water administration was established. Information is made public in time. At the meantime, shortcomings existing in public participation in water management are still stated.

Key words: water management, public participation, experience, proposal

Water governance is to promote directly or indirectly the rational exploitation and utilization of water resources through political, economic, social, legal and administrative measures.

Stewardship refers to social action with goal in the scope of rights and obligations of the masses, public organizations, units or individuals.

Nowadays, with the water resources running short and water crisis becoming severe, nations are seeking solving methods and innovating each management philosophy and mode, of which is to encourage the public and stakeholders involving in water governance in order to improve the performance of water governance and meet the appeals of stakeholders.

As a large river under the governance of the Central Government of People's Republic of China, water governance includes not only the scientific management of water resources but also water administration, flood control administration. It needs diligent work of specialized department and even stewardship.

1 The basic information and challenge of stewardship and water governance in Yellow River basin

(1) The basic information of stewardship and water governance in Yellow River basin.

In history, water resources has owned to the state, which has implemented centralized management. Water resource of the basin has been governed by the basin agency—dispatched agency of the state. The Yellow River has been governed directly by the Central Government due to its significance in history. So stewardship and water governance has a long history.

(2) The challenge of stewardship and water governance in Yellow River basin.

As the second largest river of China, flood control security, water resources supply security, ecological security and waterways control of the Yellow River would be very important to social and economic development of the country.

The imbalance between supply and demand of water resources of the Yellow River would be more and more serious in the future.

The large floodplain exists in the lower Yellow River, which is different from any other river. Human competing for land with the river on the floodplain becomes intense. Water governance environment turns complicated in Yellow River basin.

With the development of economy and society and the progress of urbanization, a large quantity of manpower in the countryside moves to towns and cities, which makes flood control by the masses meeting unprecedented difficulty.

2 Actions to promote the public involving in water governance of the Yellow River

(1) To arouse the consciousness of the public partaking in water governance of the Yellow River through publicizing.

Multiform publicizing campaigns have been launched for years to enhance the understanding and legal sense and foster the consciousness of stewardship.

(2) Making government information public is promoted to build the platform of the public partaking in water governance.

Making government information public is an important way to impel water governance to come out into open.

(3) Mechanism of stewardship is set up to guide the stakeholders to partaking in water governance.

One is water affairs management linkage mechanism; the second is the mechanism of rational utilization of water resources; the third is flood control consultation decision mechanism.

(4) System is perfected to guarantee the rights of the public partaking in water governance

3 Issues and shortcomings existing in public partaking in water governance

(1) The concept of the public partaking in water governance should be set up.

(2) The system of the public partaking in water governance would be made sound.

(3) The mode of the public partaking in water governance would be further made perfect.

(4) The platform of the public partaking in water governance would be structured.

(5) Publicity would be enforced more to enhance the consciousness of the public partaking in water governance.

Study on Potential Application and Implementation Strategy of Decentralized Wastewater Reuse in Northern China

Simon Spooner[1], *Guo Wei*[2], *Wen Huina*[3] and *Zhang Yi*[4]

1. Atkins China, Beijing, 100022, China
2. Shaanxi Provincial Academy of Environmental Science, Xi'an, 710061, China
3. Yellow River Water Resource Protection Bureau, Zhengzhou, 450004, China
4. Mott MacDonald China, Beijing, 100029, China

Abstract: Water resource scarcity in northern China can be attributed to low levels of total precipitation and extreme seasonal variation in rainfall patterns. The severe water pollution exaggerates the water resource shortage. A series of measures involving water saving, water pollution control and wastewater reuse have been proposed to alleviate water resource scarcity in northern China. Wastewater reuse has been considered more and more as an important way of supplementing dwindling water sources. This paper discusses the potential application of decentralized wastewater reuse (DWWR) in northern China and formulates the implementation strategy based on the analysis of domestic policy supporting, water reuse standards, available technologies, construction cost and recovery, operation and maintenance, financing, monitoring and so on. It concludes that the cost of constructing the pipe network for transporting the reclaimed water to the point of reuse is the biggest impediment to greater uptake of wastewater reuse, simplification of the pipe networks for the reclaimed water or seeking feasible solutions to the pipe distribution networks maybe the crux issue for water reuse and recycling. Decentralised wastewater reuse seems to be an effective way to reduce the main impediment to greater reuse of reclaimed water. The most important driver for the promotion of decentralised wastewater reuse is the government policy. Several suggestions to the government are proposed based on the key findings from the research.

Key words: decentralized, wastewater, reuse

1 Background

To help address water scarcity and pollution, China has set targets for wastewater treatment, reuse and recycling. But reuse and recycling can be costly, due to the need for long transmission pipelines to a wastewater treatment plant and back. Under certain circumstances it can be more effective to collect wastewater from a community or enterprise, treat it on site and then reuse the water directly. The EU-China River Basin Management Programme (RBMP) Research project on Decentralized wastewater treatment and effluent reuse in Northern China analysed the steps and techniques for reducing water use, collecting and making dirty water re-usable. The study conducted a review of available technologies around the world, the roles of different stakeholders in planning, financing, implementing and operating such schemes, and then prepared feasibility studies and preliminary designs for two pilot projects. This paper summarizes the key findings and recommendations on potential application and implementation strategy of decentralized wastewater treatment and effluent reuse in Northern China.

On potential applications and impact, the study concludes that a decentralised approach to wastewater treatment is suitable for areas where population density is too low for centralised sewerage and treatment to be economic, and yet too developed for rural technologies such as biogas units or composting latrines to be applicable. There are many examples of small-scale wastewater treatment technologies in the EU utilising the latest advances in membranes, anaerobic treatment, packaged plants and natural wetland systems. These technologies may be particularly applicable in Northern China, because of its water shortages, current lack of wastewater treatment infrastructure, and geographically dispersed communities. Although decentralized treatment is typically more expensive per unit of treated water, and requires a higher level of skills to operate, there are many advantages, such as more flexibility for local communities, water savings, and lower costs of wastewater collection and disposal. When payment for wastewater treatment and stricter criteria for discharge of wastewater to surface waters are introduced in China, decentralised wastewater treatment is likely to become even more attractive.

2 Key issues of decentralized wastewater reuse

2.1 National and local policies

There is clear government policy support in both central and local levels in China for centralised and decentralised effluent reuse. However, the specific law and regulation for implementation is still at an early stage now in China.

2.1.1 Lack of the special law system

Except for the water laws including the requirement of waste water resource protection and pollution control, it seems that there is not a special law or regulation for waste water reuse and recycling. The No.1 policy document of 2011, and No.3 Document of 2012 set out the requirements for water saving, water efficiency and pollution reduction but do not set specific targets for wastewater reuse and are policy directions rather than legal documents.

2.1.2 Lack of the detailed requirements for implementation

The detailed requirements for implementation are not complete and specific and do not cover the planning, design, construction, management, utilization, supervision, and encouragement and punishment aspects about the reclaimed water. These regulations should be developed over the coming years.

It is recommended that the special law and regulation systems should be established to facilitate the process of planning, design, construction, management, utilization, supervision, and encouragement and punishment.

2.2 Criteria for water reuse

2.2.1 International criteria about wastewater reuse

There are generally different priorities placed on the reclaimed water's parameters, dependant on its reuse purpose. These are described in the Tab. 1.

Tab. 1   Consumptive users of reclaimed wastewater and qualitative water quality requirements

Consumptive uses	Removal of pathogens	Chlorine residual or other disinfection	Removal of suspended solids & turbidity	Presence of dissolved oxygen	Removal of BOD & COD	Removal of nutrients	Removal of taste, odour and colour	Removal of trace organics & metals	Removal of excess salinity
Landscape & forest irrigation	x	–	x	x	x	–	x	x	0
Irrigation of restricted crops (Groups A & B*)	x	–	x	x	x	–	x	x	0
Irrigation of limited crops (Group C*)	xx	xx	xx	xx	xx	–	x	xx	x

Continued Tab. 1

Consumptive uses	Removal of pathogens	Chlorine residual or other disinfection	Removal of suspended solids & turbidity	Presence of dissolved oxygen	Removal of BOD & COD	Removal of nutrients	Removal of taste, odour and colour	Removal of trace organics & metals	Removal of excess salinity
Irrigation of crops & produce (Group D*)	xxx	xxx	xxx	xx	xxx	–	xx	xxx	x
Industrial Reuse	xx	xx	xxx	xxx	xxx	xxx	xxx	xx	xx
Dual Urban Systems – Toilet Flushing & Gardening	xxxx	xxxx	xxxx	xxxx	xxxx	xxx	xxxx	xxxx	xx
Potable Reuse	xxxxx	xxxxx	xxxxx	xxxx	xxxxx	xxxx	xxxxx	xxxxx	xxx

(–) no need; (0) usually not essential; (x) slight need; (xx) moderate need; (xxx) strong need; (xxxx) stringent requirements; (xxxxx) very stringent requirements

Source: SHELEF 1991

Note: 1. Unrestricted crops (Group D) requires high quality irrigation water, as they include vegetable deem safe to be eaten raw.

2. Limited crops (Group C) are crops that will require processing before consumption.

3. Restricted Crop (Group A&B) is based the Group C's criteria, and further assessment based on type of soil, proximity to potable aquifer, irrigation method etc.

For industrial uses, there are frequently no firm or widely recognised common guidelines dictating the water standards. The water standards are mainly considered for their impact on the plant processes and maintenance (e. g. scaling, corrosion etc) as well as its effect on the effluent discharge. In contrast, for water reuse purposes which have the potential to affect human health (i. e. potable, irrigation etc), there are usually strict guidelines by local authorities. In many cases, standards can be more stringent than what is expected of treated raw water.

For irrigation reuse:

(1) The WHO provides guidelines on using wastewater in agriculture.

(2) Similarly, the United Nations Food & Agricultural Organisation, (FAO) publishes various reports on water reuse guidelines and good practices (which in some parts, reference to WHO standards).

(3) The United States Environmental Protection Agency (US EPA) also has comprehensive guidelines and regulations for water reuse in irrigation and other reuse purposes.

2.2.2 Water reuse standard applied in China

A summary of the Chinese standards relating to effluent reuse and the sectors to which these standards apply is included in the following Tab. 2.

Tab. 2 Summary of water reuse category in Chinese standards

No.	Category	Application	Standards
1	Agriculture, forestry, animal husbandry, fishery	Agricultural irrigation, Forestry, Husbandry, Fishery	The reuse of urban recycling water-water quality for farmland irrigation(GB 20922—2007, 2007 – 10 – 01)
2	Miscellaneous water	Landscaping, Toilet flushing, Road cleaning, Car washing, Construction, Fire fighting	Reuse of recycling water for urban-water quality standard for urban miscellaneous water consumption(GB/T19820—2002, 2003 – 05 – 01)

Continued Tab. 2

No.	Category	Application	Standards
3	Industry	Cooling, Washing, Boiler, Technical process, Product	The reuse of urban recycling water-water quality standards for industrial uses (GB/T 19923 — 2005, 2006 – 04 – 01)
4	Environment	Recreation, Scenery/land-scaping, Wetland	The reuse of urban recycling water-water quality standard for scenic environment use (GB/T 18921 — 2002, 2003 – 05 – 01)
5	Water recharge	To surface water, To ground water	Not issued yet

Note: The standard limitation value can be referred to the standard texts.

(1) Comparison with the discharge standards of pollutants for municipal wastewater treatment plant.

Different water reuse standards require wastewater to be treated to a different level and most of these standards are not as stringent as Class 1 A, the urban wastewater discharge standards. Class 1 A requires achieving BOD of 10 mg/L, COD 50 mg/L, Ammonia Nitrogen 5 mg/L, SS 10 mg/L, TN 15 mg/L, TP 0.5 mg/L, and coliform less than 1,000 /L. For all the reuse water standards, only GB/T 18921—2002 and GB/T 18920—2002 requires more stringent consents.

GB/T 18921—2002 requires achieving BOD of 6 mg/L, which is achievable by certain tertiary treatment technologies; GB/T 18920—2002 requires achieving no more than 3 coliforms per litre water whilst Class 1 A requires achieving no more than 1,000 coliforms per litre water. Effective disinfection measures will be required to meet the requirement of GB/T 19820—2002..

(2) Comparison with other standards.

In addition to discharge standards of pollutants for municipal wastewater treatment plant listed above, there are some other standards related to the water reuse standards. Irrigation of crops for example is often carried using surface water and a different standard. GB 3838—2002 applies for the quality of water that is permitted to be used. Under this standard, up to Grade V surface water is permitted for agricultural irrigation. A comparison between this standard and GB 20922—2007 as well as new GB 5084—2005 is included below for information.

As shown in Tab. 3, there are fewer items for water quality assessment in GB 3838—2002 compared with GB 5084—2005 and GB/T 20922—2007, while it could be more stringent than the other two. However, there are specific standard values set for irrigating different kind s of vegetation.

Tab. 3   Comparison of water Quality Standards Related to Irrigation

No	parameter	Environmental quality standards for surface water (GB 3838—2002)	Standards for irrigation water quality (GB 5084—2005)			The reuse of urban recycling water-Quality of farmland irrigation water (GB 20922—2007)			
			Crop types			Crop types			
		V (mainly for irrigation water use and certain landscaping water body)	Wet farmland	irrigated	vegetable	Fiber crops	Dry grain and seed crops	Wet grain	Open-air vegetables
1	Temperature								
2	pH	6 ~ 9		5.5 ~ 8.5			5.5 ~ 8.5		
3	DO	≥2					≥0.5		
4	COD$_{Mn}$	≤15							

Continued Tab. 3

No	parameter	Environmental quality standards for surface water (GB 3838—2002) — V (mainly for irrigation water and certain landscaping water body)	Standards for irrigation water quality (GB 5084—2005) Crop types			The reuse of urban recycling water-Quality of farmland irrigation water (GB 20922—2007) Crop types			
			Wet farmland	irrigated	vegetable	Fiber crops	Dry grain and seed crops	Wet grain	Open-air vegetables
5	COD	≤40	≤150	≤200	≤100a, 60b	≤200	≤180	≤150	≤100
6	BOD$_5$	≤10	≤60	≤100	≤40a, 15b	≤100	≤80	≤60	≤40
7	NH$_3$ – N	≤2.0							
8	TP	≤0.4 (0.2 for lake and reservoir)							
9	TN	≤2.0							
10	Cu	≤1.0	≤0.5	≤1.0	≤1.0		≤1.0		
11	Zn	≤2.0		≤2.0			≤2.0		
12	Fluoride	≤1.5		≤2(normal area), 3(high Fluoride area)			≤2.0		
13	Se	≤0.02		≤0.02			≤0.02		
14	As	≤0.1	≤0.05	≤0.1	≤0.05	≤0.01			≤0.05
15	Hg	≤0.001		≤0.001			≤0.001		
16	Cd	≤0.01		≤0.01			≤0.01		
17	Cr^{6+}	≤0.1		≤0.1			≤0.1		
18	Pb	≤0.1		≤0.2			≤0.2		
19	Cyanide	≤0.2		≤0.5			≤0.5		
20	Volatile phenol	≤0.1		≤1			≤1		
21	Petroleum	≤1.0	≤5	≤10	≤1.0	≤10		≤5	≤1
22	Anionic surfactants	≤0.3	≤5	≤8	≤5	≤8			≤5
23	sulphide	≤1.0		≤1			≤1		
24	coliform	≤40,000	≤4,000	≤4,000	≤2,000a,		≤40,000		≤20,000
25	SS	≤80	≤100	≤60a, 15b		≤100	≤90	≤80	≤60
26	Total salt content			1,000c(normal area)					
27	Ascaris egg			≤2	≤2a,1b		≤2		
28	chloride			≤350			≤350		
29	benzene			≤2.5			≤2.5		
30	chloral		≤1	≤0.5	≤0.5		≤0.5		
31	acrolein			≤0.5			≤0.5		
32	TDS						Normal area 1,000, saline area 2000		1,000
33	Residual Chloride						≥1.5		≥1.0
34	Be						≤0.002		

		Environmental quality standards for surface water (GB 3838—2002)	Standards for irrigation water quality(GB 5084—2005)		The reuse of urban recycling water-Quality of farmland irrigation water(GB 20922—2007)			
No	parameter		Crop types		Crop types			
		V(mainly for irrigation water use and certain landscaping water body)	Wet farmland	irrigated vegetable	Fiber crops	Dry grain and seed crops	Wet grain	Open-air vegetables
35	Co				≤1			
36	Fe				≤1.5			
37	Mn				≤0.3			
38	Mo				≤0.5			
39	Ni				≤0.1			
40	Bo				≤1.0			
41	V				≤1.0			
42	Formaldehyde				≤1.0			

Notes: 1. a: With the process or procedure of peel removing.

    b: Raw vegetables, melon and herbaceous fruits.

    c: For some places the employed 'Figures' can be correspondingly higher, such as places with water infrastructure where drainage and groundwater runoff routes are in place and places with sufficient fresh water to clean and reduce soil salinity.

2. Gray highlight: Potential conflicts between GB 5084—2005 and GB 20922—2007.

3. * Other requirements: ① Forfiber crops and dry grains, strengthened primary treatment is required for urban wastewater treatment; and for wet grain and open-air vegetables secondary treatment is required. ② The main water conveyer for irrigation should be treated to be impermeable to prevent contaminating the aquifer. The water quality in the nearest water abstraction point has to meet the requirements in this standard. ③ Before the reclaimed water can be used for irrigation, irrigation experiments should be implemented under specific climatic conditions, crop patterns and soil types to determine proper irrigation schedules.

4. Red figure: basic parameter

5. Blue figure: alternative parameter

(3) Summary.

A series of water quality standards in terms of various reuse purposes have been enacted, however there are some items that disagree with some related standards. Therefore, it is necessary that research should further be conducted on the unification and specification of the reclaimed water standards. The application scope and rules of the standards need to be further identified, which will facilitate the implementation of these standards.

2.3 Treatment technology

Currently wide range of treatment technologies are used internationally. Technology selection is dependent on both the source of wastewater and also the desired reuse application. China also has considerable diverse technologies successfully applied for wastewater treatment and reuse. With the enforcement of more stringent urban wastewater discharge standards in recent years, advanced, effective and reliable tertiary treatment technologies have been imported from other countries, and most of them have been localised.

Processes such as Oxidation Ditches, MSBR, MBR, Contact-Oxidation including Aerated Fixed Bio-film and filtration including sand filters are commonly adopted in China and are able to meet different standards of water reuse. Some of these processes are capable of effectively removing TN and TP.

In terms of the ability of a particular technology to meet required discharge standards there is little difference between domestic technology and overseas technology. When making a choice between processes and technologies for decentralised wastewater treatment, the main factors to be considered include investment and operational costs, site availability, discharge standards to be achieved and the available management capabilities. To determine the most appropriate solution a formal assessment system is required.

2.4 Cost

2.4.1 Construction cost

Same as all the water construction projects, the construction cost estimations are generally to be done by the project owner initially, and then the contracted design institute gives a detailed breakdown cost.

Based on an indicative reference case from the project, typical cost break down could include the following contents: Civil work, facility, facility installation, design fees, facility testing and trainings, contingency. Projects may be divided into a number of phases for investment .

In addition to construction costs, most of Chinese projects have the following additional costs to include in estimation:

(1) Land acquisition, resettlement and greening compensation cost.

(2) Costs for particular feasibility study, EIA, Geo-technical survey etc.

(3) Construction organization management fee.

(4) Engineering quality inspection fee.

(5) Construction supervision cost.

(6) Office furniture and other relevant domestic costs.

(7) Test and monitoring.

(8) Staff training cost.

(9) Design document reviewing cost.

(10) Commissioning cost .

(11) Power connections fee.

(12) Equipment Import relevant cost.

(13) Relevant tax.

(14) Bank loan interests during construction phase.

2.4.2 Cost recovery

The project cost recovery could be different, according to the various types of water re-using and beneficiaries. In the study, major project cost recovery could be achieved through the following aspects:

(1) The reused water sold to the industrial user, with reasonable tariff arrangement.

(2) Through the collection of the wastewater treatment tariff and selling of the reused water.

(3) Through savings in the use of water from the municipal supply.

(4) Economic benefits can also be realised as a result of the reduced environmental impacts of using reused water and the promotion of environmental protection

(5) Operating costs designed to be covered by the management fee paid by the individuals and enterprises for the general operating expenses of the village.

In some of projects fully or partly financed by government, the cost recovery was not major part of this project initiative. Certain environmental benefits could be generated which can promote sustainable development and therefore bring economic benefits. For instance in one case study, a village enterprise had to pay a pollution discharge levy related to the amount of pollutants discharged. By undertaking a decentralised wastewater project the levy paid by could be reduced.

2.4.3 Characteristics for water reuse project cost estimation

In the study, it was noted that decentralized water reuse projects may have their particular characteristics in capital cost preparation. The following points may be useful to think about in the earlier stages of project preparations:

(1) Government organized financing, official financing platform and channels, and Government subsidies will be very important for proposed projects, as government finance will still share quite a big part in the coming few years. Therefore, together with relevant government assistance, public participation, stakeholders' assistance would be critical for the project design.

(2) Investment for pipeline system (ie raw water lines, product water lines, pumping stations), the pipeline system investment can be a significant portion of the total project investment.

(3) Staff training. Depending on the project nature, in a lot of cases, the lack of construction and operational staff capacity could be a major challenge for the project owner. Therefore relevant staff training is very important to the success of a project.

(4) Safety and monitoring. Need to consider relevant requirement for safety and monitoring in a project's early stages. Including clear system separation from normal tap water and risk of potential wrong connections, etc. Relevant regular monitoring for the raw water and product water will be important during operation. It is important to include budgeting considerations for these early on.

(5) Construction contracting management. As the nature of the investment project could be quite small individually, it might be useful to remind project owners to pay particular notice on the contract, which includes the contents or guarantee period, spare parts and maintenance

2.5 Operational and maintenance

Currently in China most owners of decentralised wastewater treatment and reuse projects operate by themselves, and the BOT mode is not the common one nor is it common for large private companies to operate multiple sites. The third party, the special organization, is more and more welcomed to participate in or be in charge of the operation of works through public private partnerships and concessions etc.

2.6 Project financing

2.6.1 Major project financing sources

In North China, there are some choices for the project financing. In most cases, it could be multi-source options, it was expected that in immediate coming years the majority will be from major government channels. However, some government financing methods which have particular requirements or timing limitations may not always be available. That means the project owners need to make efforts to obtain relevant information and to understand these details as early as possible.

The project financing could be assisted from international finance organizations, or bilateral government loans. The international finance organizations could be the World Bank, Asian Development Bank (ADB), or other global or regional organizations.

Different from international financing organizations, private sector investment could be more flexible for reasonable small individual projects financing.

A positive development in the Chinese water policy is the greater use of market forces in the water sector so that projects become more attractive for private sector investment. This should attract investment in water reuse in areas of water scarcity due to the comparatively high water tariffs involved. This area could also be further explored by considering a number of models whereby private sector financing can be utilised.

Private Sector Investment for water project may include: BOT, TOOT, TOT, concession agreement, BOOT, DBO, contract of operation, etc. In the last 2 ~ 3 years particularly, Chinese water sector is not unfamiliar with these PPP projects. Detailed contents could be found though various local channels

2.6.2 Water rights allocation and trading

Water trading does not have a long or extensive history in China although recent advances such as the publishing of "Regulations on water licensing", issued by the State Council in 2006 are helping to advance the sector. Some ad hoc arrangements have been developed in different locations and these could be investigated to determine whether any suitable models for water trading could be developed and replicated elsewhere.

In particular water rights trading has been piloted in the Yellow River Basin with the support of the Yellow River Conservancy commission. In 2011 and 2012 there were dozens of pilots projects where water users requiring access to additional water resources were denied access until they could demonstrate that they had made investments in water savings for existing users such that the existing user could transfer a part of their water rights. For example, a new coal powerstation would invest in lining the irrigation channels of a nearby regional irrigation scheme, so preventing leakage and would then be allowed to apply for a licence to withdraw up to the amount of water saved such that the total water use in the region remained constant within the agreed allocation.

Given the serious conditions of water resources shortage and water environment degradations, water rights trading will gradually become a core part of water resources management and institutional reform. The process has promoted water resources transferring from lower (extensive agriculture) towards higher-value utilizations (Urban supply, industry or high intensity agriculture). It showed relevant benefits for socio-economic development and sustainability of water using with limited resources.

If the water savings from a decentralised wastewater treatment and reuse scheme can be clearly demonstrated then it may be possible that neighbouring water users will be willing to invest in the project in order to obtain water rights access. This could require the coordination of multiple water users but could provide novel financing sources. This will require further research to develop.

2.6.3   Topics on financing to be studied

Funding is an important constraint for the implementation of decentralized wastewater treatment and reuse projects that cannot be ignored. For decentralized wastewater treatment and reuse projects, because they are small scale, and the direct stakeholders are not government agencies, it is difficult for them to get the financial support from local government's finances resulting in financing difficulties. There is an urgent need to broaden the channels of financing available to such projects in the future. Some ways mentioned as follows should be considered and studied Government set up special funds, subsidies(bonus), or issue stock to support the decentralized sewage treatment and reuse project.

(1) Government saves as a guarantee for banks to get low interest or interest-free loans for the decentralized sewage treatment and reuse project.

(2) To encourage the third party, specialized water treatment service companies undertake the decentralized sewage treatment and reuse projects. Are there economic means to inspire BOT models.

(3) To exempt sewage discharge tariff and return water resource fee to the decentralized wastewater treatment reuse project.

(4) Develop ways to involved private companies and individuals financing in decentralized sewage treatment reuse projects.

(5) To include decentralised wastewater treatment and reuse schemes in pilots for water rights trading.

2.7   Monitoring

Monitoring through all the treatment process is of importance for the wastewater reuse projects. It is helpful for the operational control, and is the main way to ensure the reclaimed water quality that is safe for usage.

Multi-sectoral coordination and supervision mechanisms need to be introduced and established, preferably under the wastewater reuse regulation and enforcement system. The sanitary safety of the reclaimed water is ascribed to the responsibility of the local epidemic prevention department. Efforts should be paid particularly to avoid the reclaimed water being wrongly connected to the drinking water supply system or entering the food chain indirectly. The supply of the surface water bodies by using reclaimed water as the water source should be supervised by the local environmental department, while the routine monitoring of the reclaimed water quality needs be clearly defined to a responsible body. In order to make sure the reclaimed water quality is under control, all of the operational data needs to be appropriately collected and entered into a sound record-keeping system, including the information of sources, treatment, directions, deliveries, users, etc.

3 Conclusions

Implementation of reclaimed water projects involves a series of items involving national policy and regulation, water sources, standards, treatment technologies, pipe network, buffering facility, safety ensurance, operation and maintenance, operation cost and financing. The cost of constructing the pipe network is the biggest impediment to greater uptake of wastewater reuse. Therefore, simplification of the pipe networks for the reclaimed water or seeking feasible solutions to the pipe distribution networks maybe the crux issue for water reuse and recycling. Decentralised wastewater reuse seems to be an effective way to reduce the main impediment to greater reuse of reclaimed water.

Buffering (storage of the water to be reused to balance supply and demand) is a necessary element of the wastewater reuse system. Besides the pipe networks, the reclaimed water users are limited and required water at different times. As a result, a contradiction emerges between balance of the production and utilization of reclaimed water. Accordingly, buffering units are necessary and should be considered in the design of the wastewater reuse projects.

Artificial surface water bodies (lakes and reservoirs) could be the simplest and most effective buffering unit. It can be used as landscaping water supply source, also be useful for buffering and storage. Once the surface water is considered as a buffering unit, the water quality has to be noted. These storage units can also be used to allow for the addition of surface runoff water. The velocity of the water is required for the water quality can be maintained by self-purification process of the flowing water and aeration. Care must be applied to the water quality, especially in open water bodies as any excessive algal growth may make the water unfit for intended reuse purpose. Depending on the intended reuse further tertiary treatment of the stored water may be required.

Consideration can also be given to storage of water in artificial wetlands (sealed or open to groundwater) or in underground storage units, either sealed tanks or in aquifers, recharging the natural storage structures.

4 Suggestions for government

(1) A systematic plan for reclaimed water system promotion should be organized, as it will be the basis for wastewater reuse projects. A municipal wastewater reuse and recycling plan needs to be made as the basis of projects. This Plan could be done by cooperation between environmental protection bureau and construction departments of the local government. The Plan will be comprehensive, and covering objectives, general principles, willingness, targeted water reuse efficiency in different phases. The plan will also address major technology options, the industries and areas preference or compulsory utilization of reclaimed water, wastewater tariff principal, community facilities and constructions for wastewater reuses, governmental funding or subsidies for the pipeworks, financing encouragement policy and preferential policy for enterprises.

(2) A formal document for implementation details needs to be organized for "Reclaimed Water Technical Policy" issued by the State Government in 2006···The document needs to have detailed explanations for the effective implementation.

(3) Relevant researches for the technology, standards, applications, demonstrations and disseminations of DWWR could be initiated by Development and Reform Commission (DRC), and participated by other related government departments, ie Science and technology, Urban Construction, Environmental Protection and Finance departments. Relevant government departments above mentioned may have their own dedicated plans or annual plans to cover these items regarding technologies, demonstration, standards, investment budget, which can ensure the successful implementation of DWWR program.

(4) For the construction of pipe networks and buffering/storage unit, a new standard for consistent reclaimed water might be useful.

(5) A requirement can be specified that major urban landscaping features and watering green spaces in urban areas should be sourced from a certain proportion of reclaimed water. This could be used as one of the contents for local government performance review. This would be most effective in driving forward investment in such water efficient infrastructure

Acknowledgements

We gratefully acknowledge the significant contributions to research study on Decentralized Wastewater Reuse in Northern China from the team members, from Shaanxi Provincial Academy of Environmental Science, Mr. Si Quanyin, from Shaanxi Architechture University, Mr. Yuan Linjiang, from Mott MacDonald China, Mr. Fang Songchuan and Mr. Chris Preston, from COWI Denmark Mr. John Sorensen and from Atkins China Mr. Robert Santiago, from CREAS, Professor Meng Wei and Professor Song Yonghui.

The research study on Decentralized Wastewater Reuse in Northern China was undertaken as part of the EU-China River Basin Management Programme which was funded by the European Union.

References

Zhang Liping, Xia Jun, Hu Zhifang. Situation and Problem Analysis on Water Resource Security in China [J]. Resources and Environment in Yanngtze River, 2009, 18 (2):116 – 120.

Wang Yuan, Sheng Lianxi, Li Ke, et al.. Analysis of Present Situation of Water Resources and Countermeasures for Sustainble Development in China [J]. Journal of Water Resource and Water Engineering, 2004, 18 (3):10 – 14.

Recent Scientific Development and Requirement on the Yellow River

Xue Songgui

Yellow River Conservancy Commission of the Ministry of Water Recourses,
Zhengzhou, 450003, China

Abstract: This paper summarizes the major scientific development in recent years on the Yellow River: ① on flood control and disaster reduction, mainly including satellite – based Yellow River Basin water monitoring and river forecast system, theories and technology for wandering river harnessing, technology for underwater footstone detection, removable non – emergency submerged dam technology, technology for dyke safety and danger prevention and technology of ice monitoring on the Ningxia – Inner Mongolia reaches, etc; ② on water and sediment regulation, mainly including theory and practice of water and sediment regulation, operation mode of the Xiaolangdi Reservoir during the late stage of silt detention, water and sediment regulation system and joint operation technology and so on; ③ on water resources management and regulation, mainly including unified Yellow River water resources management and regulation, theory and practice of water right transfer, etc. ; ④ on water resources protection, mainly including the goal of ecological protection and ecological water demand in the Yellow River basin, water pollutant migration and transformation process and forecast technique under the condition of hyperconcentration flow, etc; ⑤ on water and soil conservation, mainly including recent water and sediment change response on the middle Yellow River, technology of evaluation, prediction and simulation of soil and water loss on the Loess Plateau, and functional mechanism and optimized arrangement of soil and water conservation measures, etc. ; and ⑥ breakthrough progress has been achieved in index system of the Yellow River healthy life, construction of mathematical simulation system, and theory and technology of physical models. The paper also presents scientific requirements for keeping of the Yellow River healthy life and social and economic sustainable and harmonious development in the basin, mainly including: ① prediction of change process and tendency of the Yellow River water and sediment situation; ② theories and technology of water and sediment regulation in the basin; ③ support of the strictest water resources management technology; ④ the Yellow River sediment disposal and resources use technology; ⑤ river training and floodplain harnessing on the lower Yellow River; ⑥ anti – erosion and growth promotion technology in soft sandstone areas on the Loess Plateau; ⑦ major technologies for the Guxian Key Water Control Project and the west line of south – to – north water transfer project; and ⑧ construction of the Yellow River mathematical simulation system with the river natural process – ecological environment process – social – economic process coupling.

Key words: scientific requirements of the Yellow River, prediction of water and sediment change, water resources allocation, wandering river training, water – ecology – economy coupling

1 Recent major scientific development

1.1 Flood control and disaster reduction

1.1.1 Satellite – based water monitoring and river forecast system of the Yellow River Basin

The system consists of such subsystems as meteorological satellite (FY 2C/2D) data receiving and processing, large aperture scintillometer (LAS), energy and water balance monitoring (EWBMS), drought monitoring, the Yellow River source region runoff forecast (WRFS), the lower Weihe River flood forecast (WHFS) and result issue, etc. For the special geographic environment of the Qinghai – Tibet Plateau, an elevation modification model for satellite estimated rainfall and evaporation in high – elevation areas has been set up; spatial and time continuous monitoring of cloud, rainfall and real evaporation, real – time river runoff forecast with satellite – data – based distributed hydrological models for large river source regions, and the coupling of remote monitoring and ground monitoring to hydrological forecast have been realized in the Yellow River Basin. The system has laid a foundation for drought monitoring and evaluation in the Yellow River Basin.

The WRFS and WHFS in the system are both large region distributed hydrological models, with spatial and time rainfall and evaporation data derived by the EWBMS as the main input, which, through distributed runoff producing and confluence calculation, carries out runoff flood process simulation and forecast for the stations such as Tangnaihai, etc. The spatial resolution is 5 km × 5 km, and the time resolution is 1 d.

1.1.2 Theory and technology for wandering river training

Since 2000, around training programs of the lower Yellow River wandering reaches after the Xiaolangdi Reservoir was put into operation, we have carried out theoretical research, data analysis, model test and mathematical model computing. At the theoretical level, from the angle of microscopic turbulent vertex and secondary circulation, they have revealed the mechanism of "river tending to meander", discussed the sediment incipient motion condition on the side wall, proposed the critical incipient motion condition for sediment particles on the side wall; revealed the internal hydraulic mechanism of "large flow tends to follow straight path while small flow forms bends"; have set up wandering river training flow path equations, applied the phase lag theory to the study of training works layout, and given theoretical explanation of the present works layout experience; and have established the index system of training parameters such as flow sending distance and discharge of works, works radius, flow touching length and inflow angle, etc. At the technical level, we have systematically studied the formation mechanism of abnormal river bends; have evaluated training programs and effectiveness of domestic and foreign typical rivers, and have proposed that "node works" construction should be focused on in the lower Yellow River training; have put forward the new idea for training of segmented training with 'node works' focused on to achieve organic unity of wandering river training, and determined 9 sets of "node works"; and have work out implementation program for further training of wandering river of the Henan reaches, and proposed corresponding engineering measures.

The study has revealed the evolution law and mechanism of the lower Yellow River's wandering channels, and has achieved great progress in river training parameters, works layout optimization, new ideas for training, implementation program and result of further training. The result has directly supported the new round of the Yellow River's wandering channel training started at the end of 2007.

1.1.3 Underwater footstone detection technology

We have introduced the U. S. X – Star sub – bottom profiler, made localization transformation to the software and hardware of the control unit, and achieved the user interface localization; have developed data post – processing software MiniSeis, and developed an over – water test platform suitable for the lower Yellow River's water regime. The transformed profiler can locate the top boundary of underwater footstones, and can indicate the footstone thickness distribution. Cooperating with the 715 Institute of China Shipbuilding Industry Corporation, we have carried out sub – bottom profiler localization study, developed the engineering prototype, and carried out prototype test in the lower Yellow River. The main function parameters of the equipment have reached or exceeded the introduced one, and the equipment is obviously much better than the imported one in system miniaturization, ease of use and maintainability.

1.1.4 Removable non – emergency submerged dam technology

At the bottom of the removable non – emergency submerged dam are a line of pipe piles arranged at a certain interval, and at the top of the dam are the coping beams, both of which are connected by pins to form a diversion dam, see Fig. 1. "Removable" means that the dam is assembled with piles and can be taken apart, transported and re – assembled according to the need; "non

– emergency" indicates that the pipe piles are inserted to the stable depth once and for all so that no danger of instability may happen when the dam is lashed against by the flood; and the "submerged dam" means that the new structure has a good adaptability to overflow, and it can not only control river regime of small flow, but also can meet the requirement of flood discharge and water –

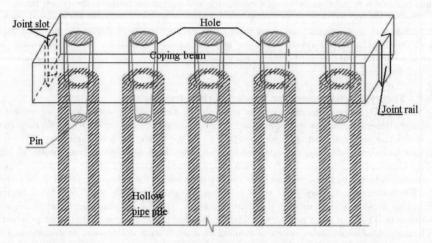

Fig. 1 Structure of the removable non – emergency submerged dam

sediment exchange between the floodplain and the channel. The dam can be disassembled, re – assembled and moved flexibly, which enables it to effectively adapt to the uncertainty of river regime changes of wandering channels. Using it together with the existing medium flow channel training works, while controlling river regime changes of large or medium flood, we can further control small flow river regime changes, thus provide a technical means for containing the lower Yellow River abnormal river regime evolution, stabling the main channel and flood control and emergency tackling.

The core technology of the achievement is the method of pile extraction with high – pressure water jetting, namely the technology and process that destroys the soil and positive circulation protection principle around the pile with water jets, reduces or eliminates the friction resistance and downward suction to the pile body during pile extraction, so as to realize quick and no – damage extraction of the thick and long piles buried in the soil.

1.1.5 Dyke safety and danger prevention and control technology

We have developed a bunching DC resistivity method detection system, which, by increasing the current density near the target, has improved the capacity of recognizing dyke hidden troubles and dangers; the mobile dyke safety monitoring system based on high – density resitivity method has solved the problem of high input for the hardware of the dyke monitoring system by means of mobility, has primarily formed a resistivity image data based early – warning mode, and has further improved the ease of use and timeliness of the testing data; we have established the corresponding relation between dyke typical dangers and electric measurement data, proposed a set of screening modes to judge the reliability of the original electric measurement data and a display mode that facilitates intuitive understanding; and, at the same time, we have carried out dyke safety evaluation and technology of grout filling at the soil – stone binding site.

1.1.6 Ice monitoring technology on the Ningxia – Inner Mongolia reaches

We have developed a continuous automatic river ice condition monitoring system, which can obtain real – time monitoring of ice melting, ice thickness during the freezing period, temperature gradient within the ice and water stage under the ice at four observation points, namely Sanhuhekou, Huajiangyingzi, Putanguai and the Wanjiazhai Reservoir area; we tried using the H – ADCP automatic monitoring system to measure flow velocity under ice during the freezing period all weather without interruption, we have set up a wireless video remote monitoring system for the major reaches in the Inner Mongolia to realize video monitoring at 7 fixed points (Sanhuhekou, Huajiangyingzi, Sanshenggong, Zhaojun Tomb, Deshengtai Road Bridge, Toudaoguai and Lamawan) and 2 mobile points. We have proposed a FM continuous wave radar (FMCW) technology based ice condition real – time monitoring system, which can scan all weather without interruption to quickly obtain parameters such as floating ice density, floating ice velocity, etc.

The Yellow River ice condition UAV remote emergency monitoring system that has been built mainly consists five parts – flying operation sub – system, ground control sub – system, communication and transmission sub – system, consultation and directing sub – system and on – line processing and analysis sub – system, and it can obtain real – time ice spot video, splice remotely – sensed images on line, transmitted them to the ice control headquarters at various levels to achieve two – way interaction and remote consultation between the ice spot and the headquarters. The system can also quickly process digital remotely – sensed images on line to achieve comparison with satellite remotely – sensed images, which may help improve the analysis and decision – making level.

1. 2 Water and sediment regulation

1. 2. 1 Theory and practice of water and sediment regulation
(1) We have revealed the water and sediment transport law of the lower Yellow River under the current situation. On the basis of mechanics research and data analysis, we have established the sediment transport expression representing the lower reaches scour – and – fill adjustment relation:

$$Q_S = kQ_{\text{下}}^{\beta}\left[\frac{\sqrt{B}}{h}\right]^{ml}\ J^{m2}V^{m3}\left(\sum P_i d_i^2\right)^{m4} \tag{1}$$

The formula expresses the effect of factors such as incoming water discharge, sediment content, suspended load grain composition, and the longitudinal and lateral shapes on the sediment transport capacity, and the influence mechanism of mutual influence and feedback mechanism between the upper and lower reaches.

Close to the equilibrium of sediment transport, there is the optimized sediment transport result on the whole. With the sediment content, discharge, and percentage of sediment particle smaller than 0. 025 mm in the suspended sediment at Huayuankou cross – section as the main factors, we have set up the water – sediment relation of no – silting in the lower channel:

$$S = 0.308QP^{1.5514} \tag{2}$$

(2) We have established the mathematical relation between the Xiaolangdi Reservoir's density current transport and muddy water reservoir settlement process. In it, the function relation of density continuously running to the dam is:

$$S = 980e^{-0.025d_i} - 0.12Q \tag{3}$$

The expression for the muddy water reservoir dynamic settlement velocity is

$$u_H = \frac{\Delta h}{\Delta t} - \varphi' u_S + \frac{\varphi' u_S S_0}{S_S} \tag{4}$$

The relationship of sediment delivery ratio for the reservoir density current is

$$\eta - 4.45\exp(0.06Q_s^{0.3})Q_{out}^{0.4}J^{1.7}H^{-1.8} \tag{5}$$

(3) We have established the index system for water and sediment regulation. The indices that can keep scour for the whole river course of the lower Yellow River are: when the incoming water sediment concentration of the lower reaches is smaller than 20 kg/m³ and the fine sand make up over 70% of the corresponding suspended load, at Huayuankou the discharge should be controlled at 2,600 m³/s and the flood duration should be kept over 9 d; when the incoming water sedimentt concentration is about 40 kg/m³ and the suspended sediment is mainly made up of fine grains, at Huayuankou the discharge should be around 3,000 m³/s, and the flood should last for over 8 d. On the basis of the above mentioned indices, during the water and sediment regulation, dynamic control and adjustment should be made according to the composition of the suspended sediment, reservoir storage condition, the river bed boundary condition and change process of the lower reaches.

(4) For different water and sediment conditions and regulation objectives, we have established three kinds of basic modes: the Xiaolangdi Reservoir based mainly single reservoir mode, the large – scale spatial water – sediment docking based mode and the trunk stream reservoirs joint operation based mode.

(5) We have formed a complete set of water and sediment regulation technology, including the coordinated water and sediment process molding technology, the test process control technology, reservoir density current molding technology and hydrological monitoring and forecast technology and so on. The innovative technology of density current molding with trunk stream reservoirs joint operation has made a new way for reservoir sediment ejection.

(6) We have achieved systematic integrated innovation. We have developed and integrated hydrometeorological forecast system, consultation system for runoff, works and danger conditions, forecast – regulation coupling system, real – time regulation and monitoring system, and, through comprehensive use, we have achieved fine adjustment and control of the Yellow River's water and sediment process.

1.2.2 The Xiaolangdi Reservoir's operation mode in the late stage of silt detention

The "late stage of silt detention" refers to the whole period from the time when the sedimentation of the Xiaolangdi Reservoir amounted to 2.2×10^9 m^3 to the time when the high floodplain and deep channel formed and the floodplain elevation in front of the dam reached 254 m.

Through research, we have proposed the operation mode of flood control and silting reduction in the late stage of silt detention, whose main objectives include extending the service life of sediment storage capacity of the reservoir, reducing siltation downstream of Gaocun and maintaining the medium – water channel of the lower reaches, and whose core idea is "multi – year sediment regulation, scouring with lowering water level at the proper opportunity, sediment trapping and water and sediment regulation". Through comparison between different programs, it was proved that the opportunity for the reservoir scouring is that the accumulated sedimentation reaches 4.2×10^9 m^3, the upper limit of regulated discharge is 3,700 m^3/s (the corresponding regulation storage capacity is 1.3×10^9 m^3), and the lower limit of the regulated discharge is 800 m^3/s.

With use of the recommended silt reduction programs for the late stage of sediment trapping, the reservoir has increased to some extent the storage degree of non – over – floodplain's hyperconcentrated flow to reduce siltation in the lower channel by the non – over – floodplain's hyperconcentrated flow; when discharging water to scour, the reservoir has increased the end discharge to reduce siltation in the lower channel during scouring with discharged water, thus has further improved the discharge capacity of the lower main channel and increased the siltation reduction effect of the lower channel.

1.2.3 Joint regulation and control technology of the water and sediment regulation and control system

In the light of the characteristics of water and sediment of the Yellow River and the requirements of basin social and economic development, coordination of water – sediment relation, reasonable allocation and optimized regulation of water resources, we have studied and proposed the general layout of the water and sediment regulation and control system; we have systematically analyzed water and sediment transport law of different Yellow River reaches, the requirement of river flood control and ice flood control, change of water resources in the basin and the relation between water consumption for social and economic development and for ecological environment, and have established binding and guiding indices system for joint water and sediment regulation and control; we have studied the internal link and interrelation between various projects of water and sediment regulation and control system, proposed the joint operation mechanism of the Yellow River water and sediment regulation and control system and the principles, mode and instructions for reservoirs joint operation of different project combination programs; we have evaluated the important role of water and sediment regulation and control in coordinating the Yellow River water – sediment relation, maintaining the flood and sediment discharge capacity of the medium flow channel and lowering

Tongguan elevation by scouring; on the basis of comparison of the development task, economic result and social effect between various to – be – built projects, we have demonstrated the development order and construction opportunity of Guxian, Heishanxia, Qikou and other to – be – built key projects, thus provided important technical support for the State's macro – decision.

1.3 Water resources management and regulation

1.3.1 Unified Yellow River's water resources management and regulation

In December, 1998, the State authorized the Yellow River Conservancy Commission to implement unified Yellow River's water resources management and regulation. Since the unified regulation began to be implemented, with administrative, scientific, engineering, legal, economic and other means, no flow drying up for successive years in the Yellow River has been achieved, and the level of the Yellow River water resources management and regulation has been improved steadily. The following main results have been achieved.

(1) We have established the unified water resources management and regulation mode "State centralize allocation of water, basin organization organized implementation, provinces (autonomous regions) responsible for water distribution, dual control of total water consumption and cross – section flow, and unified regulation of major intakes and key reservoirs", and the water resources regulation manner of "combination of annual plans, monthly and ten – day regulation programs and real – time regulation appointments".

(2) We have developed a high precision basin water regulation model with total volume control, online tracking and dynamic adjustment, which can be used for preparation of regulation programs; have developed a low – water regulation mode with forward and inverse calculation, early warning and automatic parameter optimization.

(3) We have constructed the lower Yellow River diversion sluice gate remote monitoring system that is extensive coverage, large – scale, multi – level and with strong control ability.

(4) We have developed and constructed the sediment – laden water quality automatic monitoring station, developed five – degree adjustable water sample pre – processing technology and equipment suitable for sediment – laden rivers, obtained national patents for three results, and also reformed the low water test facilities.

(5) We have developed the Yellow River water regulation decision support system that integrates information collection, program preparation, service treatment, comprehensive consultation, emergency response, remote monitoring and other functions, and have constructed the Yellow River Water Regulation Centre.

(6) We have established the unified Yellow River water management and regulation system that includes emergency administration, water right transfer and total volume control, etc.

1.3.2 Theory and practice of water right transfer

The implementation of the Yellow River water right transfer began in 2003, and, during the implementation process, we have gradually built the management system, technical system and monitoring system of the Yellow River water right transfer. The main results achieved are as follows:

(1) On management system, we have made clear the concept of water right, proposing that water right means the right of water extraction from the Yellow River, and the water right transfer refers to the transfer of the right of water extraction from the Yellow River, and defined the transferor and the transferee. We have issued the "Implementation Method of Yellow River Water Right Transfer Management", which normalize the water transfer behaviour of the local government, the transferor and the transferee; we have issued the "Verification Method of Yellow River Water Right Transfer Water Saving Projects (Trial)", which normalizes the verification work of Yellow River water right transfer water saving projects.

(2) On technical system, we have made clear that the water transfer term is 25 years, and made clear the fee composition and fee calculation method of water right transfer, water right transfer water saving method, and the relation between water saving volume and transferable water volume, compensating of various stakeholders and other key technical problems.

(3) On monitoring system, considering the requirement of measurement and monitoring of grade water right, dynamic groundwater monitoring and ecological environment monitoring, we have specified detailed layout of monitoring cross – sections and monitoring points in water right transfer project areas, monitoring content, collection, transmission and processing of monitoring data, and have initially established water right transfer monitoring system for water right transfer project areas.

Through research and practice of the Yellow River water right transfer, we have clearly proposed that the concept of water right means the right of water extraction, which broadened the connotation of the water right concept of our country; have specified the term of water right transfer, the fee composition of water right transfer, the relation between water saving volume and transferable water volume, compensating of stakeholders and other technical problems; have formed relatively perfect Yellow River water right transfer management system, which effectively directed the smooth implementation of the Yellow River water right transfer.

1.4 Water resources protection

1.4.1 Ecological protection objectives and ecological water requirement of the Yellow River Basin

We have systematically analyzed ecological function determination, water ecological protection objectives and ecological water requirement of different reaches of the Yellow River trunk stream, established the Yellow River health index system that includes such aspects as hydrology, water environment, river bed configuration and water ecology, proposed environmental flow indices of major Yellow River's control cross – sections, and its restoration objectives and technology. With the Sino – Dutch scientific cooperation project, we have proposed the proper scale, ecological water requirement and layout schemes of the freshwater wetland of the Yellow River Delta, and began ecological water replenishing practice in the delta wetland in 2008; have carried out the study of water – related ecological compensation mechanism in the Yellow River basin, set up the general framework of the Yellow River ecological compensating, and presented the compensating subject, object, compensating standard, method and system guarantee; with health assessment of major rivers and lakes in the Yellow River Basin launched in 2011, we have carried out pilot evaluation of health condition of key Yellow River reaches from aspects of hydrology and water resources, channel physical form, water environment, water ecology and social service function, etc.

1.4.2 Transport process and forecast technology of water pollutants in the environment with hyperconcentration flow

We have mainly studied the distribution in various phases of water environment and change law of soluble and particulate toxic organic pollutants (COD, ammonia nitrogen, etc.) in Lanzhou, Huayuankou and other key reaches, and initially mastered the monitoring methods of typical pollutants such as polycyclic aromatic hydrocarbons (PAHs), nonylphenol (NP), etc., revealed the characteristics of distribution in water phase and in various sediment phases and transformation mechanism of typical toxic organic pollutants such as PAHs in the water of the middle and lower Yellow River and NP in the water of the upper Yellow River; with the Sino – Euro basin management project, we have initially constructed the multi – pollutant – factor one – dimension water quality forecast model and water quality forecast system for normal flow condition and sudden water pollution incidents in the Longmen – Sanmenxia reaches and reaches downstream of Xiaolangdi, basically realized the forecast and dynamic display of dynamic evolution, concentration evolution and pollution process of river pollutant water masses, and, with the data synchronization technology, initially realized water quality real – time correction forecast with emergency monitoring data in sudden water pollution incidents.

1.5 Water and soil conservation

1.5.1 Response to recent water and sediment change of the middle Yellow River

The result of the "11th Five – Year" national scientific support subject "Study of the Yellow

River Basin's Water and Sediment Change Condition Evaluation" showed that: compared with the years before 1970, during 1997 ~ 2006, human activities such as water and soil conservation and comprehensive control reduced 8.65×10^9 m^3 of runoff and reduced 0.55×10^9 t of sediment yearly on average in the middle Yellow River. Among the figures, water and soil conservation measures reduced 3.84×10^9 m^3 of runoff and 0.42×10^9 t of sediment every year on average; recent water and soil conservation measures reduced 1.88×10^9 m^3 of runoff and 0.29×10^9 t of sediment yearly on average in the corresponding Hekouzhen – Longmen Reaches.

It is found through study that there exists a fixed relation between the effectiveness of sediment reduction benefit and the storage capacity per unit area of the reservoir, the reservoir is an important measure for sediment detention, and maintaining storage capacity per unit area is the key; the sediment reduction benefit has a positive correlation with control degree of soil and water loss in the tributaries of the Hekouzhen – Longmen Reaches, which can be clearly divided into high sediment reduction benefit value area and low value area. There exists a control degree threshold value in the low sediment reduction benefit value area. In recent years, there has been a sharp decline in runoff and sediment in the main tributary, the Kuyehe River, where, human activities amounted about 67% of the impact on the sharp decline in runoff and about 41% of the impact on the sharp decline in sediment.

1.5.2 Technology of evaluation, prediction and simulation of water and soil loss on the Loess Plateau

To meet the requirements of rapid prediction of water and soil loss on the Loess Plateau, water and soil conservation benefit evaluation and the Yellow River's water and sediment regulation and control, we have carried out study of soil loss evaluation and simulation on the Loess Plateau. We have established basic modeling theory with multi – dimension basin "dynamics – vegetation – control" system erosion and sediment yield process coupling mechanism as the core; constructed multi – function, complete heterogeneous distribution multi – dimension basin soil loss model system, and formed a complete set of simulation technology that can meet multiple objectives such as water and soil loss prediction, harnessing program optimization, harnessing benefit evaluation, flood and sediment forecast, etc; proposed the "rainfall index" evaluation method of harnessing benefit and evaluation precision judgment method based on the maximum sediment detention capacity of the water and soil conservation measures; presented methods of spatial scale transformation erosion factors and quantitative test of model uncertainty; created heterogeneous distribution soil loss model support system, and also achieved release on the web.

1.5.3 Acting mechanism and optimized layout of water and soil conservation measures

Through study, we have obtained the critical values of water and soil conservation benefit under different water and soil conservation measures optimized layout modes and their optimized layout method, clarified the effect of different coverage of slope vegetation under different spatial layout on sediment yield on slope surface and in slope ditches, and summarized typical harnessing modes of different areas; with radionuclide tracer technology, we have traced and played back sediment deposit process, characteristics and its dynamically changing relation with erosion sediment yield intensity and sources after construction of silt dams in small basins, and re – built sediment depositing speed at different evolution stages of silt dams, providing scientific basis for the correct evaluation of silt dam sediment reduction benefit; by means of indoor simulated rainfall experiment, three – dimension laser positioning scan and field survey and sampling, etc., we have established the complex relation between slope sediment yield under water and sediment conservation measures and the runoff feature parameters, and revealed the action and geoscience process of water and sediment conservation measures and their driving mechanism.

Around detention of flood sediment by the water and sediment conservation measures layout, the research shows that the sediment reduction ratio of silt dams has close relation with layout ratio. To reduce sediment entering the Yellow River effectively and quickly, for water and sediment conservation measures of the Hekouzhen – Longmen Reaches, we should adopt the overall layout mode of combination of engineering measures with silt dams as the main part and the slope surface measures; if the layout ratio of silt dams is kept over 2%, the corresponding sediment reduction ration can reach more than 45%. The degree of flood sediment detention of the water and sediment conservation measures has a close relation with measures layout, and there exists a measures layout ratio with maximum flood and sediment reduction effect.

1.6 River health indices and simulation technology

1.6.1 Yellow River health life indices system

With respect to the definition, intension, objectives, index system and way of realization of the Yellow River healthy life, we have done a lot of innovative work.

(1) We have presented the scientific intension and indicator of the healthy river. A healthy river is a river whose social function and natural function can develop in a coordinated or balanced way, and the specific health indicator includes smooth and stable channel, adequate surface runoff, good water quality and sustainable river ecological system.

(2) We have presented the determination method of river environmental flow, and presented the evaluation indices and standards of the Yellow River water ecological health.

(3) We have systematically analyzed structure, species composition and succession process of ecological system of different reaches of the Yellow River trunk stream, initially found out kinds and ecological habits of important protected species, and proposed the ecological protection objectives of the Yellow River at the present stage through analysis of the response relation between river runoff condition and typical biological reproduction in the river ecological system.

(4) We have proposed the "bankfull discharge" as the evaluation index indicating the health of the lower Yellow River and the Ningxia – Inner Mongolia Reaches, demonstrated and proposed the proper standards of channel health of the above – mentioned two reaches from the aspects of meeting the requirement of high – efficient sediment transport, meeting the requirement of flood safety and adapting to the water and sediment condition, demonstrated and proposed the water and sediment condition required for maintaining the channel health of the lower Yellow River and the Inner Mongolia Reaches from the analysis of channel scour – and – fill effecting factors.

(5) We have made an in – depth analysis of the main factors affecting the Yellow River water quality and the affecting manners, and proposed the calculation method of water need for the Yellow River self – purification with water quality objective as the constraint.

(6) Based on the water requirement for the Yellow River natural function and social function and the hydrological characteristics of the river, we have put forward environmental flow recommended programs for each typical cross – section of the Yellow River.

1.6.2 Yellow River mathematical simulation system construction

Since 2003, breakthrough progress has been achieved in the mathematical simulation system of the "Digital Yellow River" project with respect to architecture design, model development, model evaluation, model integration and application, etc.

(1) We have worked out the "Yellow River Mathematical Simulation System Construction Plan on 'Digital Yellow River' Project ", and proposed the new idea of "river natural process, ecological environment process and social and economic process coupled simulation".

(2) We have developed six kinds of models: water and sediment loss model of the first sub – region in loess hilly – gully region on the Loess Plateau, river two – dimension water and sediment mathematical model, river one – dimension flow – sediment – water quality model with unsteady flow, reservoir one – dimension water and sediment model with steady flow, reservoir three – dimension water and sediment model with turbulent flow and estuary two – dimension tidal current and sediment transport model.

(3) We have constructed the Yellow River mathematical simulation system support platform, which can basically meet the high – efficient integration and running of various professional models from different sources.

(4) We have built the Yellow River supercomputing centre, which has been providing all – weather, stable and efficient service for digital weather forecast, spatial large – scale water and sed-

iment simulation.

(5) We have presented "Yellow River Mathematical Model Development Guideline" and "Yellow River Mathematical Model Evaluation Method", which normalized develop process and provided evaluation standards.

(6) The mathematical models have been widely used in major Yellow River harnessing affairs such as soil erosion prediction and forecast of typical basins, flood risk mapping, study of reservoir operation manner, test of water and sediment regulation and production operation, warping test in Xiaobeiganliu and study of river training programs, etc.

1.6.3 Physical model theory and technology

On the simulation theory, we have proposed reservoir density flow plunging similarity conditions, sediment carrying similarity and continuous similarity conditions, thus formed complete law of similitude for reservoir models on heavy sediment – laden river, simulated successfully the density flow transport and transformation process of the Xiaolangdi Reservoir; proposed the method of simulating river bed layer silting in the Yellow River "bottom tearing scouring" reaches with different graded fly ash, successfully simulated the "bottom tearing souring" process in the flume test, and collected a large number of water and sediment factors that are difficult to obtain in the field; carried out study in the aspects of model sand characteristics, similarity rate and tide generating technology of the Yellow River estuary, and the inlet clear and muddy water automatic control technology, and presented the tide generating modeling device and control system that are suitable for the intense alongshore current and small tide range of the Yellow River estuary.

On control technology, we have developed the inlet discharge automatic control system and the tail gate automatic stage control system suitable for hyperconcentration flow mobile – bed model test; with CAM and digital control technology, we have developed HHZM – I – 1225 digital control model machine, which can make the model with the maximum size 1,200 mm × 2,500 mm × 100 mm; have introduced or developed devices such as automatic measuring bridge, muddy water topographic meter, river regime flow field control system, sand content meter, etc.

2 Major scientific requirements in the near future

2.1 Yellow River's water and sediment condition change process and trend prediction

Since the late 20th century, the policy of closed forest and grazing prohibition, various water and sediment conservation measures, human activities such as mining, road construction and sand mining have significantly changed water and sediment yield process in the Yellow River basin (especially in the middle Yellow River), and then deeply impacted the situation of water and sediment into the Yellow River. Sharp drop in runoff was mostly notable in the Toudaoguai – Sanmenxia reaches of the middle Yellow River: at a similar rainfall situation, the natural runoff in the Toudanguai – Longmen reaches and in the Fenhe River is less than 55% and 45% respectively. The sediment reduction is more remarkable. For example, during the period 2003 ~ 2010, the real annual average incoming sediment at the Tonghuan Hydrological Station was only 0.277×10^9 t, 1.3×10^9 t, less than the 1.6×10^9 t of annual mean during the period of 1919 ~ 1960.

Since the factors impacting the Yellow River water and sediment yield and transport are very complex, the Yellow River basin is very vast, the study means and methods do not adapt to the present social background, and field observation and survey are insufficient, there is still a great divergence on the understanding of the impact of human activities on water and sediment yield and flow confluence and sediment transport in the basin, and there remains a series of problems to be solved. The problems mainly include: Loess Plateau natural erosion background value; rainfall change cycle of the Yellow River basin; main driving force and driving mechanism of sharp drop in runoff and sediment in the middle Yellow River in recent years; the contribution of runoff producing environment and change in water diversion and consumption in the middle Yellow River to the sharp reduction of river runoff; present yearly total sediment reduction volume by human activities and the sustainability of sediment reduction by human activities; Yellow River Basin runoff and sediment yield characteristics under extreme climate conditions, and climate change - based prediction of the Yellow River sediment situation in the future 30 ~ 50 yr.

2. 2   Basin water and sediment regulation and control theory and technology

At present and in a period to come, we will focus on the study of the issues of key water and sediment regulation and control technology with maintaining a certain scale medium - flow channel in the lower Yellow River and the Yellow River reaches in the Inner Mongolia as the objective, the issues mainly including regulation and control key technology of the middle and lower Yellow River hyperconcentration flood, the impact of the sediment from Kongdui (flood ditches) on the scour - and - fill on the Yellow River reaches in Inner Mongolia and its control measures, and optimization of water and sediment regulation and control mode of existing Yellow River reservoirs, etc.

(1) Study of the middle and lower Yellow River hyperconcentration flood regulation and control key technology. The aim is to find possible ways to alleviate or eliminate hyperconcentration flood silting hazard of the middle Yellow River, and what need to study include the floodplain and channel scour - and - fill control indices of the lower Yellow River hyperconcentraion flood, key technology of hyperconcentration flood regulation and control by the reservoir group of the middle Yellow River, tributary hyperconcentration flood transport and control in the Hekouzhen - Longmen reaches, and hyperconcentration flood silting reduction control measures.

(2) Study of the impact of Kongdui sediment on silting in the Inner Mongolia reaches of the Yellow River and silting reduction technology. The aim is to find the way to alleviate the impact of the Kongdui hyperconcentration flood on the silting of the reaches, and what need to study are characteristics and hazards of Kongdui hyperconcentration flood, the impact of Kongdui hyperconcentration flood on the silting of the Inner Mongolia reaches, the middle and lower Kongdui flood and sediment control means, the flood and sediment effect of the upper and middle Kongdui water and sediment loss control measures, and the Kongdui control direction.

(3) Water and sediment regulation and control mode optimization for the Yellow River existing reservoir group. With the exiting key reservoir group of the Yellow River's water and sediment regulation system as the background, what need to study are silting reduction objective of reservoirs and channels in the water and sediment regulation and control system, with focus on the analysis of water and sediment control operation mode of the exiting reservoir group for flood from different sources, reservoir operation mode, scheduling control indices, etc.

2. 3   Supporting the strictest water resources management technology

The three red lines of strictest water resources management are related to the three major links of water extraction, consumption and drainage, and involve water resources evaluation, water resources allocation and regulation, water - saving and drainage - reduction technology, basin ecological protection objectives. ecological water requirement, etc. With respect to the basic support requirement of formulating and implementation of the three lines, we need to carry out the following three aspects of research work.

(1) On water resources survey and evaluation, we need to carry out upper and middle reaches ground water exploitation survey and evaluation, mine water - gushing utilization survey and evaluation, present social and economic water consumption coefficient survey and evaluation, present irrigation water consumption survey and evaluation for major tributaries, evaporation and seepage loss survey and evaluation for the existing water storage projects, etc.

(2) On water allocation and regulation, we need to carry out study of water allocation programs of cross - province tributaries such as the Huangshui River, the Kuyehe River, the Zulihe River, the Qingshuihe River, the Hunhe River, etc, study of ecological regulation and regulation programs without functional flow drying up, study of basin productivity layout and water resources restraint, reasonable storage order and water control level adjustment in flood season of large reservoirs, etc.

(3) On water saving and drainage reduction, we need to do study of water resources quantity

- quality combined allocation technology based on total amount control, total water extraction control relating to surface water and ground water, total irregular water utilization amount control, total ecological environment water consumption control, entering - sea (lake) water amount control, e-conomic ET control amount, total pollutant emission control and total sewage emission to river control.

(4) On basin ecological protection objectives and ecological water requirement, we need to strengthen the study of ecological system value and asset assessment, study of wetland vegetation and aquatic living things water requirement law, study of river living things habitat change and coastal ecological water requirement.

2.4 Yellow River sediment disposal and resources use technology

In a period to come, through overall control measures such as water and soil conservation, the Yellow River annual sediment yield may be further reduced, yet the Yellow River's basic property of being sediment - laden river can not be changed in a short period of time. Therefore, it is necessary to change ideas, make good use of the sediment coming endlessly, and, through organic combination of sediment utilization and river training, put an end to the sediment hazards faced by the river and realize Long-term stability of the Yellow River. In the near future, we need to carry out the following researches.

(1) Carry out study of Yellow River sediment resources utilization theory framework system by using gelatinization theory, architecture theory, fractal theory, etc.

(2) Ascertain the specific quantity, mineral composition, characteristics and spatial distribution of the Yellow River sediment resources, and carry out study on industrialization general layout and plans of sediment resources utilization, and study of related policies.

(3) Study of sediment resources utilization key technology, including study of mechanism of organic combination of Yellow River sediment disposal and utilization, sediment series gelling consolidation technology, technology of dam construction with modified soft sandstone, study of key technology and equipment of making dry mixed mortar, hollow block, sialon composite materials and aerated concrete blocks with the Yellow River sediment.

(4) Carry out pilot study in the Yellow River sediment artificial reserve stone and ecological unburned brick technology transfer, and pilot study in reservoir ejection, sediment transport in pipelines and sediment utilization.

2.5 Channel reconstruction and floodplain harnessing of the lower Yellow River

The lower Yellow River floodplain is very special, it is not only the passage of flood discharge and an important region of flood and sand detention, but also a main place where 1,895 thousand floodplain inhabitants live and multiply. In recent years, with steady improvement of China's national economic strength, the gap between inside and outside the floodplain has been increasing in terms of social and economic development, and the contradiction between floodplain sustainable social and economic development and flood control has been becoming more and more serious. The Xiaolangdi Reservoir and the to - be - built reservoirs such as Guxian and Taohuayu will remarkably enhance control capacity over the lower reaches flood, therefore, it is necessary to carry out advance study of the feasibility of the lower Yellow River harnessing programs, especially the newly - built flood control dykes programs.

The so - called "newly - built flood control dykes" programs mean that, while constructing the Taohuayu Reservoir to further control flood, the dyke distance between two dykes on both sides of the river will be narrowed to 3 ~ 5 km for example in wide reaches of the Lower Yellow River so as to release most floodplain in the lower reaches. The main study contents are as follows.

(1) The lower Yellow River newly - built flood control dyke program study. This includes study of newly - built flood control dyke standards, newly - built flood dyke program design, study of newly - built floodplain flood detention area layout and classified harnessing, and study of floodplain safety construction, resettlement programs and the newly - built dyke management.

(2) Impact of the newly – built flood control dykes on the lower Yellow River channel scour – and – fill and flood control situation. This includes the Xiaolangdi Reservoir operation mode optimization and the process of runoff and sediment into the lower reaches, the scale check and operation mode of the Taohuayu Project, the impact of new flood dykes construction on the scour – and – fill and flood control situation in the lower reaches channel and the estuary, and on the present flood detention.

(3) Study of measures to help improve the newly – built flood dyke program sediment transport capacity. This includes analysis of the Xiaolangdi Reservoir optimized operation result, the new floodplain – channel relation establishment and its impact on sediment transport capacity, impact of measures such as local river dredging on river stability and sediment transport capacity, impact of river training measures improvement on sediment transport capacity, and the impact of flood detention and discharge in the newly – built flood detention areas in the floodplain on sediment transport capacity of the river channel, etc.

2.6　Anti – erosion and growth – promoting technology for sandstone regions on the Loess Plateau

The Loess Plateau soft sandstone region features low diagenesis degree, very bad geologic condition, very low structural intensity, strong water and sediment loss and very fragile ecological environment, and the region is also the main source of coarse sediment that cause the Yellow River serious siltation. Anti – erosion and growth – promoting technology for soft sandstone regions in the middle Yellow River and its demonstration mainly include the following.

(1) On erosion lithologic character mechanism in soft sandstone regions, we need mainly study differential features and geologic and topographic features of soft sandstone, analyze the physical and chemical features, engineering mechanics features and hydrological environment features of soft sandstone, and reveal the soft sandstone erosion dynamic force and erosion lithologic character mechanism.

(2) On soft sandstone consolidation and growth – promoting technology study, we need develop a number of high and new tech consolidation growth – promoting materials suitable for soft sandstone harnessing, develop soft sandstone monoblock surface modification key technology, improve the physical and chemical features of soft sandstone through surface modification so as to make it have stable mechanical property and growth – promoting property.

(3) Study of modified soft sandstone dam construction technology, develop soft sandstone modification technology and form silt dam construction material making technology mainly through study of soft sandstone modification, and form modified soft sandstone dam construction mainly through construction technology study.

(4) On anti – erosion and growth – promoting measures three – dimensional configuration technology integration and demonstration, we need to carry out material – engineering – biological measures optimized combination 3d configuration mode study, carry out dynamic monitoring of anti – erosion and growth – promoting overall benefit in demonstration areas and its impacting factors, and develop demonstration multi – factor overall benefit analysis and evaluation model to scientifically evaluate the overall benefit in the demonstration area.

2.7　Major technology of the Guxian Key Project

Located in the lower reaches of the Yellow River Xiaobeiganliu Reaches, the Guxian Key Project is one of the seven key projects on the trunk stream of the Yellow River determined in the "Yellow River Training and Development Planning Guideline", is an important composite part of the Yellow River water and sediment control and regulation system, and it has the strategic status of connecting link between the upper and lower reaches. Its major technical problems are as follows.

(1) On runoff and sediment transport characteristics in the reservoir area, we need to study the flow pattern change and sediment transport effect of reservoir flow under different water and sediment conditions and different reservoir operation modes, study the impact of river bed coarsening

on sediment transport during reservoir scouring process.

(2) On siltation feature change of reservoir trunk stream and tributaries, we need to study the response relation between siltation features and runoff and sediment of the trunk stream and reservoir scheduling process, study the impact of tributary hyperconcentration flood on the trunk stream siltation features and its water and sediment transport, study the transformation of cross – section shapes during the trunk stream and tributary silting and scouring process, the floodplain and channel formation process and changes of effective reservoir storage capacity, etc.

(3) On tributary's runoff and sediment movement changes and its possible effect on the maintenance and utilization of effective reservoir storage capacity, we need to study siltation conditions of typical tributaries, the interaction between tributary's back flow siltation and itself siltation, the impact of incoming water and sediment from tributaries on siltation in trunk steam of the reservoir, and the effect feedback mechanism between flow and sediment process of the trunk stream and tributaries, reservoir regulation process and topography of the trunk stream and tributaries under different incoming flow and sediment conditions and different reservoir operation stages, and analyze its possible hazards to maintenance and utilization of effective reservoir storage capacity and control measures, etc.

(4) On joint operation mode of the Guxian, Xiaolangdi and other reservoirs, we need to study the Guxian Reseavoir and the Xiaolangdi Reservoir joint operation mode, opportunity, indices, comprehensive effects, etc. before and during flood period and in the multi – year sediment regulation term of the Xiaolangdi Reservoir.

(5) On environmental effect prediction, we need to study and predict the Guxian Water Control Project construction on the macro – environment of the Yellow River basin and related regions, on the national scenic spot Hukou Waterfall, Hukou and Shequ National Geological Parks, nature reserves, fish germplasm resources protection areas, etc.

2.8　Major technoogy for the south – to – north water transfer west line project

The south – to – north water transfer west – line project is of arduous task, large volume of water and complex problem between water source and receiving areas. Though a lot of research has been done, there still remain many problems needed to be further study scientifically so as to provide technical support for the final decision of the project. The major technical problems are as follows.

(1) On water – receiving area selection and the need of transferred water quantity, we need to analyze water shortage degrees of different areas, the water supply pattern and degree of difficulty of supporting project construction in the receiving areas and bearing capacity of water price, and present water – receiving areas and their need of transferred water quantity.

(2) On effect of water transfer and transferable water volume, we need to study the water resources supply – demand relation and transferable water volume in the water source area, analyze its economic, social and ecological effect on the water transferring river, and study its compensating countermeasures.

(3) On water transfer volume allocation strategy, we need to study the optimized water resources allocation programs, present the benefit calculation method suitable for the west line project and the effectiveness of the project for guarantee of the country's energy, grain, drinking water and ecological safety.

(4) On project financing and operation management mode, we need to present project construction financing program, project operation and maintenance cost compensating mechanism, project construction and management mode, project water price, related policy, etc.

2.9　Three – field coupled Yellow River mathematical simulation system construction

The "'Digital Yellow River' Project Yellow River Mathematical Simulation System Construction Planning" required, directed with the new idea of river natural field, social – economic field and ecological system field coupled "digital basin", developing a water conservation, macro – economic society and ecological system model, built a virtual simulated Yellow River basin, and, by means of it, achieve pre – warning, forecast and program generation in the Yellow River training, development, protection and management process. With respect to the basin mathematical simulation system construction, the present hot and difficult problems are as follows.

(1) On water cycle and multi – material transport process simulation methods, we need to study Loess Plateau slope ditch coupled erosion process and its simulation methods, similarity between slope sediment yield and retrogressive erosion and its simulation methods, flow movement – sediment transport – river bed deformation fully coupled simulation theory and technology, process of settlement, resuspension, adsorption, desorption and degradation for different genus pollutants under hyperconcentration flow environment and its simulation methods, Ice jam, ice dam change process and simulation method, simulation of reservoir group's joint operation and optimization of coupling technology, etc.

(2) On mathematical model uncertainty and evaluation methods, there are the application of Monte Carlo simulation uncertainty methods, model error sources, probability distribution and its quantatitive expressing method, criteria and elements of mathematical model quality evaluation, mathematical model evaluation criteria use case construction, etc.

(3) On coordinated large – scale and interdisciplinary model integration and cloud service technology ther are study on different flow pattern / spatial scale efficient digital format application, water – ecology – economy multi – temporal – spatial scale model two – way coupling technology, development of public development environment of integrated software components, digital basin platform / software cloud service technology, etc.

Managing Wetlands and Waterways in Regulated Rivers

Jennifer Collins, *Peter Kelly* and *Louise Searle*

Mallee Catchment Management Authority (CMA)

Abstract: The Mallee Catchment Management Authority (CMA) undertakes the coordination of natural resource management for over four million hectares of private and public land in the North West region of Victoria, South Eastern Australia.

Based on key strategies and planning documents, the Mallee CMA has focused on improving the health of over 760 km of the River Murray as it passes along the northern border of the region. The River is supported by a series of anabranches and wetlands, each with their own unique characteristics which provide various functions toward overall river health. These strategies and plans focus on the riparian landscape as well as the anabranches and wetlands along the southern bank of the most regulated section of the River Murray.

The focus of the works undertaken over the past five years has been to improve the health and management of the wetlands and waterways as well as to provide opportunities for further development and enhancement of the natural values in key locations along the system. Ensuring these natural systems are supported and protected allows the rare and threatened species of the region, including native fish to have the greatest opportunity of survival and potential to flourish.

The works include environmental regulators to allow wetting and drying of wetlands within weir pools through additional regulation and using pumps to deliver environmental flows and improved habitats. The screening of structures in waterways allows for management of fish populations and opportunities to control pest fish species. Delivery of environmental water through pumps allow for management of native fish recruitment by directing larvae from River Murray spawning events into wetlands that function as nurseries for fish growth. These works and operations allow the river to maintain the social and economic values established with river regulation while returning the natural processes, which in turn will enhance the natural environmental values.

To date, the works include six environmental regulators at River Murray wetlands. Margooya Lagoon is a key wetland which had an environmental regulator installed to re – instate an approiate wet – dry cycle to improve environmental condition. Since the regulator was installed and the wet – dry cycle commenced the Lagoon has functioned as a native fish nursery, supporting Golden and Silver perch populations. Lateral movement of the fish back into the River Murray is an important aspect to native fish conservation. The Mallee CMA has undertaken investigations to understand and manage lateral fish movement at Margooya Lagoon and seek to apply this knowledge throughout its 760 km reach of Murray River.

Key words: Australia, wetlands, environmental infrastructure, regulated waterways, Murray River

1 The Murray-Darling Basin

The Murray – Darling Basin is one of the most socially, environmentally and economically significant areas of Australia. The basin is the catchment for the Murray and Darling rivers and their many tributaries. It covers an area of more than 1×10^6 km^2, which is equivalent to 14% of Australia (MDBA, 2008). In total there are 23 river valleys in the basin, as well as important groundwater systems. The basin's average annual rainfall is 530,618 gal; a total of 94% of this rainfall evaporates (MDBA, 2008). The Murray Darling Basin generates 39% of Australia's income derived from agricultural production.

In 2002, in response to the declining health of the river system, the Murray Darling Basin Authority (MDBA), along with member states, established The Living Murray (TLM) program to help restore the health of six environmentally significant sites along the River Murray. The project is a partnership of the Commonwealth, Queensland, NSW, Victorian, South Australian and Australian Capital Territory Governments (Mallee CMA, 2010). Complementary to this program is the Mallee Catchment Management Authority's (CMA) wetland restoration program. It addresses wetland health issues along the Mallee CMA's 760 km span of the River Murray by identifying and implementing water management works. The objective of the works is to improve the health and functionality of the River Murray floodplain wetlands. The program complements the TLM program by addressing River Murray reaches beyond the realm of the Icon Sites.

2　Murray River: natural to regulated

Murray River is 2,530 km long from its source in the Australian Alps to its mouth at Encounter Bay in South Australia (Crabb, 1997). For some 1,880 km of its length, the river marks the boundary between New South Wales and Victoria. The Mallee CMA manages the northern bank of Murray River for 760 km (from downstream of the Loddon River inlet to upstream of lock 6 (Fig. 1), this section of River is also known as the lowland region of the River Murray.

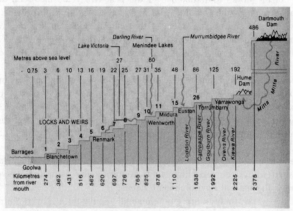

Fig. 1　Longitudinal profile of the River Murray (*Source*: MDBC)

The lowland region of the River Murray has an extensive floodplain environment surrounding its main channel. Natural flows within this system are highly variable (Puckridge et al. , 1998), with parts of the River Murray having ceased to flow on numerous occasions over the last century (Young,2001), and major flooding occurring throughout the catchment on many occasions (Walker,1985). Due to the perennial nature of the River Murray, the floodplain formed throughout the lowland tracts is characteristically broad and forested, with many anabranches, wetlands and backwaters. Organisms native to these floodplain environments have adapted to the highly variable and stochastic hydraulic regimes that occur naturally. Regulation has reduced the variability and stochasticity of the hydrological regime, which has led to a loss in productivity and the removal of environmental cues needed by many native species to sustain resilient populations (Walker,1985, Puckridge et al. , 1998; Humphries and Lake, 2000; Jenkins, 2002; Kingsford, 2000; McCasker, 2004).

Since the 1920s, development of the River Murray has seen the construction of dams, weirs and levee banks for water storage, irrigation supply, boat navigation, flood mitigation, and domestic and commercial water supply. The regulation of rivers within the Murray-Darling Basin has changed natural flow regimes and caused significant reductions in flow magnitude. As a result, the connectivity between the main river channels and their floodplains has been altered dramatically, with many floodplain areas now completely isolated from the river channel, while others have become permanently inundated by weir pools. Regulation and the resulting loss of connectivity are considered key threatening processes contributing to declines in native fish populations throughout the Murray-Darling Basin (SKM, 2003b; McCasker, 2004).

By world standards, the volume of water that travels down the River Murray is very low. It takes less than one day for the Amazon River to discharge the same amount of water as the Murray's average annual flow. The Murray's flow is also highly variable. The long-term average annual run-off reaching the river is 12,600 GL, but this may vary from around 3,000 GL in a very dry year to 57,000 GL in a very wet year (MDBC, 2003). To ensure a reliable supply of water, storages have been constructed and the flow of water in the river regulated. The long-term average diversion from the River Murray is 4,408 GL or 34% of the total flow in the river. Of this, irrigation accounts for 93% of the diverted water with the remainder being used for domestic (rural and urban), industrial and stock supplies (MDBC, 2003; ISMP, 2006).

The River Murray is valued for the range of human activities it supports, such as irrigation, hydroelectricity, tourism and recreation, and also for its cultural heritage. A number of factors, not all of which are related to flow management, have impacted the condition of the River Murray and in turn threaten these values (ISMP, 2006).

2.1 Impacts of river regulation

River regulation has had a significant impact on the natural flow regimes of the River Murray. It has reduced the normally high winter flows while enhancing normally low summer flows, reduced overall water discharge, and decreased flow velocity and variability both within and between years (Maheshwari et al., 1995; Young, 2001; Puckridge et al., 1998; Boulton and Brock, 1999). As a result, wetlands once undergoing inter-annual and annual wet-dry cycles have been effectively cut off from the main channel and left dry for many years. Additionally, the construction of weirs a-long the lower River Murray from the Mouth to Torrumbarry has produced a series of large, slow flowing pools for more than 1,100 river kilometres (Walker, 1985). Many naturally ephemeral wetlands located upstream of these weirs are now permanently inundated due to the elevation of water behind these structures. (McCasker, 2004).

River flow variability, in response to climatic variability, is a characteristic of the natural flow regime in many dryland rivers, including the River Murray (Young, 2001). River regulation alters aspects of the flow regime that are important from an ecological perspective. For example, the operation of a large dam which captures high spring inflows reduces the frequency of small to medium-sized floods downstream. The effects come about principally because river biota (fish, plants, etc.) are adapted to the pre-regulation flow regime (Boulton and Brock, 1999).

The progressive regulation of the River Murray has changed it from a variable and somewhat unpredictable system to a reliable source of water that has enabled development of inland towns. Collectively, the regulating infrastructure (dams, weirs, etc.) on the River Murray have changed its hydrological and ecological character. While the economic benefits have been substantial, there has been an associated ecological cost. The hydrological, ecological and geomorphic impacts of flow regulation are well documented (Thoms et al., 2000; Young, 2001; Erskine et al., 1993; ID & A, 1997; Gippel and Blackham, 2002; Walker, 1985; Maheshwari et al, 1993; Walker and Thoms, 1993; Rutherfurd, 1991).

Flow regulation and resource use has impacted on the environmental condition of the River Murray. Norris et al., 2001 evaluated the aggregate impacts of resource use on the environmental condition of rivers in the MDB. The results indicated that overall biological and environmental condition was degraded right along the River Murray, with most degradation towards the Murray mouth. The following summary of condition is from Norris et al. (2001):

(1) fish populations are in very poor to extremely poor condition throughout the river;

(2) macro - invertebrate communities are generally in poor condition and declining towards the river mouth;

(3) riparian vegetation condition along the entire river was assessed as poor, with grazing and alterations to the flow regime being the major causes;

(4) wetland quality is significantly reduced, with most wetland loss attributed to permanent inundation of areas previously intermittently flooded;

(5) hydrological condition in the River Murray Channel is poor for all zones, with the extent, timing and duration of floodplain inundation all significantly impacted;

(6) riverine habitat was found to be poor or very poor through all zones with connectivity, riparian vegetation and bed-load all affected by regulation;

(7) the condition of floodplain inundation has been assessed as very poor in all zones;

and nutrients and suspended sediments are poor and worsening towards the Murray.

The Mallee CMA's wetland restoration program has focused on the ecological impact of River Murray wetlands permanently inundated by weirs. In the lower River Murray the presence of weirs has extended the area of permanently flooded wetlands, so that 70% of lower Murray wetlands (backwaters, sidearms, anabranches, lakes and billabongs) are now connected to the river at weir pool level (Pressey, 1986). Many of these wetlands were formerly subject to larger, more frequent water level changes, and some would have dried periodically.

The wetland restoration program has additional focus on the ecological impact of weirs and other regulating structures on fish movement. Weirs and other regulating structures are significant obstacles for fish movement along the River Murray, and removal of these barriers or installation of fishways is an important step toward improving habitat access for native fish. The program seeks to address the impacts associated with flow regulation by implementing works and measures to reinstate the characteristics of variable flow with the objective of improving wetland health and the role of wetlands in native fish conservation.

2.2 Restoring the balance

In response to continued degradation of River Murray condition, the Mallee CMA employed a strategic approach to address the problem. An extensive investigation of the Mallee CMA 760 km of the River Murray's northern embankment was undertaken; the investigation assessed deficiencies in the water regime of floodplain ecosystems and identified water management options to address them.

The investigation identified many wetlands with a deficient water regime that could benefit from water management works. Margooya Lagoon is an example of one of these wetlands. Margooya Lagoon is a River Murray floodplain wetland situated in the Beggs Bend River Reserve, located 12 km south-east of Robinvale. The Lagoon is approximately 30 hectares in size and up to 2 metres deep at full supply level. It is situated upstream of the River Murray lock 15 weir pool, permanently inundating the Lagoon since constructed of the weir in 1937, due to the elevation of water behind the structure. Naturally, Margooya Lagoon is ephemeral undergoing inter – annual and annual wet-dry cycles, a vital component to wetland health that has been absent since 1937. Drying of a wetland promotes nutrient transformation and the consolidation of sediments, and allows for the establishment of terrestrial plant species which provide habitat and nutrients for aquatic biota upon re – inundation (Ellis and Pyke, 2010). The water management works seek to re-instate these processes, improving the ecological function and condition of Margooya Lagoon. The long and short term effects of the changes to wetland hydrology are still largely unknown, however it is now recognised that the functioning and productivity of the boarder floodplain and main channel are adversely affected (McCarthy et al., 2004; Sparks, 1995; Tockner et al., 1999; McCasker, 2004).

A condition assessment of Margooya Lagoon's fish community was undertaken in 2004. The fish community consisted of small bodied species and carp only, with no recordings of large bodied fish species (Ellis et al., 2004). Typically, these wetlands would have supported healthy and diverse native fish populations and been important juvenile fish habitat, the survey results indicate that the wetland is not functioning in this capacity. Humpheries and Lake (2000) propose that fish are affected by river regulation in two ways; firstly, by the removal of appropriate conditions needed for reproductive efforts and secondly, that regulation has decoupled the occurrence of fish larvae and the environmental conditions needed to sustain them until they become juveniles (Humpheries and Lake, 2000; McCasker, 2004).

Considering this, an environmental regulator was identified as the preferred water management work to reinstate an appropriate water regime to the Lagoon. The regulator would enable the Lagoon's water regime to be managed independent of the Lock 15 weir pool, enabling reinstatement of a wet-dry cycle to the Lagoon without impacting on the water level or function of the lock 15 weir pool (in Fig. 2). The need to manage the exotic fish population required the regulator to be fitted with a carp exclusion screen (mesh width of 35 mm), to restrict passage through the regulator to fish of a girth generally less than 35 mm.

Lock 15 weir pool

Lagoon

Fig. 2　The environmental regulator isolating Margooya Lagoon from the lock 15 weir pool

In 2009 the regulator was installed and the wetland was dried for the first time in nearly 72 years, completely eradicating the carp population and triggering ecological processes associated with a drying phase. The Lagoon was partially refilled late in 2009 by opening the regulator, while the carp exclusion screen remained closed. Due to low river levels during the refilling event, the regulator was closed and Margooya Lagoon was surcharged via pumping directly from the River Murray into the Lagoon.

Upon the Lagoon becoming full, fish surveys indentified juvenile golden perch (Macquaria ambiguα) and silver perch (Bidyanus bidyanus) in the Lagoon. Silver perch is listed under the Victorian Flora and Fauna Guarantee Act 1988 (FFG Act 1988). The presence of juvenile golden perch and silver perch indicated that larvae or early juveniles of these species had entered via the pump or through the carp exclusion screen prior to pumping. No large bodied predatory fish which could eat juvenile golden perch and silver perch were present in Margooya Lagoon at this time (Ellis and Pyke, 2010).

It is generally recognised that elevated flows and discharge induces spawning in golden perch and silver perch (King et al. , 2010; Mallen – Cooper and Stuart, 2003). The River Murray demonstrated elevated (but within bank) flows during late November and early December 2009 (Fig. 3). These elevated flows appear to have stimulated these species to spawn in the river channel, with larvae and/or juveniles subsequently transported into Margooya Lagoon (Ellis and Pyke, 2010) during the refilling event.

A following fish survey four months after the refilling, again detected juvenile golden perch and silver perch in the Lagoon. The growth of these juvenile fish (a rate of approximately 1mm per day since December 2009) is indicative of the suitability of Margooya Lagoon as a nursery habitat for native fish. This is consistent with other hypothesis and research findings which suggest off-channel habitats provide important nursery environments for fish species (Junk et al. , 1989; Lyon et al. , 2010; Closs et al. , 2005), and demonstrates that managed watering events in floodplain wetlands can contribute to the conservation of native fish biodiversity (Ellis and Pyke, 2010). The carp exclusion screen appears to have prevented adult Carp from entering the wetland, thus allowing aquatic macrophytes to emerge and create dense cover for native fish species. The absence of Carp also promoted the consolidation of wetland sediments, and emergence of abundant phytoplankton and zo-

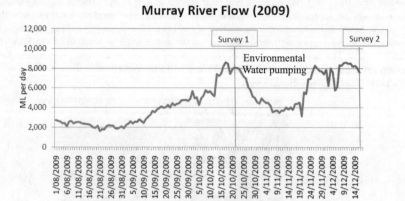

**Fig. 3   River Murray flows from June to December 2009,
with two fish surveys and pumping of water indicated**

oplankton, providing excellent nursery habitat for larval and juvenile fish.

Lateral movement of the fish back into the River Murray is an important aspect to native fish conservation and the Mallee CMA has been investigating how to stimulate this movement. Lyon et al. (2010) suggests that lateral fish movements approximated water level fluctuations. That is, as water levels rise, fish leave the main river channel and move into newly flooded off-channel habitats and on falling levels, fish move back to the permanent riverine habitats. Several exceptions to this trend were detected throughout the surveys of Margooya Lagoon. Assessments of fish movement between the River Murray and Margooya Lagoon detected movement of Golden perch and Silver perch towards the River Murray from the lagoon on filling events but not on drainage events. Thousands of juvenile Carp were; however, recorded entering the wetland during a drainage event.

The Mallee CMA seeks to continue to understand and refine lateral fish movement at Margooya Lagoon and apply this knowledge throughout the River Murray floodplain wetlands, in particularly to the other six environmental regulators already installed at wetlands. In summary, the Mallee wetland restoration program has been successful in achieving environmental outcomes, particularly for native fish, without impacting on the irrigation or economic requirements of the River Murray. Alternatively, the program has been beneficial to these industries by supporting a healthy River Murray which is fundamental in long term irrigation and economic sustainability.

References

Boulton A J, Brock M A. Australian Freshwater Ecology: Processes and Management[M]. Gleneagles Publishing. 99.

Closs J, Balcombe S, Driver P, et al. The Importance of Floodplain Wetlands to Murray-Darling Fish: What's There? What do We Know? What do We Need to Know? Murray Darling Basin Commission—Native Fish and Wetlands in the Murray – Darling Basin,2005.

Crabb P. Murray – Darling Basin Resources. Murray – Darling Basin Commission, Canberra, 1997.

Ellis I, Pyke L. Assessment of Fish Movement to and from Margooya Lagoon upon Re – connection to the Murray River[R]. Report to the Mallee Catchment Management Authority. The Murray-Darling Freshwater Research Centre,2010.

Ellis I, Ho S, Mc Carthy B, et al. Distribution of Aquatic Vertebrates within the Mallee Region [R]. Final Report Prepared for the Mallee Catchment Management Authority by The Murray-Darling Freshwater Research Centre, 2004.

Gippel C J, Blackham D. Review of Environmental Impacts of Flow Regulation and Other Water Resource Developments in the River Murray and Lower Darling River System[R]. Final Report by Fluvial Systems Pty Ltd, Stockton, to Murray – Darling Basin Commission, Canberra, 2002.

Humphries P, Lake P S. Fish Larvae and the Management of Regulated Rivers[J]. Regulated Rivers: Research and Management,2000(16):451 – 432.

Jenkins K. Zooplankton Community Responses to Flooding and Drying Cycles[R]. Murray Darling Freshwater Research Centre Technical Report, 2002.

Junk W J, Bayley P B, Sparks R E. The Flood Pulse Concept in River – Floodplain Systems[J]. Canadian Special Publication in Fisheries and Aquatic Sciences,1989(106):110 – 127.

King A J, Tonkin Z, Mahoney J. Environmental Flows Enhance Native Fish Spawning and Recruitment in the Murray River, Australia[J]. River Research and Applications, 2009 (25): 1205 – 1218.

Kingsford R T. Ecological Impacts of Dams, Water Diversions and River Management on Floodplain Wetlands in Australia[J]. Austral Ecology,2000(25): 109 – 127.

Lyon J, Stuart I, Ransey D, et al. The Effect of Water Level on Lateral Movements of Fish between River and Off-channel Habitats and Implications for Management[J]. Marine and Freshwater Research,2010(61): 271 – 278.

Maheshwari B L, Walker K F, McMahon T A. Effects of Regulation on the Flow Regime of River Murray, Australia[J]. Regulated Rivers: Research and Management, 10: 15 – 38, 95.

Mallen Cooper M, Stuart I G. Age, Growth and Non-flood Recruitment of Two Potamodromous Fishes in a Large Semi-arid? Temperate River System[J]. River Research and Applications, 2003(19): 697 – 719.

McCarthy B, Gawne B, Meredith S, et al. Effects of Weirs in the Mallee Tract of the River Murray. Murray-Darling Freshwater Research Centre, Mildura. Report to the Murray – Darling Basin Commission, Canberra. 2004.

McCasker N. (2004). Factors Determining Zooplankton Response to Floodplain Inundation and Changes to In – channel Flow Regime. Murray-Darling Freshwater Research Centre, Mildura. Report to the Mallee Catchment Management Authority.

MDBA. Explore the Basin[EB/OL][2012 – 02 – 03]. http://www. mdba. gov. au/explore – the – basin/about – the – basin.

MDBC. Murray – Darling Basin Water Resources Fact Sheet[R]. Murray – Darling Basin Commission, 2003.

Mallee CMA Hattah Lakes Icon Site Environmental Watering Management Plan Prepared for the Murray Darling Basin Authority,2010.

Pressey R L. (1986). Wetlands of the River Murray below Lake Hume. Prepared for the River Murray Commission. RMC Environmental Report 86/1.

Puckridge J T, Sheldon F, Walker K F. et al. Flow Variability and the Ecology of Large Rivers [J]. Marine and Freshwater Research,1998(49): 55 – 72.

Rutherfurd I D. Channel Form and Stability in the River Murray: A Large, Low Energy River System in South Eastern Australia [D]. PhD Thesis, Monash University. Vols. 1 & 2. Lower River Murray, Australia. Regulated Rivers: Research and Management, 8:103 – 119.

Sinclair Knight Merz, Roberts J. Assessment of Water Management Options for Lindsay and Wallpolla Islands[R]. Final Report, 2003.

Tockner K, Schiemer F, Baumgartner C, et al. The Danube Restoration Project: Species Diversity Patterns Across Connectivity Gradients in the Floodplain System. Regulated rivers: Research and Management ,1999(15): 245 – 258.

Walker K F, Thoms M C. Environmental Effects of Flow Regulation on the Lower River Murray, Australia[J]. Regulated Rivers: Research and Management,1993(8):103 – 119.

Walker K F. A Review of the Ecological Effects of River Regulation in Australia[R]. Hydrobiologia,1997,34(7):41 – 49.

Young W J. Landscapes, Climates and Flow Regimes[J]. Rivers as Ecological Systems: The Murray Darling Basin, 2001: 135 – 171.

Strengthening Integrated River Basin Management in order to Realize the Long-term of Stability and Security in the Yellow River

Zhao Yong

Yellow River Conservancy Commission, Zhengzhou, 450003, China

Abstract: On the basis of analyzing and summarizing the meaning and function of integrated river basin management, the current status of river basin management, achievements and objectives, river basin management institutions and operation mechanisms, elaboration is made on how to strengthen the integrated river basin management of the Yellow River, from such aspects as river basin planning management, integrated water resources management, the construction of river basin flood prevention and drought relief system, the construction of policy and regulation system, and the managing capacity building, etc. The management mechanism of the Yellow River is river basin management combined with regional management, through establishing the following mechanisms as river basin management consulting mechanism, the management and regulation mechanism of key water conservancy projects in its mainstream and tributaries, the unified water resources management and regulation mechanism, the responsibility supervision mechanism of water quantity and quality at provincial boundary sections, the managing mechanism of construction projects within the scope of river channel management, and incidental water emergency and response management mechanism, etc, so as to realize benign, efficient and standardized management of the Yellow River Basin. Integrated river basin planning is the main approach and means to realize integrated river basin management. The policy and regulation system construction based on the actual situation of the Yellow River is the foundation to realize water control and management according to the laws. The capacity building of the administrative law enforcement on water affairs, supervision, project management, science and technology, and human resources, etc. is the guarantee for realizing effective management. Strengthening the integrated river basin management is to ensure the flood prevention security, better support the sustainable development of the river basin economy and society with limited water resources, so as to realize the long period of stability and security of the Yellow River, and better benefit the Chinese Nation.

Key words: river basin, integrated management, institutional mechanism, planning management, water resources management, Yellow River

1   Overview of the Yellow River Basin

As the second longest river, with a total length of the main stream of 5,464 km and drainage area of 795,000 km^2 (including 42,000 km^2 interior drainage area), the Yellow River passes through 9 provinces (autonomous regions) and is the important water source of Northwest China and North China. The Yellow River Basin enjoys rich land and mineral resources, especially the energy resources, which are of great significance in the strategic pattern of social and economic development of China.

Meanwhile, the Yellow River is a river characterized by its complicated natural conditions and extremely special water regime; there is serious water shortage within the river basin and other prominent issues such as flooding, sedimentation and water and soil erosion, etc. , which bring about many challenges to the governance, development, protection and management of the Yellow River Basin.

2　Current situation, achievements and objectives of the integrated management of the Yellow River Basin

2.1　Current situation

The Yellow River Conservancy Commission (YRCC) is a water administration authority subordinated to the Ministry of Water Resources, exercising the water administration functions within the Yellow River Basin and inland river basins in Xinjiang, Qinghai, Gansu and Inner Mongolia on behalf of the Ministry of Water Resources.

The YRCC is responsible for the planning management, unified management and regulation of water resources, flood control and drought relief management, supervision and management of water resources protection, river course management, construction and management of water conservancy projects, water and soil conservation management, water administration enforcement and other management responsibilities in accordance with the *Water Law*, *Flood Control Law*, *Law on Water and Soil Conservation and Law on Prevention and Control of Water Pollution of the People's Republic of China*, *Regulations on Water Regulation of the Yellow River*, *Measures for the Yellow River Estuary Management and* other laws and regulations. With the authorization of water administration issues to the YRCC by the state, a system integrating the Yellow River Basin management with region management has been basically established, the river basin water administration functions of the YRCC have been enlarged and strengthened and various management systems are being constantly established and improved.

2.2　Achievements

(1) For the planning management, fruitful achievements have been made in preparation of various plans, so as to enrich comprehensive planning, strengthen professional planning and deepening special planning via constantly strengthening the top-level design and a preliminary water resources planning system of the Yellow River basin centering on the governance of the Yellow River has been established following the thought of water resources for people's livelihood and sustainable development. Besides, the system of approval on the construction planning of water projects is implemented to strengthen the implementation of plans.

(2) For the water resources management, the control and regulation and the overall benefits of the Yellow River water resources have been improved by proposing the distribution plan of available water supply of the Yellow River, implementing in full scale the water abstraction permit system and the approval system of water resources assessment report on project construction, developing water regulation plan and methods, implementing unified management and real-time regulation of water quantity of the mainstream and its major tributaries and strengthening water planning and water saving.

Through the unified regulation, the Yellow River has achieved no zero flow of 13 consecutive years, which effectively guaranteed the water use for economic and social development and ecological environment of the river basin and relevant regions.

(3) For the flood control and drought relief, flood control and drought relief organization system, mobilization system, flood treatment and regulation system have been established, achieving the joint flood control of eight provinces and regions, and flood prevention regulations of major reservoirs on the mainstream and tributaries, and expanding the drought relief function of the river basin. In order to combine pre-flood reservoir emptying and sediment reduction of river courses, the YRCC has conducted 14 times of water and sediment regulation tests and production runs since 2002, which have achieved good results.

(4) For the water resources protection, the licensing system on setting effluent discharge outlets has been implemented according to the verified assimilative capacity of the water function area and suggestions on limiting pollutant discharge, forming a preliminary emergency response mechanism to major water pollution.

(5) For the management of water and soil conservation, a preliminary management system of

unified planning, administrative approval, law enforcement inspection, supervision and monitoring and public announcement has been formed.

(6) For the river course management, the approval and supervision management of projects within the range of river course management has been strengthened and the flood impact assessment system has been applied.

(7) For the management of water conservancy projects, the post responsibility system, monthly report system and subject-oriented office meeting system concerning construction management, and the management-maintenance separation system have been energetically implemented, which has enhanced the construction and operation management level and the maintenance level.

(8) For the management of water administration enforcement, the order of water affairs of the Yellow River has been properly safeguarded by strengthening the foundation of water conservancy enforcement team, investigating and handling cases of water violations and positively preventing and mediating inter-provincial water disputes.

2.3 Objectives

For the integrated management of the Yellow River Basin, the construction of river basin management system and operating mechanism aiming at the shortcomings in management and operation should be further strengthened and new management modes should be created following the rules of nature and social and economic development with reference to the advanced management experience of other river basins and the actual conditions in China and the specific characteristics of the Yellow River Basin; policies and legal systems should be improved to have the water control and regulation conducted in accordance with relevant laws; scientific and technological innovations should be made to support capacity building and improve the public service and social management level of the river basin; water conflicts between economic development and governance, development and protection of the river basin, and between regions and departments should be properly coordinated and settled to improve the social, economic and ecological benefits of the governance and development of the river basin; the integrated management capacity of the river basin should be constantly improved to realize the long-term stability of the Yellow River and support the strategic objective of sustainable development of economic society and ecological environment of the river basin and relevant regions.

3 Future focuses on integrated management of the Yellow River Basin

3.1 Management system and operating mechanism

The unified planning, management and regulation of the river basin will be promoted and sound Yellow River Basin management system characterized by definite duties and responsibilities, standard operation and authenticity and efficiency will be established under the management framework integrating river basin management with regional management. River basin management and regional management organizations at various levels will be established according to the actual requirements, and the duties and responsibilities of river basin management and regional management will be divided following the principle of unity of power and responsibility. The management of public water affairs by and the public service capacity of river basin management organizations will be strengthened. An open and public service information platform concerning administrative management of water resources will be established to ensure that the public can fully participate in the river basin management.

A standard, coordinated, efficient and beneficial river basin integrated management operating mechanism will be established through further improvement of river basin management negotiation mechanism, strengthening of unified management and regulation mechanism of water resources, improvement of water quantity and quality supervision system of provincial sections, tightening the management mechanism of construction projects within the range of water course management and building sound water emergency warning and management mechanism.

3.2 Planning management

The planning of water resources is the most important basis to guide the governance, development and protection and promote the integrated management of river basins. In order to improve the planning system of the Yellow River Basin, the preparation and submission of plans of key areas should be properly carried out, the top-level design should be strengthened, the plans should be prepared in a scientific way, the *Suggestions on the Framework of the Planning System for Water Resources of the Yellow River Basin* should be supplemented and improved and the construction of planning system for managing the Yellow River should be accelerated to support sustainable social and economic development of the river basin.

The first is to accelerate the approval process of prioritized plans. Primarily, the importance should be attached to the review and approval of important plans such as the comprehensive planning of the Yellow River Basin, the construction and management planning of flood storage and detention areas in the Yellow River Basin, the planning of comprehensive treatment of the Yellow River Estuary, the planning of comprehensive treatment of floodplain in the lower reaches of the Yellow River, the planning of safety construction of floodplain in the lower reaches of the Yellow River and the construction planning of water and sediment control system of the Yellow River.

The second is to accelerate the preparation of a number of plans. The preparation of comprehensive plans for Huangshui River Basin and the water resources of Hongjiannao Lake should be completed as soon as possible by strengthening the examination and supervision; the preparation of comprehensive plans for river basins of such important tributaries as Wuding River, Kuye River, Yiluo River, Taohe River and Qinhe River and the preparation of medium-and long-term water supply and demand plan of the Yellow River Basin should be carried out in an all-round way; and the preparation of water and soil conservation plans and plans for treatment, development and management of estuaries, shores and tidal zones should be properly carried out.

The third is to start the preparation of plans for the protection of water resources in the Yellow River Basin as soon as possible.

Besides, in order to strengthen the planning and management of water resources and make the planning more instructive and have more binding force, the management of water administration permit should be strengthened by properly implementing the planning approval system and reviewing the planning approval of water projects according to the authorization of the Ministry of Water Resources.

3.3 Unified management and regulation of water resources

The Yellow River Basin is subject to severe water shortage; the drainage area accounts for 8.3% of the national land area while the runoff only accounts for 2% of the total national runoff; the water supply per capita within the river basin is 473 m^3, accounting for 23% of the national water supply per capita; the water supply for farmland is 220 m^3 per mu, accounting for only 15% of the national water supply for farmland per mu.

In the future, the requirements of the most rigid water resources management system should be applied and further importance should be attached to the unified management and regulation of water resources. The water distribution plan should be adjusted and water distribution indicators should be detailed according to the available water resources of the Yellow River and changes in supply and demand. The coordination between the control of total basin water quantity and total regional water quantity should be strengthened in the management of water abstraction permit; the unified management and regulation of water quantity and quality of the mainstream and important tributaries of the Yellow River should be strengthened; the duties and responsibilities of river basin management and regional management should be clearly defined; and the quantity and quantity indicators of provincial sections and estuaries of tributaries to the Yellow River should be subject to strict control by strengthening the measurement. Vigorous efforts should be put into the water right transfer within and between provinces (regions) and the promotion of water saving in production and living. The negotiation, coordination and communication on water regulation of the river basin should be further strengthened by hydropower generation regulation agencies and ecological protection agencies participating in the negotiation on the water regulation of the Yellow River. The water of cascade hydropower projects on the mainstream and the reservoirs on main tributaries of the Yellow River is subject to unified regulation and the water for hydropower generation is subject to water regulation. The groundwater of the river basin is subject to the control of total exploitation quantity and the water level control and the management in seriously over-exploited area.

3.4 Flood control and drought relief management

Due to the special water regime of the Yellow River, the current situation of flood control and ice flooding prevention is still tense, the conflict between water supply and demand in dry and low-water years is very sharp and the irrigation water is subject to limitations.

The organization, instruction, coordination and supervision of flood control and drought relief of each province (region) in the river basin by the Flood Control and Drought Relief Office of the Yellow River should be furthered strengthened, as well as the unified regulation of major reservoirs on the mainstream and important tributaries, the administrative supervision and control of flood control and drought relief of the river basin, and also the overall planning and coordination between flood control and drought relief. Accountability systems of the governments and departments where the chief executives are the center role should be implemented. The capacity building of departments at various levels for flood control and drought relief should be strengthened via establishing specialization-socialization integrated emergency rescue teams, and sound systems concerning flood control and emergency rescue and reserves of drought relief materials, improving flood control and drought relief emergency plans and enhancing modern management level of flood control and drought relief. Flood insurance should be encouraged and supported to perfect the non-engineering measures for flood control.

3.5 Water administration

A water administration system integrating river basin management and regional management and being characterized by well-behaved conduction, coordinated operation, fairness and transparency and honesty and high efficiency should be established. Proper management mode should be determined according to the characteristics of river reaches under direct jurisdiction of the river basin management agency, inter-provincial boundary river reaches, trans-provincial (regional) tributary river reaches, river reaches within important reservoir areas and other river reaches. The water administration power of the river basin management agency and local water administration agencies and relevant law enforcement agencies should be clearly defined so that they can assume their own respective roles on the basis of definite duties and responsibilities and work together in case of major water cases. The water administration functions and enforcement power of the river basin management agency should be further strengthened according to the laws and regulations and also the authorization of the Ministry of Water Resources; meanwhile, a sound system of joint law enforcement by the Yellow River conservancy public security team and water administration enforcement team should be established.

3.6 Policies and legal systems

The existing laws and regulations should be actively carried out. Studies should be implemented to develop laws and regulations and policies which are in urgent need and build sound water law system of the Yellow River according to the needs of the Yellow River governance and river basin management, in combination with the actual conditions of the Yellow River, successful experiences and mature policies. Mainly including: *Yellow River Law* should be formulated legally; Rules of Water Resources Protection of Yellow River Basin should be formulated administratively; and with regard to the department rules and regulations, stipulations concerning the management of the

Yellow River source region and Dongping Lake should be made; water right transfer system of the Yellow River and ecological compensation policies for the river basin should be developed, and the possibilities of systems and policies concerning water sediment replacement, emissions trading and sand excavation in river courses should be studied according to the state laws and regulations and managing requirements of the Yellow River.

4 Conclusions

The Integrated River Basin Management is an important means to solve the water problems in a systematic way, exert the multiple functions and achieve the maximum comprehensive benefits of the river basin. The special water regime of the Yellow River makes the governance and integrated management of the Yellow River protracted, arduous and complex. In the future, the importance will be attached to the exploration of integrated management mode of the Yellow River Basin, the integrated management and construction of the river basin centering on management system and operating mechanism, planning management, unified management and regulation of water resources, flood control and drought relief management, the constant improvement of integrated management level, the normalizing of governance and development activities, the protection of ecological environment, and the settlement of special conflicts and problems to guarantee the long-term stability of the Yellow River and support the strategic objective of sustainable development of economic society and ecological environment of the river basin and relevant regions.

A. Advanced River Basin and Water Resources Management with Social and Economic Development

Discussion on Zoned Harnessing of the Lower Yellow River Floodplain

Niu Yuguo, *Duanmu Liming*, *Geng Mingquan* and *Li Yongqiang*

Yellow River Henan Bureau, YRCC, Zhengzhou, 450003, China

Abstract: With the development of economy and society, the livelihood and production security of the masses in the floodplain need to be gradually improved on the condition of flood control security of both banks. Meanwhile, with the improvement of regulating and storage capacity of reservoirs on the Yellow River and its tributaries and flood forecast capability, the flood process of the lower reaches can be quickly forecasted. In this context, in the light of the current status of the floodplain harnessing and the existing problems, this paper presents a general idea, namely, forming closed zones with the existing Yellow River dykes, dangerous sections and control dams and newly built flood control embankments to carry out zoned harnessing of the floodplain, and discusses the relationship between flood management and utilization and security construction and flood diversion warping, and based on this, proposes the direction of future work.

Key words: the lower Yellow River, floodplain, Zoned Harnessing of the lower Yellow River floodplain, flood control embankments

1 Overview of the lower Yellow River floodplain

1.1 Natural geography

The special incoming water and sediment process has resulted in the general uplifting trend of the lower Yellow River channel due to siltation, and years' struggle between man and flood has helped form the current special channel pattern of the lower Yellow River. At present, over 120 large or small floodplain lands are distributing within the Lower Yellow River channel, whose area is more than 4,000 km² with a population near 1.9 million. There are 7 floodplain lands larger than 100 km², 9 ones with an area of 50 ~ 100 km², 12 ones of 30 ~ 50 km² and more than 90 ones smaller than 30 km². The floodplain lands are mostly distributing upstream of Taochengpu, whose area accounts for more than 78% of the floodplain in the lower reaches.

1.2 Effect of the floodplain in the Yellow River flood control system

Floodplain is vast in the Lower Yellow River. When flood larger than bankfull discharge occurs, the water flows into the vast floodplain. The floodplain becomes a large natural flood detention area. Take the section between Zhengzhou Railway Bridge and Dongbatou. The reach is 120 km long. The distance between the levees on both banks ranges from 5.0 km to 14.0 km and the channel area is 1,156.4 km², of which 983.4 km² is floodplain, accounting for 85.04% of the channel area. According to the analysis of nine flood events with peak flow larger than 8,000 m³/s at Huayuankou, the average peak cut ratio of the section is 6.99%, and the largest peak cut ratio is up to 11.3%, of which the channel's largest detention volumes were 0.736×10^9 m³, 1.01×10^9 m³ and 0.785×10^9 m³ respectively in 1954, 1958 and 1982, and the peak cut ratios were 11.33%, 8.07% and 5.22% respectively, with 8.21% as the average.

The lower Yellow River floodplain has not only the role of detention, but also of desilting. According to statistics of observed data, from June 1950 to October 1998, 9.202×10^9 t of sediment deposited totally in the lower Yellow River, of which 6.370×10^9 t deposited in the floodplain, making up 69.1% of the total sedimentation of the whole cross - section. Take the flood in 1958 as an example, during the flood period of 22,300 m³/s at Huayuankou, the largest flood detention volume was 1.35×10^9 m^3 in the reach from Huayuankou to Gaocun, the flood peak was cut by 4,300 m^3/s. In the flood process, 1.07×10^9 t of sediment deposited in the floodplain, while 0.86 $\times 10^9$ t of sediment was washed away, forming the high floodplain and deep channel. During the period of flood in August, 1996, due to water and sediment exchange between the floodplain and the channel, the channel was scoured deeper, while the floodplain lifted with siltation. In the flood process, 0.533×10^9 t of sand settled on the floodplain, while 0.178×10^9 t was scoured away from the main channel, and a total of 0.355×10^9 t of sediment deposited in the whole cross – section. The function of peak cutting, flood detention and desilting of the lower reaches floodplain has played an important role in ensuring the lower Yellow River flood control safety.

1.3 Flood disasters in the floodplain

According to incomplete statistics, the floodplain has suffered flooding more than 30 times to different degrees since the foundation of New China. 9 million person – times were accumulatively affected, 13,000 villages were stricken and 2.6×10^7 mu (1 mu = 1/15 hm^2) of farmland was flooded.

The floodplain was seriously flooded in 1958, 1976, 1982 and 1996. In 1958, 1976 and 1982, the low floodplain downstream of Dongbatou were almost all flooded, and some were flooded upstream of Dongbatou. In 1996, the peak flow of Huayuankou was 7,860 m^3/s. Duo to the seriously silted channel, all the other stations saw the largest discharge volume except Gaocun, Aishan and Lijin since there have been observed record, and the floodplain areas were almost all flooded, even the high floodplain areas in Yuanyang, Kaifeng and Fengqiu that had not been flooded since 1855 were submerged. During the period of the flood in August, 1996, it was investigated that average water depth was about 1.6 m, with the largest depth of 5.7 m. 1,374 villages were flooded and a population of 1.118 million was under the suffering. 2.476×10^6 mu of farmland was stricken, 265.4 thousand houses were collapsing, 409.6 thousand houses were damaged, and 560 thousand people were evacuated. The direct economic loss was up to 6.46 billion yuan according to the year's price.

Since the Xiaolangdi Reservoir was put into operation, in some river sections, the floodplain areas where the main channel bottom elevation higher than that of low floodplain area still exists in large quantities though bankfull discharge has increased to some extent. Once the river regime changes, these low floodplain areas may still be flooded. Take the floodplain areas in Lankao and Dongming as an example. Affected by the "West China autumn rain" in 2003, the river discharge maintained at 2,500 m^3/s, nine natural floodplain areas in the Dongbatou – Taochengpu section were flooded, with an area of $4,98 \times 10^5$ mu, of which 3.50×10^5 mu was farmland; the population affected by the disaster was 148.7 thousand, of which 40 thousand was evacuated.

1.4 Current harnessing and the existing problems in the floodplain

At present, the masses of the lower Yellow River floodplain mainly rely on security construction to ensure flood control safety. The floodplain security construction mainly adopts three measures: village relocation, temporary evacuation and on – the – spot flood avoidance. The above measures had some shortcomings: investment is slow, flood – avoidance works construction lacks investment and the construction is slow, so the safety of the masses cannot be fully protected; some already built flood – avoidance works are of low standard; and village relocation is difficult to implement.

Besides, problems still exist in the overall floodplain harnessing and development at present: the current floodplain security construction is not sufficiently linked up to local social and economic development (such as new countryside construction); the floodplain security construction cannot be organically integrated with the secondary—suspended river harnessing; whether the production dykes should be kept or removed; under the present floodplain circumstances, when the floodplain is flooded, the policy on compensation is difficult to implement; and the economic development of the floodplain is not connected with the general flood control strategy of the lower reaches.

The existence and solutions of the above problems are both closely related to the lower Yellow River flood control strategy. Especially, flood control of the floodplain is related to the lower Yellow River flood control safety on the one hand; but on the other hand, it is also related to production and development of nearly 1,900 thousand people in the floodplain. Therefore, it is urgent and necessary to research countermeasures to harness lower Yellow River floodplain.

2 Different ideas on the floodplain harnessing

2.1 Ideas on the lower Yellow River flood control harnessing and existing problems

Training strategies of the lower Yellow River are "stabilizing the main channel, regulating water and sediment, widening the channel to reinforce dykes and compensating in accordance with policies". Among them, "widening the channel to reinforce dykes" has always been one of the bases for defending against large floods in the lower Yellow River and ensuring flood control safety.

In accordance with this training strategy, the life safety of the masses is guaranteed by floodplain security construction during certain reappearing period of floods, while no corresponding measures to guarantee the production safety of the masses. Under the circumstances, the economic condition of the floodplain masses has obviously fallen behind other areas. To protect their own interests, the masses constructed production dykes in the floodplain to avoid the unfavorable situation where they would be stricken whenever bankfull flood occurs, but this practice reduced the chance that the floodplain is flooded, and prevented the floodplain from retarding the flood and desilting, thus resulting in steady development of the "secondary—suspended river", and endangered the dyke safety. To ensure flood control safety of the lower Yellow River and carry out the strategy of "widening the channel to reinforce the dykes", we have insisted upon removal of the production dykes for years, but this has always been questioned and opposed by the floodplain masses.

2.2 Current different ideas on floodplain harnessing

How to use the floodplain for flood detention and desilting, reduce the losses of the floodplain as far as possible and promote social and economic development in the floodplain? Experts have proposed a number of harnessing programs. There are mainly the following three representative programs for floodplain utilization mode of flood diversion and detention.

Program 1: Whole floodplain operates with gradual removal of production dykes. This program continues to use the present way for floodplain harnessing, namely, the whole floodplain is used for flood diversion and detention, the production dykes are treated as illegal works and must be demolished, and the life safety of the floodplain masses relies on floodplain security construction works.

Program 2: Whole floodplain operates through constructing production dykes with certain standard. Based on the existing production dykes, the production dykes shall be rebuilt or reinforced so that they can defend floods of certain standard. When the flood discharge is smaller than the standard, the flood will discharge through the passage restricted by the production dykes; and when the discharge is larger than the standard, the flood will overflow and the floodplain will automatically be used for flood detention and desilting. No flood inlets or recession facilities shall be set in the production dykes.

Program 3: Zoned floodplain operates with flood control dykes of certain standard. Dykes of certain standard shall be built in the floodplain, and flood inlets and recession facilities shall be set in the upper and lower ends of the flood control dyke. In addition, appropriate re – division shall be made within the floodplain areas according to their size to form smaller zones with grid dykes, and, according to flood situation, different floodplain zones shall be selected for flood diversion, detention or desilting.

2.3 Focus of different harnessing ideas

Program 1 is a floodplain harnessing measure supported with the lower Yellow River General

Flood Control Strategy. In the circumstances, the floodplain can give full play to flood detention and desilting, constituting an important part that makes the lower Yellow River Flood Control Strategy give full play to flood control effectively. However, the interests of the floodplain masses cannot be guaranteed. At present, due to large investment in the floodplain security construction and slow construction process, neither life nor production safety of the floodplain masses can be guaranteed. The floodplain masses built a large number of production dykes. Though the production dykes are illegal at present, it is very difficult to demolish them completely. In addition, in case of flooding of medium – sized floods, though threat is little to the dykes, and the quantity of the sediment that they carry and depositing in the floodplain is limited, flooding large area will seriously threaten production safety of the masses in the floodplain and livelihood safety of the masses in the low floodplain areas. It causes large economic losses, which either local government or the State is difficult to bear, and which may also produce huge social impact.

Program 2 is that the existing the Yellow River embankment is used to guarantee flood control safety on both banks. The production dykes shall be rebuilt or reinforced to defend against floods of certain standard in order to ensure flood control safety of the floodplain masses and not to threaten flood control safety of dykes on both banks. Under this program, the livelihood and production safety of the floodplain masses will be guaranteed to some extent, and the effect of large floods on the embankment of lower reaches is limited. However, the existence of the production dykes will cut off water – sediment exchange between the channel and the floodplain, which is one of the main causes of secondary – suspended river; and once the flood is larger than the flood control standard of the production dykes, the production dykes will naturally be overflowed or burst, in this case, transverse river, rolling river or discharging along dykes are likely to occur, which seriously threatens flood control safety of the embankments. Besides, after production dykes are changed to flood control dykes, their standard determination, operation and management manners are all issues that need studying and dealing with earnestly.

Program 3 is to add flood diversion inlets and recession facilities based on forming closed zones with control works, dangerous section linking dams and newly built flood control dykes together with the embankments. Therefore, Program 3, on one side, has the advantages of Program 2, can prevent the situation where the floodplain is stricken as soon as bankfull flood occurs, and can guarantee life and property safety of the floodplain masses; on the other hand, it can help realize selective flood diversion according to flood magnitudes. In addition, under the proper incoming water and sediment condition, and through appropriate operation manner, it enables the floodplain to fulfill the function of dealing with some of the sediment.

3　Exploring the feasibility of zoned harnessing program

3.1　Incoming water and sediment condition change and social development increases the feasibility of zoned floodplain harnessing

Since the Xiaolangdi Reservoir was put into operation, through joint operation with other three reservoirs of Sanmenxia, Guxian and Luhun, the ability of artificial control of downstream floods has greatly increased, which has provided a basic condition of implementation of zoned operation in the floodplain of the lower reaches in a planned way.

(1) Control, guiding and floodplain protective works have come into play. River regime of most river sections have been effectively controlled, the existing newly – formed floodplain, low, middle and high floodplain areas, production dykes and village platforms has also provided a certain controlled boundary condition for zoned harnessing and operation of the floodplain.

(2) Shape a normal water channel of certain discharge magnitude through water and sediment regulation. The situation of harnessing for normal water in different river sections, together with the demand of determined flood discharge width, has provided a basis for division of flood diversion and detention zones.

(3) The masses living in the floodplain are in large number, according to years' experience, it is very difficult to move them all out of the embankments. The floodplain masses should be main-

ly helped settle down on the spot, and concentrated residential areas shall be built combined with ongoing "dismantling villages to build towns" and the policy on new countryside construction. This can not only reduce the cost for relocation of the masses and the resistance effect on floods, but can also provide conditions for future rational utilization of the floodplain resources and development of the tertiary industry, so it can easily get understanding and support of governments of different levels and the floodplain masses.

(4) With the increasing of China's comprehensive strength, the State issued policies on floodplain compensation, which has provided policy support for guidance of classified floodplain management. Implementation of classified floodplain management can easily coordinate with relevant laws and regulations, and can help fulfillment of policies on compensation for floodplain inundation.

(5) Increase of regulating ability of the reservoirs on the Yellow River and its tributaries and the improvement of flood forecast ability have provided a vigorous technical support for zoned harnessing. With the development of science and technology and improvement of weather forecast technology, the ability of flood forecast has improved greatly. Improvement of reservoir operation and hydrologic forecast makes that the flood routine process can be well predicted, and the peak discharge and flood volume of the lower reaches can be adjudged scientifically and quickly. In the circumstances, zoned harnessing of the lower Yellow River and active flood detention and desilting have become feasible.

3.2   General idea for zoned harnessing

The general idea for zoned harnessing is as follows:

(1) Make full consideration of the changes of the lower Yellow River incoming water and sediment and the improvement of flood forecast technology.

(2) Flood control standard is raised under the condition of ensuring flood control safety of both banks and insisting on the strategy of "wide river and reinforcing the embankment" in the lower Yellow River.

(3) Make full consideration of the natural flood detention and desilting function of the lower Yellow River floodplain.

(4) The masses in the selected zones must be helped settle down properly, and the policy on floodplain compensation should be easily implemented.

(5) Make full use of the existing vulnerable spots and control, guiding works and production dykes.

(6) Actively carry out flood diversion and detention.

3.3   General engineering layout pattern of zoned harnessing

Take the floodplain in Qingzhuang – Nanxiaodi reach of Puyang as an example, the general engineering layout pattern of zoned harnessing is given in Fig. 1. The existing Yellow River embankment, Qingzhuang vulnerable spots, linking dam, flood control embankment (namely the original production dykes that have been adjusted) and the Nanxiaodi linking dam together form a closed zone, a flood inlet shall be placed at the proper position of Qingzhuang vulnerable spot, and the recession facilities shall be placed at the proper position of Nanxiaodi works, which makes the general layout pattern of zoned harnessing. For active flood diversion to protect the floodplain masses' life from flood threat, security construction shall be made in the floodplain so that the life and property safety of the floodplain masses can be guaranteed to some extent.

3.4   Flood management manner of zoned harnessing

According to the arrangement for the lower Yellow River flood control, it is primarily decided that the standard of flood control dykes is to defend against flood of 10,000 m^3/s at Huayuankou. Based on this, the flood management manner of zoned harnessing is: ①when the incoming water is less than the bankfull discharge, the water shall discharge through the main channel of the existing river; ②when the incoming water is larger than the bankfull discharge but less than 10,000 m³/s, the flood shall discharge through flood control dykes; ③when the incoming water is larger than 10,000 m³/s, according to prediction of peak discharge and flood volume, floodplain zones should be actively selected in good time to divert flood.

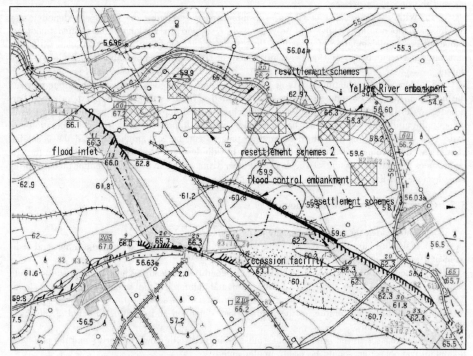

Fig. 1　Layout of zoned harnessing engineering

3.5　Relationship between zoned harnessing and floodplain safety construction

In order to avoid large loss of the floodplain by choosing the initiative flood diversion as well as the loss can be accepted by the masses on the floodplain, flood safety construction should taking into account in the zoned harnessing according to the general zone harnessing layout. Currently, Three different views on the floodplain masses' resettlement lead to three different resettlement schemes: The first scheme is to build connection platform avoiding flood next to the dike which can hold all the people who live in the zoned harnessing area. The second scheme is to build big flood – avoiding platform separately depending on the existed village in accordance with the past safety construction mode of the floodplain. The third scheme is to move all the people who live in the zoned harnessing area to the adjacent floodplain. All the schemes can be seen in Fig. 2.

These three schemes all can ensure the safety of the masses on the floodplain when flood diversion is taken under the zoned harnessing mode. Also masses on the floodplain can obtain reasonable compensation by using the floodplain submerging compensation policy. Compare with scheme two, the advantage of the scheme one is that the threat of the secondary – suspended river to the dike can be eliminated after the implementation of the floodplain safety construction, but the investment is relatively larger. The scheme three is not easily to be accepted because the people are far away from their farmland. But the life and property are easily guaranteed under the situation of zoned harnessing is implemented.

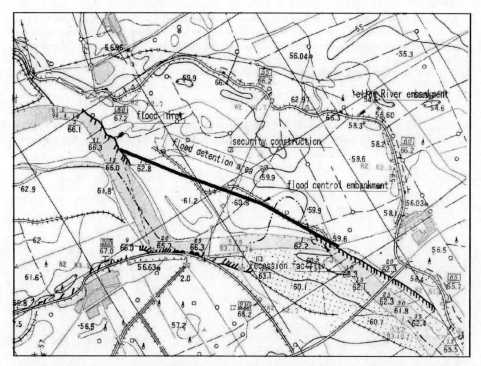

Fig. 2   Project layout of floodplain safety construction under the zoned harnessing mode

3. 6   Relationship between zoned harnessing and flood diversion warping

Under the zoned harnessing mode, masses living on the floodplain are properly settled. Production and detention area is formed between floodplain safety construction and flood control dykes, which would not affect the life and safety of the masses on the floodplain when it is flooding. Meanwhile, inlet and recession facilities are built in the zoned harnessing projects. Thus, during the condition with appropriate incoming water and sediment, even the flood discharge is less than $10,000 \ m^3/s$, flood diversion warping can be implemented initiatively in order to avoid the situation of mainstream silting continuously and the unfavorable watercourse forming. When the flood discharge is larger than $10,000 \ m^3/s$, flood diversion as well as sediment silting can be implemented at the same time when the inlet facilities are in use.

4   Conclusions and suggestions

Zoned harnessing of the floodplain is a strategy on the management of the lower Yellow River in the current situation of reservoir regulation capacity enhancement, flood forecasting improvement and high protection requirement for the masses on the floodplain with development of the society. This strategy aims to improve the flood control safety of the floodplain masses under the premise that the flood control safety of both banks is not endangered. By properly arranging the floodplain masses, this strategy cannot only guarantee the life safety but also the production safety of the floodplain masses. In addition, the compensation measures of this strategy can be easily implemented when the floodplain is used for flood diversion, flood detention and sand settling. Because the research on the zoned harnessing of the floodplain is still in the initial stage, the suggested future research includes: ①arranging the projects on the lower Yellow River according to the strategy of the zoned harnessing of the floodplain; ②analyzing the flood control influences to the design standards of embankment, outlet layout and the flood routing process in the lower Yellow River; ③developing the proper management mode for the zoned harnessing of the floodplain. Therefore, it is significantly meaningful to carry out in – depth study and enrich the thoughts of floodplain management for the flood control safety on both banks and economic development.

References

Hu Yisan. Yellow River Flood Control[M]. Zhengzhou: Yellow River Comservancy Press, 1996.
Liu Yun. Exploring on Floodplain Safety Construction and Development Mode[J]. China Water, 2012(6):30 – 32.

Sustainable Water Resources Management:
Users' Participation

Jean – François DONZIER[1] and *Christiane RUNEL*[2]

1. The International Network of Basin Organizations (INBO), The International Office for Water,
21, rue de Madrid – 75008 Paris(France)
2. Publishing Director of 'INBO Newsletter" and INBO Communication Director,
INBO PTS S/C International Office for Water, 21, rue de Madrid – 75008 Paris (France)

Abstract: Experience has shown that, regarding water management, involving the "civil society" inside mechanisms of decentralized water resources management was needed, as this civil society is at the cross – roads of institutional steps and support for development. This requires raising awareness and educating all the stakeholders on the principle of sustainable water resources management.

Thus, necessary means are to be mobilized to constitute water – related information systems, to develop the capacity of decision – making, to promote the establishment of representative organizations, in providing the possibility for them to benefit from the know – how and the means which are necessary for participating to public water management policies and for creating the appropriate legal structures, etc.

As information is essential, to be useful, it must not remain in the form of raw data, but be retrieved in the form of easy – to – understand data which can be handled by all the different categories of users. In addition, if the data are to be utilized, they must be made available in the most appropriate forms. Common standards must also be defined to gather the comparable information produced by the different parties.

Key words:participation of the civil society, water information systems, water resources management

1 Involvement of the civil society: a requirement for better water management

The experience that has been acquired for several decades, regarding water management, emphasized the need for an institutional association of the "civil society" inside mechanisms of decentralized water resources management, in order to allow an optimum and adapted meeting of growing and diversified needs.

Indeed:

(1)Administrations and public bodies in charge of water management must decentralize their actions, while relying on partnerships that enable a real participation of Local Authorities and users' representatives (households, irrigation users, industrialists, fishermen,...) in decision – making.

(2)Improving public services, such as drinking water supply, sanitation or irrigation, will only be possible if mechanisms are set up for recovering costs from the users. This will only be accepted by the users if they are given the guarantee that water is of good quality, services are permanent, management methods are transparent and that they will participate more and more in management.

(3)Decision – making will have to become progressively "democratic", widely opening possibilities for expressing counter – opinions in order not to sink into theoretical and fruitless debates, and have an independent and sound expertise capacity and access to transparent and complete information.

(4)Many needs will not be met by way of the traditional channel of Public Authorities but by individual or community field initiatives, which will not necessarily be spontaneous and will imply adequate skills and know – how.

(5) Water saving, preventing wastage, protection of the aquatic ecosystems, pollution control, imply an initial awareness – raising of all the users and thus of the inhabitants, especially children and in many countries, the womenfolk;

(6) An important part of the installations and development is carried out by the riverside property owners or individual users whose combined initiatives do not necessarily correspond to the general interest, in the absence of a global policy to the elaboration of which they would have been associated.

2　The civil society: at the cross – roads of institutional steps and support for development

On the one hand, there is the impetus given by bi – and multi – lateral Cooperation Agencies and by Governments to set up some institutions and procedures to allow for global and integrated management of water resources, which are based on:

(1) A partnership with the local communities and the users, associated more and more with decision – making, particularly within new Basin organizations;

(2) Some contradictory instruction procedures of the projects and the authorizations.

It is then absolutely essential for the Authorities to be surrounded by interlocutors who are sufficiently informed and competent, to be capable of assuming the role which is expected of them.

A widely spread movement towards the decentralization of the State's role in the organization of water supply and sanitation services to municipalities and of the tasks of collective irrigation to irrigation users' communities does exist: this will require that they become real managers of "industrial and commercial services" and thus acquire quickly a good knowledge of techniques and management to assume their responsibilities.

On the other hand, there are the field efforts displayed by the Public Authorities and Non – Governmental Organizations, aimed at the most unprovided for people in remote rural areas or underprivileged urban areas, in order for them to have access – through education and the appropriate organizational methods – to health care, to a minimum of essential services, to the development of activities, in the sectors of agriculture and fishing, in particular.

There again, it is with the support of the Local Authorities, village communities and local associations that one can obtain positive results, some of which, are very spectacular.

Whatever the method of approach, it is clear that improvement of the quality of services related to water, the development of the principle of integrated management of the resources aiming at optimum satisfaction of the needs and simultaneously, preventing pollution and rehabilitating the environment and the aquatic ecosystems, will not be possible unless organizational systems are created in order to federate interests and initiatives which could, along with the Public Authorities, introduce and impose legitimate and representative interlocutors in order to:

1) Give a stimulus to the ideas and disseminate them.

2) Thwart bureaucracy.

3) Ensure also direct responsibilities, without expecting all the solutions to come from elsewhere.

That implies the existence of local people in charge, and the forming of teams of volunteers or professionals, which intervene in the procedures, organize and manage the new structures and the projects they generate, with all the required know – how.

3　Which means for this structuring

Evolution of this kind should be axed on three priorities:

(1) Training and awareness – raising campaigns, especially for those who will have to share or assume decision – making in the field.

(2) Access to information, which implies both:

①The possibility of access to the data and to the files, to understand and to analyze them;

②The need for appropriate dissemination, through the media and the educational systems, of knowledge which is essential to understand, decide and act.

(3) Structuring of the initiatives, within competent and representative organizations which can become capable of gathering the interested partners, of speaking on their behalf and of representing them in the procedures, of developing their own capacities of expertise and expression and themselves, be the supports for collective actions.

That implies fundamental, thorough work which entails considerable time and efforts and which must be canalized by modern vectors for support, coordination and intervening, requiring important means of organization, information, professionalization and action.

The established fact shows that even today, the means are still provided mainly through the aid to development channels organized by the NGOs which often, because there is no other alternative, are limited to "small" programmes which are isolated one from another.

The "important means" from Public Aid to Development and the Authorities are concentrated primarily on the investments, leaving only a minor place even if recent evolution can be observed for specialized education, institutional organization, means for a real user partnership – association and the appearance of Local Authorities.

However, rapid progress could be made if the awareness of the "facilitating" role of the "Civil Society" improved, if the necessary means were assigned to really enable its involvement.

Thus, the necessary means must be mobilized:

(1) To constitute integrated information systems related to water.

(2) To develop the capacity of decision – making in this sector:

①Elected officials of the local communities.

②People in charge of the village communities and cooperatives of irrigation users, of fishermen, ...

● *representatives of the industrial branches and economic activities linked to water (tourism, fishing ...),*

● executives of non – governmental associations/organizations? (expertise training ...).

(3) *To encourage the creation of representative organizations, in providing the possibility for them to benefit from the know – how and the means which are necessary for participating to public water management policies and for creating the appropriate legal structures, search for financing, manage the budgets, lead the projects, have access to information, communicate, etc.*

(4) *To raise awareness and educate all the stakeholders on the principle of sustainable water resources management.*

All these actions can be classified under the term of "social engineering".

4   Institutional organization for users' participation

All direct or indirect users should be involved, in one way or another, in the decision – making process, so far as they are concerned.

4.1   Who is a "User"

A "user" utilizes water (industrialists, electricity producers, farmers, population).

The users may be participants in organizations which represent their interests.

This notion can be extended to people using water for recreational purposes (fishermen, leisure, etc.) and to associations for the protection of nature.

4.2   Why involve the users?

Acceptation and thus the feasibility of a Long-term project and its successful completion require the following steps:

(1) Approval of project objectives.

(2) Sharing of the Long-term vision.

(3) Definition of priorities.

(4) Getting the human, technical and financial means necessary to achieve the objectives.

An active participation of the users is the best means to solve possible conflicts on water use: "Dialogue is the beginning of wisdom".

4.3 Which official framework for dialogue

A framework for dialogue should take into account the impact of the decisions to be made. The more ambitious the project, or far reaching, the more widespread dialogue should be. On the contrary, a project of local interest will need a more reduced and precise dimension.

Dialogue must be organized in the most decentralized way possible while taking local constraints and specificities into account.

In a general manner:

(1) The extent of public participation in all decision – making processes must be unanimously approved.

(2) Representatives of local elected officials, communities and of all users concerned must participate in the works regarding the formulation of master plans for water development and management, priority action plans, projects,..., with the help of specialists from the Administration and specialized consulting firms.

(3) Information must be clearly distinguished from dialogue. In the first case, the administration shares information with the public, it is a one – way process. Dialogue implies a two – way process: the administration listens and takes the formulated comments into account.

(4) The public participation process must be accessible to a wide range of people concerned: it is an open process that takes the diversity of the interested parties into account (representativeness).

(5) NGOs, when they are well established in the field, can become efficient partners in programmes involving an active participation of the population.

(6) The comparison of different points of view amongst reflection groups is the fount of progress towards an integrated vision of water resources.

Of course, at the conclusion of this information and association process, the final decision is left, at adequate level, to the competent Public Authorities, subject to the eventuality of recourse before the jurisdictions.

The organization of such procedures is relatively expensive for preliminary studies, dissemination of documents, meetings, travelling, exhibitions, for experts and facilitators: it is essential that the corresponding budgets be planned for.

The approach to integrated management of water resources through river basins offers a favourable context for this participation within the Basin Committees for the important rivers, and the Local Water Commissions, for their tributaries.

5 Information is essential

In order to attain an overall management of water resources, at river basin level in particular, it is to be emphasized the prime importance for decision – makers (Directors of River Basin Organizations and Administrations, Basin Committee members, representatives of the Local Authorities and associations of users), and usually for all the users and the population, to have easy access to complete, representative and reliable information on the following:

(1) The state of surface and groundwater resources, from both a quantitative and a qualitative viewpoint, also the seasonal and yearly variations.

(2) Land uses and the development of activities.

(3) The situation concerning biotopes and the aquatic ecosystems and their degrees of sensitivity.

(4) Water uses (withdrawals), in particular drinking water supply for the population or irrigation for the farmers.

(5) Pollution sources (discharges) whether point or non – point.

(6) The risks of recurring extreme phenomena such as floods or drought and accidental pollution.

(7) Investment programmes and costs for water harnessing.

... and, in a general manner, to have access to studies, documentation, information on experiments and the availability of services and equipment, etc.

But this information is often dispersed, heterogeneous and incomplete ... and is not always comparable and adapted to the prerequisites for objective decision – making and awareness raising. Moreover, it is a fact that public, para – public and even private organizations can have access to this information but lack of sufficient means for exchanging, gathering, standardizing, summarizing and for capitalizing it amongst them or for its dissemination to other people interested...

In each situation and considering all the national and local characteristics, special attention should be paid to the organization of documentation centres, of monitoring networks and data banks, to their financing and operation, as well as to a suitable role for specific basin organizations as compared to other possible stakeholders.

It is absolutely necessary to examine:

(1) The nature of useful information.

(2) The means used for collecting, monitoring and analyzing, as well as for controlling the quality of data produced, of their transmission (in real – time, when necessary, for major risks forecasting) and for their storage.

(3) Forms in which information should be made accessible to decision – makers (data banks, reports, maps, diagrams, ...) or to technicians and scientists.

(4) Broadcasting and dissemination means (remote processing, publications, dissemination to the general public, ...).

5. 1 Real and complete "systems" must be designed, used and organized to constitute "global observatories"

The exact definition of each participant's role as well as of the issue of financing and its sustainability is of prime importance.

Gathering this information, requires a complex and consistent organization of monitoring networks, analyses laboratories, data transmission and their checking and monitoring, management of data banks, their accessibility and their "products" and a documentation management, etc. For this, permanent means must be made available and their optimizing ensured, in order to obtain at minimum public cost, all the relevant information, limiting this however, to the strict necessary.

It should be pointed out that investment costs for obtaining appropriate information (monitoring stations, laboratories, teletransmission, automatization, studies and research ...) are high.

Moreover, the qualification of intervening experts (training) and operating costs are, by far, the most important and recurring items of expenditure in the medium and Long-term.

Thus, it appears unreasonable to invest without ensuring positive means for the optimum and continuous operation of the systems over a long period of time which, of course, requires substantial, appropriate and sustainable financial resources.

It is important to avoid using excessive sophistication relying on advanced technologies instead of reflecting on a sound organization and straightforward solutions that usually are the most efficient.

Information systems only operate when skilled operators are in charge.

5. 2 Moreover, if the information is to be useful, it must not remain in the form of raw data, but be retrieved in the form of easy – to – understand data which can be handled by all the different categories of users

Information should be organized according to requirements, whether it be for the study of a "white book", master plans for water management and development, for action programmes, budgetary simulations or the basis for water charges, for delivering administrative authorizations or studying projects, for regulation of public works, warning systems or even for evaluating the results of applied policies and monitoring the environment, finally for informing the general public...

If it is generally considered that Public Authorities must be the contracting authorities for documentation centres, monitoring networks and associated information systems and that from then on, access to them must be open and free for the various users. However, due to additional costs for processing and circulating the information, it would appear quite normal that the processed data be paid for when the people making the request can afford it.

5.3 In addition, if the data are to be utilized, they must be made available in the most appropriate forms

It is clear that the information is not meant to remain confidential.

On the other hand, it does not always arrive at its destination.

(1) Most of the water studies throughout the world are to be found in the "corporate literature" format, available as unique reports which are more often than not non – referenced and never published.

(2) Many data bases can only be consulted on – site, on the premises of the organization which manages them.

It is necessary to facilitate the access to information and to organize its dissemination according to the most appropriate techniques in order to reach the various "targeted – publics".

It is interesting to have recourse to professionals of media and to use the information relays such as the elected officials, civil servants, professional or associative leaders, journalists, teachers, facilitators, etc .

The broadcasting and dissemination means must be budgeted because they are often as expensive as those for the production of information (printing, mailing, multimedia broadcasting, exhibitions, events, etc .).

Some mass communication systems can be very efficient.

5.4 Common standards must also be defined to gather the comparable information produced by different parties in order to organize real observation systems at the level of national or transboundary river basins and to centralize the summarized information necessary for formulating governmental policies and for informing the public.

Information systems for shared rivers and aquifers would be improved by being designed in a global and consistent way on the watershed scale, within the framework of agreements between riparian countries.

It is then necessary to:

(1) Consider that setting up complete information systems, corresponding to the above – mentioned specifications, is a prerequisite.

(2) Clearly define which institutional bodies are responsible for the permanent organization and operation of such systems.

(3) Guarantee not only sufficient means for corresponding investments, but also the compulsory financial techniques which will secure their Long-term operation.

(4) Encourage the development of means and specific engineering proficiency in this field.

(5) Support the works that aim at defining common standards and nomenclatures for data or documentation management in order to exchange, compare and summarize the information between partners at all relevant observation levels.

(6) Promote the setting – up of observation systems for water resources and their use, at the river basin level in particular, and the organization of national and coherent information systems that can:

①become resource centres for the various users of data and documents;

②integrate the national, regional and international Networks permitting useful exchanges, comparisons and syntheses, as for example "Aquadoc – Inter" or in the Mediterranean region, "EMWIS".

6 Awareness raising and training of decision – makers and information relays

More and more actors are thus involved in water management:

(1) To participate in dialogue bodies or in the procedures organized by the Public Authorities.

(2) To realize investments either individually or collectively and to ensure their management.

(3) To make better use of water, thus combatting and preventing wastage and pollution, and better maintain the aquatic media and the bed of the basin.

(4) To organize risk prevention and warning systems, ...

Thus, new parties are coming into the scene to mingle with the water professionals (engineers, technicians, civil servants ...). Their direct or indirect role will become more and more important:

6.1 The decision – makers

(1) Individuals: heads of industrial enterprises, farmers, fishermen, waterways representatives ...

(2) Collective: local elected officials, heads of village communities, heads of trade unions or cooperatives, associations' representatives.

6.2 Information relays

They are mainly journalists, teachers, facilitators of associations, popularizing bodies, health care staff ... and whom play an "interactifve" role in broadcasting both information and knowledge, but also in carrying the problems and the opinions of the users and of the population.

It is extremely important to implement specific means to raise their awareness, and provide them with the information they require, in and with the appropriate forms and supports.

They all have in common, on one hand, that water is not their profession and that they have not been prepared to play a role and on the other, they are often geographically dispersed, even isolated sometimes, especially in rural areas.

In France for example, the International Office for Water has developed some awareness – raising programmes with the support of the Ministries, the Water Agencies and Associations of elected officials or professionals. These programmes are particularly intended for the mayors of rural communities (more than 10,000 participants in the "Water information days for local elected officials") or the people in charge of professional agricultural organizations (European LIFE – RIVER – Water sharing programme). The Water Agencies also produce teaching materials for teachers organizing "Water classes".

In Poland, the Gdansk Water Foundation (GFW) has organized with the RZGWs, the seminars of the people in charge of all "Voï vodships".

In Hungary, the awareness – raising days for elected officials have also been organized with Hungarian facilitators who were specially trained in France.

With the fulgurating development of Internet, new "intelligent" on – line services are developing and allow responding in real – time to the most frequently asked questions of the various categories of managers.

Services of this kind are being experimented within the framework of European programmes intended for managers of Small Industries or for mayors of rural communities.

Of course, projects like in the Mediterranean region, "EMWIS", or AWIS in Africa will provide, direct access to international data banks, open to all potential users.

Idea on Achieving Long-term Stability of the Lower Yellow River

Wang Weijing[1] and *Wang Haiyun*[2]

1. Yellow River Henan Bureau, Zhengzhou, 450005, China
2. Zhengzhou Foreign Language School, Zhengzhou, 450003, China

Abstract: That there is no sufficient water for transporting sediment in the lower Yellow River is the basic cause of the situation that the channel siltation is difficult to contain. Due to limitation of the dykes and the defence water level, the sediment that the river can accommodate is limited. Once the sediment cannot be accommodated, burst or course change may still occur. To achieve Long-term stability, we have to find new ways to dispose of the sediment. The paper presents the idea of constructing canals or pipelines outside of the existing river channel to transport sediment directly to the other sea areas. This way can not only solve the problem of channel siltation completely to achieve Long-term stability, but also can alleviate the serious shortage of water resources in the Yellow River Basin. The past river harnessing strategies had two shortcomings: firstly, they could not solve the problem of channel siltation completely; secondly, treating the sediment as a "burden", they cost too much, thus formed a heavy social and economic burden. The paper proposes that we should use sediment as resources, and by learning the experience of Dubai and Caofeidian, construct artificial islands and deep water ports in the seas near Lianyugang with the sediment transported into the sea, so as to open up a huge space for the Yellow River harnessing and economic development.

Key words: the lower Yellow River, Long-term stability, sediment utilization, use as resources

The lower Yellow River is famous for its frequent floods. In the recorded history of more than 2,500 years there have occurred breaching more than 1,500 times, bringing heavy disasters to the people on both banks. The Chinese Nation has struggled for a long time harnessing the Yellow River. Though some effectiveness of harnessing was achieved in different historical periods, the flood threat of the Yellow River has not been eliminated completely so far. Why the floods of the lower Yellow River cannot be eradicated after going through thousands of years of harnessing? Are there any means and ways to bring the Yellow River floods under permanent control and achieve Long-term stability? On this we would give some understanding and ideas.

1 Shortage of water to transport sediment is the root cause of frequent occurrence of silting disasters on the lower Yellow River

In history, the disasters of the lower Yellow River were mainly flood disasters, but the root was the sediment-related disasters. Flowing through the world's largest loess plateau, the Yellow River carries huge amounts of sediment into the lower reaches. The channel for flood discharge was constantly occupied by deposited sediment, the river bed was raised to form the suspended river, hence the attendant breaching disasters. The root cause of Yellow River sediment deposited in the lower reaches in huge quantities is the shortage of water to transport sediment. The incoming water in the lower reaches, even all used for sand flushing, cannot meet the need of sediment transport. Then, how much water dose the lower Yellow River need to transport sediment? According to statistics of observed hydrological data, the annual sediment concentration to maintain non – sedimentation of the lower reaches is $20 \sim 30$ kg/m^3, that is to say, to transport 1 of sediment into the sea needs 40 ~ 50 m^3 of water. The mean annual sediment runoff of the lower Yellow River is 1.6×10^9 t, to keep the channel from deposition needs $6.40 \times 10^{10} \sim 8.00 \times 10^{10}$ m^3 of water, while the mean annual runoff of the lower Yellow River is only 4.70×10^{10} m^3, far from satisfying the need of sediment transport. This is why the river bed of the lower reaches has been constantly silted and raised.

Due to the effect of climate change and human activities, since the 1950s the observed runoff and sediment runoff coming to the lower Yellow River have been decreasing constantly. Then with decrease of incoming water and sediment, was the river siltation improved? The observed data show that, since the magnitude of floods decreased and the sediment carrying capacity of the flow was reduced, the river siltation was not improved, instead, it became more serious. According to statistics of observed data, during the 48 years from 1950 to 1997, the lower reaches' mean annual runoff was 4.163×10^{10} m^3, and the mean annual sediment runoff was 1.087×10^9 t. During the 12 years between 1986 and 1997, the mean annual runoff was $2.8.9 \times 10^{10}$ m^3, and the mean annual sediment runoff was 0.713×10^9 t, less than the average values by 31% and 34.4% respectively. Though the incoming water and sediment decreased at roughly the same rate, since the magnitude of incoming water was reduced, especially the occurrence chance of floods over 4,000 m^3/s decreased, the sediment carrying capacity of the flow declined greatly and the river siltation became more serious remarkably. From 1950 to 1997 the lower reaches' mean annual deposited sediment was 0.19×10^9 t, accounting for 17.5% of the incoming water; while during the period between 1986 and 1997 the mean annual siltation volume was 0.245×10^9 t, making up 34.4% of the incoming sediment: the siltation nearly doubled. Thus it can be seen that, with reduction of incoming water and sediment, the siltation of the lower reaches was not alleviated, instead, it became more serious.

2 Limited space cannot accommodate unlimited sediment

To solve the problem of river silatation, many measures were taken in the past, such as "widening the river and strengthening the dykes", "clearing sands with converging flow" and "dredging the river" et al. None of these measures has succeeded in containing the tendency of "siltation – river bed rising-burst and course change" of the Yellow River. Since the founding of new China, comprehensive harnessing measures have been taken, and the great success of no burst for over 60 years has been achieved. Yet, the space of the river to hold sediment is limited. If there is no new way to deal with the sediment, course change will certainly happen some day.

2.1 River's sand accommodating capacity

To ease discussion, we introduce the concept of "river's sand accommodating capacity" and give the following definition: Under a certain defense water level, the maximum value of the sand volume that the river can accommodate is the river's sand accommodating capacity. The lower boundary of the sand accommodating capacity is the river surface between both dykes or below the highest defense water level (including the estuary delta and the silting sea areas). Its upper boundary is the plane formed by the river width corresponding to the highest defense water level and the critical slope (or the balanced slope for the shallow sea). Here, the "critical slope" refers to the smallest slope required to maintain a certain sand carrying capacity, and the "balanced slope" means the slope of the coastal region of the river mouth when it reaches a balanced state under the ocean dynamics. The above – mentioned space, with channel storage deducted, is the river's sand accommodating capacity. When the river boundary condition changes, the sand accommodating capacity will change correspondingly. When the defense water level is raised, the sand accommodating capacity with expand; when the defense water level is lowered, the accommodating capacity will be reduced. With the defense water level raised, the cost of flood control works' construction, maintenance and defense will increase, and if an accident occurs, the disaster loss will be much larger. Therefore, the defense water level cannot be raised unrestrictedly. It can only be restricted to a technically possible and economically reasonable level. When the defense water level is determined, the river's sand accommodating capacity is determined, and once the space is filled up with sand, the river's life ends.

2.2 Current main ways to deal with sediment

At present, there are three main ways to deal with sediment: firstly, to intercept sediment in the upper and middle reaches; secondly, to carry the sediment into the sea through the channel; thirdly, to pile it up in the channel and its both sides.

To intercept sediment in the upper and middle reaches depends on two ways, one is water and soil conservation, and the other is constructing sand blocking reservoirs on the trunk stream and its tributaries. These measures are effective, but, from a Long-term point of view, restricted by various conditions, their effectiveness is limited. Since the founding of new China, large – scale mass water and soil conservation has been carried out and remarkable result has been achieved. Yet, water and soil conservation is a long-lasting and gradual process, besides, water and soil loss caused by mass erosion is difficult to check completely. According to the research result of geographers, even in the Shang and Zhou dynasties when Loess Plateau was covered with good vegetation and the effect of human activities was very little, the annual soil loss was up to $0.8 \times 10^9 \sim 1 \times 10^9$ t. So soil and water conservation cannot solve the problem of sediment entry into the Yellow River. Large sand checking reservoirs on the trunk stream can effectively intercept sediment. According to the Yellow River cascade development planning, the total silt detention capacity of the key projects on the trunk stream is over 4.00×10^{10} m^3, which can only solve the problem of silt detention of several decades. Constructing silt detention reservoirs on tributaries often involves flooding of large area of farmland, which will have great effect on the local economic development and people's life. In view of the basic national situation of large population with less farmland, it is not practical to construct silt detention reservoirs on tributaries in large quantities.

The second way to deal with sediment is to carry the sediment into the sea as far as possible through the existing channel. This has also long been the basic idea for sediment disposal. Though carrying sediment into the sea cannot expand river's sand accommodating capacity, since the estuarine coastal region becomes main position that the river accommodates sediment, it will make the sand accommodating capacity fully used, thus effectively extend the river's lifer. For sediment transport into the sea, there must be suitable incoming water and sediment and river boundary conditions. Harnessing the river to increase desilting capacity, regulating water and sediment to give full play to the role of sediment carrying capacity of the flow, and transfer water from other basins to increase water for carrying sediment into the sea are the main measures taken at present.

Measures to deal with sediment in the river and its both sides include strengthening dykes by warping, building relative underground river by warping, warping in the Xiaobeiganliu and in the lower reaches floodplain and artificial course change, et al. Dyke strengthening by warping in the lower Yellow River has been carried out for many years, and up to the end of 2001, a total of 0.76×10^9 m^3 of earthwork has been finished. Expanding the limit of warping further to turn the lower channel into a relative underground river will not only greatly increase the safety guarantee of the lower Yellow River, but also will find a suitable space for sediment disposal. Warping in the Xiaobeiganliu and in the lower reaches floodplain is also an effective way to deal with sediment.

Artificial course change may open up new sand accommodating space so as to make the river get new vitality, but it will cause too huge losses and too many problems, so, as a river harnessing strategy, it has never been adopted in history (except for small – scale course change in some areas). Nowadays, population grows sharply and economy develops rapidly, and the North China Plain is one of China's most densely populated areas, where infrastructures of transportation, water conservancy, energy and communication are intertwined, forming a variety of network systems like dense spider webs. In this situation, the losses caused by course change would not be compared by those of any of past dynasties. In history, every large course change would be followed by an unstable period as long as several decades. Violent scouring upstream of the entrance and serious sedimentation downstream of the entrance, the river course wandered without a fixed route, so the defense was very difficult, and the secondary losses were difficult to estimate. Therefore, course change may only be the last choice.

Corresponding to the above-mentioned sediment disposal measures (course change excluded), a rough estimate of sand accommodating capacity of the lower channel and the incoming sediment. Suppose the defense water level is raised 5 m, the existing channel between both dykes can hold approximate 2.00×10^{10} m^3 of sediment, the delta and the sea areas can hold about 6.00×10^{10} m^3, plus detention in the reservoirs on the trunk steam and its tributaries and warping in the upper and middle reaches, a total of around 1.60×10^{11} m^3 of sediment can be treated. In actual, since the silting is uneven, the river's sand accommodating capacity cannot be fully used. As for the incoming sediment, according to the water and soil conservation planning for the upper and middle reaches, from 2001 to 2010 the annual incoming sediment shall decrease from 1.3×10^9 t to 1.1×10^9 t, from 2010 to 2020 it shall reduce to 1×10^9 t, from 2020 to 2050 it shall reduce to 0.8×10^9 t, and it shall maintain at the level of 0.8×10^9 t after 2050. It can be seen that, with the existing harnessing measures, striving to avoid course change for 100 years can be achieved. However, from a Long-term point of view, if there are no new ways to deal with sediment, course change will be inevitable.

3 To achieve Long-term stability of the Yellow River requires opening up new ways to deal with sediment

Since the sand accommodating capacity of the river is limited, new ways of dealing with the sediment of the Yellow River must be opened up. The sea is a vast space to admit sediment, but its area to pile sediment up is restricted by the river. Firstly, with abundant sediment but less water, the Yellow River cannot transport all the sediment into the sea, resulting in steady lifting of the river bed; secondly, the steadily – reinforced dykes restricts the wandering of the river and the sea area to pile sediment up. Opening up a new sand transport passage of high efficiency outside of the existing river channel is a possible way to crack the hard nut.

The research in the 1970s found that hyperconcentrated flow has a strong sediment transport capacity. According to the result of the experiment carried out by the Water Resources Department of Shaanxi Province in the three gravity irrigation areas of Jinghe, Luohe and Baojixia, the highest sediment contents were 41.7% (Jinghui Canal, 1978), 42.4% (Baojixia Weihe Diversion Canal, 1979) and 60.6% (Luohui Canal, 1978) respectively, and there was no obvious siltation in the canals. Among these areas, the transport distance of Baojixia Weihe Diversion Irrigation Area was more than 100 km. Later, this method was spread to the other irrigation areas, the result was the same and obvious benefits were obtained. That hyperconcentrated flow can transport sediment over long distances has been proved by a lot of practical results, so increasing sediment transport efficiency is not a thing too far away to reach. Now the sediment transport efficiency of the lower Yellow River is low, and this is caused by the unevenness of flow and sediment process. At low water, the river channel is silted and shrinks, while at large water, the channel cannot hold the water so that it flows over bank, and after flowing over bank, the water slows down and its sediment carrying capacity decreased greatly, resulting low efficiency of sediment transport. Constructing canals or piles outside of the existing river channel may avoid the above – mentioned problem and greatly increase sediment transport efficiency. Carrying out highly efficient sediment transport in this way requires solving the problem of molding, transport and storage of hyperconcentrated flow.

3.1 Molding of hyperconcentrated flow

Hyperconcentrated flow needs a certain proportion of fine sediment, while flushed through the sluice gate, either sediment content or grain composition is difficult to control. The problem can be solved with mechanical sediment pumping in the Xiaolangdi Reservoir. The current dredging equipments are very powerful, and some are suitable for operation in reservoirs and sea areas. We can pump sediment at different positions in the Xiaolangdi Reservoir with dredging equipment and mixed the sediment to form uniform and stable hyperconcentrated flow that meet the requirement of transportation.

3.2 Transport of hyperconcentrated flow

If the hyperconcentrated flow is transported through canals, it must meet the requirement of no siltation flow velocity, boundary slope and cross-section patterns. Experts with Qinghua University have calculated four different programs, and the main parameters and results are given in Tab. 1.

Tab. 1 Hydraulic parameters and sediment transport result of four calculated programs

Program	Flow rate (m^3/s)	Sediment content (kg/m^3)	Cross – section pattern parameter	Hydraulic radius (m)	Flow velocity (m/s)
1	150	300	12	2.35	2.26
2	100	300	12	1.96	2.17
3	150	200	12	2.60	1.84
4	100	200	12	2.17	1.77

Program	Hydraulic slope(‰)	Canal bottom width (m)	Water depth (m)	Daily sediment discharge ($\times 10^6$ t/d)	Water consumption (m^3/t)
1	0.23	13.4	3.30	3.888	3.33
2	0.27	11.18	2.76	2.592	3.33
3	0.13	14.74	3.48	2.592	5.0
4	0.18	12.36	3.06	1.728	5.0

From the above calculation result, we can get the following understanding: Firstly, all the programs can meet the need of the Yellow River sediment transport. The smallest program has a annual sediment discharge of 0.631×10^9 t. Since the channel of the lower reaches will transport some sediment, all the programs can meet the need of sediment transport in terms of quantity. Secondly, there is a basic condition for implementation. The maximum canal bottom width is no larger than 15 m; there are no special difficulties in engineering layout and construction. After the Xiaolangdi Reservoir enters normal operational period, the beach face elevation of the reservoir area is around 250 m. From the Xiaolangdi Reservoir to the Bohai Sea and the Huanghai Sea, the shortest distance is less than 1,000 m and the gradient is between 0.25‰ and 0.3‰, both of which can meet the need of design gradient, so it is possible to realize gravity transport. Even in exceptional circumstances where the gradient is insufficient, pumping can be used as an auxiliary measure. Thirdly, in the programs, the largest single ton water consumption is 5 m^3, estimated at 0.8×10^9 t of annual sediment discharge, only 4×10^9 m^3 of water will be needed. This can greatly reduce water consumption for sediment transport and effectively alleviate the contradiction that water resources are becoming increasingly scarce.

Hyperconcentrated flow can also be conveyed though pipelines. In foreign countries there are successful experiences of hydraulic conveyance of coal through pipelines with maximum transport distance up to over 1,500 m. In recent years some experts have proposed a program of building additional bottom flushing openings in the Xiaolangdi Reservoir to transport the hyperconcentrated flow of the reservoir through pressure pipelines. With two pipelines of diameter of 4 ~ 5 m, 0.5×10^9 ~ 0.7×10^9 t of sediment can be transported each year.

3.3 Piling of sediment

Constructing an additional canal to transport hyperconcentrated flow is neither restricted by the existing river nor connected with day-to-day flood control problems, and the sediment can be transported either into the Bohai Sea or into the Huanghai Sea, thus helps achieve water – sediment separation of the Yellow River and provides a sufficiently vast space for sediment treatment. To solve the problem of siltation extension at the estuary of the canal, we can design several branches near the seas and use them in turn. When the delta of the river or canal is silted up fully, some part of the canal can be rebuilt to open up new sea areas for sediment piling.

Though the above – mentioned programs of hyperconcentrated flow through channels or pipes are tentative, and many problems may be encountered in the implementation process, there have been successful precedents after all, and there are no insurmountable obstructs any longer. As long as we persist in efforts, the problems can be solved gradually.

3.4 Develop sediment resources to write new chapters in the Yellow River harnessing and development

Silting in the Yellow River has troubled and puzzled numerous river harnessing workers, for this people have walked a long way of exploration, from "shunting by dredging", "sand scouring by flow contraction" to the comprehensive way of "retaining, draining, regulating, discharging and excavating". Though these measures took effect to some extent (sometimes even remarkably) in different historical periods, they had two common defects: Firstly, they were unsustainable. With steadily incoming sediment, they could not solve the problem of channel silting completely to a-chieve Long-term stability. Secondly, they all regarded sediment as a "burden". Because of this, much of manpower and financial resources were expended, thus forming a heavy social and econom-ic burden.

Sediment can cause disasters, and is also a kind of resources. The way out to achieve Long-term stability of the Yellow River is to turn the sediment into resources. At present, land reclama-tion from the sea with the sediment is an appropriate choice. It is advantageous first in large capaci-ty, and second in high efficiency, so it has a vast development prospect. In this respect, there are many successful experiences that can be learned, among which those of Dubai and Caofeidian espe-cially have value for reference.

Dubai is an emirate in the Unite Arab Emirates (UAE) (Fig. 1), where natural resources are poor, on the one side is desert, and on the other is a gulf. It was once a poor and backward place, but through mere more than 30 years' construction, it has become a modern international metropo-lis equated with New York, Tokyo and Hong Hang. One of its important practices is to create a world – class tourism and leisure and commercial and financial centre by constructing an artificial island largest in the world at present. Many of their experiences in construction and protection of the artificial island may be used in the harnessing and development of the Yellow River.

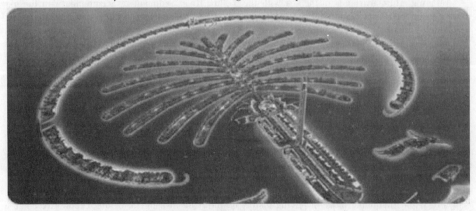

Fig. 1 Dubai Palm Island

North China lacks deep-water ports. At present, the largest deep-water port is Tianjin Port, where the maximum water-depth of the waterway is 19 m, and only at high tide 300,000 t ships can go through. It is just taking advantage of its water depth that Caofeidian achieved leap – forward de-velopment. Located between Tanggu New Port and Qinhuangdao Port, Caofeidian is a belt sand is-land, about 30 km from the shoreline. 500 m away from its head, the water is 25 m deep; and the deep channel in front of it has a water depth of 36 m, being the deepest point of the Bohai Sea. Ex-tending from Caofeidian to the Bohai Straits there is a natural watercourse with water depth of 27 m, which leads to the Huanghai Sea through the Straits. The natural combination of the watercourse and the deep channel constitutes the unique advantage of constructing large – scale deep-water port in Caofeidian(Fig. 2). It is the only natural site in the Bohai Sea to construct large berths for 300, 000 t ships without the need of excavating channels and dock basins. Approved by the State, it is planned to build four 250,000 t ore terminals, two 300,000 t oil terminals, sixteen 50,000 ~ 100,000 t coal terminals and one 100,000 t LNG terminal. Relying on this, it is planned to build an industrial zone with large docks, large steel plants, large power plants and large petroleum chemical factories as main goals, thus becoming a high ground in the "economic uplift" of the Bohai Economic Rim.

Fig. 2   Caofeidian port construction

Making artificial islands in the sea with the Yellow River sediment in accordance with the a-bove – mentioned "water and sediment diversion" has more advantages compared with Dubai and Caofeidian: Firstly, it can combine elimination of the Yellow River flood disasters and achievement of Long-term stability with utilization of sediment and promotion of economic development, thus produce huge economic and social benefits. Secondly, the Yellow River sediment is in large quantities and has broad prospective of utilization and development. Thirdly, the sediment is transported with water, so the cost is low. Fourthly, with no pollution, it is advantageous to improvement and stability of the ecological environment of the sea area. If transporting the sediment to the Huanghai Sea, we can build the world's largest artificial island with the sediment near Lianyungang, construct North China's largest deep-water port, and make it the new starting – point of the Eurasia Land Bridge and a new Far East logistics center. By learning the experience of Dubai (Palm Island), we can build it into a commercial island, a tourist island, so as to form a new economic development zone and open up a huge space for the Yellow River harnessing and economic development. Meanwhile, we can learn from Caofeidian's practice "government guiding of development direction, market – oriented capital operation, and management as enterprises", change the thousands of years' traditional mode of relying merely on the State investment for the Yellow River harnessing, so as to fill new vigor into the Yellow River harnessing and development.

Once the above assumption is realized, the Yellow River will enter a new historical era. The Xiaolangdi Reservoir will become a water and sediment regulating hub, and separation between clear and muddy water will be realized. The clear water, or the flow of low sediment content, will discharge through the existing channel to keep the channel balanced and stable. The Xiaolangdi Reservoir and other reservoirs can permanently keep the needed capacity through continuous dredging. The floodplain will have finished its historical mission, so the distance between both dykes can be greatly reduced. A good situation of dykes stable forever, green water ever flowing, harmonious between man and water and development prosperously will be achieved in the lower Yellow

River. The sediment resources will be fully utilized to produce huge economic and social benefits. The unique disaster of the Yellow River sediment will become a particular advantage of resources and will make new and greater contribution to China's economic development and national rejuvenation.

References

Editorial team of Contemporary Yellow River Harnessing Forum. Contemporary Yellow River Harnessing Forum[M]. Beijing:Science Press, 1990.

Yellow River Conservancy Commission. Collection of Rewarded Scientific and Technical Results (1) 1978 ~ 1984.

Wang Weijing. Thinking on Lower Yellow River Harnessing[J]. China Water Conservancy, 2004: 11.

Fei Xiangjun, Wang Guangqin, et al. Re-Discussion of Long Distance Transportation of Hyperconcentrated Flow-Seminar Papers of Key Technology and Equipment for Sediment Treatment of Yellow River Xiaolangdi Reservoir[M]. Zhengzhou: Yellow River Conservancy Press, 2007.

Wang Puqing. Research on Key Technology of Transporting Sediment to Sea — Seminar Papers of Key Technology and Equipment for Sediment Treatment of Yellow River Xiaolangdi Reservoir [M]. Zhengzhou: Yellow River Conservancy Press, 2007.

General Decision Support System for Water Resources Management

Karsten Havnø and *Henrik Refstrup Sørensen*

DHI Solution, Agern Allé 5 DK – 2970 Hørsholm, Denmark

Abstract: Background: The pressure on freshwater resources is rapidly increasing. Climate change and population growth is accelerating the problems further. Solutions are required for improved water management under increasingly complex conditions with competing stakeholder interests. Many organisations have started develop computerised decision support systems to aid the day-to-day management. In collaboration with multiple organisations a general decision support system has been developed, which can be tailored to specific uses. It can be applied both for planning and for real time operation.

New Methodologics: State-of-the-art simulations models and decision support systems (DSS) for water management and optimization of water allocation have been developed. The tools combine data retrieval and analyses, hydrological and hydraulic simulation models, crop models, GIS, multi criteria optimization and economic analyses. The tools are developed in partnership with water resources authorities around the world to aid their complex management.

Application experience: Under the National Hydrology Project in India, the DSS has been implemented for real time management and long term planning and optimization. The 9 riparian countries of the Nile River Basin, represented by the Nile Basin Initiative, have agreed to develop the water resources in a cooperative manner; share socioeconomic benefits, and promote regional peace and security. The development of the shared and accepted Nile Basin DSS is the first important step in achieving the joint vision. The Computer Aided River Management System being developed in Australia adds real time functionalities for optimizing the day to day operation. The DSS includes coupled surface water and groundwater models and uses the full information to meet all water demands with minimum release from the headwater storages. The paper discusses the system requirements and experience from application of this type of systems.

Key words: water resources, DSS, management

1 Increasing pressure on water resources

The pressure on water resources is increasing rapidly around the world as a result of population growth, economic growth and climate change. The project available water per capita is depicted in Fig. 1.

About 80% of the change in mean availability stems from population change. Undoubtedly the major challenge for water resources management is to enable food production for the projected 2.5 billion additional inhabitants in 2050.

Having said that, the climate change impact can well be more critical, than the mean annual figures indicate. Climate change affects the mean annual precipitation but not least the extremes (floods and draughts). Especially the draughts have major impact on the agricultural production.

Irrigation accounts for 60% ~ 80% of the world's freshwater consumption and irrigated agriculture accounts for 50% of the total production. About 30% of the water is lost due to inefficient irrigation systems and inefficient management. The increasing concern over environmental impact from agriculture production further aggravates this.

The increasing pressure on water resources has many places led to over-exploitation of groundwater resources. At the same time, many rivers suffer from poor water quality due to pollution from agriculture production, inadequate treatment of municipal and industrial effluents leading to environmental degradation.

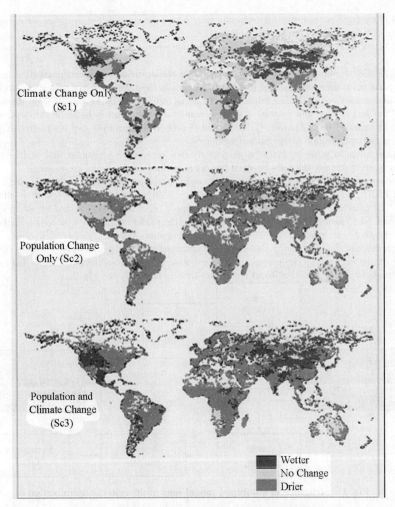

Fig. 1 Relative change in demand per discharge

Water has a key role in sustainable and economic development. Integrated and holistic approaches for water management including real time optimization and operation are key in meeting the severe challenges, we are facing. Application of these technologies for improving irrigation management practices and operations has major potential impact on the water efficiency.

2 New methodologies

Effective water management requires a comprehensive consideration of all related aspects, e. g. , technical, social, environmental, institutional, political and financial. The conventional methods of cost-benefit analysis and single-objective models are changing to multi-objective models.

Combined models and decision support systems have been developed for:

(1) Optimization of the use of water as a resource.

(2) Regulation of reliable water at the demand time.

(3) Enhanced security of supply and fulfillment of demands through correct management of water within the channel system.

(4) Enhanced possibilities for communicating future demands to the upstream water supplier. The decision support framework includes:

(1) Data Integration and Management: ①On-line sensor information. ②Demand/crop information.

(2) Modeling Manager: ①Hydraulic modeling of the upstream supplier (normally a large reservoir), the main distribution canal and important canals from the extensive system of smaller secondary and tertiary canal. In-line smaller reservoirs as well as regulating structures on the main canal. ②Data assimilation and forecast of water demands and the future behavior of the canal system.

(3) Scenario/Event Manager: ①Interactive definition of scenarios and alternatives. ②Model execution and optimization of operational strategies.

Optimization of water resources management often represents a complex and multi – purpose problem where balanced solutions between the often conflicting objectives are required.

For optimization of irrigation systems, procedures based on simulation models coupled with numerical search methods are applied. The adopted simulation-optimization framework is illustrated below. The optimization criteria (e. g. irrigation supply reliability) are defined as objective functions which are numerical measures that compare the model output with user-specified targets (e. g. impoundment storage, conveyance losses, irrigation demand, water deliveries, etc.). Based on the calculated objective functions the optimization algorithm selects new sets of control variables to be evaluated. The process is repeated a number of times until no further improvement can be made or the maximum allowed simulation time is reached, see Fig. 2.

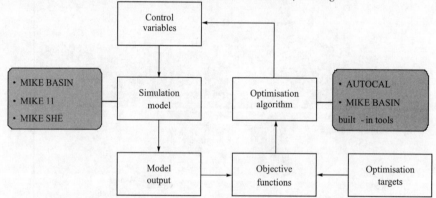

Fig. 2 Structure of combined model and multi-criteria optimization

A combined approach to irrigation system optimization is applied. It is driven by data from a network of monitoring stations established to get a reliable and accurate hydraulic and environmental state of the overall system. A collection of modelling tools is applied to predict future conditions given certain boundary conditions. A range of real-time updating routines is applied to update the models based on field data as time progresses. Control strategies are explicitly described in the numerical model of the hydraulic structures. Finally, online optimization tools are implemented to resolve conflicting demands during operation with a dynamic feedback to the control strategies.

2.1 Optimization of reservoir operation rule curves

Traditionally, rule curves are used for guiding and managing reservoir operations. These curves usually specify reservoir releases according to the current reservoir level, hydrological conditions, water demands and time of the year. Established rule curves, however, are often not very efficient for balancing the demands from the different users. Moreover, reservoir operation often includes subjective judgment by the operators. Thus, there is a potential for improving reservoir operating policies, and even small improvements can lead to large benefits.

2.2　Real-Time Irrigation Optimization

Operation of reservoir systems using optimized rule curves ensures a general optimal operation of the system. To further improve the performance, real-time optimization can be adopted, where real-time and forecast information about reservoir levels, reservoir inflows and water demands are utilized. In this case, the reservoir system is optimized with respect to the short-term operation, using both short-term and Long-term objectives. Often there is a conflict between short-term and Long-term benefits, and hence the inclusion of Long-term objectives in the optimization is important. Two different methods can be used:

(1) Combined short-term and Long-term simulation-optimization.

(2) Short-term simulation-optimization with Long-term objectives as constraints.

Combined short-term and Long-term simulation-optimization.

In this approach the system is optimized over a long period of time, and the optimal solution is implemented for the short-term operation. When new real-time data and forecast information become available, the simulation-optimization is repeated and a new optimal solution is implemented for the short-term operation. Control variables to be optimized are future releases from time of forecast to the various water users.

Short-term simulation-optimization with Long-term objectives as constraints. For complex simulation models involving several reservoirs, such an approach may not be computationally feasible in real time. Instead an approach that includes Long-term objectives as constraints in the optimization can be applied.

In this case simulation is only carried out for a short period using the available forecast data the first days after time of forecast. The control variables to be optimized consist of future releases to the various water users in the forecast period. The objective functions to be optimized include short-term objectives, which are often not included in today's operations.

3　Applications experience

The application of DSS for water resources management has demonstrated the strength in providing decision level staff with aggregated and processed information directly targeted to optimising water allocation. Examples of recent applications are given below:

3.1　Nile Basin Decision Support System

The overall objective with the Nile Basin DSS is to assist the nine riparian countries in the basin on their road towards IWRM. The dependence of the riparians on the Nile River is of huge proportions. Egypt, for example, has no other source of water than the Nile and upstream water resources developments will decrease the water availability. Historically, competition for the water of the Nile has been a key issue between the riparian countries especially for those countries downstream of Lake Victoria. The Nile Basin Initiative has recognized the need for tools and knowledge bases to assist the riparians in their assessment of water resources development scenarios and projects. The present DSS development work has resulted in two interim releases and a deployment of the NB DSS takes place in September 2012.

The riparian countries of the Nile-Burundi, Democratic Republic of Congo, Egypt, Ethiopia, Kenya, Rwanda, Sudan, Tanzania, and Uganda; have embarked on the Nile Basin Initiative (NBI). The NBI is governed by the Council of Ministers of Water Affairs of the Nile Basin states and seeks to develop the river Nile in a cooperative manner, sharing socioeconomic benefits and promoting regional peace and security. Their shared vision is: to "achieve sustainable socioeconomic development through the equitable utilization of, and benefit from, the common Nile Basin water resources". A Strategic Action Program (SAP) should translate this vision into concrete activities and projects. An important part of the shared vision is the establishment of shared and accepted water management tools and technologies. For this purpose the Nile Basin DSS was developed.

The development of the Nile Basin DSS contains two separate work packages. Work package 1 is essentially an IT project focusing on the development of the NB DSS while work package 2 is designed for independent system testing and pilot application. Key activities were elaboration of the NB DSS software requirements, software architecture and design, software development and testing, training of local staff and system deployment in the nine countries. The NB DSS software requirements are rooted in use cases developed by the NBI and further elaborated during the course of the project. The NB DSS is designed to support water resources planning and investment decisions in the Nile Basin, especially those with cross-border or basin level ramifications. The system comprises an information management system linked with river basin modelling systems and a suite of analytical tools to support multi-objective analysis of investment alternatives. NB DSS will aid the development of core national capabilities to assist in the evaluation of alternative development paths and the identification of joint investment projects at sub-regional and regional levels. The NBI has established a small, strong project management unit (PMU) staffed by DSS specialists and IT and modelling experts. In addition, IT and water resources modelling experts from all nine countries have participated in all project phases ranging from elaboration of requirements to system testing. Two interim NB DSS releases have been successfully deployed, tested and accepted by the NBI. The final NB DSS will be deployed in all countries in September 2012. A service agreement is in place ensuring that the NBI will have access to support and software updates. Training and involvement of local staff has been key. At this stage, more than 50 Nile Basin water professionals have been trained by DHI. Moreover, the NBI PMU has invested substantial resources in involving additional engineers and managers through training sessions and workshops in the NBI countries. As such a very large number of local staff have been trained or exposed to the NB DSS even before the final release.

Lessons learned:

(1) Substantial training and client involvement during the project has created a very strong feeling of ownership at the NBI.

(2) Software requirements should be based on or supported by use cases developed by the client. This process is time consuming but important to ensure and demonstrate the ability of the system to address real-life problems and key issues in relation to client involvement and ownership.

In order to sustain and further enrich the NB DSS a post project plan must be in place including staffing, institutional setup and funding.

3.2　DSS Planning, India

The "DSS Planning" in India is being implemented for Integrated Water Resources Development and Management, developed under the Hydrology Project Ⅱ (2008 ~ 2012).

Requirements for Integrated Water Resources and Water Security Planning and Management are increasing as a result of increasing population pressures and associated competing water demands from among others, agriculture, industry, domestic supplies and environment. The project was defined with the objective to develop a customized DSS – "DSS Planning", applicable for several states in India to address issues identified specifically in each state within: ① surface water planning; ② integrated operation of reservoirs; ③ conjunctive surface water and ground water planning; ④ drought monitoring, assessment and management; and ⑤ management of both surface and ground water quality. "DSS Planning" is based on DHI's generic DSS framework.

The Upper Bhima River Basin in Maharashtra was selected as a pilot area for "DSS Planning" application with the other participating states to follow. The water resources management themes in Maharashtra, for which "DSS Planning" was required, included: ① short and long term planning of reservoir management; ② conjunctive use of surface-and groundwater; ③ planning of seasonal groundwater use; ④ artificial groundwater recharge; ⑤ drought monitoring; ⑥ flood analysis and ; ⑦ water quality modelling. Decision makers needed, among others, to address issues on how to respond to increased water demands in the most efficient way, how to communicate drought indicators to the broader public and how to reduce flooding and flood damages. The "DSS Planning" assisted with detailed answers to questions on the risk of reaching critically low reservoir levels in the dry season and the likelihood of filling the reservoir in the forthcoming wet season. Based on calculated risk, decision makers could resolve that the live storage could be increased. The "DSS Planning" was also applied to analysis of droughts with low post-monsoon reservoir storage and falling ground-water levels. Mathematical models included in "DSS Planning" were used to identify a sustainable situation with addition of artificial recharge. Conjunctive use was shown to be an efficient way of supplying water and reducing the risk of water logging. On-line data presentations from the database showed the likely severity of future drought and thus remedial measures could be taken accordingly. The "DSS Planning" was further applied for flood analysis and the potential of reducing flooding and flood damage through forecasting was assessed and used by the decision makers. The continued use of the "DSS Planning" system's methodologies requires extensive training of implementing a-gency staff. The emphasis has been on modelling, data processing, model setup and calibration and model use. This was supplemented by training in more generic areas such as use of GIS and time series tools.

Lessons learned:

(1) The "DSS Planning" system can be installed as the central hub for water resources data and information in the state and access can take place through a PC, local area networks and safe internet connections.

(2) The "DSS Planning" system is very useful for long and short term planning and manage-ment of water resources and for impact assessments.

(3) The "DSS Planning" system is suitable for floods and droughts predictions against indica-tors and for issuing of emergency warnings.

Training is necessary to maintain knowledge and skills within the organizations involved.

3.3　Computer Aided River Management System (CARM), Australia

As South-Eastern Australia recovers from its worst drought on record, the experience has driv-en new innovations for achieving water efficiencies, including better river management. The Mur-rumbidgee River was first highlighted in the Pratt Water report of 2004 as an example of where ma-jor water savings could be obtained through infrastructure upgrades and smarter river operations. Murrumbidgee means "big water" in the local Wiradjuri aboriginal language. The river is one of Australia's longest at 1,600 km, and it is a major source of irrigation water to the Riverina region in NSW as well as an important source of water to the river wetlands. However, its management is a complicated task. The complex nature of the river system, coupled with the critical need to meet irrigation, environmental, and town water demands, often results in excess water being released from the headwater storages which are surplus to the actual requirements.

Now, NSW State Water, the river operator, supported by Water for Rivers (www. waterforriv-ers. org. au), have embarked on a 65 m upgrade of the river management and operational system that will set a benchmark for efficient river operations in Australia and internationally. The Murrum-bidgee Computer Aided River Management (CARM) project will make control of water flows and dam releases more precise and efficient through a combination of upgrades to river infrastructure, metering, operational modelling and information systems. Operational improvements will be realised through the integration of river monitoring, extraction metering, hydrodynamic river models and op-timisation software systems. The system will support efficient and frequent decision making by river operators to ensure that the most efficient operational settings are achieved and that irrigators, envi-ronmental and other customers receive the right amount of water at the right location at the right time. The decision support system represents world's best practise in river operations.

3.4　Current river management practice

State Water currently operates the Murrumbidgee River system(in Fig. 3), which includes the Yanco-Billabong Creek offtakes, through dam releases from the headwater storages at Burrinjuck and Blowering Dams and through an additional 10 re-regulation weirs(in Fig. 4) and two off-line storages.

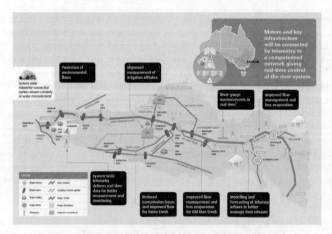

Fig. 3   Map of Murrumbidgee Irrigation System

Irrigation is by far the largest user of river water in the basin, with two of the largest irrigation supply companies, Murrumbidgee Irrigation and Coleambally Irrigation accounting for approximately 70% of water use. These are supplied by dedicated offtake canals located between 5 ~ 7 d travel time downstream of the dams. In addition to hundreds of private irrigators extract water directly from the river by pumping. Releases from the headwaters can take up to 28 d to reach some irrigators at the tail end of the system. Environmental customers are also increasingly important, and the catchment includes significant wetlands including Lowbidgee, one of the most important waterbird breeding sites in the region. Large towns that also depend on the river for their water supplies include Gundagai, Wagga Wagga, Hay and Balranald.

Current river operations rely heavily on the experience and judgement of the river operator and are based on simple water balance modelling concepts. These models do not take into account the complexities of the catchment flow processes or river flow dynamics. Providing reliable water deliveries to customers located, in some cases, many weeks of travel time downstream of the dams is particularly challenging. Consequently dam releases often exceed actual demands, which lead to significant water loss. This is because, once released, water is removed from the basin due to much higher evaporation and evapotranspiration rates in the lower part of the river system compared to the headwaters. The main drivers leading to this operational surplus have been identified as being due to unaccounted changes in channel storage, unaccounted tributary inflows and late changes in irrigation water orders. As a result, almost 12% of the annual 4,200 GL of regulated flows is currently unaccounted for.

Improving the process for the identification of unaccounted losses, and reducing operational surpluses requires modelling tools capable of reproducing the key physical behaviour of the catchment and the river system, which can then form the basis for optimal dam release and weir operation strategies.

The river operations decision support system being implemented by DHI is being built around DHI's Solution Software technology. The CARM 'engine' comprises a suite of MIKE by DHI computer simulation models that accurately reproduce the key catchment runoff and river flow processes: tributary inflows; continuously variable river flow travel times; in-channel storage dynamics; evaporation; evapotranspiration from riparian vegetation and near-river groundwater exchange.

The system will integrate the models with real time measurements from State Water's on-line data and control systems, and will provide a range of fully customised decision support user interfaces for the river operators.

The CARM will make full use of existing and new monitoring data, including rainfall measurements and forecasts from the Bureau of Meteorology, river flows and levels, as well as pumped extractions from real time metering that is being implemented as part of the project. Forecast using hydrological models, utilising both rainfall observations and Bureau forecasts. These will feed into a MIKE 11 hydrodynamic river simulation model that is being developed for the entire Murrumbidgee-Yanco Creek system, incorporating over 2,000 km of river channels and floodplains. Real time water level and flow measurements will be used to automatically update the model state so that it continuously emulates the real river behaviour exactly. Near – river bank and groundwater exchanges, which were previously unaccounted, as well as evapotranspiration along the riparian margin, will be simulated using a MIKE SHE integrated surface groundwater interaction model, fully coupled to the MIKE 11 river model. This will ensure continuous dynamic coupling between groundwater behaviour in the alluvium of the river and the dynamic water levels in the river. With these integrated models it will possible to understand the relationship between dam release operations and groundwater changes within the riparian zone.

The hydrologic and hydrodynamic models will form the basis for the optimisation of the river system's day to day operations. By utilising forecasts of river inflows and real time water orders, it is possible to optimise the operation of the dam releases and the downstream re – regulation weirs, with the objective to meet all water demands and at the same time minimising release from headwater storage.

Fig. 4　Regulation weir

4　Proven concept

In a proof of concept for the upper parts of the system (dams to Narrandera) based on historical measurements, the CARM optimised solution has been shown to significantly reduce dam releases without compromising irrigation water security. Two periods selected, October ~ November 2006 (dry) and December 2007 ~ January 2008 (wet), were previously identified as periods where significant operational surpluses had occurred. The optimized solution reduced the dry period releases from 441 GL to 351 GL, a saving of 90 GL or 20%, and in the wet period reduced releases from 181 GL to 94 GL, a saving of 87 GL, or 48%. These savings were achieved whilst still meeting all virtually all 6 day and 1 day demands for Murrumbidgee Irrigation and water orders further downstream of Narrandera. Real water savings are realised through a reduction in river operating levels, which means less water is lost through evaporation and evapotranspiration.

References

Duan Q, Sorooshian S, Gupta V. Effective and Efficient Global Optimization for Conceptual Rainfall-runoff Models[J]. Water Resources,1992, 28(4): 1015 – 1031.
Madsen H, Skotner C. Adaptive State Updating in Real-time River Flow Rorecasting-A Combined Filtering and Error Forecasting Procedure[J]. Journal Hydrology,2005, 308(1): 302 –312.

Reasonable Development and Utilization of the Water Resources of the Yellow River Realization of the Sustainable Development of Zhengzhou Economy along the Yellow River

Su Maorong[1] , *Xin Hong*[2] , *Jia Mingmin*[1] , *Jiao Haibo*[2] , *Yang Miao*[1] and *Zhang Feng*[2]

1. Water Supply Authority of Yellow River Henan Bureau, Zhengzhou, 450003, China
2. Zhengzhou Water Supply Sub – office of Water Supply Authority of Yellow River Henan Bureau, Zhengzhou, 450003, China

Abstract: The Yellow River is a river lack of water resources and the supply and demand contradiction of water resources is very outstanding. In recent years, due to the shortage of water resources, and regional economic development and population growth, the demand for water is increasing day by day, so that we should use the limited water resources of the Yellow River reasonably, promote optimal allocation of water resources, strengthen the management of water resources and scheduling standardization construction, improve water resource management and scheduling scientifically and standardly, which will be an important guarantee of effective management system, management form and management means for reasonable development, effective utilization, scientific scheduling and optical allocation of the Yellow River's water resources. In view of the contradiction between supply and demand of the Yellow River's water resources, we shall allocate the water resources of the Yellow River reasonably to alleviate the contradiction between supply and demand in Zhengzhou and promote the sustainable development and utilization of the Yellow River's water resources.

Key words: the Yellow River, water resources, economy, sustainable development

1 General

The Yellow River is the second river of China and the fifth longest river in the world, with natural annual runoff of 5.8×10^{10} m^3, occupying 2% of countrywide river runoff. The Yellow River has nurtured the Chinese splendid ancient civilization. The Yellow River basin, rich in natural resources, is the largest energy and important chemical industry base of China. Ningxia, Inner Mongolia, the Fenhe and Weihe Rivers Plain and the Huanghuaihai Plain are the main grain cotton base of China. The irrigation area of the downstream Yellow River is one of the compact areas of the Yellow River irrigation. The Yellow River irrigation has great economic benefit in fully developing farming, forestry, stock raising, sideline production and fishery in regions along the Yellow River to promote economic development and society progress. The Yellow River provides rich material basis for the survival of Chinese sons and daughters. China has a large population; the amount of water resources is 2.8×10^{12} m^3, ranking sixth in the world; but annual possession per capita is only 2,200 m^3. In the Yellow River valley, the water resources per capita and per mu (1 mu = 666.67 m^2) are only about one fourth and one second of world average level. The water resources per unit of farmland is less than one fifth of national level, ranking 109th in the world and belonging to one of 13 water-poor states.

According to statistics, the water deficit in normal year at present in the country is nearly 4.0×10^{10} m^3, of which water deficit in agriculture is about 3.0×10^{10} m^3. Because agriculture is the largest family in using water, its water consumption accounted for about 70% of total water use of China. Temporal and spatial distribution of rainfall in the Yellow River basin is extremely uneven; 60% to 70% of annual rainfall is concentrated in the period from June to September, and it has great adverse effects on the agricultural products in the basin.

Henan is a great populous province and also a great agricultural province. The water resources per capita in the province is 440 m^3, equivalent to one sixth of the average level of China, far below the warning line 1,000 m^3 of international water resources per capita. The water resources per capita, especially the regions along the Yellow River, is only 275 m³, equivalent to 1/10 of countrywide average level, belonging to regions in extreme shortage of water resources. How to ensure the sustainable development of water resources in the Yellow River has become the primary issue to promote the sustainable development of social economy in the Yellow River basin. The industrial water consumption in the city occupies about 70% of the total diverted water. It is the main industrial and agricultural production base of our province, and is also the center city of the central plain's urban agglomeration under construction and of economic uplifted zone.

Zhengzhou, the provincial capital of Henan, has total population of about 8,626,500 people, of which the urban permanent resident population is about 3,000,000; the water for industrial use and city life ranks first in the province. As the main water source crossing over Zhengzhou city, the water resources from the Yellow River accounted for nearly 90% of the total transit water resources, with 150 km of the boundary along river and a water intake from the Yellow River with gradual complete ancillary facilities, which provide a broad space for utilization of water resources to support the social and economic development of Zhengzhou. The characteristic of Zhengzhou Yellow River water supply is that the water for urban industrial use occupies the absolute dominant position; the proportion of agricultural water is small; the water for industrial use is large, but water supply to agriculture is small; the industrial water occupies about 68% of total diverted water and the income from industrial water charge accounted for more than 85% of total income from water charge. There are many local self-built and self-direct sluices in Zhengzhou, while only a few inner – managed sluices managed by the Water Supply Authority; there is a few new built water sluices, while many old water sluices; so the aging of project facilities is serious; the Huangyuankou sluice has been running for 55 years since it was built.

In recent years, although by taking measures, such as water – saving society construction, it effectively alleviate the pressure from the shortage of water resources, but with the rapid development of social economy in Zhengzhou, the total demand for water resources still maintains ascendant momentum; the contradiction between supply and demand of water resources and social economic development is still increasing; therefore, how to take effective measures to scientifically and reasonably make full use of domestic water resources, improve optimized configuration of water resources and social economic development are still key topic for development at present.

The development of the Yellow River water supply can effectively alleviate the serious water shortage contradiction in some areas of the Henan Province, improve the guarantee rate of city and industrial water, expand agricultural irrigation area and improve the ecological environment. When comparing with south-to-north water transfer and other methods, this kind of water supply cost is low; in addition, Zhengzhou has the advantage of natural environment endowed by nature and regional advantage, which determine that the Yellow River water supply of Zhengzhou has broad market prospects. By accelerating the development of Zhengzhou's water supply industry and exploring developmental potential on Henan water supply industry, it is not only has obvious social benefit and ecological benefit, but also it has significant meaning to guarantee the economic and social sustainable development of regions along the Yellow River. What's more, it is extremely essential to support well-off society construction comprehensively in the regions along the Yellow River and it also has extraordinary significance for economic development in economic uplifted zone and urban agglomeration of the central plain taking Zhengzhou as center.

2 The great social and economic benefits from the Yellow River water supply

The annual average volume of water resources is 4.55×10^{10} m³ in the province, of which the Yellow River occupies 90%. The Yellow River water resource's utilization has expanded from single agricultural irrigation to public water, industrial water, tourism, fish breeding and poultry raising. It has equal importance with grain, environment, population and energy source. It has positive role in promoting the regional economic development, improving the living and ecological environment and safeguarding people's health.

The Yellow River water is suitable for crop growth; every 1.00×10^8 m³ of the Yellow River water for agricultural irrigation will bring 0.1×10^8 Yuan of increase production efficiency, and the effect of increase production is tremendous. It has great effect in soil improvement by using the Yellow River water for irrigation; it can also recharge the groundwater, improve underground water funnel. It helps to recharge groundwater, supply water for irrigation, divert and withdraw water into local river way, enhance water dilution and self-purification ability, effectively change the local surface water quality and reduce water pollution degree by restoring and expanding the area of the Yellow River water supply.

According to relevant data, the annual water consumption per mu of the Yellow River for irrigation for rice is 1,500 m^3, garlic 800 m^3, and wheat 500 m^3. If every cubic meter of the Yellow River water charge 0.04 Yuan, their annual irrigation fee per mu separately is 60 Yuan, 32 Yuan and 20 Yuan. If use same water by well for irrigation, the water fee per every 100 m^3 used by well is 8 Yuan to 10 Yuan by motor-driven well, 12 Yuan to 15 Yuan in diesel oil; the average irrigation fee per mu separately for rice is 120 Yuan, garlic 64 Yuan and wheat 40 Yuan, 2 times higher than that of the Yellow River irrigation. The Yellow River irrigation saves time and energy, has no mechanical cost and the water from the Yellow River contains organic matter for soil and it can keep moisture, fertilizer and water in the soil and prevent the ground from hardening. If we take measures to recover 10,000 mu irrigation area, we can save more than 400,000 Yuan of irrigation cost directly, meanwhile we can restore underground water to irrigate the area of about 7,000 mu by using the recharged water.

The income list from water charge of the Yellow River water supply of the Yellow River Zhengzhou Bureau over the years is shown in Tab. 1.

Tab. 1 The income list from water charge of the Yellow River water supply of the Yellow River Zhengzhou Bureau over the years (unit: ×10^4 Yuan)

Year	2000	2001	2002	2003	2004	2005
The income from water charge	106	318	321	373	405	572
Year	2006	2007	2008	2009	2010	2011
The income from water charge	714	1,061	1,539.01	1,140.31	1,610.33	2,310

3 The development and utilization situation and the existing problems of the Zhengzhou's Yellow River water resources

With the development of social economy and the increasing water crisis, Zhengzhou's water resources condition calls for deep thought. According to water resources evaluation, the result showed that the total water resources of the city is about 13,000,000 m^3, the annual water resources per capita is 198.8 m^3 per mu for years, the annual water resources per mu is 302.2 m^3, far below the water resources per capita and per mu of the country and province. The spatio-temporal distribution of intra-area rainfall is not even; the multi-year average rainfall is 633.3 mm and the average evaporation capacity is 1,320 mm. Compared the rainfall of Zhengzhou and the water resources per capita with evaporation capacity, Zhengzhou belongs to areas serious lack of water. The Yellow River is the water source to make up for the shortage of water resources of Zhengzhou. Therefore, it is the internal factor and inevitable trend for the economic development of Zhengzhou to increase the Yellow River water supply of Zhengzhou and promote the development and utilization of the Yellow River water resources.

In the current process to boost the construction of urban agglomeration and economic uplifted zone, on one hand we are creating water-saving society vigorously in order to alleviate the pressure from shortage of water resources, on the other hand, the transit water resources from the Yellow River have not been fully utilized. The two points forms tremendous contrast, which is manifested in the following aspects.

3.1 A great contrast compared the approved water diversion quantity with the actual water diversion quantity

It has formed a great contrast when compared the approved water diversion quantity by the government with the actual water diversion quantity. There are 16 water intakes in the jurisdiction of Zhengzhou, among which there are 7 water sluices for the Yellow River Diversion Project, which is self-built and self-managed by the Water Supply Authority; 9 water-intake projects, which is self-built and self-managed by the local government; there are 10 water intakes for surface water and 6 water intakes for underground water. From 2000 to 2005, the total diversion water quantity approved by the Yellow River Conservancy Commission (YRCC) was 6.32×10^8 m^3, among which the amount of industrial and city life water was 2.85×10^8 m^3, agricultural water 3.47×10^8 m^3, while the actually average water diversion quantity was about 2.00×10^8 m^3 in recent years, just accounting for 32% of the approved water diversion quantity; the actual water diversion quantity for industrial use was about 1.30×10^8 m^3, for agricultural use about 0.60×10^8 m^3. 4.00×10^8 m^3 of water diversion target can not be fully used, which directly resulted in that the higher level of authority reduced the water diversion target. The YRCC from 2005 to 2009 reduced the total water diversion quantity to 5.41×10^8 m^3 in reply to the Yellow River Zhengzhou Bureau, among which the quantity of industrial and city life water was $3.278,4 \times 10^8$ m^3, the agricultural water $2.134,5 \times 10^8$ m^3. From 2010 to 2015, the YRCC reduced the total diversion water quantity to 4.38×10^8 m^3 in official reply to the Yellow River Zhengzhou Bureau. Such as, in 2007 the actual total quantity of diversion water was 1.84×10^8 m^3, among which the quantity of industrial water was 1.57×10^8 m^3 and the agricultural water 0.27×10^8 m^3; 3.58×10^8 m^3 of diversion water target did not get used; in 2008, the actual quantity of diversion water was 4.51×10^8 m^3, among which the industrial water was 2.39×10^8 m^3, agricultural water 2.12×10^8 m^3; 0.90×10^8 m^3 of water diversion target did not get used. In 2009, the actual quantity of water diversion was 5.42×10^8 m^3, among which the industrial water was 1.69×10^8 m^3, the agricultural water 3.73×10^8 m^3. In 2010, the actual quantity of water diversion was 6.22×10^8 m^3, among which the industrial water was 1.75×10^8 m^3, the agricultural water 4.47×10^8 m^3. In 2011, the actual quantity of water diversion was 8.46×10^8 m^3, among which the industrial water was 1.67×10^8 m^3 and the agricultural water 2.39×10^8 m^3.

The causes resulting in decrease of water diversion quantity: ①The water diversion condition is poor. The river within the Henan Province belongs to the wide and wandering channel, so that the Yellow River makes its flowing contact point of both sides unstable and the water intake changeable at any time. As a result, the work amount of the Yellow River Diversion Project is increased and the overlong channel produce deposition. ②The water conservancy facilities and canal system for farmland outside embankment is not perfect. The existing one is aged and out of repair, often with "large mouth with small intestine". For example of Zhaokou sluice, its designed diversion flow from the Yellow River can reach 210 m^3/s, while the actual annual water diversion flow only maintained at about 30 m^3/s. ③The reduction of Yellow River irrigation area led to the insufficient reinvestment for the reform and development of irrigation area, while the operating cost of the management department of irrigation area on the project is raised without restriction, so that the terminal charge for water is high, which influence the farmer's enthusiasm to use the Yellow River water. ④Due to frequent swing of the Yellow River's mainstream, in recent years water diversion and sand diversion make the mainstream riverbed down-cutting and the approach channel before channel-head sluice overlong. Such as, the length of approach channel before Yangqiao sluice is 4 km. the channel is seriously deposited, so it is difficult to draw water, the water diversion quantity is decreased year after year, the guarantee rate for using water is reduced. In 2006 the Yangqiao sluice did not draw any water and it begins to draw water again from 2008.

3.2 Irrigation area in the irrigation districts is reduced gradually

Affected by the poor management system of the Yellow River irrigated districts, the irrigation area in the irrigation districts is reduced gradually. There are two Yellow River irrigation districts in Zhongmou, Yangqiao and Sanliuzhai. Yangqiao irrigation district was built in 1970 and started to irrigate in 1975; and the beneficial range involved ten villages and towns. Sanliuzhai irrigation district was built in 1965 and its beneficial range involved 5 villages and towns. The Yellow River irrigation districts in Zhongmou, have experienced reform and expansion since they were built and started to irrigate in 1960s and 1970s, and were provided with facilities covering area of 539,100 mu in total, among which Yangqiao irrigation district covered 274,100 mu and Sanliuzhai irrigation district covered 265,000 mu. During 20 years before 1992, the average water diversion quantity was 2.46×10^8 m³ in the Yellow River irrigation district of Zhongmou and the actual irrigation area is above 450,000 mu. In 2005, the existing actual Yellow River irrigation area only remained 160,000 mu. In 2006, it was reduced to 80,000 mu; Yangqiao irrigation area stopped irrigate. On one hand, it is because of serious shortage of water resources; on the other hand, the water resources can not get fully used. The diversion water quantity of the Yellow River irrigation area of Yangqiao and Sanliuzhai respectively were 0.8×10^8 m³ and 0.25×10^8 m³, which was approved by the YRCC in 2005, while the water diversion quantity for Yangqiao was $0.173,544 \times 10^8$ m³ in 2005, accounting for 21% of the approved water diversion target. In 2006 Yangqiao sluice did not draw any water in the whole year; the water diversion quantity of Sanliuzhai was $0.199,368 \times 10^8$ m³ in 2005 and $0.244,917 \times 10^8$ m³ in 2005 and 2006, separately accounting for 80% and 98% of the approved water target.

The irrigation area's shrink even close directly lead the local water index reduction, meanwhile, it will produce a series of adverse consequences on the development of regional social economy.

3.3 Not paying enough attention to the economic and social benefits brought by diverting water from the Yellow River

Not enough importance has been attached to the economic and social benefits brought by diverting water from the Yellow River under the situation of rapidly increasing demand for urban life water, ecological and industrial water.

3.4 Irrigation district engineering benefits can not be fully played and economic effectiveness can not be reflected

As the needs of irrigation districts cannot be met in the peak period for seasonal water – use and sometime there is even no water, it happens a lot that the farmers appeal for help, put off the payment of water fees or refuse to pay. Since the water can not be ensured, farmers in irrigation districts lose confidence in diverting water from the Yellow River water and then adjust the planting structure, resulting in the increasing number of wells in irrigation districts, which leads to the decrease of water diversion amount from the Yellow River year by year so that the project benefits could not be developed and economic effectiveness could not be reflected. For instance, in 2003, owing to the inability of supplying water normally for Yangqiao Irrigation District, rice of 16,000 mu could not be planted and the paddy field had to be planted with corns, which caused 0.6×10^6 Yuan loss to the rice seedlings cultivation alone. In July 2005, because of the low water flow of the Yellow River, diverting water through sluices could not satisfy the demand of irrigation districts, which leaded to replant the upland crops like corns or soybeans instead of the already planted rice of 4,200 mu. Two irrigation districts of Zhongmu County owned the rice area of 160,000 mu in 1995 while reduced to 45,000 mu only in 2003; in 2005, the area recovered to 60,000 mu, among which the most field was changed into dry farmland due to the instability of water supply. According to the measurement and calculation, compared with the well irrigation, the Yellow River irrigation can help to save about 30 Yuan per mu annually with less time and efforts. Based on the current actual irrigation area of 200,000 mu in Yangqiao Irrigation Districts, people have to pay over 6×10^6 Yuan more each year to irrigation alone.

3.5 Problems

(1) Constraint of bottle-neck effect emerges after the rapid development of water supply cause. During the "Eleventh Five – Year", the robust growth of water diversion quantity from the Yellow River and income from water charges hit record high. However, at the time when water diversion quantity from the Yellow River and income from water charges are creating a breakthrough,

the constraint of bottle – neck effect objectively emerges in water supply cause. There is not much development space for diverting surface water from the Yellow River to Zhengzhou urban life and industries; the demand of diverting water from the Yellow River for industrial corporations has been severely affected by the international financial crisis, like China Aluminum Co. , Ltd. Henan Branch; national policy support is needed to realize the full charge of water fees by the two original local self-built sluices with self-management— Mangshan Pumping Irrigation Station and Sino-French Raw Water Co. , Ltd. after the alteration of water intaking subject.

(2) The engineering management foundation is weak. The sluices were built lone before and have worked for a very long time with seriously aged programs and equipments. Most sluices are operated with many problems.

(3) There are little financial resources to use and some objective factors are affecting the investment in engineering reconstruction and maintenance.

(4) The water supply system is not smooth, with inflexible mechanism, multi-sectoral management for part of the work and imperfect rules and regulations.

(5) There is no policy basis for charges of the groundwater intaking from the Yellow River in the beach area. As it's difficult to conduct the management and fee collection only by an agreement, the Long-term stability and firmness can hardly be realized.

(6) Front-line employees for water supply are relatively older with lower educational level and poor technological ability. Some are still holding the obsolete ideas.

4　Current status and development trend of water intaking of the Yellow River in the project of diverting water from Yellow River

Current status and development trend of water intaking of the Yellow River in the project of diverting water from Yellow River is shown in Tab. 2.

Tab. 2

No.	District	Name of water diversion institute (head works)	Actual water diversion ratio of 2009	Actual water diversion ratio of 2010	Actual water diversion ratio of 2011
1	Xingyang City	Gubaizui Pumping Irrigation Station	142%	152%	174%
2		Mangshan Pumping Irrigation Station	64%	89%	126%
3	Huiji District	Sino-French Raw Water Co. , Ltd.	69%	74%	103%
4		Huayuankou Sluice	244%	571%	862%
5		Dongdaba Pumping Irrigation Station	49%	8%	0%
6	Jinshui District	Madu Sluice	26%	28%	57%
7		Yangqiao Sluice	187%	122%	213%
8	Zhongmou County	Sanliu Village Sluice	214%	161%	174%
9		Zhaokou Sluice	190%	304%	372%

4.1　Gubaizui Pumping Irrigation Station

Annual approval intake quantity of water is $0.12 \times 10^8 \ m^3$. Under the effect of financial crisis in 2009, China Aluminum Co. , Ltd. Henan Branch (Shangjie Aluminum Factory) had to shut down in a large scale. However, its production capacity has been expected to gradually recover afterwards with an upward trend of the actual water diversion ratio. In 2011, the supply quantity of diverting water from the Yellow River all though the year was $0.208,8 \times 10^8$ m³.

4.2 Mangshan Pumping Irrigation Station

Annual approval intake quantity of water is 1.2×10^8 m³. The actual water diversion ratio of 2011 (the supply of diverting water from the Yellow River was $1.133,584 \times 10^8$ m³) was 25% higher than that of the past two years, which could basically satisfy the need of urban development in Zhengzhou over the years.

4.3 Huayuankou Sluice

Annual approval intake quantity of water is 0.05×10^8 m³. In recent years, the farmland around the city has been occupied little by little along with the development of urban economy and the expansion of urban construction so that the irrigation area is reducing year by year. As Water Conservancy Bureau of Zhengzhou started to apply for the eco-environmental water supply to Dongfeng Canal and Jialu River of Zhengzhou City from Huayuankou Sluice to improve the urban ecological environment in 2008, the actual water diversion ratio was growing at nearly three times a year and the actual water diversion quantity in 2011 was $0.431,002 \times 10^8$ m³. The current approval water quantity has been far from enough to satisfy the practical demands.

4.4 Sluice of Diverting Water from the Yellow River of Sino – French Raw Water Co., Ltd.

Annual approval intake quantity of water is 1.10×10^8 m³. It is mainly responsible for the water supply of Zhengzhou urban area with an average water diversion quantity of 0.80×10^8 m³ in recent years. In 2007, the first sluice for the canal of diverting water from the Yellow River of Dongdaba was built and put into use, with a remaining average actual water diversion ratio of 71% over the past three years.

4.5 Madu Sluice

Annual approval intake quantity of water is 0.166×10^8 m³. With an average water consumption of millions of cubic meters in recent years, the irrigation area has reduced a lot under the influence of city development (the average actual water diversion ration was 18%) with the same reason for the Huayuankou Sluice. The actual water diversion quantity in 2011 was 5,663,100 m³.

4.6 Yangqiao Sluice

Annual approval intake quantity of water is 0.8×10^8 m³. The thermal power plant of Zhengdong New Zone was built and put into use in 2007, and then it diverted water from the Yellow River through the Yangqiao Sluice with a daily water diversion quantity of 100,000 t around. Since the Phase 2 Project of the thermal power plant of Zhengdong New Zone is going to be completed, the water diversion quantity is expected to increase by near 25%. Besides, at the same time, Yangqiao Sluice assumed the responsibility of delivering eco-environmental water to Jialu River in 2008. The actual water diversion quantity in 2011 was 149,130,100 m³ and the average actual water diversion ratio over the past three years reached 140%. But it can not meet the current water demand.

4.7 Sanliu Village Sluice

Annual approval intake quantity of water is 0.25×10^8 m³. It was also responsible for delivering eco-environmental water to Jialu River in 2008; moreover, it diverted water from the Yellow River to fight a drought. In 2009, the actual water diversion ratio reached 214%, the highest over the past three years, with an average ratio of 168%. The actual water diversion quantity in 2011 was $0.435,365 \times 10^8$ m^3 and it can not meet the current water demand.

4.8　Zhaokou Sluice

Annual approval intake quantity of water is 0.80×10^8 m^3. However, the current Zhaokou irrigation district which plays a role in exerting benefits of diverting water from the Yellow River is mainly the original irrigated area of 2.3×10^6 mu in Phase 1 Zhaokou project. And the planning irrigation area in future is expected to expand to 5.72×10^6 mu, with the gradually recovery irrigation area depending on the Yellow River in districts like Kaifeng, Zhongmou, Weishi, Tongxu, Yanling and son on. Meanwhile, it is also responsible for delivering eco-environmental water to Jialu River. The actual water diversion quantity in 2011 was $3.169,854 \times 10^8$ m^3 and the average actual water diversion ratio is going up year after year. But it can not meet the current water demand.

5　Measures adopted to intensify the development and utilization management of water resources

Water supply from Yellow River is a Long-term strategic task related to the overall situation with plenty of difficulties. In order to realize the planning goal and increase the degree to ensure the sustainable utilization of water resources in Zhengzhou city, a series of protection systems and policy measures shall be established, and the following actions must be carried out: improving awareness, strengthening leadership, intensifying management, making overall plans and pursuing practical results.

5.1　Strengthening the leadership

Strengthening leadership needs to establish a high-efficiency working system, make overall plans and scientific distribution to take full advantage of water resources assigned by the superior. For purpose of strengthening the leadership over the water supply from Yellow River in Zhengzhou, the water of the Yellow River shall be fully used and a coordinated system with level-to-level administration, department coordination, upper and lower linkage and close cooperation shall be formed based on the actual needs of national economic development and conforming to the general idea of taking precedence over the city life-water, striving to develop the industrial water supply from the Yellow River and giving all-round consideration to the agricultural irrigation water.

5.2　Developing scientific scheduling to increase the water supply guarantee rate

During the agricultural water season, the management method of scientific water diversion, rational water distribution and metrological water control should be implemented to reverse the passive situation of water competition between industry, life and agriculture in the irrigation peak season and ensure the timely sufficient water supply from the Yellow River. In line with the water consumption in farming season and seasonal duty of water, increase the flow and make full use of water resources of the Yellow River, satisfying the water demand of people in irrigation districts to arise their enthusiasm of water use so that the irrigation area depending on the Yellow River would be recovered and expanded.

5.3　To change the management system of water supply from the Yellow River in irrigation districts

To change the management system of water supply from the Yellow River in irrigation districts and actively explore new ways for the compensatory transference and replacement innovation of the water right of Yellow River. Given the current severe situation of agricultural water supply from the Yellow River in Zhengzhou city, the management system in irrigation districts depending on the

Yellow River needs to be reformed to change the present disordered administration caused by the nonstandard water supply system.

5.4 To adopt the comprehensive water-saving measures with the combination of engineering constructions to improve the utilize efficiency of water resources from the Yellow River

To adopt the comprehensive water-saving measures with the combination of engineering constructions, technology support, system security and economic adjustment to improve the utilize efficiency of water resources from the Yellow River. Further strengthen the management and scheduling of water resources, reinforce the examination and approval of the water permission, carefully carry out the *Regulations for Water Distribution of the Yellow River* and ensure the safety of water supply from the Yellow River with legal means to protect the legitimate rights and interests from being infringed. Water resources of the Yellow River shall be fully explored with scientific control to maximize the benefits of water resources.

5.5 Accelerate the execution pace of inter-regional water supply projects to promote the unified and coordinated water resources allocation of the whole city.

The water resource is seriously in shortage in Zhengzhou with uneven distribution in time and area. If the inter-regional water supply project is carried out, the current target allocation of water from the Yellow River in Zhengzhou could completely solve the condition of water scarcity in counties and cites lack of water. In this process, combined with the conception of the water right replacement, the new management model of construction with specialization of labor, independent operation, high-quality water with best price and win-win purpose shall be adopted, which would make full use of water resources of the Yellow River to bring new vigor and vitality to the economic development of the whole city.

5.6 Increase input in engineering maintenance fund

To increase input in engineering maintenance fund and seize the optimum value of the river flow to provide the best opportunity for the irrigation districts, which would generate the maximum benefits with the lowest cost. Besides, measures should be taken to prevent or reduce the silt. Dredge the channel to enhance the ability of water diversion, realizing the healthy and safe operation of the project.

5.7 Further perfect the measurement system

Further perfect the measurement system of water diversion and conduct regular maintenance for the instruments to make sure the normal operation of them. Establish the water preferential policy to encourage the water use in irrigation districts, ensuring the effective utilization of water resources with economic means.

5.8 Enact matched management provisions of water supply from the Yellow River

On the basis of Water Law and related laws and regulations, to enact matched management provisions of water supply from the Yellow River to guarantee that the work of water control, water diversion and water charges would be in line with laws.

6 Conclusions

To fully explore and utilize water resources of the Yellow River, unremitting efforts like investing numerous manpower and materials to consolidate the dyke and silt area, developing land resources near the Yellow River, diverting water from the Yellow River for soil improvement and so on have been made to turn the dozens of hectares of saline beach area by the Yellow River into a land of abundance since people started the treatment on the Yellow River more than six decades ago. Water resources of the Yellow River are not solely applied in agricultural irrigation but also in various aspects including public water supply, industrial water use, irrigation, soil improvement, tourism and breed aquatics, etc. , which plays an equal role as grain, environment and energy in promoting regional economic and social development, improving life standard and ecological environment and ensuring people's health.

As the Yellow River is a stream lacking water resources, there is a prominent contradiction between supply and demand of water resources. Thus, the suppliers and consumers need to take the initiative to communicate and unify the recognition, energetically promoting the contradiction between supply and demand of water resources and importance and necessity of water scheduling to make sure that all government departments and water – use institutes truly realize the difficult of obtaining every drop of water from the Yellow River and furthest achieve their understanding, support and cooperation to create a harmonious social environment for water supply. The development of water supply from the Yellow River can effectively alleviate the intense conflict of water shortage in districts along the Yellow River, increase the agricultural irrigation area, raise the guaranteed rate of urban and industrial water use and improve the ecological environment. Especially after Xiaolangdi Reservoir has been put into operation, with outstanding achievements for unified water control of the Yellow River, the reliability and stability of diverting water from the Yellow River have been enhanced, which shows a broad prospect for the future cause of water supply from the Yellow River.

References

Chen Chaoyang. Sample Analysis on Sustainable Development of Water Resources of Yellow River [EB/OL]. Water Conservancy and Hydropower Market of China, 2008(02).

Shan Lun. Current Status and Future Prospect of China's Water – saving Agriculture [EB/OL]. People's Net, 2005(08).

Wang Hongqian. Water Supply from Yellow River Downstream[M]. Beijing: China National Photographic Art Press, 2009.

Zhao Yi. Strengthen Water Saving of Agriculture and Enhance Efficiency of Water Utilization [EB/OL]. Water Resources Journal of Henan, 2004(05).

The Use of Wastewater in Irrigation as a Way for Fresh Water Protection

Yedoyan T. V. , *Hovsepyan G. S.* and *Qalantaryan M. A.*

Yerevan State University of Architecture and Construction, 105/1a Teryan Street,
Yerevan ,0009, Republic of Armenia

Abstract:The main sources of pollution of surface waters are untreated or insufficiently treated household and industry wastewater; meteoric and other waters from polluted air basin which are polluting the surface waters with a wide range of various pollutants, thus violating their ecobalance. They can contain toxic matters that are difficult to identify and the harmful affect of which are not yet studied.

Having a good understanding of the above statements, still, the unsparing operation of water facilities goes on, thus irreversibly changing their natural rotation with different pollutants. Maintenance of the ecobalance of surface waters is a critical problem that requires an immediate solution. Use of wastewater for irrigation purposes can partially solve the problem; it involves more targeted use of the scarce stock of fresh water, while at the same time ensuring reliable supply of water rich in nutritients for farmers. Such exchange has also a positive environmental effect since it leads to decrease in wastewater runoff to the natural waters as well as to accumulation of nutrients contained in wastewater in soil.

Currently, use of wastewater is largely practiced throughout the world; and this enables to bring closer the demand existing in different spheres of water use to the existing supply in low-water areas. Motivations for wastewater use are certainly different in developed and developing countries; however the growth in the number of world population, lack of sufficient food, global warming, water scarcity, and the growing level of pollution of the nature are global problems that make the treated water a valuable resource. Therefore, by applying treated wastewaters, not only environmental problems but also problems of household and economic significance are solved.

The paper considers modernization of water distribution, water quality and standards, legal and institutional framework, and financial mechanisms for the purpose of enhancing the efficiency of wastewater use in agriculture; development of comprehensive planning system for setting up bases for wastewater using process that will enable to determine and clearly present the core standards of management system, identify the problems and clarify the ways for solution thereof, and define the functions of agencies involved in the process.

Key words: wastewater, agriculture, irrigation, planning process, development programme

The main sources of surface water pollution are raw or not sufficiently treated domestic and industrial wastewater, atmospheric water of polluted air basins, flood water submerging agricultural land, communal wastewater of suburban villages etc. These sources carry various polluters causing surface water contamination and thus giving rise to ecological disbalance. In most cases they include even such toxic substances which are difficult to reveal and their dangerous influence has not yet been studied.

The pollution of water basins in the first place negatively influences on their self – cleaning and biological balance and also on the useful microflora of the basins biocenosis. As a result the quantity of useful microorganisms is rapidly reduced in 1mL water. In addition their species composition is also changed and reduced. At that potentially hazardous microorganisms are simultaneously developing very actively in the polluted water.

Even fully realizing all this, just the same, uncontrollable use of water bodies continues causing irreversible change of their natural alternant order change by different polluters. Conservation of surface water ecological balance is an urgent problem waiting for its immediate solution. The use of treated or not sufficiently treated wastewater for irrigation purposes is a partial solution to that problem, which gives opportunity to economically use of fresh water limited resource at the same time providing the supply of water with nutritive for irrigated lands. Such exchange has its positive environmental influence as it reduces the flow of grey water to natural fresh water rivers and basins and accumulation of nutritive elements contained in wastewater in the soil.

The typical content of microelements in communal wastewater is given in Tab. 1.

Tab. 1   The typical content of microelements in communal wastewater

Microelement	Designation	Concentration	Type of wastewater flow		
			Average density	Scarce	Very scarce
Nitrogen, N (g/m^3) general	C_{Ngen}	80	50	30	20
Ammonium *	S_{NH_4}	50	30	18	12
Nitrite	S_{NO_2}	0,1	0,1	0,1	0,1
Nitrite	S_{NO_3}	0,5	0,5	0,5	0,5
Organic	S_{Norg}	30	20	12	8
Phosphor, P (g/m^3) general	C_{Pgen}	14	10	6	4
Orto – phosphor	S_{po4}	10	7	4	3
Polyphosphor	S_{pph}	0	0	0	0
Organic	C_{porg}	4	3	2	1

Note: * $NH_3 + NH_4^+$.

It is seen from the table that wastewater contains a certain quantity of nutritive elements necessary for the normal growth of plants.

Tab. 2 presents concentrations of microorganisms in communal wastewater before and after biological treatment.

Tab. 2   The change of the concentrations of microorganisms in the wastewater

Microorganisms	Concentration 100 mL. in wastewater	
	Initial	After biological treatment
E. coli	10^7	10^4
C. perfringens	10^4	3×10^2
Faecal streptococcus	10^7	10^4
Salmonella	200	1
Campulobacter	5×10^4	5×10^2
Listeria	5×10^3	50
Staphylococcus aureus	5×10^4	5×10^2
Coliphages	10^5	10^3
Giardia	10^3	20
Roundworms	10	0.1
Enterovirus	5,000	500
Rotavirus	50	5
Suspended matter (mg/100 mL)	30	2

It is obvious that after biotreatment wastewater is sufficiently clean from undesirable microelements, suspensions, etc. and can be used for irrigation purposes.

Taking into account the growing scarcity of fresh water resources the use of urban wastewater for agricultural purposes can cover essential needs especially in arid and semidry climate. Despite the fact that in majority of developing countries legal norms limiting or inhibiting the usage of raw wastewater have been adopted or are underway, the use of raw wastewater remains everyday practice.

Presently, the use of wastewater enables nearing water demand in various water using spheres to the present-day supply. Certainly, the motives of wastewater use are different in developed and developing countries, but the increase of population on Earth, lack of food, global warming, and scarcity of water resources and troubling high level of environmental pollution are problems common for all countries and make the treated wastewater a valuable resource. Hence, usage of treated wastewater solves not only environmental but also problems of household and economic significance.

It enables to save water resources, mineral and organic fertilizers as well as to raise productivity of land.

The usage of treated wastewater for irrigation is a state policy in a number of countries (Israel, Jordan, Peru, Mexico etc.). These countries have hundreds of water basins designed for collecting and reusing of wastewater of which nearly 70% is urban wastewater.

If poorly purified household and communal water doesn't meet agricultural standards and therefore doesn't provide enough safety, it would be preferable to use it for irrigation of cereals, forage and technical cultivated plants, as well as for green tree planting and lawns.

There are some regions in the Republic of Armenia where water resources are scarce; therefore wastewater treatment has become imperative. It helps not only solving some irrigation problems, but also problems of surface water purity preservation.

For instance, in several villages of the Republic of Armenia – Geghanist, Khachpar, Lake Yerevan water mixture with nearby village's domestic and communal wastewater is used for irrigation. This water contains not only nutrients, but also large amount of pathogens. Thus, it is forbidden to use this water to cultivate lands where are growing potatoes, berries and pot – herb. Instead of this, large part of population has built green houses in the yards where they use this water to cultivate roses. Comparatively pure water they use for citrus plants, particularly for lemon, onto lemon tree grafted grapefruit.

To provide reliable water supplying system and development of agriculture in the country the wastewater usage should be based on determination of some main principals for development of drainage sphere, including irrigation of agricultural land using wastewater. To this end an appropriate program should be created and implemented.

Therefore, the aim of drainage sphere development program should be designed to clarify development directions and main approaches for the following years (decades).

To define main approaches we should take into consideration:

(1) Development and improvement of drainage services, including technical equipment and staff training.

(2) Development and improvement of secure water supply and drainage services in rural regions.

(3) Registration of wastewater accumulation sources, introduction and development of wastewater leakage accounting system.

(4) Development of drainage infrastructures.

(5) Reconstruction and establishment of new structures of wastewater treatment plants.

(6) Creation of efficient mechanisms for wastewater initial treatment.

(7) Introduction of modern technologies in procedures of wastewater treatment.

(8) Improvement of drainage systems management mechanisms.

(9) Involvement of required amount of investment to organize proper work of infrastructures and future development of the sphere.

(10) Implementation and determination of an economic mechanism to promote investment of wastewater generation systems and procedures.

In setting up key approaches for agricultural land irrigation using wastewater, the following priorities should be taken into consideration:

(1) Providing irrigation water of required quality and needed quantity (including raw and treated water).

(2) Providing reliable and stable water supply for irrigation.

(3) Essential reduction of water loss in irrigation water supply systems.

(4) Applying incentive mechanisms in employing irrigation water saving state – of – art technologies.

Development of a comprehensive planning system is the most appropriate approach for establishing the base for wastewater use. The definition of realistic approaches for such a planning will make it possible to decide and clearly present the basic standards of management system, discover problems and clarify the ways of their solution, establish the functions of bodies involved in the process. Fig. 1 shows general stages of planning process.

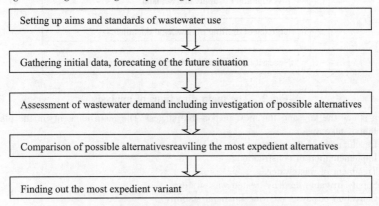

Fig. 1   The general stages of planning process

In the first stage of planning the initial data is gathered, registration of potential water users is implemented, determination of their water demand, as well as water users' willingness to use treated wastewater and pay against used water in accordance with statutory tariff. With regard to this, it is necessary to gather information both about existing water use volumes and water springs, and water users who are ready to irrigate using treated wastewater and about their water need.

Based on the above information the following data are gathered and entered the database on each water user's level:

(1) The water spring which is used for irrigation at a given moment, the kinds of irrigable crops.

(2) The surface of the irrigable land.

(3) The position of the land on the map.

(4) The need of water in recent years (at least last 3 years).

(5) The schedule of maximum need for irrigation water.

(6) The reliability of irrigation water supply.

(7) The willingness of irrigation using wastewater.

(8) The need of a local equipment installation necessary for wastewater use and the required expenses.

(9) The future trends of land – use, possibility of irrigable land expansion.

(10) The future trends of crops change.

Data having been gathered are analyzed and respective documents processing performed according to the presented in Fig. 2 standards of directions.

Now on the ground of gathered and analyzed data we can start implementation of planning by comparing possible versions and choose the most feasible one. Then the implementation stage be-

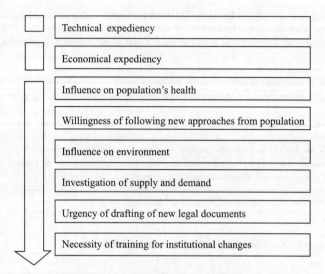

Fig. 2　The main directions of analysis

gins which is one the key and the most responsible stages. The main elements of the planning in implementation stage are:

(1) Development of the project.

(2) Implementation of construction.

(3) Treatment of wastewater.

(4) Supply irrigable land by treated wastewater.

(5) Collection of pays, securing financial stability.

(6) Rendering farmers technical assistance.

(7) Analyzing, monitoring, and evaluation.

In general, wastewater, removed through a sewage system, contains household grey water, industrial wastewater, and flood water. The wastewater from its initiation point to an irrigable land passes through a number of structures in particular a sewage system, wastewater treatment plant, treated or partially treated wastewater supplying system, and then is used for irrigation of the agricultural land.

One of the most important problems of planning process is the analysis of wastewater supply and demand and the balance analysis. At that unsteadiness of wastewater flow should be taken into account, and particularly:

(1) Municipal wastewater quantity fluctuation during a day reaching its maximum value in the daytime and the minimum at night.

(2) Rain water also flows through a sewerage network and in rainy seasons quantity of wastewater flow is increased.

(3) Possible increase of wastewater flow in periods of active tourism.

(4) Seasonal industrial branches also may discharge their waste to the sewerage.

The last stage of planning process deals with establishing of well-defined legal relationships and forming of necessary papers for clarification of involved sides functions and establishing relation among different sides.

The aim of the country drainage system development program is to make clear direction and key approaches for coming years. The strategic aim of the program is to implement reliable and safe drainage, improve environmental characteristics, reduce the environment and, in particular, water resources pollution rate through organizations producing drainage and wastewater treatment services.

The use of treated wastewater for irrigation purposes can solve a number of household, economic and environmental problems. It will enable to keep water resources as clean and save as possible. To implement comprehensive observation of drainage, wastewater treatment, and irrigation when wastewater is used, it is necessary to form new managerial and organizational ~ structural approaches. In this regard the suggested planning mechanism for the agricultural land irrigation using wastewater will enable to implement reliable and safe irrigation.

References

Odegaard H. Norwegian Experiences with Chemical Treatment of Raw Wastewater[J]. Water Science Technology, 1992, 25 (12): 255 - 264.

M. Ghence, P. Armoes, I. Lya - Kur - Yansen. E. Arvan Wastewater Treatment[J]. Moscow, 2006: 480.

Door J., Ben - Josef N. Monitoring Effluent Quality in the Hypertrophy Wastewater Reservoirs Using Remote Sensing. - Appropr. Waste Manag. Techno. Dev. Countries: Techno. Pap. Present // 3rd Int. conf., Nagpur, Feb., 25 ~ 26, 1995, T 1. - Bombey, 1995, 199 - 207.

Shuravin A. V., Vorobyova R. P., Davidov A. S., etc. Municipal Wastewater and Prospects of Their Usage in Agriculture of West Siberia[J]. Water, Ecology and Technology, 2000:595 - 595.

Lotosh V. E. Utilization of Wastewater and Sludge/ Scientific and Technical aspects of environment protection/ All - Russian Institute of Scientific and Technical Information - 2002, N6.

A Cost Benefit Analysis Based Approach on Soil and Water Conservation
—A Case Study of the Yanhe Project

Wang Feng[1] , *Yang Juan*[2] , *Niu Zhipeng*[1] and *Du Yajuan*[1]

1. Department of Soil and Water Conservation, YRCC,
Zhengzhou ,450003, China
2. Yellow River Engineering Consulting Company, YRCC, Zhengzhou, 450003, China

Abstract:Soil and water conservation is an important policy to protect environment and a-bate impoverishment in the Loess Plateau. Though it is known that there are various bene-fits from the implementation of soil and water conservation, the evaluation is mainly based on the physical terms. It is often difficult for making a trade – off between the costs and the total benefits of soil and water conservation because they are not compared using one single unit. Furthermore, the discounting costs and benefits over time are not considered in related researches. Therefore the main purpose of this thesis is to evaluate the costs and benefits of soil and water conservation using a Cost Benefit Analysis (CBA) based ap-proach. In order to conduct this study, a soil and water conservation project at the Yanhe in the Loess Plateau area of the Yellow River Basin is selected as a study case. The re-sults of analysis are described firstly in physical terms. Then, the net present value (*NPV*) of the benefits, costs and net benefits are calculated using one unit, the monetary unit. The positive value of the *NPV* of the net benefits implies that the Yanhe project is an economically feasible project.

Key words:soil and water conservation, economic analysis, cost benefit analysis, net present value,the Yanhe project,the Loess Plateau

The Loess Plateau part of the Yellow River Basin, covering a total land area of 640,000 km^2, is the most seriously erosive area in the world. The soil erosion-prone area is 454,000 km^2, 70.9% of the total land area here. Out of this total erosive area, the portion subject to intensive water erosion constitutes 64.4% in total in China, and that subject to extremely intensive water e-rosion takes up 89% of the national amount. Serious water and soil losses cause poor vegetation, frequent natural hazards and disasters, low and unstable crop yields, hinders the sustainable devel-opment of the regional society, and results in impoverishment and environmental deterioration. With a long-term average sediment yield as high as 1.6 $\times 10^9$ t, the lower Yellow River has become a world famous "suspending river" due to the deposit of sediment in channel of the downstream Yellow River, which puts the North China Plain in a severely dangerous position.

Soil and water conservation is an important policy to control soil loss. Since 1950s, soil and water conservation on the Loess Plateau has made significant progress and large areas have been controlled. Until 1997, the areas of 180,800 km^2 have been primarily controlled in which 3.08 $\times$ 10^6 hm^2 were terraced; 358,500 hm^2 of dam land and floodplains established; 10.06 $\times 10^6$ hm^2 of afforestation completed; 2.33 $\times 10^6$ hm^2 of grassland sown, more than 3 $\times 10^6$ small-scale soil and water conservation works such as water-holding cellars, ponds, check dams and gully head protec-tion works constructed (Shen et al. , 2001).

These conservation measures have significant ecological, economic and social benefits for the Loess Plateau and the Yellow River. According to the statistical data of 2005, due to these conser-vation measures, the annual incremental yield of crops and fruits are 1.0 $\times 10^7$ t and 7.7 $\times 10^5$ t re-spectively, and the annual incremental income is 2.7 $\times 10^9$ Yuan in the Loess Plateau area. These measures also have significant impacts on reducing sediment and water flow into the Yellow River, annual 3.00 $\times 10^8$ t and 3.1 $\times 10^9$ t respectively in average according to the data provided by the Yellow River Conservancy Commission (YRCC). Soil and water conservation can also improved the local (on-site) and the downstream (off-site) environmental situation, such as an increase in vege-tation cover rate and the improvement of water quality. Besides, soil and water conservation can also cut the peak flow and increase the normal flow of rivers, reduce flood hazards and increase biodiversity.

Those benefits mentioned above are presented in terms of the Code of "Method of Benefit Calculation for Comprehensive Control of Soil Erosion". In this Code, the benefits of soil and water conservation are divided into four parts: basic benefits (soil erosion control and water retention), ecological, economic and social benefits (Jiao et al. , 1996). As a part of these four benefits, the economic benefits only include the values of goods and services that can be traded directly in the market place, not include the benefits come from environmental improvement and social progress of a project area and from impacts on the downstream. Furthermore, the discounting cost and benefit over time is not considered in the analytical procedure. Those economic benefits hence could not reflect the overall benefits of soil and water conservation. Moreover, it is difficult for making a trade-off between the costs and the overall benefits of soil and water conservation because three important aspects are left out in economic evaluation. Due to these reasons, the main purpose of this study is to integrate the benefits and costs in at a broad scale and evaluate the economic efficiency of a finished soil and water conservation project by using a Cost Benefit Analysis (CBA) based approach.

1 Description of research case

In order to conduct this study, the Yanhe project is selected as a case. The Yanhe River is the first class tributary in the middle reach of the Yellow River with $7,725$ km^2 basin areas. A project of soil and water Conservation was implemented in the middle sub-catchments involved 3 counties (Ansai, Baota and Yanchang Counties), and 30 Xiangs (township), 485 administrative villages. The project area covers an area of $3,034$ km^2. $2,900$ km^2 of the area is subject to water and soil loss. The location of the Yanhe project in the Yanhe River Basin is shown in Fig. 1. Gullied loess hillyland is the main landscape. Soil erosion type is the intensive water erosion with soil erosion modulus of $10,004$ $t/(km^2 \cdot a)$. The total population is $214,500$ in 1993, including $193,400$ in the agricultural sector.

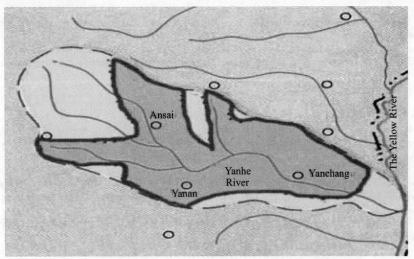

Fig. 1   Location of the Yanhe project in the Yanhe River Basin

The objectives of the Yanhe project contain three aspects. The first objective is to increase the agricultural production and farmers' incomes in the project area. The second objective is to improve the environmental situation in the project area. The third objective is to reduce sediment discharge into the Yellow River (Lei et al. , 1997). The construction took 8 years, from 1994 to 2001. In the end of 2001, the total investment is $3.922,2 \times 10^8$ Yuan; the total harnessed area is

1,118. 562 km^2.

In the end of the project, the slope land decreased 48,624 hm^2. As a result of this decrease, 16,982 hm^2 slope land is reclaimed as terrace land; 11,124 hm^2 is reclaimed as orchard land; 305 hm^2 is reformed as irrigated land (due to construction of water pumping station) ; 20,213 hm^2 is afforested as forest land, including 15,584 hm^2 arbor forest and 4,629 hm^2 shrub forest. The unused land is decreased 63,232 hm^2. As a result, 36,689 hm^2 is afforested, including 9,736 hm^2 arbor forest and 26,953 hm^2 shrub forest; 26,103 hm^2 is grassed as grass land; 439 hm^2 is reformed as dam land (the new land before a check dam or key dam formed by sediment deposition). There are 38 key dams (a large dam is built in a gully in order to impound run – off and deposit sediment) and 366 check dams (a small dam is built in a gully in order to impound run – off and deposit sediment) and 25 water pumping stations constructed. Besides, 40 hm^2 nursery is constructed for supplying fruit trees for orchard, and a 100 t fruit storehouse is built for storing fruit.

Due to the implementation of soil and water conservation, there are many effects on the project area (on-site effects) and the downstream (off-site effects) areas of the Yellow River. The on-site effects include: the reduction of soil loss; the improvement of soil properties; the retention of water within the soil and the project area; the increase in the basic farmland (terrace land, dam land and irrigated land) , orchard land, forest land and grass land; the decrease in the slope land and unused land; the increase in vegetation cover rate (arbor, shrub and grass) ; the improvement of micro – climate condition and environmental situation. Since the reduction of sediment and water flow out from the project area and the Yanhe River, the magnitude of sediment and water flow into the Yellow River is decreased too. As a result, less sediment deposited in the downstream reservoirs and channel of the Yellow River; less water provided for agriculture, industry and living in the downstream Yellow River. These are the off-site impacts of this project.

2 Research methodologies

2.1 Two scenarios

This study is based on two scenarios: without soil and water conservation measures (Scenario 1: the baseline scenario) and with soil and water conservation measures (Scenario 2). Scenario 2 is a real situation that soil and ware conservation measures are implemented in a project area; while Scenario 1 is an assumed situation that soil and water conservation measures are not implemented in the same area.

2.2 Market price method

In this thesis, the direct economic benefits or costs are those benefits or costs that can be traded in the market place, so that they can be evaluated based on the market price. This kind benefit or cost is described in Eq. (1):

$$B/C = Y \times P \tag{1}$$

where, B is the benefits of one goods that can be traded in the market; C is the costs of one goods that can be traded in the market; Y is the yield of one goods that can be traded in the market; P is the price of one goods that can be traded in the market.

In this project, the goods that can be traded in the market place include the crops, the fruits, the woods and the grasses, et al.

2.3 Replacement cost method

In this study, many benefits or costs of soil and water conservation, such as the environmental benefits and costs, are the non-market benefits or costs. They can not be traded in the market place. In order to monetize those benefits or costs, the replacement cost method is used in this study. By this method, the non-market benefits or costs is replaced by the value of goods that can be trade in the market place. As a result, the Eq. (1) is used to evaluate the non-market benefits or costs indirectly. In this thesis, the benefits of soil erosion control, the reduction of sediment deposited in the downstream the Yellow River and the reduction of water flow in the downstream the Yellow River are evaluated by this method.

2.4 Cost benefit analysis

In this project, the net benefits will be evaluated using the Net Present Value (NPV). The NPV is described in Eq. (2):

$$NPV = \sum_{t=1}^{T} \frac{B_t - C_t}{(1 + r)^t} \tag{2}$$

where, B_t is the benefits of year t; C_t is the costs of year t; and r is the discount rate.

There are T years in total. The NPV is the total benefits in the overall analysis period. The social discount rate should be used in this project because the funds come from the government.

3 Results

3.1 The on-site benefits

3.1.1 The on-site direct economic benefits

For the direct economic benefits, they can be directly evaluated in terms of market prices. In this soil and water conservation project, the direct economic benefits include the increase in crops, fruits, timbers, firewood and grasses production. These benefits come from the increase in the basic farmland, orchard land, arbor tree land, shrub tree land and grass land, and the increase in land productivity that is induced by soil loss control and water-holding. Because no data for distinguishing the benefit of the increase in land area and that in land productivity, the benefit caused by these effects will be evaluated in a whole. According to the calculation of the yield increment of grain, fruit, wood and grass, it is known that the total yield of grains, fruits, wood and grasses increased 87,469 t, 30,779 t, 133,920 t and 1,476,536 t respectively in the whole construction period (see Tab. 1). All those results are from the implementation of the Yanhe project.

Tab. 1   The yield increment of grain, fruit, wood and grass   (unit: t)

Year	Grain	Fruit	Wood	Grass
1994	1,130			43,794
1995	5,340			79,449
1996	14,065			129,116
1997	14,299	469	6,044	167,208
1998	16,631	3,283	12,896	215,893
1999	17,650	6,241	25,195	248,865
2000	17,296	8,746	35,070	262,854
2001	1,058	12,040	54,716	329,357
Total	87,469	30,779	133,920	1,476,536

For the on-site direct economic benefits of the Yanhe project, except for the first two years, they are positive and continuously increasing in the construction period. The total direct economic benefit reaches 2.8775×10^8 Yuan without discounting (see Tab. 2). However, it is lower than the total investment of the project (3.9222×10^8 Yuan without discounting). At the same time, it can be expected that the direct economic benefits will exceed the investment one day because of the lasting effects of each soil and water conservation measure. Furthermore, some measures do not work effectively, such as fruit trees and forest trees are in their infant stage, little economic benefits are gained from them. The third reason is other economic benefits are not included in the estimation of the difference of the economic benefits and the investment, such as the indirect economic benefits and the environmental benefits which could be monetized indirectly.

Tab. 2　The on-site economic benefits of the Yanhe project

(unit: $\times 10^4$ Yuan)

Year	Terrace	Key dam & check dam	Water pumping station	Orchard	Afforestation	Grassing	Fruit storehouse	Sub-total
1994	405	15	15	−527	−496	229		−359
1995	950	59	24	−804	−700	397		−74
1996	1,793	125	93	−1,213	−728	617		687
1997	1,882	135	125	−1,334	−792	762		778
1998	2,530	189	118	1,514	−1,767	937	40	3,561
1999	2,721	197	134	3,828	−1,452	1,027	40	6,495
2000	2,930	265	129	5,693	−1,802	1,029	40	8,284
2001	2,324	238	110	9,058	−3,586	1,220	40	9,404
Sub-total	15,535	1,222	747	16,216	−11,323	6,218	160	28,775

Note: " − " means "the economic benefits are negative".

3.1.2　The on-site environmental benefits

Soil and water conservation can improve local environment condition. Firstly, it can prevent surface soil loss, control soil nutrient loss and improve soil chemical and physical characteristics due to soil erosion control. Secondly, it can increase the normal water flow and decrease the peak flow in gullies due to water-holding. Thirdly, it can increase the plant photosynthesis amount. Fourthly, it can improve local micro − climate, mitigate natural disaster, and improve local air quality condition. Lastly, it can improve local water quality condition, increase local biodiversity and improve natural landscape. However, only the benefit of soil erosion control is monetized in this thesis for the lack of data.

Due to the implementation of the Yanhe project, the degree of soil erosion is mitigated. Tab. 3 shows the decrement of soil loss during the project period. The average reduced amount of soil loss is 5,640,042 t/a; the total amount is 4.51×10^7 t, and the soil erosion modulus drop from 10,004 t/($km^2 \cdot$ a) to 8,145 t/($km^2 \cdot$ a) from 1994 to 2001. This means the Yanhe project can effectively prevent surface soil from eroding.

Tab. 3　The reduced amount of soil loss

(unit: t)

Year	Terrace	Key dam and check dam	Orchard	Afforestation	Grassing
1994	205,190	1,621,620	174,903	40,131	67,830
1995	389,804	4,929,725	322,719	145,902	143,285
1996	1,359,512	8,627,018	705,155	651,431	344,049
1997	214,360	2,335,133	152,257	315,523	128,188
1998	1,316,838	2,594,592	705,531	1,577,461	567,515
1999	517,068	3,243,240	284,392	961,676	331,049
2000	738,648	324,324	487,135	1,389,619	465,851
2001	1,630,263	64,865	951,707	3,149,072	945,752
Total	6,371,683	23,740,517	3,783,799	8,230,815	2,993,519

Since soil contain large amount of nutrient, such as nitrogen (N), phosphorus (P), potassium (K) and organic matter, soil loss will result in nutrient loss. The replacement cost approach will be used to monetize the value of soil nutrient loss. The data of N, P and K is available in this study. Because potassium is the rich element in the Loess Plateau, its loss could not induce more fertilizer input for crop growth. In contrary, nitrogen and phosphorus are the limited elements in the Loess Plateau, and their losses could induce more fertilizer input for crop growth. Based on analyses above, the values of N and P are used to represent the total value of soil nutrient in this thesis. According to the data of the reduced amount of soil loss, the total N and P content of soil in the project area and the market price of urea and super – phosphate, the reduced amount of soil nutrients loss and the corresponding economic benefits are calculated as a result (see Tab. 4, Tab. 5 and Tab. 6).

Tab. 4 and Tab. 5 show that the total reduced amount of soil nitrogen and phosphorus loss are 21,659 t and 56,400 t respectively during the project period. This means the Yanhe project can saved 78,059 t nutrients in order to achieve the same level of soil fertility. As a result, it can decrease farmers' expenditure on farming or increase the productivity. From Tab. 6, it is known that the total economic benefit of the reduced amount of soil nutrient loss is $6.058,5 \times 10^8$ Yuan during the construction period without discounting.

Tab. 4 The reduced amount of total nitrogen loss in soil (unit: t)

Year	Terrace	Key dam and check dam	Orchard	Afforestation	Grassing
1994	98	778	84	19	33
1995	187	2,366	155	70	69
1996	653	4,141	338	313	165
1997	103	1,121	73	151	62
1998	632	1,245	339	757	272
1999	248	1,557	137	462	159
2000	355	156	234	667	224
2001	783	31	457	1,512	454
Total	3,058	11,395	1,816	3,951	1,437

Tab. 5 The reduced amount of total phosphorus loss in soil (unit: t)

Year	Terrace	Key dam and check dam	Orchard	Afforestation	Grassing
1994	256	2,027	219	50	85
1995	487	6,162	403	182	179
1996	1,699	10,784	881	814	430
1997	268	2,919	190	394	160
1998	1,646	3,243	882	1,972	709
1999	646	4,054	355	1,202	414
2000	923	405	609	1,737	582
2001	2,038	81	1,190	3,936	1,182
Total	7,965	29,676	4,730	10,289	3,742

Tab. 6   The economic benefit of the reduced amount of nutrient loss in soil

(unit: $\times 10^4$ Yuan)

Year	Terrace	Key dam and check dam	Orchard	Afforestation	Grassing
1994	276	2,177	235	54	91
1995	523	6,619	433	196	192
1996	1,825	11,584	947	875	462
1997	288	3,135	204	424	172
1998	1,768	3,484	947	2,118	762
1999	694	4,355	382	1,291	445
2000	992	435	654	1,866	626
2001	2,189	87	1,278	4,228	1,270
Total	8,555	31,877	5,081	11,052	4,019

3.2   The off-site benefits

Due to the implementation of the Yanhe project, much soil and water are held within the Yanhe River Basin. As a result, less sediment and water are discharged into the Yellow River. The effects of less sediment and water flowed into the Yellow River is an off – site benefits or costs from the viewpoint of the Yanhe project. In this study, the off-site benefits include: a decrease in sediment deposit, an increase in normal water flow, a decrease in the peak flow, the improvement of water quality and the improvement of waterscape in the downstream Yellow River. All they are the off-site environmental benefits. Only the benefit of the reduction of sediment deposited in the downstream Yellow River is considered in this thesis due to the limitation of data.

According to the research on boundary division of coarse sand concentrated source region of the middle Yellow River (Xu, et al. , 2006), it is known that about 62.3% of the amount of eroded soil which come from the middle Yellow River is deposited in the downstream Yellow River. Tab. 7 shows the decrement of the sediment deposited in the downstream Yellow River due to the implementation of the Yanhe project. The total decrement of the sediment deposited in the downstream Yellow River is 2.81×10^7 t during the construction period. As a result, the benefit of the reduction of sediment deposit could lower the rising speed of the riverbed and improved the waterscape of the downstream Yellow River.

Tab. 7   The decrement of the sediment deposited in the downstream Yellow River

(unit: t)

Year	Terrace	Key dam and check dam	Orchard	Afforestation	Grassing
1994	127,833	1,010,269	108,965	25,002	42,258
1995	242,848	3,071,219	201,054	90,897	89,267
1996	846,976	5,374,632	439,312	405,842	214,343
1997	133,546	1,454,788	94,856	196,571	79,861
1998	820,390	1,616,431	439,546	982,758	353,562
1999	322,133	2,020,539	177,176	599,124	206,244
2000	460,178	202,054	303,485	865,733	290,225
2001	1,015,654	40,411	592,913	1,961,872	589,203
Total	3,969,559	14,790,342	2,357,307	5,127,798	1,864,962

Because dredging is a main measure to remove the sediment deposited in the downstream channel, the replacement cost of dredging measure is used to calculate the benefit of the reduction of sediment deposited in the downstream Yellow River in this study. Tab. 8 shows the economic benefit of the reduction of sediment deposited in the downstream Yellow River according to the replacement cost of dredging measure. The total economic benefit is 3.373×10^8 Yuan from 1994 to 2001 without discounting. This implies that the same investment would be saved when dredging measure is carried out to remove the same quantity of sediment.

Tab. 8 The economic benefit of sediment reduction in the downstream Yellow River
(Unit: $\times 10^4$ Yuan)

Year	Terrace	Key dam and check dam	Orchard	Afforestation	Grassing
1994	153	1,212	131	30	51
1995	291	3,685	241	109	107
1996	1,016	6,450	527	487	257
1997	160	1,746	114	236	96
1998	984	1,940	527	1,179	424
1999	387	2,425	213	719	247
2000	552	242	364	1,039	348
2001	1,219	48	711	2,354	707
Total	4,763	17,748	2,829	6,153	2,238

3.3 The on-site costs

According to the analyses, it is known that the on-site costs equal the investment in the Yanhe project. The costs include not only the investment in each construction measure, but also the investment in the supporting measure of the project. Tab. 9 shows the on-site cost of the project. The total on-site cost is $3.922, 2 \times 10^8$ Yuan from 1994 to 2001 without discounting.

Tab. 9 The on-site cost of the Yanhe project (Unit: $\times 10^4$ Yuan)

Year	Cost	Year	Cost
1994	5,020	1998	3,008
1995	8,130	1999	3,492
1996	6,889	2000	3,740
1997	3,474	2001	5,469
		Total	39,222

3.4 The off-site costs

The implementation of the Yanhe project decrease not only the sediment deposited in the downstream, but also the water flowed downstream. The effect of the reduction of water flow downstream is a negative effect for the downstream Yellow River because it reduced water availability downstream. Therefore, this effect can be considered as the cost to the downstream according to the cost benefit analysis based approach.

Tab. 10 shows the decrement of water flow into the downstream Yellow River. The total decrement of water flowed into the downstream Yellow River is 1.43×10^6 m^3 during the construction period. As a result, the reduction of water could decrease the water use for living, irrigation and industrial production in the downstream Yellow River.

Tab. 10   The decrement of water flowed into the downstream Yellow River

(unit: m³)

Year	Terrace	Key dam and check dam	Orchard	Afforestation	Grassing
1994	1,046,180	3,891,888	890,094	892,258	0
1995	1,646,409	11,831,340	1,404,544	1,648,097	24,990
1996	3,990,256	20,704,844	2,467,811	3,728,432	17,850
1997	2,154,318	5,604,319	1,497,293	2,583,014	288,571
1998	5,155,545	6,227,021	2,937,453	6,051,711	220,925
1999	3,289,127	7,783,776	1,834,541	4,568,411	626,566
2000	4,541,898	778,378	2,973,588	6,434,820	802,667
2001	6,758,800	155,676	4,176,945	10,787,691	919,369
Total	28,582,533	56,977,242	18,182,269	36,694,434	2,900,938

Because the priority of irrigation water is lower than other two water provisions according to the situation of water use in the Yellow River. Therefore, it is assumed that the water reduced is used for irrigation. Tab. 11 shows the economic loss of the reduction of water flowed into the downstream Yellow River according to the economic loss of the irrigated crops. The total economic loss, also the cost of the downstream benefits, is 7.45×10^7 Yuan from 1994 to 2001 without discounting.

Tab. 11   The economic loss of the reduction of water flowed into the downstream Yellow River

(unit: ×10⁴ Yuan)

Year	Terrace	Key dam and check dam	Orchard	Afforestation	Grassing
1994	54	202	46	46	0
1995	86	615	73	86	1
1996	207	1,077	128	194	1
1997	112	291	78	134	15
1998	268	324	153	315	11
1999	171	405	95	238	33
2000	236	40	155	335	42
2001	351	8	217	561	48
Total	1,486	2,963	945	1,908	151

3.5   The *NPV* of the project

The benefits and costs of the Yanhe project are evaluated in physical terms and in monetary unit without discounting above. In this section, the benefits and costs of the Yanhe project will be evaluated in one unit, the monetized benefits and costs, and the *NPV* is used for evaluating the economic efficiency in this thesis. According to Eq. (2), the social discount is applied in the Yanhe project, and equalled 6% according to the Code of "Methodologies and Parameters for Economic Assessment of Construction Projects" (Yang, et al., 2006). The construction of the Yanhe project lasted eight years, from 1994 to 2001. Furthermore, the data that are collected from 1994 to 2001 are available in this thesis. Hence, the eight-years are applied as the evaluation period in this thesis. The baseline year for discounting is the first year (i. e. 1994) of the project implementation.

The NPV of the benefits, costs and net benefits are calculated and are shown in Tab. 12. The NPV of the total monetized benefits, costs and net benefits are 9. 235,6 × 10⁸ Yuan, 3. 682,5 × 10⁸ Yuan and 5. 553,1 × 10⁸ Yuan respectively. The NPV of the total monetized benefits of the Yanhe project are the sum of the NPV of the monetized benefits from the increase in grains, fruit, wood and grasses yield, the decrease in soil erosion in the Yanhe project area and the decrease in sediment deposited in the downstream Yellow River. The NPV of the net monetized benefits of the Yanhe project is positive.

Tab. 12   The NPV of the benefits, costs and the net benefits of the Yanhe poject
(unit: ×10⁴ Yuan)

Year	The NPV of the benefits	The NPV of the costs	The NPV of the net benefits
1994	3,822	5,066	-1,244
1995	10,969	8,002	2,967
1996	21,089	7,134	13,955
1997	5,824	3,251	2,573
1998	13,223	3,048	10,175
1999	12,444	3,125	9,319
2000	10,244	3,024	7,219
2001	14,742	4,175	10,567
Total	92,356	36,825	55,531

Since a lot of benefits could not be monetized according to the analyses in Section 3. 1 ~ 3. 4, the real NPV of the net benefits is underestimated. All those calculation and analyses results imply that the Yanhe project is a feasible project in economics.

4   Conclusions

By summarizing all the results and analyses above, it is known that the implementation of the Yanhe project could increase grains, fruits, wood and grass yield, improve the environmental situation in the local area and so on. The implementation of the Yanhe project could also decrease the sediment deposited in the downstream Yellow River. However, the implementation of the Yanhe project could decrease the water flowed into the downstream Yellow River. This effect is a negative effect on the downstream Yellow River. The NPV of the net monetized benefits is 5. 553,1 × 10⁸ yuan in the construction period, which implies that the Yanhe project is an economic feasible project.

References

Jiao J R, Liu W Q, Xu C Z, et al. GB/T 5774—1995 Method of Bnefit Calculation for Comprehensive Control of Soil Erosion[S]. Beijing: Chinese Standards Press, 1995. (in Chinese).

Lei Z Y, Wang X J, Feasibility Research of the World Bank's Loan Project of Soil and Water Conservation on the Loess Plateau[M]. Zhengzhou: Yellow River Conservancy Press,1997. (in Chinese)

Xu J H, Lin Y P, Wu C J, et al. Research on Boundary Division of Coarse Sand Concentrated Source Region of the Middle Yellow River[M]. Zhengzhou: Yellow River Conservancy Press, 2006. (in Chinese)

Shen G F, Wang L X. Water Resource Strategies Reports of Chinese Sustainable Development [M]. Beijing: Chinese Water Publication, 2001. (in Chinese)

Yang Q W, Wang X T, Luo G S, et al. Methodologies and Parameters for Economic Assessment of Construction Projects[M]. Beijing: Chinese Planning Press, 2006. (in Chinese)

Discuss on Management of Water Conservancy Planning in the Yellow River

Yang Huijuan[1], *Zhao Yinliang*[1], *Song Huali*[1], *Wen Hongjie*[1] and *Zhai Ran*[2]

1. Department of Planning and Programming, Yellow River Conservancy Commission (YRCC), Zhengzhou, 450003, China
2. School of Earth Sciences and Engineering, Hohai University, Nanjing, 211100, China

Abstract: The work analysis the management status of water conservancy planning in the Yellow River, and sum up the main problems, as well as discussing solutions. Water conservancy planninng classification and their interrelation are analyzed firstly. Secondly, planning for water resource devolopement of the Yellow River Basin is reviewed. Thirdly, present situation and problems of water conservancy planning management are analysised, through discussing process management and system establishment. Then, under the new situation that water conservancy reform and development progress rapidly, challenges for are researched, and how to prove management level of water conservancy planning in the Yellow River are put forward.

Key words: water conservancy planning, management, the Yellow River

Water conservancy planning, is planning about water resources developing, utilizing, protecting, managing and controlling water disasters, according to demand of national economic and social development and water resource sustainable utilization. Water conservancy planning is important means to carrying out their duty for department of water administration at all levels, is important foundtion of water conservancy public service and social management, and is important basis of making water conservancy construction plan and making water resource management system and policy.

1 Water conservancy planninng classification and their interrelation

Water Law of the People's Republic of China divided water coservancy planning into river basin planning and regional planing. River basin planning, as well as regional planning, include comprehensive planning and specialty planning.

In 2010, Ministry of Water Resources (MWR) made and published *Measures of Water Conservancy Planning Management (Trial Implementation)*, which divided water resource planning into national, river basin and regional planning. National planning include strategic planning, development planning and special planning. River basin planing inculdes comprehensive planning, specialty planning and sepcial planning. The planning for a region within a river basin should be subordinated to the planning for the river basin, and the specialty planning shoule be subordinated to the comprehensive planning.

Strategic and development planning often focus on macro targets and missions such as water conservancy reform and development direction, guiding ideology, overall scheme, strategy and so on (Sun Pingan, et al., 2008). Stategic planning period is often 20 ~ 30 years and be revised every 10 ~ 15 years. Development planning period is often 5 ~ 10 years and be revised every 5 years. Comprehensive planning focus on water resources development, utilizing, saving, protecting and preventing water disasters. Its planning period is often 15 ~ 20 years, and be revised every 10 years. Specialty planning focus on flood control, waterlogging control, irrigation, navigation, water supply, hydroelectric generation, bamboo or log rafting, fishery, water resources protection, soil and water conservation, sand control, saving water and so on. The planning period is often 15 ~ 20 years, and be revised every 10 years. Special planning focus on important water resources reform and construction, crucial projects and so on. The planning period is determined according to specific situation.

Thus it can be seen that water conservancy planning classification is made on the basis of space, time and speciality scale (Luo Chengping and Zhu Shikang, 2005). River – basin water conservancy planning management scope covers river basin planning and planning for a region within a river basin (in Fig. 1).

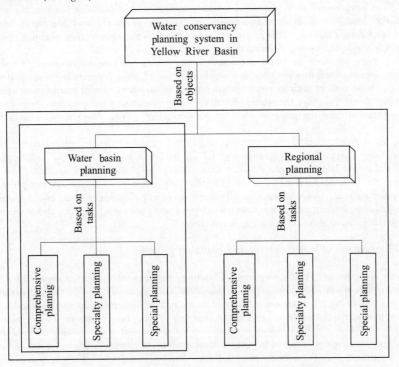

Fig. 1 Water conservancy planning classification

2 Water conservancy planning management status quo and problems in the Yellow River Basin

The Yellow River is one of the most difficult management valleys, because of less water resource, more sand, and inharmonious between water and sand. In history, Jia Rang, Wang Jing, and Pan Jixun put forward methods for the Yellow River harnessing. Since the foundation of the P. R. China, the Yellow River Basin management have improved greatly.

Four river-basin comprehensive planning have been made in the last 60 years. The first planning is *Economic and Technical Report of Comprenhensive Utilizing Planning in the Yellow River* in 1945, which is the first main river basin planning and is approved by the Second Session of the first National People's Congress in 1955. The second planning is *Outline of the Yellow River Plan for the Management and Development* made in 1997, and reviewed by former State Development Planning Commission and Ministry of Water Resources. The third planning is *Management and Development Plan in the near future of the Yellow River*, approved by the State Council in 2002. The forth planning is *Comprehensive Planning of the Yellow River*, which is going to be approved by the State Council.

Outside these comprehensive planning, many specialty planning and special planning have been made and taken great role in the improvement of the river-basin management.

2.1 Development of water conservancy planning management in the Yellow River Basin

As the water conservancy planning be made, management of planning has beening explored. On the one hand, rules of planning formulating, review, approval, and implement have been worked out (Zeng Zhaojing, 2009). On the orther hand, water conservancy planning system has been put forward, to improve planning management.

But, there are many problems about water conservancy planning management. Firstly, prospective, overall research is not enough when formulating planning. Secondly, some planning are absent, or some content such as resources allocation, environment, social management and public service are absent. Thirdly, the relationship among existing planning are not clear. Sometimes, two or more planning have the same content but not consistent (Peng Zhongfu and Huang Huajin, 2010).

In resent years, there is more attention to water conservancy planning magagement. *Measures of Water Conservancy Planning Management (Trial Implementation)*, came up with setting up a system within which different planning are harmonious, complementary in function. The *Central Document No. 1* in 2011 presented to the scientific preparation of water conservancy planning, to improve the national, river-basin, regional water conservancy planning system, and strengthen water conservancy planning on water activity management and constraint function. This marks the water conservancy management coming into a new stage.

2.2 Management of water conservancy planning process

Both *Measures of Water Conservancy Planning Management (Trial Implementation)*, and *Measures of investment plan management of pre-phase work* explicitly provide the process of water conservancy planning. A water conservancy planning can be launched through the plan design assignment formulation, submission, review and approval. Then, the department of water ministration under the State Council, basin administration, and the department of water ministration of the local people's governments at or above the county level shall be in charge of different planning formulation in accordance with the duty assigned to them. And pre-phase work, launch, public participation, demonstration, approval, publish, assessment of a water conservancy planning is strictly required (Yang Weimin, 2004). It can be seen that process management of planning is ruled preliminarily, and have main problems such as following.

2.2.1 Preparation and approval of a planning related to many sectors and their high requirements

A water conservancy planning is related to many sectors, such as relevant regioal sectors of economy, society, industry and environment. Approval of a planning should be get agreement of relevant local government, relevant ministries and commissions under the State Council. So, formulating planning should take opinion and suggestion from relevant sectors such as development and reform, land resource, immigration, urban and rural planning, forestry, environmental protection, wetland protection, cultural relics protection into account.

2.2.2 Poor executive force and weak management of water conservancy planning

Water Law of the People's Republic of China and Measures of Water Conservancy Planning Management (Trial Implementation) require that planning approval, must be strictly enforced, and highlight the important role of river basin administration. But no implement scheme to support (Zhao Pengmin, et al., 2007). Government at all levels is the enforcement principal body. And river basin administration has no effective mean of supervision.

2.3 Water conservancy planning system in the Yellow River Basin

According to the duty of Yellow River Conservancy Commission (YRCC) of the Ministry of

Water Resources (MWR), the preliminary water conservancy planning system has been set up. The system emphasis on comprehensive planning and specialty planning, special planning contained. The system include 68 planning. Among these planning, 12 planning have been approved, 13 terms have been reviewed, 21 terms have been finished formulation, and 22 terms are being formulated.

Although the preliminary water conservancy planning system of the Yellow River Basin has been set up, there is lots of work should be done to improve it. As just 4 planning have been approved by the State Council. Among these 4 planning, *Management and Development Planning in the near future of the Yellow River* is valid before 2010, and other 3 planning are specialty planning. *Comprehensive Planning of the Yellow River*, *Construction Planning for Regulation System of the Yellow River Water and Sediment*, *Planning for Integrated Management in the Yellow River's Downstream Beaches*, and *Planning for Safty Construction in the Yellow River's Downstream Beaches* should be approved by the State Council as soon as possible. Planning for soil and water conservancy in the Yellow River Basin, *Planning for Water Resources Protection in the Basin*, and some other planning should be formulated or revised.

3 Development situation of water conservancy planning management in the Yellow River Basin

3.1 Water resources reform and devolopment

Central Document No. 1 in 2011, is the first comprehensive policy on water conservancy. It present the main task of water resources reform and development in the next 5 ~ 10 years. In this situation, the Yellow River Basin management has a good opportunity to improve much, but there are many challenges as well. In resent years, situation of water and sediment has changed a lot, and economic and social development make more new demand for river-basin management. At the same time, development of YRCC make more demand for planning management.

Therefore, in the process of formulating, idea and thought of basin management should be improved continually. Water resources devolopment, utilizting, allocation should be combined with water saving, protection, maintaining river eco-environment. River-basin management pays more attention to demand of economic and social development in relevant regions. Long-term goal and typical phase goal should be taken into account together. The interest of upstream and downstream areas, the left and right bank areas, main stream and branches areas should be taken into consideration together. Engineering and non-engineering measures should be considered together. And to construct water and sediment regulation system, flood control and sediment reduction system, water resources integrated management and dispatching system, water quality monitoring and protection system, soil and water conservation system as soon as possible.

3.2 More requirement from economic and social development in relevant regions

The Yellow River Basin, having large area, has abundant enegy resources. And relevant regions within the basin have great potential. In the last years, national development strategy has made in these areas. For example, Qaidam Recycling Economic Zone, Efficient Ecological Economic Zone of the Yellow River Delta, Guanzhong—Tianshui Economic Region, Central Plains Economic Region, etc. in the Major Function Oriented Zoning Planning approved by the State Council in 2010, Taiyuan City Cluster, Hu—Bao—Er—Yu Region, Economic Region along the Yellow River in Ningxia, Lan—Xi Economic Region are proposed, too. All of these strategies will improve economy and society in relevant regions, at the same time, ask more for flood control, water supply, ecosystem conservation, etc. Therefore, in the process of formulating, ability of water resources sustainable supply for these important zones should be taken into consideration.

3.3 Demand of the Yellow River management

There are problems to be solved at present. Firstly, task of flood or ice run control is still severe. "secondary suspended river" grow rapidly in the downstream of the Yellow River and do a lot of harm. The safety precautions are incomplete yet. Secondly, the contradiction between water supply and need becomes more and more serious. Production water supply squeeze ecological environment water supply. Thirdly, soil erosion is serious. Fourthly, water pollution is serious; Fifthly, integrated management of basin is weak. Thus, in the management of water conservancy planning, main tasks are as below: the first, do anything to make sure that *Construction Planning for Regulation System of the Yellow River Water and Sediment*, *Planning for Integrated Management in the Yellow River's Downstream Beaches*, and *Planning for Safety Construction in the Yellow River's Downstream Beaches* will be approved as soon as possible. The sencond, carry out *Planning of Yellow River Water Resources*, and formulate *Yellow River Water Resource Allocation Plan*, *Planning of the Yellow River Water Supply and Demand in the Medium and Long-term* quickly. The third, formulate *Planning of soil and water conservation in the Yellow River Basin* quickly. The fourth, formulate *Planning for Water Resources Protection in the Basin quickly. The fifth*, strictly implement *Measure of Water Project Construction Consent for Planning*.

4 Working thought of improving planning management in the Yellow River Basin

4.1 Perfect planning system

Circling around water and sediment regulation system, flood control and sediment reduction system, water resources integrated management and dispatching system, water quality monitoring and protection system, soil and water conservation system, strengthen top-design, and formulate more scientific. The first of all, crucial planning must be approved and carried out as early as possible. Then, some specialty planning, such as *Planning of soil and water conservation and Planning for Water Resources Protection in the Yellow River Basin* should be worked out as quickly as possible. And some regional planning such as Hongjiannao Lake, Sanjiangyuan Region, and Heihe River Basin, Huangshui River Basin, Wudinghe Basin, Yiluohe Basin, Kuyehe River Basin, Taohe Basin, Qinhe Basin should be worked out quickly. At the same time, start work on planning for some river basins which have important problems.

4.2 Strengthen Implementation of Planning

Formulation aims to provide scientific guidance for water conservancy development. So planning need to be implemented into practice. In the present situation, guarantee system of investment, policy, law, technology, talents should be researched and improved (Liu Haijiang, 2007). River basin administration organization should work hard to make sure planning quality. Meanwhile, *Measure of Water Project Construction Consent for Planning* issued in 2007, is a good mean to make sure that water project construction conforms to the requirements of planning.

4.3 Enhance Mass Participation

In the process of pre-phase demonstration, formulation, review of a water conservancy planning, it is necessary to ask for advise from experts in relevant field, and ask for advise from the public widely. It can be done to help the public get more knowledge about planning through announcement, propaganda material, lectures, etc (Zhao Wei, et al. , 2003). Besides, administrative organization, representative of interests of relevant sectors and closely related planning, should be communicated closely and asked for opinion and suggestions.

5 Conclusions

At present, water conservancy planning system has just been set up, and management of planning has a lot work to do. Next, aiming at solving the main problems of less water resource, more sand, and inharmonious between water and sand, seek solutions to increase water, reduce sediment, and regulate water and sediment. And work hard to improve water conservancy planning system, conducive to construction of water and sediment regulation system, flood control and sediment reduction system, water resources integrated management and dispatching system, water quality monitoring and protection system, soil and water conservation system.

References

Sun Pingan, Xi Sixian, Yang Xiaoru. Research on Water Conservancy Plannings Syestem Framework in Shanxi Province[J]. Shanxi Water Conservancy and Hydropower Technology, 2008 (1):1 –5.

Luo Chengping, Zhu Shikang. Research on Water Conservancy Planning's Syestem in Pearl River [J]. Pearl River, 2005(4):11 – 14.

Zeng Zhaojing. Water Conservancy Planning in the Last 60 Years[J]. China Water Resources, 2009(19):13 – 15.

Peng Zhongfu, Huang Huajin. Discussion on Water Planning System in New Period[J]. China Science and Technology, 2010(6):194 – 195.

Yang Weimin. Four Issues to Be Solved to Innovate Planning System[J]. Water Resources Planning and Design, 2004(S1):19 – 22.

Zhao Pengmin, Liu Yan, Li Guanghua. Research on River – basin Water Conservancy Planning Syestem[J]. Water Resources & Hydropower of Northeast China, 2007(3):11 – 12.

Liu Haijiang. Discussion on Status and Role of Water Conservancy Planning in New Period[J]. Shaanxi Water Resources, 2007(2):20 – 22.

Zhao Wei, Yin Huaiting. Research on Public Participation in City Plannig[J]. Journal of Northwest University (Social Science), 2003(4):75 – 78.

Strategy Research for Water Resources Sustainable Utilization in Ordos

Peng Shaoming, *He Liyuan* and *Jing Juan*

Yellow River Engineering Consulting Co. ,Ltd. , Zhengzhou, 450003, China

Abstract: Along with the rapid development of regional economic, and the acceleration of industrialization and urbanization process in Ordos, a serious disharmony occurred between the eco-social development and the water resources utilization and protection. Ordos meets deep-seated problem in water resources and which is becoming the key bottlenecks element of regional development. With the core of improving water use efficiency and the main line of opening up new water sources and rational using water resources, this study analyze and optimize different water demand methods under a variety patterns of water consume of the regional economy departments in Ordos. On the basis of three equilibrium of water supply and demand, we put forward a rational water resources allocation scheme in planning level years, and design a new sustainable use pattern of water resources in Ordos. The strategy provide a good support for regional eco-social development, and realize the goal of supporting the sustainable, coordinated and healthy development of regional economic society.

Key words: Ordos, strategy system, sustainable utilization of water resources, three equilibrium, Configuration scheme

1 Introduction

Ordos Cis located in the southwest of Inner Mongolia, and it is situated in the upper and middle reaches of the Yellow River, with the area of $86,800 \text{ km}^2$. The Yellow River runs around west, north and east sides of Ordos City, of which the west, north and east sidesaresurrounded by the Yellow River, isadjacent to Shanxi provincein the east side, Shaanxi provincein the south side and Ningxia Autonomous Regionin the west side, which is the main source of Ordos' water supply. The north of Ordos City is separated from Baotou City which is called "Grassland Frank" and Hohhot City by the Yellow River.

In history, Ordos is major of agriculture and animal husbandry, and the output value of agriculture and animal husbandry accounted for 60% of GDP inbefore the early 1990s. Since 2000, along with the implementation of the Chinese western development strategy and regional coordinated development of economic policy, the pace of economic development is rapid increase in Ordos and it has maintained sustained and rapid development momentum in recent years. After ten years of development, Ordos has become one of the most active areas in China. The GDP of Ordos has reached 216.1×10^9 Yuan in 2009 and the per capita GDP was 1,320,000 Yuan. With the development of regional economic, the water demand of Ordos is keeping growth, and the problem of supply and demand is increasingly prominent. The sustainable use of regional water resources meets great challenges.

2 Main issues of water resources in Ordos

2.1 Abundant economic developed quickly with rich mineral resources, exuberantwater resources demand grow rapidly

Ordos is located in the upper reaches of the Yellow River, and it is the economic zone junction of Ningxia, Inner Mongolia, Shanxi and Shaanxi and the center area of the economic belt along the Yellow River. Ordos City has abundant coal resources with good quality, and it is the most important high-quality coal base in China. In the $87,006,800 \text{ km}^2$ area land of Ordos City, the coal-bearing areas account for about 70%, with the proven coal resources amount to 149.6×10^9, and total reserve is accounted for 1/6 of the total national coal reserve, with accounting for 80% of high-quality thermal coal reserves to maintain in China. The natural gas is rich in Ordos, and the natural gas proved reserves is about 1/3 of the country, with the Sulige gas field proved reserves of 750.4 $\times 10^9$ m^3, which is the largest packaged gas field in China. The richemrand Ordos is listed as one of the important energy production bases in China.

In recent years, with the industrial structure adjustment and upgrading in the country, the acceleration in the pace of construction of Western energy and heavy chemical industry base, the exploitation of coal resources, and a number of major projects which have settled in Ordos City, it has provided a good strategic opportunity and policy environment for taking advantage of resources, undertaking industrial transfer, and optimizing the industrial structure. The regional economy in Ordos has been rapid development, especially in energy and coal chemical industry. The GDP in 2009 reached 216.1 $\times 10^9$ Yuan which is 13 times of 2000, with the average annual growth rate is 24%. The rapid development industry and economic in Ordos has brought about the rapid growth of water demand, and industrial water demand has increased from 2.16 $\times 10^8$ m^3 in 2000 to 2.45 $\times 10^8$ m^3 in 2009.

On one hand, the increasing industrial water demand has brought more stringent requirements to the water supply guarantee rate and quality. On the other hand, it takes more pressure on water resources and water environment because of the increasing waste water discharge. Because of the poor water resources in Ordos City, it has become the key factor which restrictsthe economic development, and water resources sustainable useutilization in water resources has has become the key factor which restricts the harmonious development and faced challenges in Ordos City.

2.2 Because of water resources Shortage, there is sharp contradictory between supply and demand

Ordos is a typical temperate continental climate. The annual average precipitation is 265.2 mm, with the evaporation of 1,500 ~ 2,200 mm, and the climate is drought. According to evaluation, Ordos annual runoff (1956 ~ 2009) is 1.18 $\times 10^9$ m^3, and the unit area water resource is equal to 1/5 of the average level of the Yellow River Basin. The Yellow River is the main water source in Ordos, however, the mean annual watershed indicator is only 0.7 $\times 10^9$ m^3.

With the rapid development of economic and the rapid growth of population, agriculture and ecological construction, the excessive extraction of groundwater is serious and shortage of water resources has become increasingly evident.

Because of the rapid economic development, population growth and urban-scale expanding, the city's water resources development and utilization is increasing year by year. By statistics, the total water consumption in Ordos was 1.946 $\times 10^9$ m^3 in 2009, and it had reached the available water resources in general. Currently, it is very difficult to find new sources of water for the new industrial projects. Cones depression in some areas have formed due to the irrational exploitation of groundwater which lead to the water level dropped, and it is a serious threat to urban water security.

Ordos is in a critical period of industrialization accelerated developing and economic transformation in the next two decades. Along with economic development, especially accelerated construction of the energy and heavy chemical industry base, the demand for water in Ordos will show a rigid growth. The contradiction between water supply and demand has become increasingly sharp, and water shortage will restrict the development of economic and social.

2.3 Irrational use of water resources exacerbate the contradiction between supply and demand, a further optimization is needed

The water supply source in Ordos is mainly dependent on the Yellow River and groundwater, with a small portion of unconventional water, so the source of water supply is very single. The surface water supplied 0.785 $\times 10^9$ m^3 in 2009, which is close to the Yellow River water allocation indicator. The groundwater supplied 1.08 $\times 10^9$ m^3, and in some areas ground water was excessively extracted. The recycled water and other unconventional (including sewage treatment and reuse of mine water and rain) water supplied $74 \times 10^6 \ m^3$, accounting for only 3.8% of the total water supply.

The present use water structure is irrational with large agricultural water use and industrial water use take up a small percentage (compare with it output is insufficient) in Ordos. According to statistics, Ordos agricultural water use in 2009 was $1.577 \times 10^9 \ m^3$, accounting for 81% of the total water consumption, and the contribution to the GDP accounted to only 2.8%. Industrial water consumption was only $2.45 \times 10^9 \ m^3$, accounting for less than 13% of the total water consumption, and it didn't match to the positioning of the energy and chemical of the area.

The water use efficiency is not high, with the waste of water and excessive emissions, and the water consumption of per unit GDP and 10,000 Yuan of industrial added value is much higher than the advanced regions and developed countries. Some work has been done in agricultural water saving in recent years, but the standard of agricultural water-saving projects is low, and flood irrigation is still the main irrigation method. The waste of water resources is serious with the industrial water recycling rate of only 70% and urban water supply network loss rate exceeding 20%.

2.4 Fragile ecological environment and poor anti-interference ability

In Ordos, the Mu Us Sandland and Kubuqi desert takes 48% of the total area while hilly and gully region and arid hard Liang takes 48% of the total area, so the natural conditions is poor and the ecological environment is fragile. Ordos is in the interlaced zone of the climate transition zone and the ecological environment. The ecological environment has a significant transition and volatility characteristics, so it is extremely sensitive to climate change and human disturbance in response. It will cause the loss of important ecological functions and the natural recovery cycle is relatively long once by damaged.

2.5 Water management is disorder, and cannot meet economic and social development

Solving the shortage of water resources in Ordos need an overall consideration, comprehensive response and strengthened management, and implementing the most stringent water management system. Considering the current situation of water resources allocation and management in Ordos, the management of the development and utilization of water resources belongs to different departments, and the development and utilization of surface water and groundwater are managed by the departments of hydraulic engineering, mining, agriculture, urban construction respectively. The construction, scheduling and management of hydraulic engineering are controlled by different departments and levels of government. The "multi-sectors manage the water resources" model results in the fragmented ownership, unclear responsibilities, waste of manpower and the phenomenon of mutual buck-passing. Because of these factors, the limited water resources cannot be reasonable configuration and unified supervision, resulting in confusion in water resources management, predatory development and the adverse situation aggravated by water pollution. The problem of water resources in Ordos has become increasingly prominent, and it has been difficult to adapt to the needs of economic and social development, and it is contrary to the most stringent water resources requirements management proposed in current.

3 The main goal of strategy research

By reasonably allocating water resources, improving the regional water use efficiency and the water supply system, it can improve the support of capabilities water resources for sustainable economic and social development, and ensure regional economic and social development. The specific objectives can be divided into the following five aspects.

3.1　Water-Using quantity control targets

Setting regional watershed indicator of the Yellow River as control indicator of the surface water use, the exploitable volume of groundwater as upper limit of developing, establish a water management system combining total control and quota management. Under the premise of protecting the economic and social development and improving the ecological water use, the recent water consumption in 2015 is controlled in less than 2.2×10^9 m^3, and the city's water consumption is controlled at around 2.4×10^9 m^3 in 2020.

3.2　Water supply security targets

By 2015, the urban and rural drinking water safety system will be completed, water shortage situation in main areas will be effectively alleviated, urban and rural residents will use the safe and clean drinking water. By 2020, basically establish the city's water security system, the ability to resist drought will be remarkably improved, and water safety will be protected effectively.

3.3　Water conservation and efficient use targets

According to comprehensively promoting water-saving society and changing water use methods, improve water use efficiency and effectiveness, and guide the optimization of industrial structure. By 2015, the city's water consumption per 10,000 GDP will be reduced to around 40 m^3, more than 55% lower than the status quo, the proportion of water consumption of the three industries will be adjusted from the status quo 86:13:1 to 66:32:2. By 2020, water consumption per 10,000 GDP will be reduced to under 25 m^3, which is lower by 40% compared to 2015, the water structure has been further adjusted to 57:41:2.

3.4　Ecological protection and restoration targets

By suppressing over-development of water resources, transforming unreasonable use method, reasonable allocating life, production and ecological water, establish ecological water security systemto maintain the normal function of the river. By 2015, water ecological environment in main areas, the over-exploitation region and the vulnerable areas will be significantly improved, and the wetland ecological water of rivers and lakes will be basically guaranteed.

3.5　Water resources management targets

By 2015, improve integrated water resources management system, control water volume and emission volume , clear the initial water rights, allocate initial water rights, and build a reasonable price formation mechanism, water resources and aquatic ecosystem protection operating mechanism and water resources management information system, initially realize the real-time monitoring, optimizing scheduling and digital management of water resources. By 2020, establish water management system with water rights management as the core. Water right trading market will be sound, mature, and smoothly circulate. Establish an economic structure system in harmony with the water resources carrying capacity, form a reasonably water consumption method and a conscious water-saving mechanism of the whole society.

4　Main measures

4.1　Optimize industrial scale and layout

Optimize the industrial scale, structure and layout, and adjust the patterns of water consumption, and strictly control the total amount of water. Consider the shortage of water resources and sharp contradiction between water demand and supply in Ordos, we should schedule water supply according to the water volume.

In agriculture, prohibit the blindfolding expansion of irrigated area, and actively adjust the planting structure, and develop the agriculture without increasing water consumption. For industry, determine a reasonable industrial scale according to water resources conditions, actively promote the adjustment of industrial structure and layout, enhance water demand management, and control the development of high water consumption industries, strictly control the water growth rate.

4.2　Strengthen water-saving in the whole course

In agriculture, to promote the efficient water-saving agriculture development, it will take some measures, such as developing efficient ecological agriculture, taking reasonable adjustment of crop layout and optimization of crop structure, speeding up the irrigation district water saving, developing field water saving efficiency projects and promoting advanced water-saving technologies, developing pastoral areas according to local conditions, vigorously developing the water-saving agriculture, actively promoting the fruit industry and breeding industry water saving.

In industry, to develop as new path of industrialization and improve the efficiency of industrial water, it will take measures, such as determining a reasonable industrial scale, optimizing layout of regional industry and increasing the readjustment of industrial, vigorously developing the circular economy and promoting of advanced water-saving technologies and water-saving technology, strengthen the business plan water and internal water management, active use of unconventional sources of water, organization and implementation of major water-saving technology development and demonstration projects. For urban water use, strengthen the construction of urban water supply network and promote the penetration of water saving appliances.

4.3　Open water sources by more channels

For Ordos, which is located in an arid and semi-arid place, it is important to allocate and operate a variety of water resources, to achieve the virtuous circle of a variety of water resources and the efficient use of local water resources and cross-boundary water resources, to establish a reasonable allocation system of water resources which is built with the center of surface water and groundwater and increase the low-quality water use intensity.

4.4　Reasonable deployment of water

The planning establishes the rational allocation pattern of water resource and urban and rural safe water supply protection network system to guarantee reasonable needs of water resources for sustainable economic and social development, according to the water resources carrying capacity in Ordos, supporting sustainable economic and social development, maintaining the sound development of regional ecological environment as a starting point, overall considering of economic, social and ecological environment, water relations between the industry and other departments, jointing deployment Yellow River, the territory of rivers, groundwater, and unconventional water sources.

4.5　Strict management of water resources

The most stringent water management system and the integrated management of water supply is implemented on the basis of water resource allocation pattern. Improve the water permit and water proof system. Through the total control and quota management, enhance the macro-control measures to economic and social development and ecological environmental protection effected by water permit and water rights allocation. The water quota is enacted by total control of the approved flag and regional total water. The planning establishes reasonable water pricing system and achieves the flow of water to the efficient direction by economic mechanism, to achieve structural water-saving.

5 The planning effects

5.1 Promoting water-saving society

After the implementation of the sustainable use of water resources conservation program, the u-tilization coefficient of agricultural irrigation water in Ordos increased from 0.65 in 2009 to 0.79, and irrigation water consumption reduced by 5,130 m^3/hm^2 in 2009 to 4,095 m^3/hm^2 in 2030. The industrial water recycling rate increased from 70.5% at percent to 93% in 2030, and the industrial million increase value reduced to 10.52 m^3 in 2030 from 21.68 m^3 at the present year. Urban water supply pipe network leakage rate reduced to 10.9% from 21%. In Ordos, water-saving social man-agement system, economic structure system coordinated with the carrying capacity of water re-sources and water environment, engineering and technical system of water resources rational alloca-tion and efficient use, and standard system of conscious water-saving social behavior are initially es-tablished to promote water conservation and water efficiency. The water saving amount is 4.02 $\times$ 10^8 m^3.

5.2 Establish water use pattern adapted to the economic and social development

Considering of the water shortage of Ordos , the planning proposes the water resources use general idea with combination of water-saving and open-source and the joint deployment of a variety of water. The general pattern of the planning is developing in water-saving, strengthening diversion works from the Yellow River, effective regulation and storage, optimal allocation, rational use, and alleviating the contradiction between water supply and demand in a certain period of time.

5.2.1 Optimize the rational use mode of the regional water resources

The planning compares the various patterns of water use in the two levels of regional develop-ment patterns and water resources development and utilization, and optimal selects a recommended program after the water supply and demand balance analysis. The regional scale of the industry and industrial layout is defined by the "First supply and demand analysis" of water s supply and de-mand, the use of water resources scheme is proposed by the "Second supply and demand analy-sis", and the direction of the medium-term use of water resources pointed out by the "three equi-librium of water supply and demand". The planning reflects the reasonable water demands of the economic and social development in Ordos for a long period of time, and it can promote the harmo-nious development with nature. By the planning, the recent water demand will be controlled at 2.2 $\times 10^9$ m^3 in 2015 and 2.4 $\times 10^9$ m^3 in 2010, and it is consistent with the social construction require-ments of "resource conservation, environment-friendly" and achieves the objective of sustainable regional economic and social development.

5.2.2 Comprehensive analyses the water supply and demand situation

The contradiction between water supply and demand in planning level year is sharp. "First equilibrium" shows that the contradiction between water supply and demand is obvious and it can not support the economic and social development in planning level year in accordance with the high programming water demand. The water shortage rate will be 18.0% in 2015, 26.5% in 2020 respectively, and the water shortage volume will be 1.079 $\times 10^9$ m^3 in 2030. "Second equilibrium" shows that the water demand can be basically meet recently by the reasonable con-trol of the economical scale and optimize the industrial layout, and the level of water shortage rate will be 2.4% in 2015. The water shortage will be very serious in the year 2020 and 2030, with the water shortage rate 11.3% and 18.4% respectively. "Third equilibrium" takes the measures of water replacement and the development of efficient ecological agriculture strategy. The analysis shows that the water supply and demand will be alleviated to some extent and the water deficit will reduce to 2.45 $\times 10^8$ m^3 in 2020, with the water shortage rate of 9.9% , but it does not fundamentally change the resource problem of water shortage in Ordos. The water shortage will be 4.70×10^8 m^3 in 2030 with water shortage 17. 2% , and it will mainly be industrial water shortage. The severe shortage of water demand will affect the effective implementation of the regional development strategy, see Tab. 1.

Tab. 1 Water supply and demand equilibrium analysis in planning level year

(Unit : $\times 10^8$ m^3)

Scheme	Level year	Water demand	Water supply quantity				Volume of water shortage	Water shortage rate(%)
			Surface water	Ground water	Unconventional water	Total		
First equilibrium	2015	26.28	9.02	10.72	1.81	21.55	4.73	18.0
	2020	29.86	9.02	10.60	2.33	21.95	7.91	26.5
	2030	33.12	9.02	10.65	2.66	22.33	10.79	32.6
Second equilibrium	2015	22.09	9.02	10.72	1.81	21.55	0.54	2.4
	2020	24.75	9.02	10.60	2.33	21.95	2.80	11.3
	2030	27.37	9.02	10.65	2.66	22.33	5.05	18.4
Third supply equilibrium	2020	24.69	9.02	10.89	2.33	22.24	2.45	9.92
	2030	27.32	9.01	10.94	2.66	22.61	4.70	17.21

5.2.3 Propose the pattern of rational water resources development

From the aspects on the social and economic development, ecological environment protection, and water resources development and utilization, the planning put forward to the rational allocation of water resources. The total surface water depletion configured to Ordos is 7.00×10^8 m^3 , which is no more than the Yellow River water allocation indictor. The shallow groundwater extraction volume is 1.089×10^8 m^3 which does not exceed groundwater extraction.

In the parts of Dalate and Etuoke where the status quo is more than the exploitation, strictly control accordance with the total water exploiting, gradually implement groundwater diminishment, adjust the industrial structure, vigorously develop water-saving irrigation and water projects construction, to basically meet the reasonable water requirements of the economic and social development and the protection of the eco-environmental water. For some areas which have remained development potential on the current situation of water resources, such as Wudinghe Basin and Ten Tributaries, gradually increase the efforts on territory surface water resources, and protect the water resources, to ensure the development and utilization of water resources does not exceed the available volume.

5.2.4 Construct water conservancy system suited to the rational allocation of water resources

Ordos is lack of water resource. Most of the territorial rives are seasonal rivers and water resources is uneven distribution during the year, so the Yellow River is the main water supply source of the region. In the planning years, the regulatory capacity of water resources will be increased by building new regional reservoir in the territorial tributary. The planning will improve local water resources development and utilization, increase water project construction of the Yellow River, regulate the spatial and temporal distribution of water resources, and optimize the water resources allocation of the regional. It will also gradually refund the amount of excessive extraction of groundwater caused by economic and social water using, and build new groundwater wells in groundwater potential areas to improve the protection of groundwater level. According to increase the use of unconventional water resources engineering and technical inputs, improve the utilization of unconventional water sources and reduce dependence on surface and groundwater conventional water.

5.2.5 Establish the economic structure system coordinated with water resources carrying capacity

The first, second, third structure was 18. 2: 46. 4: 35. 4 of Ordos City in 2009. According to the effect of water resources distribution, adjust the three structures, promote the primary industry, accelerate the development of secondary industry, especially accelerate the upgrade of the industrial proportion in the secondary industry, and enhance the tertiary industry. The planning will adjust and optimize positively the internal structure of each industry, increase the proportion of secondary and tertiary industries and water according to local characteristics, and compress agricultural water properly. According to enhance the adjustment of planting structure and water consume, accelerate the formation of the grain, economic, grass ternary planting structure, and change the mode of agricultural production, three industries structure will be 0. 6: 53. 6: 45. 8 of Ordos in 2030.

5.3 Support sustainable economic and social development

5.3.1 Guarantee the urban and rural water security

The planning will speed up construction of rural water supply source by accelerating the protection, support and improve of urban water sources, water supply pipe network and water supply facilities. Urban drinking water security situation will be improved through the new water supply project measures and reasonable dispatch, and a well-off society of safe drinking water requirements will be maintained by the protection of the centralized drinking water sources. By implementation of the planning, it will meet urban water demand of 1.55×10^8 m^3 and rural drinking water demand of 4.9×10^8 m^3 in 2020, and it will provide water security in order to speed up the process of urbanization of Ordos City and the new socialist countryside construction.

5.3.2 Support the energy and heavy chemical industry base construction

Ordos City is one of the important energy and chemical base in China. Because of the regional water scarcity and the sharp contradiction between supply and demand, it is difficult to carry water demand to maintain sustainable economic and social development and ecological environment. In accordance with the priority to efficiency and taking into account the principle of equitable allocation of water resources, 5.73×10^8 m^3 of water supply will be increased to the Jungar Dalu Industrial Zone, Dalate Economic Development Zone, Hangjin Duguitala Industrial Zone, Etuoke Qipanjing and Mengxi IndustrialZone, Etuoke Front Shanghaimiao Industrial Zone, and Wushen Tuke Industrial Zone in planning level year. The increasing water supply will basically meet the new water demand in 2015, solvethe water demand of industrial development to some extent in 2020, and support the formation of the most competitive domestic energy and heavy chemical industry base of the Ordos.

5.3.3 Improve agricultural water supply guarantee rate

In planning level years, the contradiction between water supply and demand will become more prominent in Ordos City, and the agricultural water is often diverted to industrial and urban domestic water especially in dry years or dry season, which cause the agricultural water cannot be guaranteed. After the implementation of the planning, the water resource protection for increasing food production and improving income of peasants and herdsmen will be provided by increasing the region's available water supply, replacing and reducing agricultural water consumption which is diverted by industrial and domestic. By these measures, the irrigation area will increase 306 hm^2 on the basis of the status quo irrigation area.

5.3.4 Promote the regional coordinated development

Taking into account the development needs of each (area), co-ordination of various water sources, coordinating water relation between various departments and water users, the planning unify distribute water resources of the city in case of both fair and effective. The planning give priority to agricultural water and water supply projects of major grain producing areas, the lower economic level areas, and rural drinking water difficulty areas, which will accelerate the development of minority areas and underdeveloped areas and promote the coordinated regional development. The problem of uneven distribution of regional water resources and water shortage of tributary will be solved by building new storage projects in tributary, implementing the water intake project from the Yellow River and increasing the capacity of regional water supply.

5.3.5 Support the rapid economic development

The next 20 years is a critical period for Ordos City to accelerate the construction of the national energy base, to rapid the economic development mode shift, to accelerate the wealthy and strong city strategy and build a moderately prosperous society comprehensively. The planning will support the economy sustained rapid development of the city by the optimization of industrial structure, the rational allocation of water resources and efficient utilization, environmental protection and construction, and the rational allocation and efficient use of water resources. It can be foreseen that the implementation of the planning will promote the economy rapid growth momentum of the region, and guarantee to realize the goal of building a moderately prosperous society: by 2015, the per capita GDP is double the 2009 year, and the per capita GDP will reach $0.269,1 \times 10^6$ Yuan. By 2020, the per capita GDP will be more than 0.4×10^6 Yuan and realize the goal of wealthy and strong city.

5.4 Promote the harmonious development with nature

We overall plan and coordination the relationship between people with nature and watercourses within and pass through water, control the water resources development in strict accordance with the available water resources, and control the water consume and sewage by pollutants bearing capacity. The ecological condition of the river will be significantly improved while promoting economic development. The planning control the groundwater exploitation strictly and return the shallow groundwater of overexploitation water 38×10^6 m^3 in planning level year, which will realize the balance exploitation — supplement of groundwater. The planning focuses on the water of urban environment, the replenishment for wetlands and the ecological construction for water eco-grass construction, and provide water security for the construction of ecological environment. The planning can effectively reduce the amount of pollution discharged into rivers and lakes, recovery the river lake body function gradually, improve the ecological environment of rivers and lakes, ensure water needs of lake ecological and environmental, maintain river health and promote the development of man and nature.

5.5 Meet the demand for water resources management in the new era

We planning puts forward to the water resources management system, and establishes management system and mechanism which is conducive to the rational development and efficient utilization and effective protection of water resources, Includes: establishing and improving water resources management system, improving the total amount of water control and quota management system, establishing a scientific and rational price formation mechanism, improving the management system of water function zones, establishing a system of water recycling system, and establishing water resources of the emergency dispatch system et al.

6 Conclusion

The implementation of the planning will make the development pattern of water space to be clear, optimize the water structure, significantly improve the efficiency of water use, enhance the coordination with regional development, greatly promote sustainable development capacity, forma good support to regional economic and social development, and realize the target of sustainable use of water resources.

References

Wang Hao, You Jinjun. Advancement Sand Development Course of Research on Water Resources Deployment[J]. Journal of Hydraulic Engineering, 2008, 39(10): 1168 – 1175.

Li Heping. The Key Technologies research of Regional Water Resources Efficient Use and Sustainable Development [M]. Beijing: China Water Power Press, 2011.

Xia Jun, Zhang Yongyong, Wang Zhonggen, et al. Water Carrying Capacity of Urbanized Area[J]. Journal of Hydraulic Engineering, 2006,37(12):559 – 9350.

Hydraulic Research Institute of Ministry of Water Resource in Pastoral Areas. Planning Report of Water-saving Society of Ordos [R]. 2009.

Analysis of Coupling between Urbanization and Water Resources and Environment of Shiyang River Basin —A Case Study of Liangzhou District

Zhang Shengwu, *Shi Peiji* and *Wang Peizhen*

College of Geography and Environmental Science,
Northwest Normal University, Lanzhou, 730070, China

Abstract: Based on the evaluation indicator system concerning the coupling between urbanization and water resources and environment system and GRA (Grey Relationship Analysis) method, taking Liangzhou District as case study, this paper reveals the main factors influencing the coupling between urbanization and water resources and environment and the temporal distribution of the coupling degrees. The results show: ① Low social service level, rapid urbanization and irrational industrial structure are the main momentums of urbanization influencing water resources and environment, while water pollution, water shortage and low water management level are the main factors of water resources and environment affecting urbanization, which verify the interactive and complicated coupling between two systems. ② the curve of the change of coupling degrees present a combined and continuous "U" shape as the population urbanization grow, which verify the complexity between urbanization and water resources and environment systems. ③ According to its undulatory characteristics, the change of coupling degrees could be divided into two phases in time series. During the first phase the curve present a "W" shape from 2001 to 2008, which verifies the Law of Environmental Kuznets Curve(EKC). The coupling degrees are in decline during the second phase from 2008 to 2010, which verifies that the ecological deterioration had eased since 2007 when the Key Control Plan of Shiyang River Basin in Gansu Province was published.

Key words: urbanization, water resources and environment, coupling, Shiyang River Basin

1 Introduction

There is an interactive and coupling relationship between urbanization and ecological environment. the stress of urbanization on ecological environment and the restraint of ecological environment on urbanization in arid inland basins is much stronger than other places because of water shortage, so is the interactive and coupling relationship between two systems. Water resourcess is the lifeline of continental river basin in arid regions and the basic restraint condition of the urbanization development in inland river basins, while the orderly process of urbanization is the key point of sustainable development in arid regions, so it is with great theoretical and practical significance to research on the change rule of the coupling between urbanization system and water resources and environment system in inland river basin in arid region.

The interactive relationship between urbanization and water resources and environment system has drawn many researchers' eyes in China which are mainly focusing on three aspects. The first is concerning the restraint of water resourcess on urbanization, with representative outputs are on the connotation , classification of the restraint of water resourcess and the urbanization threshold and moderate urban population under the restriction of water resourcess. The second aspects are related to the city development under the restriction of water resourcess, such as the development mechanism and the characteristic of spatial distribution of city and towns in arid region and development models and etc. The last aspect is on the relationship between two systems. The corresponding mechanism between two systems , the process coordination and its characteristic are the main research points in theory discussion. In the respect of the analysis of the quantitative relationship between two systems, simple mathematical statistics are popular, while recently dynamic coupling models and GRA are also widely used.

There have been obvious research outputs on the coupling between two systems, however, analysis in this field in the innerland river basin has not been carried out. As far as Shiyang River Basin is concerned, water shortage has always been the main restraint on the economic and social sustainable development. The economic and social development and urbanization has made great achievement, however, it is still in slow process compared with other regions. Taking Liangzhou District as case study, based on the evaluation indicator system concerning the coupling between urbanization and water resources and environment system and GRA method, this paper reveals the main factors influencing the coupling between urbanization and water resources and environment and the temporal distribution of the coupling degrees from the perspective of time serial, which can provide theoretical reference to realize the coordinated coupling between urbanization and water resources and environment system in Shiyang River Basin.

2 General situation of study area

As one of the three largest inland river basins in Northwest China, Shiyang River Basin is located in the east of in the Hexi Corridor, north margin of the Qlian Mountain with continental temperate arid climate, low precipitation but strong evaporation. The administrative region of Shiyang River Basin is covering seven counties, three cities, which mainly belong to Wuwei City and Jinchang City. Nowadays an urbanization development pattern with Liangzhou District and Jinchuan District as its center has formed in Shiyang River Basin and the urban population are mainly in Liangzhou District, Jinchuan District, Hexibao Town and each administrative town of every county. Considering the situation and the availability of data, this paper takes Liangzhou District as case study as it has the largest population among districts of county administrative level in Gansu Province with its urban population at 45.04×10^4 and total population at 103.64×10^4 in 2010 year.

3 Indicator system and methods

3.1 Indicator system

Methods with single indicator would not be capable of revealing the mechanism and rule of the coupling between two systems because of the extremely complex coupling. Based upon the past research results and following the principles of scientificalness, availability dynamics and independence, a preliminary evaluation indicator system is constructed through the methods of theoretical analysis, frequency analysis and experts consultation. The final evaluation indicator system is formed after being analyzed on the relevance and principal component to solve the problems of information overlap by SPSS (to see in Tab. 1).

Tab. 1 Evaluation indicator system on the coupling between urbanization and water resources and environment system

	Functional group	Indicators	Indicators
Urbanization	Population urbanization (X_1)	Scale(X_{11})	Urban population(X_{111})
		Structure(X_{12})	Proportion of urban population(X_{121})
		Change(X_{13})	Growth rate of urban population(X_{131})
	Economic urbanization (X_2)	Scale(X_{21})	Output value of per urban built-up area(X_{211})
		Structure(X_{22})	Proportion of total industrial output value in the gross output value of industry and agriculture(X_{221})
		Growth(X_{23})	Growth rate of real GDP per capita(X_{231}), growth rate of tertiary industry(X_{232})
		Efficiency(X_{24})	All-personnel labor productivity(X_{241})

Continuaed Tab. 1

	Functional group	Indicators	Indicators
Urbanization	Social urbanization (X_3)	Life quality (X_{31})	Urban per capita living area (X_{311}), urban engel coefficient (X_{312}), health workers among ten thousands population(X_{313}), per capita public green space(X_{314})
		Infrastructure (X_{32})	Per capita communal facilities (X_{321}), density of road network (X_{322}), gas coverage(X_{323}), hand-free telephone coverage(X_{324})
		Science, education and health(X_{33})	Elementary and secondary school teachers among ten thousands population(X_{331}), proportion of science and education funds(X_{332})
	Spatial urbanization (X_4)	Amount(X_{41})	Amount(X_{411})
		Scale (X_{42})	Per capita urban built-up area(X_{421}), proportion of urban built-up area in urban area(X_{422})
		Density(X_{43})	Density of cities and towns(X_{431}), urban population density(X_{432})
Water resource and environment	Background condition(Y_1)	Abundance(Y_{11})	Urban water supply amount(Y_{111}), annual precipitation(Y_{112})
		Per capita amount(Y_{12})	Per capita water resources(Y_{121})
		Quality(Y_{13})	Water qualification rate of urban centralized drinking water quality compliance rate(Y_{131})
	Use and management (Y_2)	Use(Y_{21})	Per capita water use amount(Y_{211}), Water consumption per capita day life(Y_{212}), rate of tap water(Y_{213})
		Management(Y_{22})	Water supply pipe density(Y_{221}), drainage pipe density(Y_{222}), Water supply comprehensive production capacity(Y_{223}), actual water supply by water conservancy project(Y_{224})
	Pressure(Y_3)	Load(Y_{31})	Urban sewage emissions(Y_{311})
		Intensity of pressure(Y_{32})	Proportion of irrigation water(Y_{321}), proportion of flood area in total area damaged by natural calamities(Y_{322})
	Resistances to adverse conditions (Y_4)	Water environmental improvement (Y_{41})	Urban sewage treatment rate(Y_{411})
		Efficiency(Y_{42})	Water use amount per ten thousand GDP(Y_{421}), water use amount per ten thousand industrial added value(Y_{422}), water use amount per grain output(Y_{423})

3.2 Data sources and preprocessing

Taking urban areas as research target, data from 2001 to 2010 year are collected, among which urbanization data are collected from the annual construction of city and county of Gansu Province, economic and social development data conservancy statistics yearbook and water resources bulletin of Gansu Province. Concerning the different dimension of data, dimensionless treatment of the data is done by means of Standardization of Data Range before GRA.

For the positive index,

$$Z_{ij} = (X_{ij} - \min Z_{ij})/(\max Z_{ij} - \min Z_{ij}) \tag{1}$$

For the negative index,

$$Z_{ij} = (\max Z_{ij} - X_{ij})/(\max Z_{ij} - \min Z_{ij}) \tag{2}$$

In the above two equations, Z_{ij} is the standardization of one index; X_{ij} is the initial amount of one index; while $\max Z_{ij}$, $\min Z_{ij}$ are respectively the maximum and minimum of one index.

3.3 Data processing methods

GRA is suitable for dynamic and evolutional analysis as it can give a quantitative analysis on the development tendency of one system. This paper reveals the dynamic coupling between urbanization system and water resources and environment system in Shiyang River Basin through the modeling of relevant coupling put forwarded by Liu Yaobin .

$$\xi(j)(t) = \frac{\min_i \min_j |\ Z_i^X(t) - Z_j^Y(t)\ | + \rho \min_i \min_j |\ Z_i^X(t) - Z_j^Y(t)\ |}{|\ Z_i^X(t) - Z_j^Y(t)\ | + \rho \min_i \min_j |\ Z_i^X(t) - Z_j^Y(t)\ |} \tag{3}$$

In the Eq. (3), $Z_i^X(t)$, $Z_j^Y(t)$ are respectively the standardization score of urbanization and water resources and environment in t year, ρ is the resolution ratio, the value of which is normally equal to 0.5. $\zeta(j)(t)$ is the relevance coefficient in t year.

A matrix of coupling degrees, named γ with m lines and l columns, can be got by calculating the average value of those relevance coefficients according to the sample K, which here are the time serial samples. The correlations between each index in two systems could be analyzed by means of comparing the value of each relevance coefficient. γ_{ij} is between 0 and 1, and if it's value is bigger, the relevance is stronger, and so is the coupling between two systems, and vice versa. Taking for example, $\gamma_{ij} = 1$ indicates that $Z_i^X(t)$, one index of urbanization system, has a strong relevance with $Z_j^Y(t)$, the other index of water resources and environment system, further more the rule of changing of $Z_i^X(t)$ and $Z_j^Y(t)$ are identical. $0 < \gamma_{ij} \leq 0.35$ indicates weak relevance and weak coupling, and $0.35 < \gamma_{ij} \leq 0.65$ indicates moderate relevance and moderate coupling, and $0.65 < \gamma_{ij} \leq 0.85$ indicates stronger relevance and stronger coupling, while $0.85 < \gamma_{ij} \leq 1$ indicates extremely strong relevance and extremely strong coupling.

Eq. (4) can be got by calculating the average value of those relevancies according to lines and columns respectively, and the main factors of one system influencing another system according to its value.

$$\begin{cases} d_i = \dfrac{1}{l} \sum\limits_{j=1}^{l} \gamma_{ij} & (i = 1, 2, \cdots, m; j = 1, 2, \cdots, l) \\ d_j = \dfrac{1}{m} \sum\limits_{i=1}^{m} \gamma_{ij} & (i = 1, 2, \cdots, m; j = 1, 2, \cdots, l) \end{cases} \tag{4}$$

In Eq. (4), γ_{ij} is the relevance, and l, m are the index numbers of two systems respectively.

A modeling of relevant coupling is constructed based upon Eq. (3), through which we can give a quantitative analysis on the coupling between urbanization system and water resources and environment system. The following equation is the calculation formula.

$$C(t) = \frac{1}{m \times l} \sum_{i=1}^{m} \sum_{j=1}^{l} \xi(j)(t) \tag{5}$$

In Eq. (5), $C(t)$ is the relevance coupling between two system.

4 Results

4.1 Main factors that influence urbanization and water resources and environment system

Tab. 2 shows that the linkage between urbanization and water resources and environment system is of moderate and more type as almost each coupling degree of indicator is above 0.35, which verify that the close linkage between two systems.

Tab. 2   The matrix of coupling between urbanization and water resources and environment system(2010 year)

	Y_1 0.627 Y_2 0.625		Y_3 0.647		Y_4 0.607		$\bar{X}$											
	Y_{111}	Y_{112}	Y_{121}	Y_{131}	Y_{211}	Y_{212}	Y_{213}	Y_{221}	Y_{222}	Y_{223}	Y_{224}	Y_{311}	Y_{321}	Y_{322}	Y_{411}	Y_{421}	Y_{422}	Y_{423}
X_1 0.634 X_{111}	0.675	0.356	0.374	0.897	0.691	0.372	0.868	0.433	0.354	0.391	0.786	0.647	0.464	0.908	0.798	0.873	0.948	0.873 0.650
X_{121}	0.713	0.716	0.813	0.417	0.920	0.801	0.451	0.616	0.708	0.919	0.483	0.575	1.000	0.439	0.478	0.450	0.429	0.450 0.632
X_{131}	0.901	0.905	0.982	0.369	0.867	1.000	0.361	0.757	0.892	0.868	0.380	0.431	0.618	0.353	0.377	0.360	0.347	0.360 0.618
X_2 0.630 X_{211}	0.568	0.570	0.626	0.611	0.684	0.619	0.684	0.508	0.565	0.684	0.755	0.988	0.905	0.658	0.743	0.681	0.636	0.680 0.676
X_{221}	0.498	0.399	0.430	0.582	0.461	0.426	0.659	0.363	0.396	0.461	0.735	1.000	0.610	0.631	0.722	0.655	0.608	0.655 0.572
X_{231}	0.757	0.760	0.864	0.428	0.979	0.850	0.463	0.652	0.751	0.977	0.495	0.586	0.913	0.451	0.490	0.462	0.440	0.461 0.654
X_{232}	0.753	0.757	0.871	0.446	0.904	0.856	0.371	0.642	0.747	0.999	0.393	0.454	0.698	0.362	0.389	0.370	0.355	0.370 0.597
X_{241}	0.458	0.358	0.378	0.910	0.396	0.375	0.950	0.435	0.356	0.396	0.849	0.683	0.475	0.941	0.863	0.956	0.960	0.957 0.650
X_3 0.640 X_{311}	0.460	0.360	0.380	0.921	0.398	0.378	0.950	0.397	0.358	0.398	0.850	0.685	0.478	0.898	0.864	0.956	0.972	0.957 0.648
X_{312}	0.418	0.419	0.447	0.754	0.474	0.443	0.862	0.385	0.416	0.474	0.972	0.899	0.599	0.823	0.953	0.857	0.790	0.856 0.658
X_{313}	0.390	0.391	0.416	0.768	0.440	0.413	0.883	0.362	0.389	0.440	1.000	0.872	0.548	0.841	0.980	0.877	0.806	0.876 0.650
X_{314}	0.944	0.949	0.983	0.447	0.871	0.912	0.369	0.792	0.935	0.872	0.388	0.440	0.627	0.362	0.385	0.368	0.355	0.368 0.632
X_{321}	0.990	0.985	0.862	0.442	0.779	0.875	0.362	0.883	0.897	0.780	0.379	0.425	0.586	0.355	0.376	0.361	0.349	0.361 0.614
X_{322}	0.765	0.768	0.878	0.394	1.000	0.863	0.424	0.656	0.758	0.999	0.451	0.529	0.855	0.414	0.447	0.423	0.404	0.423 0.636
X_{323}	0.518	0.520	0.568	0.609	0.616	0.562	0.684	0.466	0.516	0.616	0.757	0.912	0.871	0.657	0.744	0.680	0.634	0.680 0.645
X_{324}	0.538	0.540	0.591	0.612	0.643	0.584	0.687	0.483	0.535	0.642	0.759	0.901	0.919	0.660	0.747	0.683	0.637	0.683 0.658
X_{331}	0.856	0.860	0.978	0.390	0.909	0.982	0.362	0.723	0.848	0.910	0.382	0.435	0.634	0.354	0.378	0.361	0.348	0.361 0.615
X_{332}	0.480	0.482	0.540	0.375	0.603	0.533	0.414	0.420	0.477	0.602	0.451	0.564	1.000	0.400	0.444	0.412	0.388	0.412 0.500
X_4 0.614 X_{411}	0.455	0.356	0.374	1.000	0.391	0.372	0.868	0.369	0.354	0.391	0.786	0.647	0.464	0.908	0.798	0.873	0.948	0.873 0.624
X_{421}	0.995	1.000	0.937	0.397	0.837	0.952	0.369	0.831	0.985	0.838	0.387	0.437	0.614	0.361	0.384	0.368	0.355	0.367 0.634
X_{422}	0.449	0.356	0.374	1.000	0.391	0.372	0.868	0.393	0.354	0.391	0.786	0.647	0.464	0.908	0.798	0.873	0.948	0.873 0.625
X_{431}	0.760	0.763	0.875	0.371	1.000	0.860	0.398	0.650	0.753	0.999	0.423	0.492	0.776	0.389	0.419	0.397	0.380	0.397 0.617
X_{432}	0.684	0.385	0.414	0.578	0.443	0.411	0.655	0.351	0.382	0.443	0.732	1.000	0.582	0.628	0.719	0.652	0.604	0.651 0.573
$\bar{Y}$	0.653	0.607	0.650	0.597	0.682	0.644	0.607	0.546	0.597	0.673	0.625	0.663	0.683	0.596	0.622	0.606	0.593	0.606 0.625

To uncover the characteristics of the interactive coupling among indicators and the main driving force in two systems, through simple average and sort on the results of the Tab. 2, the main factors of urbanization system influencing water resources and environment system and the main factors of water resources and environment system influencing urbanization system and the main relationship of each functional group of two systems.

(1) The comprehensive stress index of urbanization on the water resources and environment system is 0.629, and, sorted by the stress degree, the rank is social urbanization (0.640) > population urbanization (0.634) > economic urbanization (0.630) > spatial urbanization (0.614). At present the urbanization of Liangzhou District is in the transition process from I phase to II phase and the social development is in low development level, which could explain why the social urbanization influence water resources and environment system most. The urban population of Liangzhou District in 2010 grew 52.7% compared with that in 2001, which followed by the rapid increase in water use, thus the influence of population urbanization on water resources and environment system also stood out. In general the industry of Liangzhou District depend on water closely nowadays with irrational water use structure and low water use efficiency, so the economic urbanization, one key content of the urbanization development, give a big influence on water resources and environment system.

(2) The comprehensive relational degree of water resources and environment system on urbanization system is 0.626, which indicates that water resources and environment system has a strong restraint on the development speed, quality and progress of urbanization system. Concerning each functional group, the pressure of water resources and environment has the biggest relational degree with urbanization system at 0.647, and the following rank are the background condition(0.627),

water use and management (0. 625) and resistances to adverse conditions (0. 607) respectively. The irrigation water has been more than 80% in total water use amount for a long time leading inefficient water use structure. Meanwhile, the economic and social development and the quick urbanization has resulted in severe water pollution and thus the pressure of water resources and environment influence the urbanization system most, while the water shortage, backward water management and low efficiency in water use are also the main restraint of urbanization process in Liangzhou District.

4. 2 The time sequential variation characteristics of the coupling between two systems

The complicated coupling between two systems are not only displayed in the interactive intersects among factors in each system, but also are expressed by its periodicity in time serial evolution. Combined with population urbanization rate, the coupling degrees curve, which is drawn based upon Eq. (5), can show the time sequential variation characteristics from 2001 to 2010.

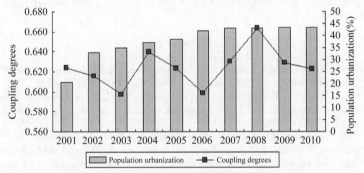

Fig. 1 The curve of the coupling degrees and the population urbanization in Liangzhou District from 2001 to 2010

The following variation characteristics of coupling degrees can be concluded from Fig. 1. :

(1) From 2001 to 2010, the coupling degrees show its obvious undulatory property as it changed from 0. 597 to 0. 663. It can be seen from the fitting curve that the variation curve of coupling degrees present a "W" (or two continuous "U") shape during 2001 to 2008, and it has declined since 2008, which demonstrates the tightness of the interactive coupling on one hand, and there are differences in intensity, key point and coordination of coupling between two systems on the other hand.

(2) There is not a linearity but a complicated relationship between population urbanization and the coupling degrees, and for more details, the change of coupling degrees present a characteristic with combined "U" shape.

(3) According to its undulatory characteristics, the change of coupling degrees could be divided into two phases in time series. During the first phase the curve present a "W" shape from 2001 to 2008. In this phase, the restraint of water resources and environment system on urbanization system declined first and increased later in general through its own feedback as the urbanization process accelerated, which also verifies the Law of EKC well. The second phase is from 2008 to 2010. The ecological deterioration has eased as the investment in ecological improvement has greatly increased since 2007 when the Key Control Plan of Shiyang River Basin in Gansu Province was published, which is being verified by the decline tendency of coupling degrees during the second phase.

5 Conclusions and discussions

Based on the evaluation indicator system concerning the coupling between urbanization and water resources and environment system and GRA method, taking Liangzhou District as case study, this paper reveals the main factors influencing the coupling between urbanization and water resources and environment and the temporal distribution of the coupling degrees.

(1) Low social service level, rapid urbanization and irrational industrial structure are the main momentums of urbanization influencing water resources and environment, while water pollution, water shortage and low water management level are the main factors of water resources and environment affecting urbanization, which verify the interactive and complicated coupling between two systems.

(2) The curve of the change of coupling degrees present a combined and continuous "U" shape as the population urbanization grow, which verify the complexity between urbanization and water resources and environment systems.

(3) According to its undulatory characteristics, the change of coupling degrees could be divided into two phases in time series. During the first phase the curve present a "W" shape from 2001 to 2008, which verifies the Law of Environmental Kuznets Curve(EKC). The coupling degrees are in decline during the second phase from 2008 to 2010, which verifies that the ecological deterioration had eased since 2007 when the Key Control Plan of Shiyang River Basin in Gansu Province was published.

References

Huang Jinchuan, Fang Chuanglin. Analysis of Coupling Mechanism and Rules between Urbanization and Eco-environment [J]. Geographical Research, 2007, 22(2): 211 – 220.

Qiao Biao, Fang Chuanglin, Huang Jinchuan. The Coupling Law and its Validation of the Interaction between Urbanization and Eco-environment in Arid Area [J]. Acta Ecologica Sinica, 2006, 26(7):2183 – 2190.

Bao Chao, Fang Chuanglin. On Concept, Significance and Strategic Framework for Water Resources Constraint Force [J]. Journal of Natural Resources, 2006(5):844 – 852.

Bao Chao, Fang Chuanglin. Water Resourcess Constraint Force on Urbanization in Water Deficient Regions: A Case Study of the Hexi Corridor, Arid Area of NW China [J]. Ecological Economics, 2007(3-4): 508 – 517.

Dong Lin, Chen Xuanxuan. Analyze on Water Resources Restriction of Urban Sustainable Development [J]. Water Conservancy Science and Technology and Economy, 2006(8):525 – 527.

Fang Chuanglin, Qiao Biao. Optimal Thresholds of Urban Economic Development and Urbanization under Scarce Water Resources in Arid Northwest China[J]. Acta Ecologica Sinica, 2005(9): 2413 – 2422.

Yang Xuemei, Shi Peiji, Dong Hanrong, et al. Study on Urban Moderate Scale in Arid Inland River Basin Restricted by Water Resources—A Case Study on Liangzhou District in Shiyang River Basin [J]. Economic Geography, 2011, 31(12):2039 – 2045.

Fang Chuanglin, Bu Weina. Competitiveness and Extent of Regional Expansion Restricted by Water Resources of Hexi Corridor [J]. Scientia Geographica Sinica, 2004, 24(5):513 – 521.

Fang Chuanglin, Sun Xinliang. Mechanism of Urban System Development and its Space Organization in Northwest Arid Area with Scarcity of Water Resources—the Case of Hexi Corridor [J]. Journal of Desert Research, 2006, 26(5):860 – 867.

Fang Chuanglin, Li Ming. Urbanization Mode with the Restraint of Water Resources in Hexi Corridor Arid Area of Northwest China[J]. Geographical Research, 2004(6):825 – 832.

Lin Xueqin, Fang Chuanglin. The Process of Urbanization in Wuwei City with Water Resources Hard Constraint and the Approach to Establishing Water-saving City [J]. Journal of Arid Land Resources and Environment, 2009, 23(1):117 – 124.

Ma Guoxia, Gan Guohui, Tian Yujun. Development Models of Zhangye Oasis Cities and Towns under Restraint of Water Resources [J]. Journal of Desert Research, 2006, 26(3):426 – 431.

Bai Qianqian, Wang Lucang, Yang Xiaomei. Development Model of Urbanization under Water Resources Carrying Capacity——Take Heihe River Basin Zhangye City as a Case [J]. Journal of Arid Land Resources and Environment, 2009, 23(4):75 – 78.

Bao Chao, Fang Chuanglin. Interaction Mechanism and Control Modes on Urbanization and Water

Resources Exploitation and Utilization [J]. Urban Studies, 2010, 17(12):19 –24.

Du Hongru, Zhang Xiaolei, Wang Bin. Study on the Adaptability between the Development of Modern Oasis City and the Use of Water Resources [J]. Chinese Science Bulletin, 2006 (51):156 – 161.

Chen Dinggui, Lu Xianguo, Wang Yanping. Study on Constraint of Urbanization of Changchun City from Urban Water Resources and Environments [J]. Journal of Northeast Normal University (Natural Science Edition), 2008, 40(2):130 – 135.

Gao Guizhi, Liu Junliang, Tian Zhiyong, et al.. Study on the Relationship between the Urban Water Use and Urbanization[J]. China Water & Wastewater, 2002(2):32 – 34.

Bao Chao, Fang Chuanglin. Study on the Quantitative Relationship between Urbanization and Water Resources Utilization in the Hexi Corridor [J]. Journal of Natural Resources, 2006(2): 301 – 310.

Qiao Biao, Fang Chuanglin. The Dynamic Coupling Model of the Harmonious Development between Urbanization and Eco-environment and its Application in Arid Area [J]. Acta Ecologica Sinica, 2005,2 5(11):3003 – 3009.

Li Na, Sun Caizhi, Fan Fei. The Coupling Relation Analysis between Urbanization and Water Resources in Liaoning Coastal Economic Zone [J]. Areal Research and Development, 2010, 29 (4):47 – 51.

Gao Xiang, Yu Tengfei, Cheng Huibo. Temporal-spatial Variation of System Coupling between Water Resources and Environment and Urbanization in Northwest China: A Case Study of Gansu Section of Western Longhai—Lanxi Economic Zone [J]. Arid Land Geography, 2010, 33(6):1010 – 1018.

Zhang Zhongwu, Yang Degang, Zhang Xiaolei, et al.. Correlation between Comprehensive Urbanization Scale and Water Resources in Urumqi [J]. Journal of Desert Research, 2011, 31(2): 536 – 542.

Liu Yaobin, Li Rendong, Song Xuefeng. Grey Associative Analysis of Regional Urbanization and Eco-environment Coupling in China [J]. Acta Georaphica Sinica, 2005, 60(2): 237 – 247.

Liu Yaobin. Study on Mechanism and Regulation Policies of Urbanization and Eco-environment Coupling [M]. Beijing: Economic Science Press, 2007: 192 -227.

Reflections on Investment Plan Management of the Lower Yellow River Flood Control Project

Wang Wanmin

Department of Planning and Programming, Yellow River Conservancy
Commission, Zhengzhou, 450003, China

Abstract: Think of the related issues of changing the investment plan management of the lower Yellow River flood control project, in accordance with the Yellow River Conservancy Commission's the changes of construction management system and the new situation of financial budget management, combined with the construction characteristics of the lower Yellow River recent flood control project. The related suggestion of annual investment plan scale of the lower Yellow River flood control project, investment plan classification management, annual investment implementation plan, investment control management and design variation management is put forward.

Key words: flood control project, investment plan, project management, the lower Yellow River

Feasibility study report of the construction of the lower Yellow River's recent flood control projects has been approved by National Development and Reform Commission in 2011 and will enter construction and implementation phase. Recently, state budget management procedures and management system of the lower Yellow River's capital construction have changed greatly. There are also some new requirements for the implementation of professional projects such as project land requisition compensation and resettlement of inhabitant, water and soil conservation, environmental protection and so on. And the investment plans' management of engineering construction is facing new situation which has undergone profound changes. This paper does some research and thinking on the issues related to the investment plans' management of the lower Yellow River recent flood control projects.

1 The construction features of the lower Yellow River's recent flood control Projects

Feasibility study report of the construction of the lower Yellow River's recent flood control projects, compiled according to *Yellow River Basin Flood Control Planning* and other regulations, focuses on unceasingly promoting the construction of the lower Yellow River standardization embankment projects and river regulation works and gives consideration to the construction of Dongping Lake flood retarding basin's and the lower Qinhe River's flood control projects. The project area, along the river channel of the lower Yellow River with zonal distribution, involving relevant areas in 12 cities and 28 counties (cities, districts) of Henan and Shandong province, mainly include the embankment projects such as reinforcement, heightening and widening of embankment, road hardening on the top of embankment and so on, the river regulation works such as continuing construction of control works, reinforcement and heightening of control works, reconstruction of dangerous sections and so on and sluice reconstruction and extension and so on.

The competent departments of state investment manage recent flood control projects of the lower Yellow River as a project and construction funds are totally appropriated by the central government. The recent flood control projects of the lower Yellow River will be in the charge of Shandong and Henan province Yellow River Affairs Bureau, Engineering Construction Bureau as project legal person, founded in 2011. The project legal person will establish site construction management structure in which city and county level Yellow River Affairs Bureaus are the main part. The project budget is managed through Ministry of Water Resources, Yellow River Conservancy Commission and project legal person three-level budget management. In principle, annual budget entirely finishes within the year.

2 The content of the lower Yellow River's recent flood control project investment plans' management

2.1 Do a good job in the item dividing work scientifically and reasonably

The determination of engineering project is the premise and basis of the investment plans' management. We should change the past small and scattered situation of single construction projects of the Yellow River flood control projects and do a good job in the item dividing work scientifically and reasonably.

Through the comprehensive consideration of factors such as construction management system, project legal person settings, requirements for preliminary work, investment plans management, project construction management and so on, the lower Yellow River's recent flood control projects can be divided into two single constructions: Shandong province's and Henan province's Yellow River recent flood control projects. The preliminary design budgetary estimate can be divided into Shandong province's and Henan province's preliminary design budgetary estimate of Yellow River's recent flood control projects and acts as the basis of annual investment plans' management. In the implementation of engineering construction and the executive management of the investment plan, similar projects with continuous construction in a county can be regarded as a unit project. The engineering construction and dynamic investment management should be on the basis of unit project. Considering in favor of assessment of the investment plan implementation, implementation of the construction management, land acquisition compensation, resettlement of inhabitant and so on, the unit projects in a county can be gathered as a implementation unit project. land acquisition compensation, resettlement of inhabitant and the overall arrangement and implementation of engineering construction are on the basis of implementation unit project.

2.2 Control the investment by the principle of "static control, dynamic management"

The investment control is important goal of investment plan management. We should follow the principle of "static control, dynamic management" and do a good job in investment control and implementation of budgetary estimate.

Under the circumstances of Shandong and Henan province Yellow River Affairs Bureau as project legal person forming two executive subjects of investment plan, the recent flood control projects of the lower Yellow River take breaking – down two single constructions , Shandong province's and Henan province's Yellow River recent flood control projects, preliminary design budgetary estimate as the basis of investment "static control". Under the circumstances of city and county level Yellow River Affairs Bureaus acting as project legal person site construction management structure, the implementation of investment plan takes unit project, that is similar projects with continuous construction in a county, as the basis of "dynamic management". According to investment control of unit project's corresponding parts in the preliminary design budgetary estimate, dynamically changing investment can be solved by reserve funds and the savings formed in the process of engineering construction.

2.3 Determine the annual investment planning scale accurately

We should increase predictability of investment planning arrangement, adjust the past mode of the Yellow River flood control projects arranging project annual investment planning according to total investment of unit project and accurately measure and calculate annual investment demand of engineering construction.

On the basis of the overall consideration of controllable factors such as land acquisition compensation and resettlement, the whole project construction scheme, general construction schedule and so on, the project legal person should organize all units involved in the construction to accurately measure and calculate annual investment demand of engineering construction. On the basis of unit project and county implementation unit project, the annual investment suggestions and plans of each project legal person should be put forward and submitted to the higher authorities for examination and approval. The Yellow River Conservancy Commission coordinates overall project and annual investment scale, takes into account conditions such as the process of land acquisition and resettlement work, engineering construction, budget payment and so on. The Yellow River Conservancy Commission also accurately determines annual investment plan scale, declares investment suggestions and plans and coordinates, implements and issues annual investment plans.

2.4 Classified management of investment plan

We should meet the laws' and policies' requirements for implementation and management of professional projects such as engineering construction, land requisition compensation and resettlement, water and soil conservation, environmental protection and so on, change our past thinking that manage all in accordance with engineering construction and do a good job in the classified management of investment plan.

According to the requirements of laws and policies such as *Large and Medium − sized Water Conservancy and Hydropower Project Land Requisition Compensation and Resettlement Regulations*, *Construction Items Environmental Impact Assessment Regulations*, *Management Method of Hydraulic Projects Directly Under Yellow River Conservancy Commission Water and Soil Conservation* and so on, on the basis of budgetary estimate of investment, the investment plans of professional projects such as land requisition compensation and resettlement, water and soil conservation, environmental protection and so on as the important part of the management of engineering construction investment plan will be classified to manage, taking linking up with independent accounting of professional projects as the way, targeting passing completion acceptance of professional projects.

2.5 Establish and perfect management system of annual investment implementation plans

The annual investment implementation plans are a management system, provided in the current *Interim Procedures for Hydraulic Capital Construction Investment Plan Management*, which is implemented in order to put into practice the annual investment plans. The lower Yellow River recent flood control projects' construction management is in the charge of Shandong and Henan province Yellow River Affairs Bureau, Engineering Construction Bureau and several site construction management structures. There are many construction units, many management hierarchies, a lot of construction contents and wide scope of construction. So we have to implement fine-grained management, change the past plan implementation methods of construction units watching for a chance to implement engineering construction and establish and perfect management system of annual investment implementation plans.

The annual investment implementation plans are specific implementing scheme of investment plan for the implementation of the annual investment plans. The annual investment implementation plans should be formed in accordance with preliminary design budgetary estimate, mainly reflect the investment to finish in the staging (month) of this year, quantities, image progress and other construction situations. The annual investment implementation plans include annual investment, construction contents, quantities, image progress, starting and ending times of engineering part, land requisition compensation and resettlement, water and soil conservation, environmental protection and other parts. The part of land requisition compensation and resettlement should make clear arrangements of project land acquisition and removal (physical quantities and relevant independent costs of permanent occupation of land, temporary occupation of land, resettlement population, production resettlement population, ground attachment and so on) and their implementation measures. Engineering part should be divided into construction engineering, electromechanical device and installation projects, metal structure equipment and installation projects, temporary projects, independent costs, basic reserve funds and other projects. The annual investment implementation plans should be compiled on the basis of unit project and in units of county implementation unit project and gather the annual investment implementation plans of Shandong province and Henan province

Yellow River recent flood control projects.

According to the whole project construction scheme, construction contract, compensation for land acquisition and resettlement agreement, the annual investment implementation plan is scientifically formulated by project legal person of organizing compensation for land acquisition, resettlement implementation management and contractors of engineering design, construction and supervision. And then the plan is reported to competent department for approval, as the basis of investment plan implementation.

2.6 Completes management work of the design variation

Design variation means the modified activities for approved preliminary design from the approval date of water conservancy project preliminary design to the date of project completion acceptance. Strengthen the variation design management and properly deal with the changes and problems in the process of engineering construction.

The construction scale, design standard, general layout, layout scheme, main building structure, the main electrical and mechanical equipment, metal structure of the important technical problems processing measures, the construction organization design, etc. of the lower Yellow River recent flood control project appear major design variation, which were checked by the Yellow River Conservancy Commission and then reported to the Ministry of Water Resources for examination and approval. As for the general engineering design variation, which is examined and approved level by level by the project legal person, Shandong of the Yellow River Affairs Bureau, Henan of the Yellow River Affairs Bureau, the Yellow River Conservancy Commission, according to the character of design variation and the project quality, safety, schedule, investment amounts and benefit influence degree. Major design variation is in line with the lower Yellow River recent flood control project as a whole project unit for consideration. General design variation is on the basis of unit project for management. The design variation of compensation for land acquisition, resettlement, water and soil conservation, environmental protection on engineering construction are executed in accordance with the relevant regulations of the State.

The project legal person must strengthen the design management work of the project construction period, further perfect, supplement and optimizing design scheme. As for the objective conditions and design a leakage, etc., the project legal person must consummate approval process for design variation, make preliminary design, bidding design, the construction drawing design, construction and implementation, etc., to meet the result requirements of the previous phase and the actual situation and needs of this stage. Making sure the tender design guide the bidding and contract, guarantee the project construction can be according to construction drawing design. For the changes within the construction scope during the project construction period, which have to be solved legally through supervision visa.

Through the design variation, working out the changing problems in each project design stage and in the constructing process, and ensuring the smooth realization of the goal of engineering construction and the effective implementation of the investment estimate.

2.7 Make overall plans and coordinate to the operation management and maintenance of completed project in the engineering construction period

The completed project via the construction contract acceptance check are delivered over project legal person by construction unit, and which enter the engineering quality assurance period. The completed projects are also handed over operation management unit by project legal person after through completion approval. Then make overall plans and coordinate to the operation management and maintenance of completed project in the engineering construction period according to the duties of project legal person and operation management unit.

Project legal person entrust the Yellow River Affairs Bureau above the county level and other relevant operation management unit with taking charge of management and maintenance work for completed project of the lower Yellow River recent flood control project. According to the operation maintenance workload and operation efficiency, the project legal person and the operation management unit raise the handling suggestion for cost allocation of management and maintenance work for completed project of the lower Yellow River recent flood control project. And the handling suggestion is reported to superior departments for approval. The project legal person can solve certain financial problem of operation management and maintenance funds from engineering reserve funds and engineering construction surplus funds.

If flooded, and other natural disasters, great danger occurs or emergencies happened, the operation management unit should timely consult with the project legal person, and properly finished the protection and maintenance work and other related work of project completed in engineering construction period.

2.8 Completed the work for construction and financial and other management cohesive work

Assist completed the cohesive work of construction management around executive management of Annual investment plan. The project legal person takes full charge of the project construction management, and organized the construction management as a whole in the unit to county implementation unit engineering.

Acceptance must in accordance with the principle of unified management and responsible grading. The project legal person conduct acceptance on the basis of county implementation unit engineering, conduct collection in accordance with the unit of county project. According to the relevant regulations, the related competent departments conduct the special acceptance, for example, the compensation for land acquisition, resettlement, water and soil conservation and environmental protection. The Yellow River Affairs Bureau of Shandong, Henan province respectively organize stage acceptance for the Yellow River recent flood control project of Shandong, Henan province in the unit to county implementation unit engineering, that is government acceptance. The Yellow River Conservancy Commission conducts the completed acceptance for the Yellow River recent flood control project of Shandong, Henan province. The ministry of water resources or The Yellow River Conservancy Commission authorized by the ministry of water resources that they conduct overall completed acceptance for the lower Yellow River recent flood control project.

Assist completed infrastructure financial management work according to annual investment plan. Carry out the financial management on the basis of two subprojects of the Yellow River recent flood control project of Shandong, Henan province. Conduct every annual budget implementation and management work according to annual investment plan. Assist completed accounting cohesive work in accordance with investment plan classification management and implementation requirements of preliminary design accounting.

2.9 Focus on the key link of inspection and supervision on investment plan's implementation

Inspection and supervision is an important measure for promoting investment plan implementation, also a system requirements of ensuring realize the goal of project construction. Taking budget implementation as the main line, the annual investment plan implementation as grips, strengthen the inspection and supervision work of the key link.

The tender design, contract management, accounting are the key link of investment plan control and budget implementation. Whether going along bidding for project design and contract management according to the preliminary design, whether the project is constructed in accordance with the construction drawing, whether conducting accounting on the basis of preliminary design budgetary estimate cost items and standards, etc., which as the focus of the budget implementation check the content. For the changed, checking whether the variation procedure is ready.

3 Conclusions

Based on the construction management system of the Yellow River Conservancy Commission and project construction characteristics of the lower Yellow River recent flood control project, change the previous investment plan management mode, conduct scientifically project method division, put the investment control into practice effectively, determine the scale of annual investment plan accurately, carry out the investment plan classification management, set up annual investment plan system, finish the design variation management work, make overall plans and coordinate to the operation management and maintenance of completed project in the engineering construction period, assist complete the construction management and financial management work, give full play to the management system advantage of the lower Yellow River recent flood control project construction, which these will lay a good foundation for the smooth implementation of investment plan of the lower Yellow River recent flood control project.

A Study on Problems and Countermeasures of Water Resources Utilization of Hubei Yangtze River Economic Belt

Liu Tao

Institute of Economy Research in Yangtze Basin, Hubei Academy of Social Sciences, Wuhan, 430077, China

Abstract: Abstract: Hubei Yangtze River economic belt is not only an important fulcrum of strategy of central China uprising, is also the megalopolis concentrated area, in which a new round of open and development needs to have safe and reliable water resources as the guarantee and support. However, Hubei Yangtze River economic belt is also facing some problems, such as the contradiction of water supply and demand increasingly prominent, health status deterioration of river and lake, threatened safety of city water supply, the functional atrophy of traditional golden waterway, transition difficulties of high water consumption industries, irrigation facilities aging, and so on. The most strict water resources management should push the regional economic development mode change, which provides a new method and way for the water problem of Yangtze River economic belt. Firstly, by the total water consumption control, it is to promote optimal allocation of water resources so as to realize the optimization and adjustment of the industrial structure on the one hand, to guarantee the basic ecological flow and free shipping on the other hand. Secondly, by water use efficiency control, it is to promote water-saving technology adoption and promotion, irrigation facilities repair and property rights reform, and to promote the construction of a water – saving society and ecological civilization. Finally, pollutant capacity of the water function area may control production and living sewage, ensure urban and rural drinking water safety, and promote health improvement of river and lake.

Key words: Yangtze River economic belt, water resources, the total volume of water consumption, water use efficiency, water function area

1 Water resources utilization and social economic development status of Yangtze River economic belt in Hubei

1.1 Water resources utilization Status

Hubei Yangtze River economic belt is belonging to the abundant rainfall area, then the average depth of precipitation is 1,351.8 mm and an annual precipitation is 7.322×10^{10} m^3 accounting for 33.5% of the total annual precipitation in the whole province. In this region, for many years the average total amount of water resources is 7.261×10^{10} m^3, accounting for more than 70% of the total water resources in the whole province. Its total amount of water resources are relatively abundant, but per capita water resources is in low level, the per capita water resources quantity is 1.718 m^3, which is 22% of the average level of the world. There are a number of rivers in Hubei Yangtze River economic belt, and the surface water resources is about more than 6.5×10^{10} m^3, and the average annual runoff of Yangtze River Crossing guest water is more than 9.0×10^{11} m^3. In 2008 the total water supply of Hubei Yangtze River economic belt is about 1.587×10^{10} m^3, then the ratio of water resources development and utilization (water supply / water resources capacity) is about 21.86%, which is slightly higher than 17.8%, in whole Yangtze River Basin still remaining to be developed and utilized reasonably for further.

The main stream river of Hubei Yangtze River economic belt is 1,061 km long, of which the main stream river of the Yangtze River upper reaches is 124 km long, after Lower-Jingjiang cut-off the main stream river of the Yangtze River middle reaches is 937 km long. The total two side shoreline of Yangtze River main stream is 1,982. 1 km long (including the Three Gorges Reservoir shoreline and part of eyot shoreline), and at present, the used Yangtze River shore is 354. 2 km long, its utilization rate is 17. 87%, then the overall extent of shoreline utilization is not high. Shoreline utilization rate is higher in Wuhan, whose total shoreline is 240 km long, which have been using the shoreline of 90. 5 km, accounting for 37. 7% of its shoreline length. The low utilization rate of is lower in the Three Gorges Reservoir area and Songzi, Jianli city and other counties of Jingzhou city, of which shoreline utilization are less than 10%. The dike construction along Yangtze River has reached the standard of Yangtze River Basin planning, and flood control capacity is significantly enhanced.

It has 25 ports in Hubei Yangtze River economic belt region, and Hubei has 4 ports belonging to 16 major inland river port of the national planning in Yangtze River, which are Wuhan port, Yichang port, Huangshi port and Jingzhou port. In 2008, the port cargo throughput of the whole province is 1.5×10^8 t, including waterway freight volume 5.278×10^7 t and container throughput nearly 390 thousand mark of Wuhan port.

1. 2 Social economic development status

Abundant water resources provides good water conditions for agricultural, industrial and city construction of areas along the Yangtze River. The land area of Hubei Yangtze River economic belt is 54,168. 5 km, accounting for 29. 1% of the province. In 2008 the population of Hubei Yangtze River economic belt amounts to 2.75×10^7 person, accounting for 48. 10% of the province, which will achieve 68.9698×10^{10} Yuan in GDP, accounting for 60. 87% of the province.

Agricultural development is more obvious by the main way of irrigation. In 2008 agriculture achieves 6.583×10^{10} Yuan of total product value in Hubei Yangtze River economic belt, accounting for 36. 98% of the province. At present, it has formed a farming to develop in the round pattern, with the high-quality grain, high-quality cotton, double low rapeseed, vegetables and aquatic products as the main body, paying attention to characteristic. Industry area, electricity, petroleum and petrochemical, steel, textile, papermaking, chemical, food and other high water industry are all advantage industries of Hubei Yangtze River economic belt. With "the Yangtze River—the mother river" as the cultural heritage, the tourism industry is also unique. In the western part of Hubei, it is characteristic of natural scenery with the landscape tourism, such as the dragon river of Badong, the Yangtze River Three Gorges sightseeing and holiday tourism area of Yichang. In Jingjiang section and the east part of Hubei, it is characteristic of the cultural landscape with the historic city as representative, such as the history ancient city of Jingzhou, the Three Kingdoms culture of Chibi, urban tourist area of Wuhan.

Relying on water resources and shoreline resources, along the Yangtze River, these Provincial key towns such as jingkou of Wuhan, fuchi of Huangshi, Xiaohekou of Jingzhou, Zhicheng of Yichang, Yanji of Ezhou, Xiaochi of Huanggang, Panjiawang of Xianning and so on, accelerate process of Urbanization, which enhance the drive ability to peripheral rural economy.

2 Potential safety problem analysis of water resources utilization in Hubei Yangtze River economic belt

2. 1 Water security problem

According to the supply and demand analysis of water resources, water resource supply and demand situation is not optimistic in different fields of Hubei Yangtze River economic belt.

2. 1. 1 The supply and demand of city water resources

In 2005 as far as the shoreline cities is concerned in Hubei Yangtze River economic belt, water supply capacity can all meet the water requirement. 14 cities of them such as the Wuhan urban district, Zhijiang, Jingzhou urban district, Huangshi, Daye, Zigui, Shishou, Gongan, Jiangling,

Huanggang urban district, Wuxue, Tuanfeng, Xishui, Qichun, Huangmei, Chibi and Jiayu, have a bigger surplus to status quo of water supply capacity, and if the water quality safety factor and the city pipe network connectivity (dividing) are not considered, their total amount of water can meet the water requirement of cities in 2015 and 2020. 9 cities of them such as Yangxin, Yichang urban district, Yidu, Ezhou urban district, Honghu, Songzi, Jianli, Xianning and Badong, have a relatively weaker water supply ability, and their total water amount can not satisfy the water requirement of cities' development in 2020 and 2015. In addition, because the material conditions, the analysis takes no account of the city pipe network connectivity and slices the supply and demand balance calculation, nor on the city water supply for classification of balance, and if the influence of these factors is not taken into account, different level years shortage of every city will be more serious.

2.1.2 Rural people living water supply and demand

Adopting rating method to predict the living water requirement, in 8 level administrative areas of Hubei Yangtze River economic belt, rural residents and livestock are all thirsty for water in the base year(2005), 2015 and 2020. Among them, Wuhan needs to add water supply 3.928×10^7 m^3 in the base year(2005), and will need to add water supply $1.231,6 \times 10^8$ m^3 in 2020. Huangshi needs to add water supply 2.985×10^7 m^3 in the base year(2005), and will need to add water supply 4.991×10^7 m^3 in 2020. Ezhou need to add water supply 8.04×10^6 m^3 in the base year (2005), and will need to add water supply 1.211×10^7 m^3 in 2020. In short, in the region of Hubei Yangtze River economic belt rural living water is in short supply, and suitable water source project to alleviate this situation must be planed.

2.1.3 Irrigation water supply and demand situation

Under the condition of current water conservancy project and normal water demand of social economic development, Hubei province has a water requirement of 3.299×10^{10} m^3 in the general low water year ($P = 75\%$ is typical year), lacking water quantity of 6.033×10^{10} m^3, and the lacking water rate is 18.3%. Within the region of Yangtze River economic belt, Huanggang is the largest city short of water amount, followed by Wuhan city. The main water shortage are in engineering water shortage and water resources shortage. Through building water conservancy facilities and promoting water - saving technologies and crop planting structure adjustment, agricultural irrigation conditions will be greatly improved, then in 2015 ~ 2020 the water shortage problem will gradually be relieved, and the average water shortage rate of the province will be dropped to 7.4%. Under the standard of the designed guarantee rate, the province has a total water requirement of 2.91×10^{10} m^3, a water supply of 2.696×10^{10} m^3, and still have water shortage of 2.14×10^{10} m^3, within the region of Hubei Yangtze River economy belt Yichang is the largest city as far as the water shortage rate is concerned.

2.2 Drinking water safety problem

The water sources of big and medium cities of Hubei province are mainly builded in Yangtze River, the main stream of Hanjiang River and reservoir in which the water quality is relatively good. The water intakes and sewage outfalls of many urban water plants are cross distribution, and many sewage outfalls are urban drainage culvert brake and pump stations, having a large amount of sewage, and bringing certain potential safety hazard to water sources. The water quality is poor in dry season in lower reaches of Hanjiang River main stream, especially the organic pollution and eutrophication problems. By nitrogen and phosphorus pollution, in lower reaches of Hanjiang River the algae such as diatoms is Mass propagation, changing water to brown, which may lead to the algae phenomenon of "water bloom". Han River "water bloom" seriously impact the cities along the lower reaches of Hanjiang River on water quality of water plants, and Wuhan is one of them, which is the important city of Hubei Yangtze River economy belt. In addition, the quality of water sources around Xiaogan is poor, which relies on groundwater to get water in many places, at the same time groundwater overexploitation is serious in Xiaogan urban district, which have formed the larger funnel of groundwater mining. As a result, groundwater water quality can not be guaranteed, and drinking water safety problem of Xiaogan urban district is very prominent in the whole province.

2.3 Water ecological security

Water environment quality was generally good in Yangtze River, but the water pollution of some tributaries is serious. Huangbo River, Juzhang River, Tongshun River, Yunshui and Changgang have been polluted to a certain extent, and the lakes controlled by nitrogen and phosphorus pollution in the province began to show a trend of eutrophication, and urban lake pollution is more serious, all which fully exposed the serious situation of water environmental protection in Hubei Yangtze River economic belt. The agricultural non-point source pollution controlled of ineffectively, and the fertilizer breed caused the reservoir water pollution seriously. Around the sewage outfall of cities along Yangtze River it exists of pollution belt unequal length, which seriously affected the quality of urban water sources. Within the water source protection zones, there are a lot of wharfs for many years construction, such as Wuhan Pinghu-door water plant, crane-mouth water plant, Jingzhou-liulin water plant and so on. At freight dock there are many ships, therefore, petroleum pollution and especially accident such as chemicals, oil products leakage in the process of loading and unloading will make water security threat. Many high energy consumption and easy pollution industrial projects are distributed along Yangtze River, then the project layout and environmental protection measures are not reasonable for more concerned with economic benefits, which also caused adverse effects on the prevention and control of water pollution.

3 Water resources utilization countermeasures of Hubei Yangtze River economic belt

The most strict water resources management should push the regional economic development mode change, which provides a new method and way for the water problem of Yangtze River economic belt. Firstly, by the total water consumption control, it is to promote optimal allocation of water resources so as to realize the optimization and adjustment of the industrial structure on the one hand, and to guarantee the basic ecological flow and free shipping on the other hand. Secondly, by water use efficiency control, it is to promote water – saving technology adoption and promotion, irrigation facilities repair and property rights reform, and to promote the construction of a water-saving society and ecological civilization. Finally, by pollutant capacity of the water function area, to control production and living sewage may ensure urban and rural drinking water safety, and promotion health improvement of river and lake.

3.1 To establish system of total water consumption control for promoting the industrial structure optimization and adjustment

It is to establish index system of total water consumption control, to issue the Yangtze River economic belt related documents of total water consumption control management and strict water license approval management. It is to establish strict and perfect water resources assessment system and program, to establish water resources argumentation of expert database and water resource demonstration report filing system, to promulgate the water resources demonstration program and the water resources demonstration implementation rules of planning and construction project, and to promote water resources demonstration of major planning such as economic development plan and urban overall plan of Yangtze River economic belt. It is to establish the supervision system of water plan and water license, to achieve unified management of water resources in Hubei Yangtze River economic belt, to promote the industrial structure transformation of Yangtze River economic belt.

3.2 To establish the efficiency of water control system for promoting the construction of water-saving society

It is to establish the index system of water efficiency control, to determine water ration process management system, and to reinforce water quota management of water resources argumentation of construction projects, water license approval, water planning and water-saving level evaluation. It is to establish "three simultaneous" management system of water saving facilities of construction projects, to guide the development of water-saving measures to the new, alteration and expansion of construction projects, and to carry out water-saving assessment. It is to establish unconventional water resources utilization system such as sewage treatment reuse, to promote water-saving technological transformation of every walk of life, vigorously promoting water-saving society construction and system and mechanism construction, to focus on building a number of industrial and urban living water-saving demonstration project.

3.3 To establish limit sewage system of water function zone for promoting the ecological civilization construction

Ecological civilization is the advanced civilization form after the industrial civilization, and it emphasis more on the coordinated development of social economy and ecological environment. Hubei Yangtze River economic belt should systemically establish limit pollution control index system of water function zone, supervision and management system of outfall into the river, health evaluation system of key lakes, and protection and supervision system of drinking water source, and reinforce the emergency response ability of sudden water pollution incident.

3.4 To establish water resources management and assessment system for ensuring the most strict water resources management system implementation

Hubei provincial government can sign form of responsibility of water resources management with municipalities (state) government along Yangtze River, promulgate the assessment methods of implementing the most strict water resources management system, establish indicators evaluation system of water resources management, take the evaluation results as the important content assessing science development level of the assessed region, as well as an important criterion for local leading cadres comprehensive assessment.

References

Weng Lida. Protection and Development, Opportunities and Challenges—30 Years Review of the Yangtze River Water Resources Protection[J]. Yangtze River, 2008,39(23):4 –7.

Qin Zunwen, Liu Tao. Water Resources Utilization and Protection of Hubei Yangtze River Economic Belt[A]. In: Qin Zunwen, Peng Zhimin. A New Open Development Round of Hubei Yangtze River Economic Belt[M]. Wuhan: Hubei Changjiang Publishing Group of Hubei People's Publishing House, 2010:167 –176.

Peng Zhimin, Liu Tao. Hubei Yangtze River Economic Belt Planning of Environmental Protection Research[A]. In: Qin Zunwen, Peng Zhimin. A New Open Development Round of Hubei Yangtze River Economic Belt[M]. Wuhan: Hubei Changjiang Publishing Group of Hubei people's Publishing House, 2010:187 –200.

Chen Wenke. Open Development Research of Hubei Yangtze River Economic Belt[M]. Wuhan: Hubei Changjiang Publishing Group of Hubei people's Publishing House, 2010.

Hong Yiping. To Fully Enhance the Supervision Ability and Realize the New Span of Yangtze River Water Resources Protection[J]. Yangtze River, 2011,42(2):8 –11.

Urban Expansion and Spatial Governance in the Shiyang River Basin

Shi Peiji and *Wang Zujing*

College of Geography and Environmental Science, Northwest Normal University,
Lanzhou, 730070, China

Abstract: The Shiyang River Basin has the most populous and the least per capita water resources in Chinese inland river basin. The contradiction between water supply and demand in the middle and lower reaches is extreme prominent. The environmental degradation seriously. And the basin is a typical regional of unsustainable developed economic and social. This article systematically analyses the land evaluation and prediction of urban spatial expansion on the basis of the socio-economic, regional transportation and ecological environment. We use land suitability evaluation method to identify the high-value area of urban expansion possibilities and townships suitable development. We proposed a program of spatial planning based on this. Firstly, we must identify the environmental protection areas and guide the reasonable expansion of urban. Secondly, it in order to strengthen the sand-fixing, soil and water resources conservation. Thirdly, it is to establish the local ecosystem in harmony with the coordination of arid zone to protect the system of forest, shrub, grass and desert. Fourthly, it is to promote the coordinate development of planning of the main functional areas and urbanization in Shiyang River Basin.

Key words: basin development, sustainable, spatial governance, Shiyang River

1   Introduction

Regional spatial governance is very important for regional development planning. While in the past 3 decades, the urbanization of eastern coastal areas taking into account the suitability of spatial planning around the development and protection. And scholars have its large number of studies. The choice of ecological protection factor varies due to the region difference. but few studies have attention the western region. The Shiyang River Basin is a northwest arid area and has great difference with the eastern coastal areas. The recent studies in regional planning only consider the road network . Without taking into account the factors of terrain slope. Because of the western region is a complex terrain and has harsh natural conditions. Terrain and traffic space divisions are equally important for a relatively large area.

This article is supported with ArcGIS. First of all, we take grid as evaluation units. Combined with regional ecological and environmental protection. We use spatial analysis methods of vector data and raster data to identify ecological function planning impact factor of the Shiyang River Basin. Then we superimpose on each factor, and divide suitable zone for urban development of the main functional areas. In order to provide a scientific basis for adjust the regional spatial structure and prepare the relevant town and county planning.

2   Study area and method

2.1   Study area

The Shiyang River Basin is located in the northeastern section of the Hexi Corridor and Qilian Mountain. It between 101° 22 ' E ~ 104° 16 ' E and 36° 29 ' N ~ 39° 27 ' N. It contains mountains, hills, desert and alluvial plains. The terrain is higher in south and north. Its area is about 41,600 km². The administrative divisions of the basin including Liangzhou, Gulang, Minqin and some areas of Tianzhu County (Wuwei City). And it includes Yongchang, Jinchuan County (Jinchang City) and some areas of Sunan City (Zhangye City). The total population is about 2. 6

million in 2010. And the urbanization level is about 36.5%.

2.2 Study method

The underlying data of the Shiyang River Basin is from the National Natural Science Foundation of Environmental and Ecological Science Data Center for West China (http://westdc. westgis. ac. cn). And the 2010 land use data is interpreted from Landast TM images. The economic data is get from socio-economic statistical data of each county in 2010. Finally, we considering the terrain slope factors combined with the underlying data of the basin convert into the analysis of the base map.

In order to make the results more accurately. The evaluation factors are divided into regional homogeneous factors and spatial diffusion factors. Administrative divisions or natural unit is used as homogeneous factor units. It is arable land and GDP per capita. The diffusion factors divided into vector and raster diffusion factors. The raster diffusion factors are divided into homogeneous and requirement diffusion. We use vector diffusion factors to calculate the gravity of the Basin's central counties. Using requirement diffusion factors to calculate the accessibility of provincial capital. Using homogeneous diffusion factors to calculate the stream distance and industrial enterprises concentration degree (in Fig. 1).

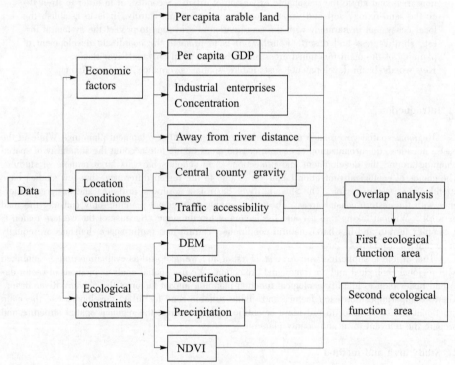

Fig. 1　Follow chart of ecological – function regionalization

3　Process analysis

3.1　Economic factors

According to each county's socio-economic statistics of the Shiyang River Basin in 2010. We use township as the basic unit to generate the per capita arable land impact factor (in Fig. 2(a)). According to *Gansu Development Yearbook 2010*. We extract population data and GDP of each county. Using counties of the Basin as the basic unit to generate the per capita GDP impact factor (in Fig. 2(b)). Urban construction must near water, especially in arid areas. We use the Minimum Euclidean Distance Algorithm to generate away from rivers and reservoirs impact factors gradient map. And overlay them (in Fig. 2(c)). The mineral resources have a decisive role in the formation of a city. So we extract the mining site plaques in construction land of 2010 land use data. It stands for the industrial enterprises. And we use the same method in Fig. 2(c) to generate industrial enterprises concentration factor gradient map (in Fig. 2(d)).

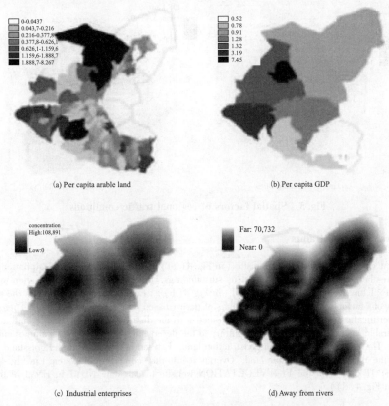

(a) Per capita arable land   (b) Per capita GDP

(c) Industrial enterprises   (d) Away from rivers

Fig. 2   Spatial factors of economic conditions

3.2   Location conditions

First of all, we calculate the central urban gravity of the basin. Overlay the terrain and road network. According to the spatial analysis method of vector data. Use the gravity model to estimate the gravity between the central urban and the county on other areas of the Basin (in Fig. 3(a)).

It can be seen in Fig. 3(a). With the land coverage farther away from the road network. The gravity of the central urban became smaller. By topographic slope of the southern Qilian Mountains and flat of the northern desert region. In the case of equal road transport system, the urban gravity of the former is bigger than the latter. Because of four cities of the Basin (except Minqin) are in the central plains area. And the road network is complex. So the urban gravity is bigger than the other two regions. And it formed a "star" structure along the road network.

Because the traffic accessibility of provincial capital Lanzhou is also very important. The data of highway, railroad and terrain is based on the national databases 1 : 4 million data of National Fundamental Geographic Information System. We extract the part of Gansu Province. Combining with the terrain, land and water of traffic conditions of the Shiyang River Basin. Then combined with DEM of Gansu Province to generate the cost weighted raster map of Lanzhou. Calculating Lanzhou traffic accessibility in the basin, As shown in Fig. 3(b).

It demonstrates the isochronous rings reflect spatial tightness of the city and its hinterland. If the city close to the high – grade traffic system and the terrain is flat. The traffic accessibility will be good. The Lanzhou accessibility shows the location advantages between the basin and provincial capital.

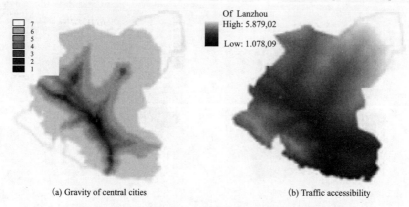

(a) Gravity of central cities　　　　　　　　　(b) Traffic accessibility

Fig. 3　Spatial factors of regional traffic conditions

3.3　Ecological constraints

DEM data is from the SRTM website (in Fig. 4(a)). The annual average precipitation data is from 7 weather stations of the Basin and its surroundings. Using the Kriging interpolation to generate 1 km × 1 km average precipitation grid (in Fig. 4(b)). Ecosystem service value is the regional homogeneous factor. We extract the elements of desert based on 2010 land use data. Then we extract desertification map of arable land. Combined with the distribution data of arable land to generate the arable land protection system. Contrary to the desert land. The lower the value of cultivated systems. The better it can be urban construction land. Extraction of woodland and grassland data as same as the classification of arable land. Overlay it into the eco-constraint system (in Fig. 4(c)). We get SPOT data from the Free VEGETATION website to calculate NDVI in ENVI of the year 2010 (in Fig. 4(d)).

3.4　Overlap analysis

We convert all the above vector data into raster data by using Albers Conical Equal Area projection and GCS_Krasovsky_1940 geographic coordinate system, which combined with all GRID data re-classification. According to the principle in Fig. 1, we calculate in ArcGIS. Then we will get main functional areas of the Shiyang River Basin (in Fig. 5).

The Shiyang River spatial governance function zoning divided to 2 levels. The first level ecological function areas is using location conditions as the dominant factor. It reflecting the overall pattern of ecological functions. And it named as geomorphic units + ecological functions. The second level ecological function areas is using land use as the dominant factor. It reflecting the difference between the first level ecological function areas. And it named as intermediate geomorphic units + vegetation type + ecological function. The Shiyang River Basin is divided into 3 landscape functional areas accordance with the above principles. It namely: the suitability region of urban de-

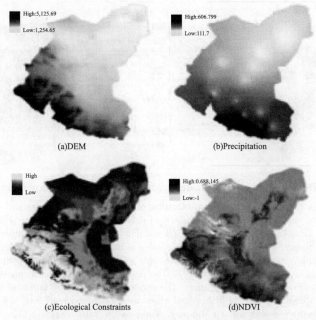

(a)DEM (b)Precipitation

(c)Ecological Constraints (d)NDVI

Fig. 4　Spatial factors of ecological constraints

velopment (I) , the protection region of ecological agriculture (II) , the unused region of mountain and desert (III) (in Fig. 5(a)).

On this basis, we divide the second landscape functional region into 9: the high-suitability sub-region of urban development (I 1) , the middle-suitability sub-region of urban development (I 2) , the low-suitability sub-region of urban development (I 3) , the develop sub-region of oasis agriculture (II 1) , the protection sub-region of soil and water conservation (II 2 ,) , the marginal sub-region of urban expansion (II 3) , the governance sub-region of desertification (III 1) , the sand fixation sub-region of the Gobi desert (III 2) , the unused sub-region of desertification (III 3) (in Fig. 5(b) , Tab. 1).

Tab. 1　The statistics of the landscape function regions

First ecological function area	Second ecological function area	Area (km^2)	Proportion (%)
The suitability region of urban development	The high-suitability sub-region	293.45	0.71
	The middle-suitability sub-region	1,368.29	3.28
	The low-suitability sub-region	2,593.67	6.22
	The develop sub-region of oasis agriculture	3,382.89	8.11
The protection region of ecological agriculture	The protection sub-region of soil and water conservation	5,326.05	12.77
	The marginal sub-region of urban expansion	7,670.86	18.39
	The governance sub-region of desertification	5,751.62	13.79
The unused region of mountain and desert	The sand fixation sub-region of the Gobi desert	7,283.72	17.46
	The unused sub-region of desertification	8,046.64	19.29

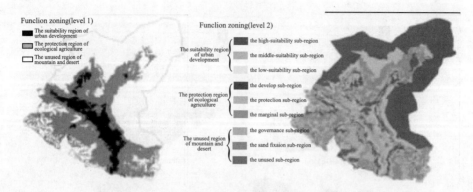

Fig. 5   An overlay of model predictions about urban expansion

It illustrates the comprehensive consideration of arable land protection and ecological constraints. With the land far away from the main road network, the suitable areas of urban expansion became smaller. Thanks to the terrain slope and woodland, grassland area of unused area in southern mountains (Sunan, Tianzhu). The possible value of urban expansion is small. The road transportation is underdeveloped of urban development and ecological agriculture region in middle area (Liangzhou, Laguna, Jinchuan, Yongchang). The possible value of urban expansion is the highest. But due to sufficient water for irrigation. The region mostly are basic farmland. So the possible value of urban expansion along the arable land class distribution and road level decrease. The unused region in northern desert is flat (Minqin). The possible value decrease along the distance of river. Because of four cities of the Basin (except Minqin) are in the central plains area. And the road network is complex. So the urban gravity is bigger than the other 2 regions. And it formed a "star" structure along the road network.

4   Conclusions

Due to the local conditions in regional planning are very important factors, which restricts the supply of resources and intelligence flow. This article taking local conditions as emphasis. Dividing spatial governance zoning combined with arable land protection and ecological constraints. Finding out the suitability region of urban expansion in the Shiyang River Basin.

This article starts from the socio – economic, regional transportation and ecological environment to find out the land evaluation and prediction of urban spatial expansion. We use land suitability evaluation method to identify the high – value area of urban expansion possibilities and townships suitable development. Finding that along the G30 Lianhuo Highway and 312 State Road, surrounding the urban land in the Liangzhou City and Yongchang County is the best suitable area. Providing reference to guide the sustainable development and planning of the basin.

Acknowledgments

This work was supported by the National Natural Science Foundation of China (40971078).

References

Wang Chuansheng, Zhao Haiying, Sun Guiyan, et al. . Function Zoning of Development Optimized Area at a County Level: A Case Study of Shangyu, Zhejiang[J]. Geographical Research, 2010,29(3):481 –490.

Li Yonghua. Method of Determining Urban Growth Boundary from the View of Ecology: A Case Study of Hangzhou[J]. City Planning Review,2011,35(12):83 –90.

Chen Jingqin, Chen Wen, Sun Wei, et al. . Rational and Coordination Evalution of Land Use in

Planning Based on Regionalization of Regional Potential Development—A Case Study of Wuxi City[J]. Resources and Environment in the Yangtze Basin,2011,20(7):866 – 872.

Lu Yuqi, Lin Kang, Zhang Li. The Methods of Spatial Development Regionalization: A Case Study of Yizheng City[J]. Acta Geographica Sinica,2007,62(4):351 – 363.

Sun Wei, Chen Wen. Regionalization of Spatial Potential Development and Distribution Guidance: A Case Study of Ningbo City[J]. Journal of Natural Resources,2009,24(3):402 – 413.

He Dan, Jin Fengjun, Zhou Jing. Urban Construction Land Suitability Evaluation in Resource Based Cities: Taking the Grand Canal Ecologic and Economic Area as an Example[J]. Geographical Research,2011,30(4):655 – 666.

Xu Xibao, Yang Guishan, Zhang Jianming. Simulation and Prediction of Urban Spatial Expansion of Lanzhou City[J]. Arid Zone Research,2009,26(5):763 – 769.

Jiang Haibing, Xu Jiangang, Qi Yi. The Influence of Beijing-Shanghai High-speed Railways on Land Accessibility of Regional Center Cities[J]. Acta Geographica Sinica,2010,65(10): 1287 – 1298.

Zhang Li, Lu Yuqi. Studies on Spatial Analysis Method of the "Pole & Axis System": A Case Study of the Yangtze River Delta[J]. Acta Geographica Sinica,2010,65(12):1534 – 1547.

Fan Yu, Yang Guishan, Tu Xiaosong. Forecast of Quantity of Land Reserve Base on Urban Extension—A Case Study of Nanjing City[J]. Scientia Geographica Sinica,2010,30(1):53 – 59.

Water Regulation and Water Resources Management Selections for North China: Using Beiyun River Basin as an Example

Chen Minghong

College of Water Resources and Civil Engineering, China Agricultural University,
Beijing, 100083, China

Abstract: Most rivers in North China suffer serious water shortage and water quality deterioration nowadays, especially for Haihe River. Methodology and strategy of water regulation and water resources management in Beiyun River which is one of the biggest branches of Haihe River are studied as an example to alleviate the contradiction. An integrated broadcasting and predictive system consisting of a distributed hydrological model, a hydrodynamic model coupled with sluice hydraulic model, a water quality model, as well as an optimal model is developed. Using this system, the operation mode of reservoirs and sluices can be implemented directly referring to flow discharge process instead of water elevation operation mode which is used at present. Optimal operation schemes are explored considering water resources with different qualities (runoff, reclaimed water and wastewater), in order to achieve the multiple objectives of minimum disaster loss, maximum ecological water and great improvement of water quality. Optimal regulations are calculated for daily flow discharge processes of Beiyun River. The calculated results show that the optimized water regulation strategy can change the present situation of zero flow in non-flood season and improve water quality during the whole year. The research will provide technical support for maintaining river health and sustainable economic development in Beiyun River basin and could be applied to other basins with large numbers of reservoirs and sluices in North China.

Key words: Beiyun River, mathematical model, broadcasting, water regulation, water quality improvement

1　Introduction

In recent years, the water resources of North China have decreased significantly. The most significant ones are the Yellow River, Huaihe River, Haihe River and Liaohe River. The amount has decreased by 17% for surface water resources and 12% for total water resources. In the Haihe River, District surface water resources have decreased by 41% and total water resources have decreased by 25%. At the same time, the regional environmental deterioration is evident. The worsening of the water environment is caused by over exploitation of groundwater, continuous drought, industrial and domestic pollution. In order to alleviate the plight of this water shortages and deterioration of water quality, we must change the water use patterns, and positively reform the water resources management system. The quality-quantity-joint scheduling, considering the efficient use of water resources and ecological environment protection and restoration, is an effective water resource management measure.

Before the 1980s, scholars only concerned about the water resource scheduling configuration, and theoretical research matured day by day. Since the 1980s, water quality and water environment deterioration trend has exacerbated and has already damaged aquatic ecosystems and threatened human life and health. The water environment has become a hot research. During this period, many countries began to develop water quality standards for surface water and groundwater, adopt national water resources management, establish various types of water quality model for rivers, lakes and reservoirs, and tried to unite water quantity with water quality in order to achieve water quantity and quality of the scheduling management of water resources in the unified description and joint regulation. Many achievements has been achieved in water quality-quantity-joint scheduling models, which increase the reliability of the models. Some researches have been carried out in the joint scheduling and control of water quantity and quality in our country. In the 1990s, Fan and Li

(1996) established a stratified reservoir water quantity and quality joint model, which is multi-objective integrated optimization scheduling dynamic nonlinear mathematical model, and obtained some achievements in the analysis of reservoir water temperature, water quality and reservoir water intake structure on the basis of relations. Using the multi-target weight coefficients to form an overall objective, step-by-step optimization method with a successive approximation, Heihe reservoir, for example, was solved with calculation. In recent years, water resources scheduling has been used in Shanghai plain rivers to improve water quality study, in the Huaihe River Locks and Dams to give the impact of scheduling on water quantity and quality, and in Wenyu River basin for water quality and quantity regulation. However, these studies are not deep in water resources management and also lacking integration of scheduling on the watershed area. Based on the hydrological model, the models, used for the river flood routing and transport of pollutants are very rough.

This article intends to build a set of model system including flood forecasting, hydrodynamic and water quality modeling, multi-objective optimization and taking the Beiyun River for the case, studies water distribution and water management methods and strategies and explores management models to alleviate the contradiction of water resources in north China.

2 An integrated system of forecasting and prediction

An integrated forecasting and prediction system is aimed to complete the hydrological simulation, flood routing and transport of pollutants, Locks and Dams controlling and other multi-objective dynamic optimization tasks. The system contains distributed and non-point source model, hydrodynamic and water quality model (including sluices and weirs hydraulic model), and multi-objective optimization model. Time scale of the simulation can be the flood event or long continuous series.

2.1 Distributed hydrological and non-point source model

The distributed model can be SWAT model and DWSM model based on the simulation time scales.

SWAT (Soil and Water Assessment Tool) model is developed by Agricultural Research Service (ARS) of the U. S. Department of Agriculture (USDA), which is distributed hydrological model having a 30 years history of application in large watershed scale. It is built on SWRRB (Simulator for Resources in Rural Basins) model and with a combination of the characteristics of the U. S. Agricultural Research Center several models (CREAMS, With GLEAMS, EPIC, ROTO, et al.). It is based on physical processes and continuous time simulation. It emphases hydrology simulation and uses day as time unit, and can be used to simulate different land uses, different soil types and a variety of agricultural management practices on Long-term effects of the basin's water, sediment, chemicals, and can be used to predict the total runoff, sand loss and nutrient loading of a watershed less than 100 years. The model has a wide range of applications at home and abroad, which are not discussed here.

DWSM (Dynamic Watershed Simulation Model) model is a distributed hydrological model developed by the Illinois State Water Survey, U. S. Bureau. The initial concept and structure of the model is SEDLAB of model which is developed Mississippi State University and National Sediment Laboratory of the state Ministry of Agriculture in 1979. After years, the dispersion of pollutants caused by the storm flood, soil runoff, underground pipe flow and reservoir simulation modules are added into the modules. And the original model turns into the DWSM model. It can effectively simulate surface runoff, subsurface flow, flood waves, soil erosion, sediment, and migration from agricultural areas, urban pesticides and other pollutants caused by a single rainstorm or short-term continuous rainfall. DWSM model is suitable for simulation of basins from a few square kilometers to several thousand square kilometers, using the time step from a few minutes to several hours. Also it also has some applications in typical small watersheds.

2.2 Hydrodynamic and water quality model

According to the computational efficiency and accuracy of forecasting system, generally the one-dimensional model is chosen as the river water dynamics and water quality model. One-dimensional river model for complex river systems is chosen for the limb rive river.

The one-dimensional flood routing model is working by solving the basic equation of water movement for the Saint – Venant equations, namely:

Continuity equation:
$$\frac{\partial A}{\partial t} + \frac{\partial Q}{\partial x} = q_l \qquad (1)$$

Equation of motion:
$$\frac{\partial Q}{\partial t} + \frac{\partial}{\partial x}\left(\frac{Q^2}{A}\right) + gA\frac{\partial Z}{\partial x} + g\frac{Q|Q|}{C^2 AR} = \frac{Q}{A}q_l \qquad (2)$$

where, Q is the average flow of the cross section; C is the Chezy resistance coefficient, $C = \frac{1}{n}R^{1/6}$; R is the water section of the hydraulic radius; Z is the water level.

One-dimensional water quality model uses the convection-diffusion equation, namely:

$$\frac{\partial}{\partial t}(AC) = \frac{\partial}{\partial x}\left(AE\frac{\partial C}{\partial x}\right) - \frac{\partial(QC)}{\partial x} + A\frac{dC}{dt} + S \qquad (3)$$

where, C is the concentration of pollutants; E is the longitudinal dispersion coefficient; $\frac{dC}{dt}$ is biochemical reactions; S is the source and sink terms.

Idealized analytical solution can be obtained only for a very small number of cases in the non-linear partial differential equations. In the actual cases generally it must be solved by numerical methods, that is, to approximate the infinite points in a contiguous area with a finite number of discrete grid points, using these nodes on the discrete approximate solutions to approximate the exact solution. Preissmann four eccentric formats, with convergence and high efficiency and good stability, are used to obtain discrete equations. Water quality model of convection-diffusion equation can also be discretized into corresponding differential equation which can be organized into a three-diagonal matrix equation using an implicit iterative method.

For complex river systems, the three-order solution is used. River network is divided into three micro-segments, reaches, branching points. The relationship between the micro-segment reaches are used to obtain the recursive equation of the end and the first section relationship through quality. Coefficient matrix is set up by conservation and energy equation. The matrix dimension is equal to the branching points, and the number of unknown variables is equal to the branching point of the water level. First the river equation is solved, and then the branching point of the water level is substituted into the recursive equations. And all the section level and flow are obtained.

2.3　Sluices and weirs hydraulic model

Regulatory flood sluices and weirs of the open channel are mainly used for flood regulation, so the reasonable use of the sluices and weirs is an important part of turning flood into resources. In order to maintain the calculation of continuity, sluices and weirs are generally treated as an internal boundary in the simulation by one-dimensional flow model. According to the gate opening ways, flow in four ways maybe happen, i. e. free weir flow, submerged weir flow, free flow in gate hole and submerged flow in gate hole.

(1) Free weir flow:

$$Q = mB\sqrt{2g}H_0^{3/2} \qquad (\frac{e}{H_0} \geqslant 0.65, \frac{h_s}{H_0} \leqslant 0.72) \qquad (4)$$

where, Q is the flow discharge through gate holes (or weirs); m is the integrated flow coefficient for free weir flow; B is the gate width; H_0 is the total head on gates, including the velocity head; e is the gate opening; h_s is the top depth of downstream weir.

(2) Submerged weir flow:

$$Q = \varphi_m Bh_s\sqrt{2g(H_0 - h_s)} \qquad (\frac{e}{H_0} \geqslant 0.65, \frac{h_s}{H_0} > 0.72) \qquad (5)$$

where, φ_m is the integrated flow coefficient for submerged weir flow.

(3) Free flow in gate hole:

$$Q = \mu Be \ \sqrt{2gH_0} \qquad\qquad (0 < \frac{e}{H_0} < 0.65, h_s \leqslant h''_c) \qquad\qquad (6)$$

where, μ is the integrated flow coefficient for free flow in gate hole; h''_c is the downstream water depth after jump, which is calculated as follows:

$$h''_c = \frac{h'_c}{2}\left(\sqrt{1 + 8\frac{q^2_c}{gh'^3_c}} - 1\right) \qquad\qquad (7)$$

where, h'_c is the downstream water depth before jump, that is, the water depth at vena contraction; h_c is discharge per unit width through gates.

(4) Submerged flow in gate hole:

$$Q = \mu Be \ \sqrt{2g(H_0 - h_s)} \qquad\qquad (0 < \frac{e}{H_0} < 0.65, h_s > h''_c) \qquad\qquad (8)$$

where, μ_0 is integrated flow coefficient for submerged flow in gate hole.

2.4 Multi-objective optimization model

As for scheduling characteristics for river basin water resources, the four kinds of targets of a multi-objective optimization model are considered, namely, the economic objectives, social objectives, environmental objectives and ecological targets.

Economic goals are calculated by the maximum direct economic benefits of river basin water supply:

$$\max f_1(\mathrm{x}) = \left\{ \sum_{k=1}^{K} \sum_{j=1}^{J(k)} \left[\sum_{i=1}^{I(K)} (b^k_{ij} - c^k_{ij}) x^k_{ij} \alpha^k_i + \sum_{c=1}^{M} (b^k_{cj} - c^k_{cj}) x^k_{cj} \alpha^k_c \right] \beta^k_j \omega_k \right\} \qquad (9)$$

where, x^k_{ij}, x^k_{cj} are the water supply from independent water sources i and public water source c for user j in subregion k ($\times 10^4$ m^3); b^k_{ij}, b^k_{cj} are the benefit coefficient of water supply from independent water resources i and public water source c for user j in subregion k (Yuan/m^3); c^k_{ij}, c^k_{cj} are the cost coefficient of water supply from independent water resources i and public water source c for user j in subregion k (Yuan/m^3); $\alpha^k_{ij}, \alpha^k_{cj}$ are the order coefficient of water supply from independent water resources i and public water source c for user j in subregion k; β^k_j is the water fair coefficient of user j in subregion k; ω_k is the weight coefficient of subregion k.

Social goals use the minimum regional total water shortage to reflect the social benefit indirectly.

$$\max f_2(x) = -\min\left\{ \sum_{k=1}^{K} \sum_{j=1}^{J(k)} \left[D^k_j - \left(\sum_{i=1}^{I(k)} x^k_{ij} + \sum_{c=1}^{M} x^k_{ij} \right) \right] \right\} \qquad (10)$$

where, D^k_j is the water demand of user j in subregion k.

Environmental goals use the minimum excessive emission of important pollutants to reflect the environmental benefits.

$$\min f_3(x) = \sum_{l=1}^{L} \left[Q_\lambda - \sum_{k}^{K} \sum_{j}^{J(K)} x^k_{ij} d^k_{\lambda j} \right] \qquad (11)$$

where, Q_λ is the water environmental capacity of pollutant l; d^k_{lj} is the emission of pollutant l for user j in subregion k.

Ecological goals are chosen from the eco-environmental water demand to meet the highest level.

$$\max f_4(x) = \max\left\{ \sum_{k=1}^{K} \sum_{j \in E(k)} \sum_{i=1}^{I(K)} x^k_{ij}/EX_j \right\} \qquad (12)$$

where, $E(k)$ is the collection of ecological environment water-use sectors in subregion k, EX_j is the upper limit for the water sector of ecological water.

The minimum cost maximum flow principle (max min method, the standard network solver) is chosen for the solution of joint scheduling model of water quantity and quality. The solution is different from the general linear programming simplex method. And the details won't be given here.

2.5 Model coupling and integrating

In order to maintain the continuity of the calculation, the Locks and Dams are regarded as the inner boundary of the river. Locks and Dams hydraulic model and river hydrodynamic model are coupled during calculation. The appropriate over – current formula is chosen according to hydrological conditions of the river upstream and downstream and the regulation of Locks and Dams. This makes it easy to feedback the regulation of Locks and Dams and river. According to the rainfall amount, distributed hydrological and non-point source model can calculate producing convergence process of the main tributary of the River Basin and the process of non-point source loads. The results can be used as the outer boundary conditions of the river (or river net) hydrodynamic and water quality models. According to schedule objectives to optimize the scheduling the way of the sluices and weirs, the multi-objective optimization model can be used to evaluate the pros and cons of the hydrodynamic and water quality model results.

3 Study case

3.1 Beiyun River Basin

Beiyun River is the major drainage channel in Beijing, which is responsible for flood control, drainage and dual function. It is originated in the territory of Yanshan Mountain of Beijing. The upper reaches are located in mountainous and hilly region, and downstream are located in the plain area. The terrain tilts from northwest to southeast. The river above Beiguan gate of Tongzhou District is called Wenyu River. And the rest is called Beiyun River. The river receives flows from Tonghui, Liangshui, Fenggangjian River along the way, and meets Yongding River at Qujiadian, and then meets the Haihe River at Zibeihui. The length of main stream is 143 km and the drainage area is $6,166$ km^2 in which the mountain area is 952 km^2 accounting for 16% of the total area, and the plain area is $5,214$ km^2 accounting for 84% of the total area as shown in Fig. 1 (a). In Beiyun River Basin, river network is developed and there are numerous buildings. There are more than 60 rivers above level 3 totally in the watershed. And there are 6 gates on the river including Yangwa, Beiguan, Yulinzhuang, LuTuan, Xinbao, Weigou and 4 rubber dams. And on tributaries there are 140 Locks and Dams, shown in Fig. 1 (b). Mean annual rainfall is 643 mm in Beiyun River basin and mainly concentrated in June to September, accounting for 84% of the full year rainfall; annual average of runoff (1956 to 2000 series) is 4.81×10^8 m^3, in which runoff in the mountain area is 1.29×10^6 m^3, and runoff in the plain area is 3.52×10^8 m^3. Annual water flow at the exit (including wastewater) is 9.03×10^8 m^3, in which the city floods and waste water are the main part. With the rapid economic and social development, discharges of industrial wastewater and urban sewage continue to increase in Beijing.

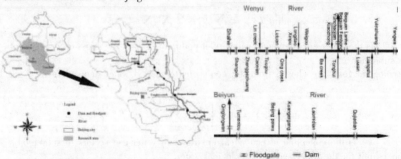

(a) Drainage network   (b) The main stream of sluices and weirs

Fig. 1   Sketch of Beiyun River basin

3.2 Model calibration

In order to test the reliability of the model system, the rainfall, flow, water level and water quality data in Beiyun River basin in 2007 is selected for model calibration and validation. The simulated and measured COD, NH$_3$-N concentrations are listed in Fig. 2 and Fig. 3. The simulation accuracy of the model for the concentrations is acceptable.

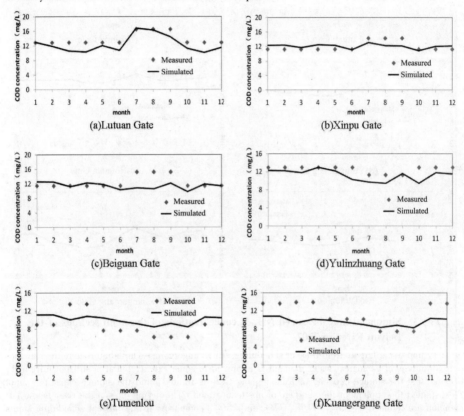

Fig. 2　Simulated and measured COD concentrations of the main sections of the Beiyun River in 2007

3.3 Scheduling calculations and analysis

The model system is used for schedule calculation of discharge and water quality of Beiyun River in 2007, the wet year, normal year and dry year, respectively. The results of ecological scheduling (optimal scheduling) and present scheduling are compared. The satisfaction conditions of ecological water of each river reaches under different schemes are listed in Tab. 1. It can be seen from the table that in the three reaches of Shahe Gate—Beiguan Gate, Beiguan Gate—Yangwa Gate, and Yangwa Gate—Tumenlou, all the ecological water demands can be satisfied under different hydrological patterns and different scheduling, and the basic heath of river reaches can be maintained. But under the present schedule scheme, Tumenlou—kuangergang and Kuangergang-Qujiadian reaches often dried up, the ecological water demand cannot be fully met. After ecological scheduling, in Tumenlou—Kuangergang reach the ecological water demand throughout the year can

188

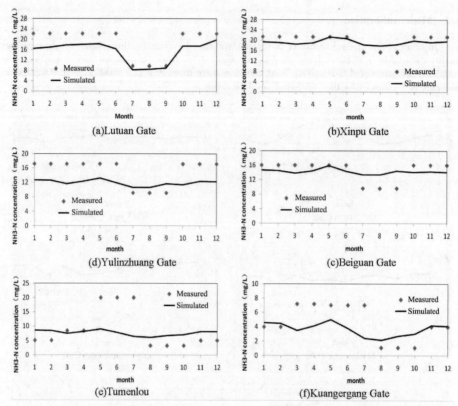

Fig. 3   Simulated and measured NH₃-N concentrations of the main sections of the Beiyun River in 2007

be met, improving largely from present scheduling. In Kuangergang – Qujiadian reach, the satisfaction conditions of ecological water demand is improved to some extent, but it is still not fully met, because that the storage capacity of Kuangergang Gate is limited, and the storage capacity cannot be sustained to supplement the shortage of upstream runoff. The drying days of the river below Tumenglou are reduced by 80%, 71%, 77% and 75%, comparing to the present scheduling status in the base year (2007), wet years, normal year and dry year respectively. By discharge analysis, the annual discharged water of the river reaches under different scheduling schemes is not large, but there is a regulatory function of the water distribution in the year.

Take wet years for an example. Fig. 4 shows the corresponding monthly average COD concentration variation process of the monitoring sections of the control reaches including Beiguan Gate, Yulin Zhuang, Kuangergang Gate and Laomidian under the present scheduling and ecological scheduling. Fig. 5 shows the corresponding monthly average ammonia concentration variation process of the monitoring sections of the control reaches under the present scheduling and ecological scheduling. It can be seen by the results comparison, in wet year, Kuangergang and Qujiadian drying up occurs under present scheduling, while ecological scheduling can avoid this from happening. Ecological scheduling can reduce the COD and ammonia concentrations of each reach to some extent, but the rate of reduction is very limited, especially in the Beiguan Gate and YulinZhuang Gate sections, compared with the current year, the improvement of water quality under ecological scheduling is also greatly weakened. To achieve water quality goal in water function zone planning, it must focus on upstream pollution control in wet year and taking into account the optimal allocation of reclaimed water in the flood season and non-flood season. The water quality improvement by the sluices and dams scheduling in flood season (June to September) is slight, water quality is even deteriorated in July and August. The main reason is that in flood season the flow discharge is large and the water quality from upstream is poor, therefore after ecological scheduling, large flood still discharges directly, basically no change in the discharge of the sluices. In addition, sluices and dams scheduling improves the downstream reaches of Beiyun River significantly. The water quality improvement at Laomidian section is most obvious. The improvement, however, is not very obvious in Beijing reach.

Tab. 1   The satisfaction condition of the Beiyun River ecological water demand under different annual scheduling scheme

Hydrological year	River reach	Days satisfied ecological water demand under different scheduling	
		Present scheduling	Ecological scheduling
2007	Shahe Gate—Beiguan Gate	365	365
	Beiguan Gate—Yangwa Gate	365	365
	Yangwa Gate—Tumenlou	365	365
	Tumenlou—Kuangergang	312	365
	Kuangergang—Qujiadian	284	338
Wet year	Shahe Gate—Beiguan Gate	365	365
	Beiguan Gate—Yangwa Gate	365	365
	Yangwa Gate—Tumenlou	365	365
	Tumenlou—Kuangergang	291	365
	Kuangergang—Qujiadian	283	320
Normal year	Shahe Gate—Beiguan Gate	365	365
	Beiguan Gate—Yangwa Gate	365	365
	Yangwa Gate—Tumenlou	365	365
	Tumenlou—Kuangergang	310	365
	Kuangergang—Qujiadian	276	332
Dry year	Shahe Gate—Beiguan Gate	365	365
	Beiguan Gate—Yangwa Gate	365	365
	Yangwa Gate—Tumenlou	365	365
	Tumenlou—Kuangergang	294	365
	Kuangergang—Qujiadian	274	324

According to the analysis method of multi-factor index method, calculation of the Beiyun River under the present scheduling and ecological scheduling multi-factor index is shown in Tab. 2. It can be seen that optimal scheduling can improve the Beiyun River water quality to some extent, the whole river water quality in 2007, in wet year, normal year and dry year is increased by 9.20%, 11.81%, 11.08% and 12.32%.

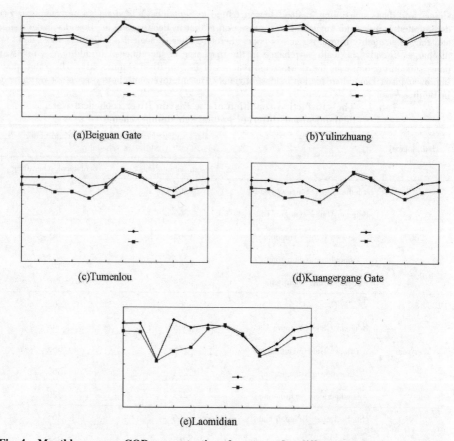

(a)Beiguan Gate　　　　　　　　(b)Yulinzhuang

(c)Tumenlou　　　　　　　　(d)Kuangergang Gate

(e)Laomidian

Fig. 4　Monthly average COD concentration changes under different modes in wet years

Tab. 2　I values under different scheduling modes in different hydrological years

Year	Present scheduling I_1	Ecological scheduling I_2	$(I_1 - I_2)/I_1$
2007	2.32	2.10	9.20%
Wet year	2.40	2.11	11.81%
Normal year	2.45	2.18	11.08%
Dry year	2.52	2.21	12.32%

4　Conclusions

　　Regarding to the serious problems of water resource shortages and water quality deterioration of rivers in northern China, we develop a predicting and forecasting integrated system which is comprised of distributed hydrological model, the hydrodynamic and the Locks and Dams hydraulic model, water quality model and optimization model. The system considering the whole watershed can be used to schedule the watershed flow process dynamically, according to the spatial and temporal distribution of the basin-wide water resource with diverse quality. By scheme optimization, it can achieve the multiple goals of minimizing flood losses, maximizing ecological water and improving wa-

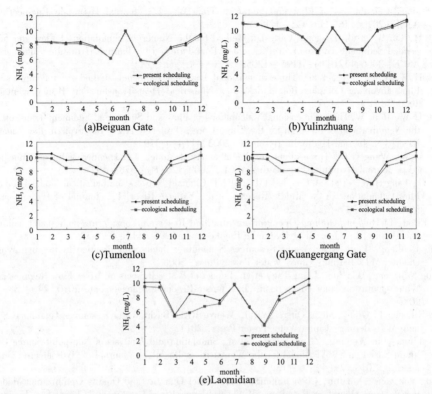

Fig. 5 Monthly average NH₃ – N concentration changes under different modes in wet years

ter quality mostly. Choosing Beiyun river watershed as a study example, we carry out calculation of optimizing the flow process scheduling on Beiyun River. The results indicate that under the premise of flood control, the drying days of the river below Tumenglou are reduced by 80% , 71% , 77% and 75% , comparing to the present scheduling status in the base year (2007) , wet year, normal year and dry year respectively. The mean index of whole river water quality is reduced by 9.2% , 11.8% , 11.1% and 12.3% , comparing to the present scheduling status in the base year (2007) , wet year, normal year and dry year respectively, and the river water quality below Tumenglou is improved more obviously. Due to limited regulation and storage capacity of the river – type gates and dams, the change of the annual discharged water of the river reaches under different scheduling schemes is not large, but there is a regulatory function of the water distribution in the year. The optimized water scheduling scheme can change the status of non – flood season drying and be a guide to the watershed water management.

References

Tecle A, Fogel M, Duckstein L. Multicriterion Selection of Wastewater Management Alternatives [J]. Journal Water Resource Planning Management ASCE, 1988, 114(4): 383 –398.

Brendecke C M, Deoreo W B, Payton E A, et al. Network Model of Water Rights and System Operations[J]. Journal Water Resource Planning Management ASCE,1989, 115(5): 684 – 696

Chen Z, Fang H, et al. Numerical Simulation of Wind-induced Motion in Suspended Sediment transport[J]. Journal of Hydrodynamics, 2007,19(6): 698 –704.

Fang H, He G, et al. Influence of Vertical Resolution and Nonequilibrium Model on Three-dimensional Calculations of Flow and Sediment Transport[J]. Journal Hydraulic Engineering. ASCE, 2010,136(2): 122 –128.

Fang H, Liu B, et al. Diagonal Cartesian Method for the Numerical Simulation of Flow and Suspended Sediment Transport over Complex Boundaries[J]. Journal Hydraulic Engineering ASCE, 2006,132(11): 1195 –1205.

Fang H W, Chen C J, et al. Three-dimensional Diagonal Cartesian Method for Incompressible Flows Involving Complex Boundaries [J]. Numerical Heat Transfer Part B-fundamentals, 2000,38(1): 37 –57.

Fang H W, Rodi W. Three-dimensional Calculations of Flow and Suspended Sediment Transport in the Neighborhood of the Dam for the Three Gorges Project (TGP) Reservoir in the Yangtze River[J]. Journal Hydraulic Resource, 2003,41(4): 379 –394.

Fang H W, Wang G Q. Three-dimensional Mathematical Model of Suspended-sediment Transport [J]. Journal Hydraulic Engineering ASCE, 2000,126(8): 578 –592.

Han D, Fang H W, et al. A Coupled 1-D and 2-D Channel Network Mathematical Model Used for Flow Calculations in the Middle Reaches of the Yangtze River[J]. Journal of Hydrodynamics, 2011,23(4): 521 –526.

Fan Erlan, Li Huaien. Optimal Operation for Stratified Reservoir Considering both Water Volume and Quality[J]. Journal of Hydraulic Engineering, 1996 (11): 33 –37.

Ruan Renliang. Plain River Water Resources Scheduling theory and Practice to Improve Water Quality[M]. Beijing: China Water Power Press, 2006.

Zhang Yongyong, Xia Jun, Liang Tao, et al. Impact of Water Projects on River Flow Regimes and Water Quality in Huai River Basin[J]. Water Resource Management, 2010(24): 889 –908.

Liao Rihong, Li Qijun, Meng Qingyi, et al. Wenyu River Basin Water Resources Guarantee System[M]. Beijing: China Water Power Press, 2011.

Tang Lihua, Lin Wenjing, Zhang Sicong, et al. Simulation and Analysis of Non-point Source Pollution Based on SWAT Model for the Wenyu River Basin[J]. Journal of Hydroelectric Engineering, 2010, 29(4): 6 –13.

Zhang Yongyong, Xia Jun, Chen Junfeng, et al. Water Quantity and Quality Optimization Modeling of Dams Operation Based on SWAT in Wenyu River Catchment, China[J]. Environment. Monit Assess, 2011(173):409-430.

Borah D K, Bera R M,Xia. DWSM-dynamic Watershed Simulation Model[M]. Colorado: Water Resource Publication, 2002.

Zheng Yi, Fang Hongwei, Han Dong. Principle and Application of Runoff Yield Module of Dynamic Water Simulation Model (DWSM)[J]. South-to-North Water Transfers and Water Science & Technology,2008, 6(3): 28 –31.

Liu Tong. Olympic Forest Park Stormwater Runoff and Sediment Loss Simulation Study[D]. Beijing: Tsinghua University, 2008.

Yang Guolu. River Mathematical Model[M]. Beijing: Ocean Press, 1993.

Cao Fangping. Study on 1-D Water Quality Simulation and Visualization for River[R]. Hunan: Central South University, 2008.

Li Yitian. A Junctions Group Method for Unsteady Flow in Multiply Connected Networks[J]. Journal of Hydraulic Engineering, 1997(3): 49 –51.

Lu Yongsen. Environmental assessment[M]. Second Edition. Shanghai: Tongji University Press, 1999.

Analysis on the Utilization and Management of the Yellow River Water Resources of Shanxi Province

Liu Xu[1] , *Dong Liguo*[1] and *Li Wei*[2]

1. Yellow River Shanxi Bureau of YRCC, Yuncheng, 044000, China
2. Shanxi Xiaolangdi Yellow River Diversion Project, Yuncheng, 044000, China

Abstract: It is rich in coal resources, but short for water resources in Shanxi Province with nine drought years out of ten. The water shortage has been a major "bottleneck" restricting the economic and social development in Shanxi Province. With the growing shortage of water resources, the rational development and effective utilization of the Yellow River are becoming more and more important for the economic and social development in Shanxi Province, especially for the areas along the Yellow River. This paper analyses the main reasons for the low efficiency of water resources utilization of the Yellow River in Shanxi Province.

On the condition that in the "12th Five – Year" period Shanxi Province will improve the water project construction, while industrial and agricultural water consumption of the Yellow River Basin will significantly increase, we should further strengthen the development and utilization of the Yellow River water resources to guarantee the sustainable economic and social development in the basin by the greatest extent on the premise of maintaining the healthy life of the Yellow River.

Key words: water resources, contradiction between supply and demand, management measures

It is rich in coal resources, but short for water resources in Shanxi Province with nine drought years out of ten. The water shortage has been a major "bottleneck" restricting the economic and social development in Shanxi Province. According to the per capita water resource, the index of water resources of Shanxi Province is the second last in the whole nation. The per capita water resource is only one sixth of the national average level, and the distribution of water resources is extremely unbalanced. For Shanxi Province, the Yellow River is the largest and the most stable and reliable water resources. With the growing shortage of water resources, the rational development and the effective utilization are very important for the economic and social development in Shanxi Province, especially for the development of the counties and cities along the Yellow River.

1   The basic situation

Shanxi Province is located in the east of Loess Plateau, and its climate is semi-humid and semi-dry. More than 80% of the whole area is mountain area and hilly region, and the landscape on the whole is the mountainous plateau covered by the loess widely. The special geographical environment forms the unique water conservancy conditions. The multi-year average precipitation is 508.8 mm and the multi-year average total amount of water resources is 12.38×10^9 m^3 cubic meters, which occupy 0.4% of the national water resources. The per capita possession of water resources is 381 cubic meters, which covers 17% of the nationwide per capita possession of water resources on the same period. The Haihe River basin area covers 38% in Shanxi Province and the main tributaries are Sanggan River, Yongding River, Hutuo River, Zhang River, etc. The Yellow River basin area covers 62% in Shanxi Province and the main tributaries are Fenhe River, Qin River, Sushui River, Sanchuan River, etc. The Yellow River goes into the Shanxi—Shaanxi Gorge from the Laoniu Bay of Pianguan County by way of Inner Mongolia, from the north to the south flows through 19 counties and cities of Xinzhou city, Lvliang city, Linfen city, Yuncheng city, and goes into Henan province in Yuanqu County.

2　The current situation and analysis of development and utilization of the Yellow River water resources in Shanxi Province

2.1　The situation of development and utilization

According to the 61[#] file from the general office of the state council(1987), "The notice for the allocation plan of the Yellow River water resources", the distribution plan assigned Shanxi Province is 4.31 $\times 10^9$ m^3 of water resources quota. (Details refer to the Tab.1).

Tab.1　Comparison between actual water consumption and distribution quota in Shanxi Province for the years from 1999 to 2010　　　　　　　　　　(Unit: $\times 10^8$ m^3)

Year	Surface water withdrawal	Surface water consumption	The annual water consumption distributed	Surplus water consumption quota
1999	10.83	9.59	36.06	26.11
2000	10.86	9.94	34.08	24.44
2001	11.48	10.46	30.13	19.67
2002	11.38	10.43	27.57	17.14
2003	11.17	9.60	31.44	21.84
2004	11.67	10.07	35.93	25.86
2005	13.41	11.81	38.17	26.36
2006	14.75	12.90	40.01	27.11
2007	15.65	13.58	37.70	24.12
2008	16.95	14.47	38.40	23.93
2009	17.53	15.08	38.19	23.11
2010	21.07	18.17	37.47	19.30
Average(1999~2010)	13.90	12.18	35.43	23.22

2.2　Analysis of the current situation

The reason for the low efficiency contains the following two factors. Firstly, engineering water shortage problems are prominent. Because of the natural conditions, the section in Shanxi Province of the Yellow River is mostly advanced the canyon area, the flow path is short, the gradient is steep, and the geographical elevation is high with the water level is low. The most people of the areas along the Yellow River can not utilize the Yellow River water because of the special landscape of "a river between two mountains". In spite of the fact that in recent years the government has increased the development and utilization of water resources of the Yellow River in order to solve the water shortage problems, for a variety of reasons the quantity of water conservancy projects are few and almost disrepair besides Wanjiazhai Yellow River Diversion Project costly built at the beginning of this century and the Yellow River Water Diversion North Main Stream Project being built. The construction of the Yellow River water diversion and supply project lags behind and the ability of water supply is seriously insufficient, which have restricted the regional economic and society development. Secondly, water supply structure is not reasonable. In these days, the people use underground water as the surface water is out of use. More and more farmers discover their wells are sere, when the industrial consumption needs more and more underground water. According to the survey, the average annual total water utilization in Shanxi Province is about 6.5×10^9 m^3, in which surface water utilization is 2.5×10^9 m^3, groundwater extraction reaches 4×10^9 m^3, accounting for two-thirds of the total provincial water supply. On the one hand this shortage situation is related to the destruction of the groundwater resources caused by the coal mining. On the other hand, due to the insufficient capacity of surface water supply, most areas in Shanxi Province must rely on groundwater over-extraction, which can barely keep equilibrium of the supply and demand. Groundwater exploitation degree is as high as 77%, forming many underground funnels whose size and depth are different; among them the situation is most serious in Taiyuan, Linfen and Yuncheng city. Because of groundwater over-extraction, shallow layer groundwater has dried up and deep layer groundwater level drops from 2 m to 30 m one year. The depth of the most wells reaches 500 m, part of them over 1,000 m. There are 19 large karst springs in Shanxi province. Three of them have been completely dried up and two of them have been dried up basically, which are all locate in the Yellow River basin. At present it is difficult to see the springs in Shanxi Province, which is named "province with thousand springs" at one time.

Groundwater is the most valuable strategic resources for economic and social development, so it is should be kept maintenance completely and protected effectively in the general year, and be used for emergency situation in the drought year. From the structure of water utilization, we should utilize surface water primarily and exploit groundwater as supplemental way. Quite the contrary, groundwater utilized accounts for two-thirds and surface water one – third in Shanxi Province. If such situation can not be reversed at all, we would not ensure the economic security and social stability by groundwater resources supply once the circumstances occur such as a drought for continuous years, river dried up and no groundwater extracted.

3　The growing tendency of development and utilization of the Yellow River water resources and the problems in the management

With the continuous development of economy and society in Shanxi Province, the demand to the Yellow River water resources will also greatly increase, which intensifies the contradiction between supply and demand of the Yellow River water resources. On the one hand, the government strives to develop the construction of water conservancy in Shanxi Province, and water use index will increase in the near future. In the water for life and agricultural irrigation aspects, in the period of "11th five – year plan", Shanxi province had carried out the implement "the strategy of sustainable development for water resources", the constructions including Wanjiazhai Yellow River Diversion Project had been built; In the period of "12th five – year plan", Shanxi Province actively raised funds and built "Shanxi Big Network of rivers" project. At present, the central Yellow River diversion project and Xiaolangdi Yellow River diversion project are successively started. For the industrial use, Shanxi combined with "the reform pilot for the transformation with the economic development of comprehensive", and vigorously promoted industrial structure adjustment, a large number of industrial project will be built. As the result, the demand for the industrial use will increase. At the end of "12th five – year plan", the water supply capacity will exceed more than 8.6×10^9 m^3, in which the surface water will reach 6.1×10^9 m^3, while groundwater will be controlled in 2.5×10^9 m^3; On the other hand, the most stringent water resources management system will be carried out, Yellow River Conservancy Commission will control water total amount and section flow. How to properly handle the relationship between our life and water basin, reasonable distribution of production, ecological water, the supply and demand balance points, express the comprehensive benefits of the Yellow River water resources. This problem needs higher requirements.

4　Countermeasures and suggestions to strengthen water resources management of the Yellow River

4.1　Further improving the water utilization efficiently

We suggest, according to the requirements of the strictest water resources management system, Shanxi Province should overall consider the quota of the local surface water, groundwater and the water supplied by the Yellow River mainstream and tributary and the water distribution of south-to-north water transfer middle line project, optimize the allocation of water resources, strengthen to save water, improve water utilization efficiency and benefit and promote water-saving society construction. On the base of produce on water and demand on supply, Shanxi province should adjust industrial arrangement reasonably, not expand farmland irrigation areas and increase massive agricultural water utilization blindly in the areas along the Yellow River. We recommend that for those cities without the quota of water utilization, Shanxi Province should extend the successful experience of transferring water rights in the Ningxia Hui Autonomous Region and the Inner Mongolia Autonomous Region on the middle reaches of the Yellow River, organize and form the general plan of transferring water rights, and analyze the potential of the Yellow River division water utilization and saving and the feasibility of water right transfer according to "Measures for the Administration of Transferring Water Rights of the Yellow River".

4.2 Strengthen reservoirs construction in the middle reaches, and develop the water regulation system

At present the mainstream and tributary reservoirs are few in the middle reaches of the Yellow River, and there is the only one reservoir of Wanjiazhai Reservoir, having the regulating ability of water yield to some extent. The regulating ability of the reservoirs is not strong and water resources are in short of supply. We recommend that, in the Yellow River administration and development plan, the rational allocation of water resources in upper, middle and lower river reaches should be considered as emphasis, the construction of Guxian, Qikou, and Yumenkou reservoirs be sped up on the middle reaches of the Yellow River, the water and sediment regulation system on the north mainstream of the Yellow River be formed, and engineering methods for the middle reaches of the Yellow River be increased.

4.3 To speed up and realize the fine management of water quantity

At present, the remote monitoring system has been built dozens of culverts and sluices in the lower reaches of the Yellow River and the remote control management have been realized. But some controlled key projects with many large reservoirs on the upper and middle reaches of the Yellow River are still under the artificial monitoring and control, and it is hard to ensure water-take time and utilization by the index. So we should speed up the remote monitoring system construction of the water intakes in the upper and middle reaches of the Yellow River mainstream, strengthen monitoring and control, effectively control the water-take index of water intakes, realize the fine management of water quantity, ensure the water utilization reasonably, realize no drying up of the river, and maintain the sustainable development of the Yellow River.

4.4 Further improve and establish the water regulation management system of the Yellow River in Shanxi Province

According to "Water Regulation Ordinance of the Yellow River", the long-effect mechanism of the Yellow River water regulation is set up, which greatly promotes the optimal allocation of the limited water resources of the Yellow River, is in favor of improving the utilization efficiency, eases the contradiction between supply and demand of the Yellow River water resources, solves the existent problems about the water regulation, reduces and eliminates the serious consequences caused by the Yellow River drying up, and provides a powerful legal protection for local people's lives and work and Long-term development. Therefore, we also should improve water regulation management system and the annual water regulation plan, and increase the consultation mechanism in the future. Firstly, it is to handle the relationship between left and right bank of the upper and lower reaches correctly, adjust the water regulation quota timely, and deal with the relationship between the local interests and the national interests. In accordance with the Yellow River water distribution solutions, annual plan of water regulation and month, ten-day plan of water regulation, further intensify the provincial consultation mechanism and procedure, strengthen the supervision, ensure smooth implementation of the ordinance, and reduce water disputes events. Secondly, it is to continue to set up and perfect the Yellow River water distribution system, further standardize and improve the total amount control index (refined to city) plan about Yellow River water intake. Thirdly, it is to optimize the emergency regulation system and give full play to comprehensive benefits of water resources in the Yellow River.

References

Zhao Haixiang, Xing Xinliang, Jia Xiaokai, et al. Ten Years Effect Evaluation of the Yellow River Water Regulation and the Construction of Functional Continuous Flow Regulation Index System [R]. Shanxi: Yellow River Shanxi Bureau of YRCC , 2011.

Wang Ling, Chen Yongqi, Pan Qimin, et al. The Yellow River Water Resources Communique [R]. Zhengzhou: YRCC, 2010.

Kang Meixiang. Develop Rationally and Utilize Effectively the Yellow River Water Resources, Resolve the Water Shortage Situation [N]. Shanxi: Shanxi Daily, 2010.

Study on Turning Jindi River Flood into Resources

Zhang Shuhong, *Chen Chen* and *Song Ning*

Yanggu Yellow River Engineering Bureau, Liaocheng, 252300, China

Abstract: The current study on turing Jindi River flood into resources is responding to the spirit of this year's first central document of implementing the most stringent water management system. We can conclude from the analysis that middle and upper reaches of Jindi River have water resources that can be utilized to irrigate Yanggu County in normal years and that drainage from Zhangzhuang gate alone is able to meet the storage capacity of the river channel in the end of Jindi River. Also that turning the downstream flood of Jindi River into resources will not affect waterlogging drainage; neither will it affect the safety of flood control of Yellow River and Jindi River.

If we can find out the balance between the Jindi River flood safety and utilization, it will have a significant influence in supporting the stable development of the society and the economy with limited water resources, maintaining the health of Yellow River, actively carrying out the construction of water – saving society and taking fully use of water resources.

Key words: Jindi River, flood, resources

1 Origin

The first central document in 2011 emphasized on implementing the most stringent water regulation system. At present, China has a big population but not enough water resources. This situation is especially typical in Yellow River, Huaihe River and Haihe River area, with only 7% of the country's water resources and an amount of water resources per capita less than 450 m^3. How to save water and take good advantage of the available water resource is now a priority. However, when we look at Jindi River, every year the end of Jindi River has flood storage of 1.2×10^8 m^3, which turns into stagnant water in the water course going downstream from Liuhai in Northern Jindi (114 +000). The 1.2×10^8 m^3 of flood water is not only not turned into resource, but poses great pressure on flood control projects in the end of Jindi River, which ruins 1,500 hm^2 farmland of Shiwuliyuan, Zhangqiu and EcRiverng in Yanggu County and affects the local people's life and agricultural activities.

According to the analysis, since 1989 Taochengpu irrigation district in Yanggu draws an average of 1×10^8 m^3 of water from Yellow River. If the flood of Jindi River can be effectively utilized, it will supply enough water to people's life as well as industrial productions of Yanggu County; plus it will save the Yellow River water resources and keep Yellow River healthy. Therefore, how to achieve the harmonious coexistence of humans and water, turn flood water into resources in Jindi River area and find out the balance between safety and utilization of Jindi River flood remains an urgent issue.

2 Sources of Jindi River flood

2.1 Overview of Jindi River

Jindi River is a tributary on the left bank of the lower reach of the Yellow River basin with an area of 5,047 km^2 crossing Henan and Shandong Province, involving 12 counties and 5 cities including Xinxiang, Hebi, Anyang and Puyang in Henan Province and Liaocheng in Shandong Province. The flood in the mainstream of Jindi River is formed by the following process: water is discharged into Yellow River through Zhangzhuang gate and with Yellow River's bed becoming higher and higher, diverting through Zhangzhuang gate becomes more and more difficult. The situation has become even worse after the Jindi River management, with the river channel lowered by 1 ~ 3 m, northern and southern dyke elevated by 2 ~ 3 m, reaching the same level as Northern Jindi flood water level; And after the renovation of Zhangzhuang gate in 1999, the base board was elevated by 3 m, causing the increase of water coming downstream. Therefore, the 1.2×10^8 m^3 of flood water in Jindi River was stored in the watercourse going downstream from Liuhai, and became stagnant water in the end of Northern Jindi River(after $114 + 000$), posing great pressure on flood control projects in the end of Jindi River. And it ruins 1,500 hm^2 farmland of Shiwuliyuan, Zhangqiu and Echeng in Yanggu County and affects the local people's life and agricultural activities.

2. 2 Overview incoming water of Jindi River

The runoff of Jindi River is as follows: the main stream Wu Yemiao station has an average yearly runoff of 1.108×10^8 m^3, Puyang station 1.64×10^8 m^3, Fanxian station 2.22×10^8 m^3 and the tributary Kongcun station 0.405×10^8 m^3. The runoff has a wild change during the year. For example, annual maximum runoff of Puyang station is 7.044×10^8 m^3, 53.8 times of that of the minimum, $0.131,3 \times 10^8$ m^3. So the runoff distribution is extremely uneven. During flood season (July to October) Puyang station accounted for 68.3% throughout the year while Fanxian station accounted for 75%. Since 1968, the average yearly runoff of Fanxian station is $1,905,36 \times 10^8$ m^3. Now let us look at the monthly runoff, the runoff of Yanggu station in 1956 ~ 1987 is relatively small (except that of August and September). And in some particular year such as 1959, 1965, 1979 and 1981, zero flow happened in three months, while the same situation also happened in Fanxian station in 1979.

3 Impact of Jindi River flood on the flood control project

3.1 Overview of sluice gate project in the end of Northern Jindi River

Northern Jindi flood detention basin is an important part of the flood control system of the lower reaches of the Yellow River with a planed storage and detention volume of 2.7×10^9 m^3 (2×10^9 m^3 of the Yellow River water and 7×10^8 m^3 of waterlogging of Jindi River). The flood control project in the flood detention basin governed by Liaocheng is located in the lower end of the basin, which is not only charged with the task of defense of the lower reaches of the Yellow River flood, but also acts as a barrier against Jindi River flood. There are eight sluice gates used for flood discharge in the lower reaches of Northern Jindi River, most of which were built in the sixties and seventies. In 1984 and 2000, Zhangqiu gate and Mingdi gate were renovated. Gaotikou gate was built in 1996, originally under the jurisdiction of Jindi River Bureau, after the abolition of which, it was classified under the jurisdiction of Liaocheng Yellow River Engineering Bureau.

3.2 Overview of Zhangzhuang Gate

Zhangzhuang Gate, situated in Wuba village, Taiqian County of Henan Province (No. of the stake on the left bank of Yellow River is $193 + 981$), is a multifunctional gate for water recession, water retaining and flood intrusion when Jindi River flows into Yellow River. The nearby area of Zhangzhuang Village was originally an estuary where Jindi River flowed into Yellow River. When floods happen in Jindi River, they can be discharged into Yellow River; and when the Yellow River floods, water flows back into Jindi River. When Min Nianpei was built into Yellow River embankment in 1949, the estuary was intercepted, so people use Zhangqiu Gate in the upper reaches of Jindi River to discharge water from Jindi River into the Grand Canal. When larger floods happened in Jindi River, Yellow River embankment would be dug to drain water into Yellow River. When Northern Jindi flood detention basin was built in 1951, there was no recession project. When Northern Jindi detention basin was temporarily abandoned, the State Planning Commission agreed to build Zhangzhuang gate until 1962. When the water level of Jindi River is higher than that of

Yellow River, the sluice is opened and water is released; otherwise, the sluice is closed to prevent water intrusion.

The Zhangzhuang gate was built in 1963, with a total of 6 holes. After the first phase of reconstruction in 1999, the sluice bottom is elevated from 37.0 m to 40.0 m, and the capacity of flood discharge or flood intrusion and diversion is 1,000 m^3/s. It has four functions, that is, flood drainage, flood prevention, flood discharge and flood diversion. It can drain the flood in Huaxian, Changyuan, Puyang, Fanxian, Taiqian County into the Yellow River, and drainage water level are 43.39 m, and the corresponding maximum discharge capacity of 620 m^3/s. To compensate for the Zhangzhuang gate weak capacity of flood discharge, in the north end of the sluice is a 400 m temporary blasting entrance. After reconstruction and extension in May 2008, Zhangzhuang drainage scale reaches 104 m^3/s.

Having been harnessed, Jindi River with its rapid converging, high speed and strong scouring force, poses a threat to north Jindi flood control project. In recent years, several floods have taken place in Jindi River; especially the one which happened in September 2010 is the biggest flood in 36 years, and at 2:00 on September 10th, the flow rate of Fanxian station was 359 m^3/s. The drainage of Zhangzhuang gate is 362 m^3/s, with a total drainage of 6.27 × 10^8 m^3, up to the highest point in the history. In the past 10 years, the average drainage have reached 1.1 × 10^8 m^3.

There is a 10 km channel from Zhangzhuang gate to Zhangqiu in the Yanggu County. The bottom slope is inverse slope. The channel bottom elevation near Zhangqiu is about 36.5 m, while the channel bottom elevation near Zhangzhuang gate is about 39 m, differing by 2.5 m. In the Zhangzhuang gate the estuary into the Yellow River is more than about 900 m, and the channel with a width of more than 100 m is formed by drainage, the spillway and artificial excavation, on both sides of which is the Yellow River Beach with the elevation of 43 m or so. When Jindi River runs into the Yellow River at the flow rate of 400 m^3/s, the corresponding water level is approximately 40.0 m, the same as the bottom elevation of Zhangzhuang gates; when the flow rate is more than 400 m^3/s, it is helpful for Jindi River to flow into the Yellow River.

3.3　The situation of Jindi River end

When waterlogging in Jindi River and high water level of the Yellow River happen at the same time, it is bound to cause poor drainage. Regardless of the flow, as along as the water level reaches the corresponding level, flood control comes into a "state of alert". Suppose at the end of North Jindi the water level of Mingdi is more than 45.70 m, Shouzhang more than 44.50 m, Zhangqiu more than 44.50 m, Dongdi more than 44.41 m. At this point, the water level of Jindi River is the same as that of Nanxiaodi, as to the depth of water of the embankment, the water depth of Mingdi above is 4.7 m, that from Mingdi to Liugai is 5.0 m, Liugai to Dongdi is 7.2 m. In this case, under high water pressure, North Jindi especially the embankments below Liuhai may appear water seepage (107 + 000 ~ 107 + 800 Meng Dikou,) and piping danger, due to strength deficiency and flat terrain opposite the river; unqualified gate appears serious leakage and the debris junction may occur seepage; vulnerable spot may make the embankment roots go slip and scour dams, dam body will sink and flood control project will enter a "critical state", which will seriously affect the safety of flood control projects.

4　The overview of diverting water from Yellow River over the years in irrigation district in Taochengpu in Yanggu County

The irrigation district in Taochengpu in Yanggu County, located in southern Liaocheng City is lack of water resource, involving 18 townships of Yanggu: Jindouying, West Lake, Litai, Gaomiaowang, Dabu, Shouzhang, Boji Bridge, Qiaorun, Shizilou, Shifo, Shiwuliyuan, Yanlou, Dingshui Town, Zhangqiu, Guotun, Echeng, Qiji, Anzhen, with a total area of 1,048.5 km^2; and three townships of Shenxian: Chaocheng, Xuzhuang, Shibalipu. Since irrigation area run in 1989, the cumulative water diverting from Yellow River is 2.037 × 10^9 m^3; the average annual water diversion is nearly 0.926×10^9 m^3; effective irrigation area is 106.5 hm^2. With the expansion of the scale of the industrial and agricultural production and improvement of people's living standards, water has become the "bottleneck" of economic and social sustainable development in the irrigation area in Taochengpu. Facing increasingly water resources of Yellow River, water resources dispatch is becoming more and more difficult. Especially during peak period, the Yellow River irrigation areas "fight" against each other over water, which adds to the difficulty of the water dispatch.

5 The flood water quality of Jindi River end

Around 2000, water quality test results show that water quality of Jindi River is inferior V, which not only has serious impact on the surrounding environment, but also affects the emissions to the Yellow River and water storage.

In recent years, as most of the small paper plants and small chemical plants located upstream has been closed, the water quality is significantly improved. In May 2011, after staff of full-time environmental detection in Liaocheng University tested water and soil quality, according to "Environment Quality Standards for Surfacer Water" (GB 3838—2002), it shows that Jindi River has changed completely from waste water into water resources, which is able to meet agricultural irrigation and ecological needs, and creates a favorable environment for the use of water resources of Jindi River.

The test shows the following result, Zhangzhuang section: dissolved oxygen, ammonia, volatile enzyme meet national standards for Class I; total phosphorus meets the national Class II standard; the permanganate index meets national standards for Class IV; COD and BOD meet Class V standards; only TN does not meet national standards.

Zhangqiu section: volatile enzyme meets national standards for Class I; dissolved oxygen, ammonia nitrogen, total phosphorus meet the national Class II standard; COD, permanganate index meet national standards for Class IV; BOD meets national standards for Class V; total nitrogen does not meet national standards.

Jiagai section: volatile enzyme meet national standards for Class I; dissolved oxygen, ammonia nitrogen, total phosphorus meet the national Class II standard; the permanganate index meets national standards for Class IV; BOD meets national standards for Class V; chemical oxygen demand and total nitrogen do not meet national standards.

Ming embankment section: volatile enzyme meet national standards for Class I; dissolved oxygen, total phosphorus meet the national Class II standard; the ammonia meet national standards for Class III; COD and permanganate index meet national Class IV standard; BOD meets national standards for Class V; total nitrogen does not meet national standards.

6 Balance analysis

Jindi River is a seasonal river, and the second large cross-boundary water resources at the end of North Jindi. Effective measures can be taken in the irrigation district in the Taochengpu in Yanggu to take full advantage of flood water of Jindi River, in order to safely and effectively achieve harmony between people and water and maintain ecological balance.

6.1 Water from the upper-middle reaches of Jindi River is available water resources

According to the results of the water quality and soil sampling test conducted by School of Environmental Protection of Liaocheng University, water impoundment of downstream Jindi River(except a few indicators) meets national level 3 and above thus meets the water demand for agricultural irrigation. This test result referred to the "Environmental Quality Standards for Surface Water" (GB 3838—2002).

6.2 Water quantity of downstream Jindi River is ample, and drainage of Zhangzhuang and Mingdi sluice gate alone can meet the demand of agricultural irrigation in this area

The run-off quantity is as follows: yearly average of the station of mainstream Wuye Temple is 1.108×10^8 m^3, while Puyang station is 1.64×10^8 m^3, and Fanxian station is 2.22×10^8 m^3. Based on the above analysis, the yearly average run-off of Fanxian is 2.22×10^8 m^3, so it can qualify as usable water resources in lower reaches of Jindi River. With available water storage of 1.00×10^8 m^3, it should be taken good advantage of. Yearly water volume diverted from Yellow River is 0.90×10^8 m^3 of water and is able to meet Yanggu's water demand for irrigation.

According to the statistics, water drainage volume of Zhangzhuang gate for the past 10 years has an average of 1.10×10^8 m^3 and that of Mingdi gate is more than 5.00×10^9 m^3. Going downstream from North Jindi chainage(number 104 +000), the watercourse of Jindi River is 20 km long with an average width of 550 m, and it has a stagnant water area of 11 km^2 with an average depth of 5.5 m, so with the help of downstream Jindi River channel, it has a storage volume of 1.10×10^8 m^3. According to some other statistics, for nearly 20 years accumulated runoff of Fanxian Station is 3.682×10^9 m^3, while cumulative flood discharge of Zhangzhuang gates and Mingdi sluices is 2.423×10^9 m^3, and Taochengpu's water intake from Yellow River has an accumulative amount of 1.933×10^9 m^3. The graphic below (Fig. 1) shows the relationship between accumulated runoff (—), discharge volume(– – –) and water demand(······) for the last two decades. X represents year and Y represents water volume (1.00×10^8 m^3).

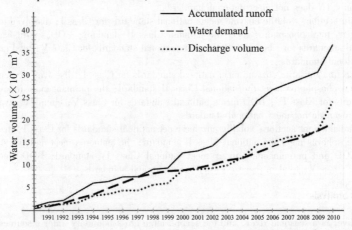

Fig.1 The relationship between accumulative run-off, drainage and water demand

6.3 Turning the downstream flood of Jindi River into resources will not affect waterlogging drainage

Water storage of downstream Jindi River comes from the waterlogging of the upper and middle reaches. The terrain is low-lying, plus main stream treatment has accelerated flood convergence speed, so the duration of the high water level in flood area is prolonged. If the flood happens once every three years, the water level of Zhangzhuang gate is 43.22 m, going downstream from 104 + 400, water level of the embankment will reach a depth of 4 m. If the flood happens once every 20 years, the water level of Zhangzhuang gate will reach 45.6 m while embankment of water will reach 6.5 m. Thus, it is a natural water storage place and will not affect Jindi River flood drainage.

6.4 Turning the downstream flood of Jindi River into resources will not affect the safety of flood control of Yellow River and Jindi River

The upper reaches of Jindi River is wide while the downstream part is narrow, and with a relatively flat terrain, a wide watercourse as well as many lowlands, it can store and slow down the flood, which explains why the process line of Jindi River flood is relatively fat. The flood duration is generally more than eight days, and bimodal flood due to two consecutive rainfalls can last up to 13 d. For Example, during the floods of August 1963, which lasted 22 d, the peak flow of Zhangzhuang station is 735 m^3/s, and the flood volume is 6.50×10^8 m^3.

We have the following findings after analysis : when the millennium flood in Huayuankou station happened, the corresponding flood volume of jindi River on the 12th is only 1.04×10^8 m^3, only slightly larger than that of 1.01×10^8 m^3 of Zhangzhuang station in 3 d in the once – every – three – year assumption. In case that catastrophic flood happens in upper and middle reaches of Yellow River, we should on the one hand divert water into Yellow River through Zhangzhuang gateway, and on the other hand open Zhangqiu and other gateways drain water into Tuhai River as well as the Grand Canal. In addition, water intake from Jindi River during the North Jindi flood regulation drill is 7.00×10^8 m^3, while now the water storage is only 1.00×10^8 m^3, far less than that of the drill. At the end of North Jindi River is built Zhangzhuang electrical pumping station, so storing water at the end of Jindi River does not affect flood control.

7 Application situation and benefit analysis

The current study on turing Jindi River flood into resources is responding to the spirit of this year's first central document of implementing the most stringent water management system. We can conclude from the above analysis that middle and upper reaches of Jindi River has water resources that can be utilized to irrigate Yanggu County in normal years and that drainage from Zhangzhuang gate alone is able to meet the storage capacity of the river channel at the end of Jindi River. Also that turning the downstream flood of Jindi River into resources will not affect waterlogging drainage; neither will it affect the safety of flood control of Yellow River and Jindi River. If we can find out the balance between the Jindi River flood safety and utilization, it will have a significant influence in supporting the stable development of the society and the economy using limited water resources, maintaining the health of Yellow River, actively carrying out the construction of water – saving society and taking fully use of water resources.

If this study is applied into practice, assuming that the irrigation rate is raised by 10% , that is to say, the irrigation of 7.1×10^4 hm^2 land in Taochengpu irrigative area is assured; assuming that per mu yield has an added value of 50 yuan, the increased profit per year amounted to $50 \times 106.5 \times 10\%$ = 5.355×10^6 Yuan. This is significant for the agricultural development of the area along the Yellow River, and economic benefits and social benefits are evident.

8 Key suggestions

After the data analysis we can see that an annual 1.00×10^8 m^3 flood of Jindi River can be turned into resources, which could meet the irrigation demand and ecological demands of Yanggu County. However, studies need to be furthered on how to better store the 1.00×10^8 m^3 water; reinforcement of existing projects or the construction of the reservoir are questions worthy of discussion by experts and scholars coming from different domains of society.

References

Management of Water Intake, Decree No. 34 of the Ministry of Water Resources of the People's

Republic of China. April 9 , 2008.

Restatement of Fanxian Station Yearly Runoff Statistics. Hydrological Stations Compilation Committee of Fanxian Station.

Water Regulation of Yellow River in Shandong. No. 4 of Shandong Yellow River Regulation. March 15 , 2007.

Yanggu County Annals. County Chronicles Codification Committee. ISBN 7 – 101 – 00939 – 5/Z. 105.

Yanggu County Water Management Annals. Yanggu County Water Management Compilation Committee, Chinese Literature and History Press.

Yellow River Water Regulation, Order No. 472 of the State Council of People's Republic of China. July 24 , 2006.

B. Ensuring Water Right of the River's Demand with Strategy and Measures of Keeping Healthy River

D. Ensuring Water Right of the River's
Demand with Strategy and Measures of
Keeping Healthy River

Foreign Experiences with Horizontal Cooperation Mechanisms for Trans-boundary Water Environment Management and Their Implications

Li Pei, *Zhang Fengchun* and *Zhang Xiaolan*

Foreign Economic Cooperation Office, Ministry of Environmental Protection,
Beijing, 100035, China

Abstract: Foreign cases and experiences of horizontal cooperation mechanism for cross-border water environment management will be introduced in this paper, including the joint implementation of the Convention on Water Quality in Great Lakes by the United States Government and the Government of Canada, the EU Water Framework Directive by EU member governments and the cross-border administrative cooperation within the basin, the cross-border watershed pollution control by the nine countries in Rhine Basin, and the inter-regional coordination and cooperation in the Thames River Basin. Besides, by taking reference of the horizontal coordination and cooperation between the relevant national departments for biological diversity in Norway, analysis and summary were made about these cases in terms of the coordination mechanisms, policies and legal coordination, overall planning, operation process, the scientific and technical support, and so on. The foreign successful experiences in trans-boundary water environment management include: ① a strong coordinating institution which is able to exert influence on the economic decisions by governments involved with trans-boundary water bodies; ② strict enforcement of the laws and regulations developed by all the governments involved; ③ coordinated planning which takes the trans-boundary water body as an ecosystem, unify management, and clear responsibilities and obligations for all related parties; ④ the need for an effective technology institution to provide scientific support for rational and sustainable trans-boundary water environment management; ⑤ continuous innovation and improvement of management technologies. These experiences have important implications for the domestic cross-border water environment management. For the trans-boundary water environment management , the establishment of and innovation in the trans-boundary water management must be compatible with existing economic and social structures and mechanisms; the management of trans-boundary water is a process of constant modification, improvement and complementation; strict enforcement of laws is the guarantee of effective management; academic counseling and support is extremely important for the management; the fully integrated river basin management is the only route leading to success; the international and inter-sectoral cooperation under the same overall framework is an effective way for the trans-boundary water management; the transboundary water management is a long-term, arduous task.

Key words: trans-boundary water environment, water resource management, ecosystem, horizontal cooperation, foreign experiences

1 Introduction

More than 260 cross-border rivers in the world cover 45% of the earth (Bernauer, 2002). And there also occurs trans-boundary water environment problems that have physical and chemical influence on nature, surface water, groundwater and atmosphere. These problems are relevant to different institutions and organizations and have something to do with policies, laws and regulations, politics as well as other factors such as technology, concepts, coordination and cooperation of institutions involved, and chronological factor. At the same time, new problems such as climate change are constantly emerging. The fluidity of water makes the trans-boundary water environment management, zoning of the rivers, coordination of obligations and interests between nations involved far more complex than the united management of trans-boundary land resources. That explains why trans-boundary water environment management has always been a global problem (Gooch and Stalnache, 2006). It is evidenced that effective coordination and cooperation between parties is the fundamental condition for the solution of trans-boundary water environment management. The study of joint management of trans-boundary water environment starts much early abroad, especially in European and American nations. Many achievements and experiences in coordination and cooperation mechanism for trans-boundary water environment management have been gained with one-odd century's constant practice and gradual improvement (Kliota et al. , 2001). Therefore, this paper analyzes some successful foreign cases and summarizes practices that China can copy and learn from with an aim to improve China's trans-boundary water environment management.

2　Experiences from the horizontal cooperation mechanism for trans-boundary water environment management

2.1　Trans-boundary water environment management of the great lakes on the canada-u-nited states border

The great lakes are a collection of freshwater lakes along the Southern Canada and the Northern US, including Lakes Superior, Michigan, Huron, Erie, and Ontario, located in eight states in US and one province in Canada. Though the five lakes reside in separate basins, they form a single, naturally interconnected body of fresh water. Four of the five lakes form part of the Canada United States border; the fifth, Lake Michigan, is contained entirely within the United States. The five lakes form the largest group of freshwater lakes on Earth, containing 21% of the world's surface fresh water, and 84% of the freshwater reserves in North America. Their total surface is 245,660 km^2 with basin area covering 766,100 km^2 (OECD, 2005).

Over 100 years ago, with the rapid increase of population in the Great Lakes Basin, urban expansion, and its industrial development, water in the basin began decreasing in quantity and being polluted. The water supply and pollution in the lakes were even deteriorated after coal, iron, copper, limestone and many other minerals are found and mined in large scale. What's worse, the deforestation caused by logging since early. Irrational agricultural activities and over farming and inappropriate water environment management, coupled with climate change and other natural and human factors, all accelerated water pollution and habitat loss, which put a threat to the wildlife in the basin. The trans-boundary water environment problems in the lakes are characterized by lack of water resources, water pollution, biodiversity loss, and invasion of alien species. And the degradation of water environment exacerbates the contradiction of the United States and Canada on trans-boundary water environment management.

Both governments later realized the importance of cooperation to protect water in the basin, which is line with the fundamental interests of both nations. Thereafter, both started their cooperation and joint efforts in water management in this basin for over 100 years.

The first agreement was signed by the United States and Canada in 1909—Boundary Water Resources Treaty, which was a great step for their joint legal system management of water environment. This agreement aims to solve all issues concerning the rights, obligations and interests of the great lakes between both nations to avoid trans-boundary water utilization conflicts. It was the behavioral framework for addressing new possible problems, but didn't touch upon underground water management. According to the treaty, an international Joint Committee was founded. Composed of three people from each nation, this committee was dedicated to the investigation, management and resolution of conflicts between the United States and Canada on the basin and its tributaries. According to the treaty, without consent from the committee, neither nation can change the level of trans-boundary waters, but both can reserve the autonomy of the management of their own water resources (Treaty Between the United States and Great Britain Relating to Border Waters Between the

United States and Canada, 1909). The committee consists of the Water Quality Board and the Scientific Advisory Board, which are responsible for the decision-making and technical consulting on water pollution. A branch office was set up in each nation to provide management support for the committee.

In 1968, another agreement—The Great Lakes Basin Compact was implemented as a complement of Boundary Water Resources Treaty, which included all basins of the Great Lakes and their underground water as well as surface water.

The third agreement—The Great Lakes Water Quality Agreement was concluded in 1972. For the first time, the Great Lakes basin was regarded as a whole ecosystem in the agreement, which stipulated each nation's obligations to restore and maintain the Great Lakes basin ecosystem. Apart from an emphasis on the obligations and rights stated in Boundary Water Resources Treaty, this agreement oriented to areas that hadn't reach their targets and areas that had already threatened or could possibly threaten aquatic organisms as "key areas" that needed focus and attention in future. Invasion of alien species, point source pollution and non-point source pollution caused by industries along the rivers were the major problems in the key areas.

In 1985, an integrity type agreement was signed—The Great Lakes Charter, which was a massive basin water withdrawal management program aimed at water supply in the lakes. It stated that each nation's withdrawal of the basin water should be informed of, consulted, and consented by both sides. Yet, little projects needed its approval because of the too high standard of consumption quantity clarified in the charter (Engler, 1993; Chester, 2003).

In 1986, The Water Resources Development Act was signed. The act legalized the terms in The Great Lakes Charter and stated that the basin water supply should be consented by all members of the committee. However, the act had its flaws, without stipulation on water consumption and whether it was applicable to the ground water (Klein, 2006).

In 2001 Annex to the Great Lakes Charter provides a legal framework for the basin water cooperation and co-management between the states of the United States and the Canadian provinces.

The Great Lakes-St. Lawrence River Basin Sustainable Water Resources Agreement signed in 2005 between U. S. states and Canadian provinces improved the planning put forward in 2001 Annex to the Great Lakes Charter and gave a trans-boundary water resources joint management process for each member state/province. Each member state/province in the lakes could join the agreement by legislation. According to the agreement, the states/provinces have the right to decide to the appropriate amount of water utilization but are prohibited to increase new water diversion from the Great Lakes.

The Great Lakes play a significant role in regional economic development and ecological environment maintenance. Their waters connect the U. S. and Canada so closely that both nations find that cooperation becomes their best single choice. And their consensus lays a foundation for their successful cooperation and integration of trans-boundary water environment management of the Great Lakes that endures over 100 years and is of great importance in pollution control and water resources allocation of the Great Lakes. At present, water quality in the five lakes has been significantly improved; biodiversity is restored; and the ecosystem service function is strengthened. The 100 odd years' cooperation between these two nations on trans-boundary water environment management of the Great Lakes is a history of constant exploration and improvement of relative laws and legislation. Their successful experiences in trans-boundary water environment management mode of the Great Lakes, which they have always been proud of, provide precious assets for trans-boundary water environment management worldwide.

2.2 Cooperation under "EU Water Framework Directive"

With an area of only about 10×10^6 km^2, Europe has 37 countries and regions, and about 70 international rivers (Anonymous, 2002). 40% of the world's population lives in trans-boundary river basins, and 55% of them are in Europe. Meanwhile, EU has lots of members, each with its own language, culture, laws, government, and institution. All these make it inevitable that those member nations to occur disagree on how to share the border water resources and how to divide obligations as well as rights. For example, the Rhine was once considered the European Sewer, and the Dutch had to "consume" sewage from its neighboring countries. In recent decades, trans-boundary water environment management has always been listed as one of the major environmental problems of EU. Cooperation is their only choice in order to effectively manage trans-boundary water environment. That's how EU Water Framework Directive came in place, signifying the start of trans-boundary water environment planning and management with basin as basic unit in Europe (Woltjer et al. 2007).

The EU Water Framework Directive implemented in 2000 is one of the most important environmental laws that EU has so far issued. It is also proved to play a decisive role in EU's water environment management.

However, the introduction of The EU Water Framework Directive is a long process. It is the joint efforts of the coordination of member countries and themselves within, and the repeated discussion of differences and common grounds, and the determination of correct laws and mechanism among involved nations, governments, institutions, and the public, so that it can satisfy both different human needs and the needs of the animal and plant communities.

In 1973 ~ 2000 period, the EU's water policies were implemented through five environment action plans, which determined many priority actions to reduce pollution and improve water quality, and accordingly introduced a large number of special directives. To the end of 1990, it was found that those special directives usually led to management division and sometimes were in conflict with the EU's overall water policies. Therefore, the EU decided to adopt a more comprehensive and integrated system, namely EU Water Framework Directive, to manage water environment in its border. The new directive replaces many special directives, and utilizes a more integrated and overall means, which helps promote the unified EU trans-boundary water environment management laws rather than local and regional ones, pollution control of the whole basin rather than partial controls, ecosystem management mode rather than different management styles in their own way, and focus on biological and ecological conditions rather than physical and chemical ones (Apitz et al. , 2006).

The EU Water Framework Directive is a joint decision made by both the Council of Ministers and the European Parliament, taking the parties' suggestions extensively, which gave NGO more opportunities to involve in the decision making process (Kaika and Ben Page,2003)

Some important contents and actions in the EU Water Framework Directive are as follows: all EU member countries and candidate countries are required to include the directive in their legislation and put it into effect. To support its implementation, the member countries and the European Commission made joint implementation strategies as guidelines, among which is the Best Practice of River basin Management Planning. To achieve the overall target, the EU requires its member countries to determine their management areas of the basin (including underground water, coastal water and rivers) and allocate capable departments to manage the areas before December, 2003. Each member country should grasp each sub-basin's characters and formulate regional river basin management plan based on the spatial management units of the basin.

The directive even sets an international river basin area in particular for river basins that cross international borders, and stipulates that all countries involved in trans-boundary rivers should submit joint river basin management plans, while for those rivers which cover areas outside the EU boundary, member countries involved are encouraged to cooperate with non-member countries to ensure managing rivers according to units clarified in the directive (Nilsson et al. , 2004).

To guarantee scientific decisions and actions and advanced technical methods, the directive emphasizes the importance of science and technology in decision-making, planning, management, and implementation of the project (Quevauviller et. al. , 2005), and encourages scientific research projects, which aim to directly serve the decision-making and implementation of directives by solving some scientific and technical problems and developing tools and methods through public relations (Mostert, 2003). Science and technology play a great part by introducing innovation and new technologies for management. Yet, the EU is aware that their next problem to solve is the conversion rate of scientific research (Willems and de Lange, 2007).

As the result of reform of economy, politics, and the society that are related with water resources management, The EU Water Framework Directive requires all member countries started integrative management on their own water environment, and each responsible department align their actions with the EU standards (Weiß et al., 2008). The directive is definitely playing an increasingly important role in regional, national, and European levels (Kaika, 2003).

To ensure the smooth implementation of the directive, the EU made lots of efforts in law enforcement and correspondent support mechanism. For example, it combines Strategic Environment Assessment Directive and the EU Water Framework Directive, and carried out strict environmental impact assessment of plans and strategies according to river basin management plan evaluation requirements (Carter and Howe, 2006). To ensure strict law enforcement, the directive adopts discrete principle to decrease decision-making errors, and punishes the polluter to pay for the restore costs (Correljé et al., 2007). To strengthen monitoring, the directive develops various monitoring index system and monitoring guidance for the fixed level (Ferreira et al., 2007), and will give them legal support.

Some people think the success of the EU Water Framework Directive is partly attributed to the full consideration of the fitness level between current departmental structure and actions and the policies and actions to be made (Moss, 2004), which promotes the EU's diversion to more diversified water resources management mode (Page and Kaika, 2003).

Since the implementation of the directive, all member countries made a fundamental change—basing all management decisions upon water ecology. Under this directive, all basin management plans should be made on the condition that living organisms can adapt to environmental stress (Hering et al., 2010).

Just like other environmental laws, the EU Water Framework Directive also has its flaws. How to enforce the cooperation between EU member countries and non-member countries (Nilsson and Langaas, 2006), and how to improve diversion rate of scientific research are issues to be solved in next steps (Willems and de Lange, 2007).

2.3　Joint pollution control of the Rhine

The Rhine is the largest river in West Europe, at about 1,232 km, covering a basin area of over 220,000 km^2. It originates from Switzerland, and flows through Liechtenstein, Austria, France, Germany and other countries, and then runs into the North Sea in the Netherlands. As an important international river and international shipping channel, the Rhine plays a significant part in the economy, society and ecological environment of the countries that it flows through, including flood control, navigation, ecology, water and electricity, agriculture, industry, and drinking water for man and animals (Middelkoop and Kwadijk, 2001).

The major problems of the Rhine's trans-boundary water environment are water pollution, lack of water resources, coordination among countries of resources allocation, and pollution control rights and obligations. In the past 170 years, especially during 1955 ~ 1977, the development and water projects along the Upper Rhine brought tremendous changes to its landscape, hydrology, vegetation, and fauna (Dister et al., 1990). Besides, construction of dams along the river, excessive water withdrawal for agriculture, and industrial and domestic water pollution are also problems that the Rhine is confronted with.

Since 1950, the International Committee of Protection of the Rhine (ICPR) had started coordinating countries of the Rhine to jointly protect the basin, but its efforts didn't pay off. Not until the Sandoz Disaster in 1986 that came the turning point of the Rhine's management mode. The countries along the Rhine gradually changed their mindsets, and decided to restore the Rhine's ecological environment. Consequently, the eco-system method progressively replaced the traditional simple management mode (Huisman, 1995; Wieriks and Schulte-Wülwer-Leidig, 1997). Under the new mode, the integrative management of the whole basin is promoted instead of local pollution control; ecological protection of the basin is emphasized rather than protection with flood control and shipping as the core. Countries along the river cooperated to build a fish corridor (van der Kleij et al., 1991). The Sandoz event also drives countries along the Rhine to jointly formulate The Rhine Action Plan with an aim to create conditions suitable for animals and plants return, guarantee the safety of drinking water, reduce pollution and implement the North Sea Action Plan (since the Rhine finally flows into the North Sea) by 2000, and finally realize consistency of the local, national, and international targets (Cals et al., 1998). The countries also work out and implement Ecological Master Plan for the Rhine to restore the habitats for migratory fish in the mainstream and tributaries (van Dijk et al., 1995).

The integrated management of the Rhine also attaches much importance to scientific support, in favor of joint research carried out by more than one country. For example, IEMA-SONGE program is a united research project by more than 30 institutions in many countries, with an aim to seek sustainable flood management method for the Rhine (Hooijer et al., 2004). Now since climate change is predicted to aggregate the supply and demand conflict of water resources, a research project on countermeasures to deal with climate change has begun as a precaution (Middelkoop and Kwadijk, 2001).

Located in the most downstream of the Rhine, the Netherlands is representative in participating in all kinds of activities of joint management of the Rhine. The treatment of the Rhine by the Netherlands is a history of constant improvement of original laws and actions. Though it has now fully joined the Integrative Management System of the Whole Basin, this kind of improvement has not yet stopped. An interactive water resources management is put forward, with more emphasis on interaction, ecosystem approach and sustainable management (van Ast, 1999).

The trans-boundary water environment management mode for the Rhine is often seen as an international model of successful management of trans-boundary water basins. Its successful experience is that water environment improvement is not the outcome of the change of one single factor, but rather the joint result of all relevant factors, which should be fully taken into account when policies and action plans are made. In fact, the management of the Rhine is a history of constant improvement and update of one management system.

2.4 The inter-regional cooperation on River Thames

With a total length of 402 km, and the basin area of 13,000 km^2, River Thames is the famous "Mother River" of Britain. It originates from the Cotswold Hills in southwest England, and flows through Buckinghamshire, Berkshire, North Hampshire, Surrey (and Gloucestershire and Wiltshire with other sources), along which it gathers many streams in England, and finally flows into the North Sea through Noel Island. Along the Thames, there are more than ten important cities such as the capital city London, and Oxford. River Thames Basin is one of the richest regions in EU. Since 2002, it has always been the zone with the fastest economic growth in the UK.

River Thames plays an important role in the UK's economic and social development, and ecological environment. The rise of the industrial revolution at the latter half of the 18th century brought prosperity to London and River Thames. Meanwhile, London's population rocketed from 1 million in 1800 to 2.75 million in 1850. At the end of the 1930 s, it was over 8 million. Together with the rapid industrial growth and population boom came the grief environment pollution of River Thames, with the discharge of a large amount of domestic sewage and untreated industrial wastewater, coupled with the garbage along the coast, which once made the Thames the London's sewage ditch.

The problems that River Thames is faced with include water pollution, flood and tide, lack of sustainable use of water resources, waterway, biodiversity loss, etc.

Though the lower reaches of River Thames—the Greater London is the major source of water environment problem, the UK regards the trans-boundary basin as one, and carries out integrative management on the Thames' whole basin through laws and administrative measures. A unified and integrative basin management mode is gradually formed, which completely broke the administrative boundaries of the counties.

Specific measures of River Thames' integrative management include: strict rules are made about discharge of waste water and sewage into River Thames through legislation; by improving infrastructure, waste water and sewage discharged into the Thames should be treated carefully; customers and polluters are charged according to ecological compensation mechanism; inspectors will strengthen law enforcement and carry out spot-checking from time to time; factories that discharge substandard sewage and disobey supervision will be prosecuted and get punished by fines or even closure.

In order to integrate management of the basin in UK, the government had issued a series of environmental laws related with water from the late fifties to the seventies of the last century since it issued Land Drainage Act in 1930. According to those laws, the UK's National Rivers Authority merged and reorganized about 1,400 institutions related to water based on basins, and founded 10 Water Authorities in England and Wales, and clearly defined their nature, tasks, and responsibilities through legislation. The Water Authorities are responsible for management and protection of the water resources of the whole basin, including preparation for long-term water resources strategic plan, development of new water sources, water diversion and supply, water quality management, licensing, and fee collection. They were supposed to realize integrative management of water treatment, aquaculture, irrigation, livestock, shipping, flood control, and basin ecosystem monitoring. The water authorities' decision-making body is the Board of Directors, consisting of representatives from fields of environment, agriculture, fisheries, food, and local representatives within the basin.

National Rivers Authority requires each partition to work out their river basin management plan. For example, Water Authority for River Thames is in charge of treatment of River Thames and its water resources management. Specifically, it includes hydrological network business management, hydrology monitoring and forecasting, industrial and domestic water supply, sewage pine system, sewage treatment and discharge, water quality control, flood control, fisheries, water tourism, etc.

Thames River Authority established a modernized water quality and hydrology monitoring network in the whole basin. Within the cover of 23,100 km^2, 57 water quality and hydrology monitoring stations and over 300 sampling points were distributed, which constituted automatic water quality and hydrology monitoring network to forecast the conditions of water environment of the whole basin. The Authority also made Thames River basin Plan and Flood Control Strategic Plan for Thames River Basin, which cover not only flood control and disaster decrease strategic plan of the whole basin, but also the sub-plans for each county, and constitute a complete system to better coordinate the overall and local interests and responsibilities.

Just like other successful management systems, the UK always attaches great importance to science and technology for support. Founded in 1963, the National Water Council, like a ministerial advisory body, offered technical assistance to the Water Authorities, and guided in their work. The council was also engaged in developing new management technology and methods, such as their achievements of environment assessment method and basin assessment method. Meanwhile, it also published River Engineering and Basin Protection: Overall Assessment Manual (Handmer, 1987).

Another feature of Thames River basin management is participatory and partnership. For example, through participating in Thames 21, the National Rivers Authority and other organizations are closely connected; some technical and policy analyses are made and specific suggestions are put forward to the national government. Thames 21 once launched an activity of cleaning up the dead ends of waterways to return the wildlife to their habitats with thousands of volunteers' participation (Bruce Mitchell, 2005).

Dating back to 120 years ago, the UK's water environment treatment has gone through constant practice and improvement. The experiences of Thames River basin can be concluded as follows: integrative management and overall consideration; unified planning, and strong law enforcement; science reliance, and massive participation of the public. A clear proof to Thames River's successful treatment is the return of over 100 fishes, many invertebrates and various birds to River Thames (Castro et al. , 2003).

2.5 Norway's cooperation of related departments in biodiversity

The protection and sustainable use of biological diversity is a typical trans-boundary environment problem. The natural and socio-economic characters of biodiversity and its management requirements have much in common with trans-boundary water environment. Therefore, some successful practices in biodiversity management can directly or indirectly provide reference for trans-boundary water environment management. The formation and implementation of Norway's Biodiversity Strategy and Action Plan is a successful case of inter-departmental coordination and cooperation.

Norway has about 40,000 species, of which 2/3 are among the forest. That's why Norway gives top priority to forest biodiversity and forest ecosystem protection. Currently, the main threat of Norway's forest biodiversity come from habitat fragmentation, road cutting, clear cutting, weak awareness of protection and sustainable use, lack of inter-departmental coordination and cooperation, etc. (Sverdrup, 1997; Selvik, 2004). In addition, in recent years, the influence of the expansion of villas to the forest areas caused much concern. Yet, there still lacks the environmental influence assessment of such move (Kaltenborn et al. , 2007).

In 2001, Norway's Biodiversity Strategy and Action Plan (2001 ~ 2005) is submitted in the form of white paper to the Congress for approval. In order to fulfill the obligations of strengthening cross-departmental coordination and cooperation in the Convention on Biological Diversity (CBD), this plan put coordination and cooperation between departments as the core, and set in particular one chapter about cross-departmental responsibilities and cooperation, which took up much space in the paper. This chapter is, in fact, a joint action policy and plan of biodiversity management across 17 departments.

To improve the implementation efficiency of Norway's Biodiversity Strategy and Action Plan, its government established, in particular, a new biodiversity management system based upon existing problems of cross-departmental cooperation and coordination. With legislation and economy as ties, this system aims to simplify the management processes and improve efficiency through delegating powers to local authorities, to make decision-makers easier to balance the interests of all parties, to make the planning more economic and effective, and to increase predictability of land management plan.

The new system firstly determined priority fields and priority regions as the focus of concern in future, and then made seven major tasks for the 17 departments according to the plan's overall requirements. Based upon those seven task and their features, each department determined their priority actions. Altogether 300 actions are gathered, and the government organized the 17 departments to implement them. The 17 departments are Ministry of Agriculture, Ministry of Children and Family Affairs, Ministry of Culture and Religious Affairs, Ministry of Defense, Ministry of Education and Scientific Research, Ministry of Environment, Ministry of Finance, Ministry of Fisheries, Ministry of Foreign Affairs, Ministry of Health, Ministry of Labor and Government Administration, Ministry of Petroleum and Energy, Ministry of Social Affairs, Ministry of Industry and Trade, Ministry of Transport and Communication, etc. To guarantee the operation of the new system, while to play the supporting role of science and technology, various committees are founded, including legislative assessment committee, and construction assessment committee.

In the implementation process, the government divided the above 17 departments into three categories: the first category is those that directly manage biodiversity resources; the second category is those that directly utilize biodiversity resources; and the last category is those that indirectly manage and utilize biodiversity resources. Moreover, it also clarified the responsibilities of each department: firstly, based upon the overall plan, all departments should assess the environment influence of activities carried out within the scope of their area, and worked on biodiversity baseline survey and monitoring; secondly, in principle, all departments should be responsible for the administration and financial management of activities arranged within their work scope, and any cooperation framework agreement or responsibilities division should be bound with finance; thirdly, all departments should actively practice cross-departmental cooperation in order to carry out more effective dialogue; fourthly, if possible, the responsibilities of the plan should be delegated to the local, which determine the choice and focus according to the overall targets and focus; fifthly, all departments should, within their own scope, provide their environment trends and impact, as well as list of expenses needed for all the activities in the annual plan.

This knowledge management based new system is actually a policy tool, which bases on scientific survey and analysis, and requires all related departments to integrate their performing responsibilities with the administrative management, including not only integration within the department, but also integration between departments. Through this tool, the government organized all departments together to work in cooperation in a due division of labor, and made administrative arrangements when necessary. This management system proved to be an efficient way for multiple departments to cope with trans-boundary environment problems, further evidenced by Norway's Barents Sea Ecosystem Integrative Management implemented in 2006, which has already achieved desired performance (Olsen et al. , 2007).

There are a lot of practices in Norway's integrative management of biodiversity that are worthwhile to learn from. For example, to encourage more extensive participation and mobilize the enthusiasm of all parties, the government fully delegated authority to the lower department, and achieved a balance of interests and rights between the central government and all departments (Fallethand Hovik, 2009). Norway made a connection between science and policy, one of whose product is Norwegian Species Red List (Jørstad and Skogen, 2010). A third lesson is the establishment of public-private partnerships as an auxiliary way to help integrative management of biodiversity (Hovik and Edvardsen, 2006).

3 Implications of the horizontal mechanism in trans-boundary water environment management in foreign countries

3. 1 Trans-boundary water environment management needs compatible innovation

Since trans-boundary water environment management is only one of environment problems in human society, innovation or improvement of any management method or coordination and cooperation mechanism should fully take into account its compatibility with existing mechanism and regulations. Adoption of a brand new way without any consideration of present mechanism can barely succeed. A good horizontal mechanism of trans-boundary water environment management should not just introduce new concepts and technology, but take fully into consideration current conditions of the region, departments, laws and regulations, policies, society, economy, and environment to integrate and make full use of available resources to the maximum. The match level between the new horizontal mechanism of trans-boundary water environment management and existing mechanisms always determine its effectiveness, which is one of the reasons that the EU Water Framework Directive's success is mostly attributed to. The formulations of all policies and actions in the Directive have taken into account their fitness level with existing department structure and actions, including their coordination with present laws, policies, and institutions, which is reflected within each member country, between member countries and EU. Their experience tells that this kind of compatibly innovative mechanism can not only fully mobilize the enthusiasm of all parties and consider their interests, but also promote the diversion to diversified water resources management mode in trans-boundary water environment management. It doesn't mean that compatibility doesn't have its management idea. In fact, it is a balance between kinds of interests, rights and obligations, which is also reflected in Norway's management mode of biodiversity protection and sustainable utilization.

3.2 Trans-boundary water environment management is a constant revising, improving and complementing process

It is the basic requirement for scientific and developmental management mode that even the best management system needs constant improvement and updating. To horizontal mechanism of trans-boundary water environment management, constant improvement and complement are particularly much more important, for the reason that the timely introduction of new concepts and technology can ensure the adopted management technology is always the most advanced one to cope with new problems that keep coming up and to update or complement out-of-date technologies that were once most efficient. Moreover, timely adjustment is also needed in line with the change of political and economic environment or policies. The joint management of the Great Lakes is a case in point. In the past almost 100 years from 1909 to 2005, laws related to the joint management of the Great Lakes water basin had gone through seven great amendments, each updating or improving the lost and unreasonable or outdated terms with new and more perfect laws issued, which enabled the Great Lakes joint management to be lead in the forefront. The management history of the Rhine is also a history with constant improvement and complement of management.

3.3 Strict law enforcement is a guarantee of effective management

The establishment of horizontal mechanism in trans-boundary water environment management calls for solid laws as its code of conduct or norms of action framework. Those laws include international conventions, regional conventions, bilateral or multi-lateral treaties. In terms of content, it includes laws and regulations in real sense, such as the Convention on Biological Diversity, EU Water Framework Directive, the UK's Water Act, River Act, Pollution Control Act, etc., and rules and documents constrained by moral, such as the US-Canada Great Lakes' Great Lakes Charter. It has been proved that law making is not the most difficult in trans-boundary water environment management. According to Rogers' (1992) research, there are already 286 conventions on international river management worldwide, among which 2/3 were from Europe and North America with an early start in trans-boundary water environment management. Apart from that, there are also numerous regional law documents in this field. Law enforcement after legislation is the hard nut to crack. How to ensure the strict implementation of laws, treaties, or agreements is the key to success. All the above successful cases attached great importance to this regard with strict law enforcement systems. To get the recognition from and implementation of all parties, all governments involved in the trans-boundary waters should participate in the formulation of laws and regulations. To assist law enforcement, many management systems establish a series of reward and punishment systems. For example, under the EU Water Framework Agreement and the management of River Thames, countries or governments that violate related agreements will get serious penalties or punishments.

3.4 The supporting science and technology should always be emphasized

The above mentioned successful cases all highly value the importance of the supportive science and technology. The functions of science and technology in trans-boundary water environment management can be concluded as follows: ①it gives scientific and technical support in decision making, which ensures the decisions made scientific and rational; ② through research and development activities, it can constantly develop and improve new management technologies, methods, and measure, and lead to innovation in policy and technology; ③it is used to cope with and solve various problems that newly occurred in the management process; ④ it supports capacity building through training, communication, etc. ; ⑤ it assists in publicizing education and raising public awareness; ⑥ it supports information and technical communication. Scientific study includes theoretical study and practical research, namely exploration on the level of theory, and specific study on problems that occur in management. Generally speaking, mature management system always has a subsidiary academic or research institution. For example, the EU Water Framework Directive, the Great Lakes management, and Norway's biodiversity management systems all have the specialized standing committee in charge of scientific and technical problems in management. Meanwhile, all those cased also encourage NGO and scientific research groups in the society to participate in the management's theoretical study, scientific investigation, technical development, information management, etc. .

3.5 Integrative management of the whole basin is the inevitable route to take

Based on the knowledge that it is of great danger to take measures to manage only some parts of the trans-boundary water basin or some reaches of rivers (Kliota et al. , 2001), all successful cases in trans-boundary water environment or biodiversity management all take integrative and comprehensive management methods without exception. This management method's idea is as follows: firstly, according to the multi-dimensional and multi-regional characters of trans-boundary water environment, it emphasized integrative treatment and management, which includes all concrete contents, institutions and interest parties concerned; secondly, according to water's mobility feature, it discards all constraints of boundaries, regarding the whole basin as a whole or complete ecosystem, and carries out overall planning and integrative management of the trans-boundary basin with ecosystem methods. Each country or government is seen as the sub-system of the whole ecosystem. In order to maintain or restore the function of ecosystem, partial policies and actions should be in accordance with the overall interests, while the overall target should also take the partial interests into account. With this idea, integrative management emphasizes the inseparability of the basin. Therefore, the EU Water Framework Directive also encourages the coordination and cooperation of non-member countries who also share the waters, apart from its member countries. The management mode of the Rhine is another successful case in this regard.

3.6 Division of labor under the overall framework is the pathway to cross-departmental cooperation

Regarding the trans-boundary water environment management object as one complete ecosystem is a common practice of today's successful management, which requires not only overall and integrative management policy and planning, but also an overall target and detailed actions to realize it. Those actions must be allocated to all countries and departments involved in the trans-boundary basin, which relates to the division of labor under the overall framework, and guarantee of joint and consistent actions under the common targets and strategies. In this aspect, Norway's biodiversity protection and sustainable utilization is a representative case. The government works out the overall plan and target, coordinates the cooperation between departments, and allocated the priority actions in the action plan to related departments according to their business characters. Under this circumstance, the work of each department is part of the whole target realization process. With such a good link and bind of all departments, Norway's biodiversity protection is well addressed. Similarly, the overall planning of River Rhine and River Thames also has such programs. Countries along River Rhine and counties along River Thames all work together to do their own parts and make their contributions for the realization of the common overall objective. However, it should be noted that such kind of division of work works under one condition, namely it needs a forceful coordinating institution, which can influence economic decisions and legislation of the local government related with trans-boundary waters.

3.7　Trans-boundary water environment management is a long-term task

The successful experiences worldwide tell us that trans-boundary water environment management, which needs joint action plans under the common strategy, is a long-term task. Specifically, only through long-term and arduous negotiation and coordination to overcome obstacles from each member country's government, institution, laws, cultures and other factors, can arrangements acceptable to all parties be made in terms of the rights, obligations, and interests of each party concerned, zoning of the basin, and the relationships of one region within the jurisdiction of one country and its surrounding areas. And they can eventually reach political consensus. In addition, with various complex, interactive, and interdependent factors influencing the trans-boundary water environment management, as well as the political factor increasingly more involved in environmental protection, a long-term and constant coordination is called for. It is faithfully reflected by the formation process of EU Water Framework Directive. The Rhine Action Plan takes 20 years in total from its primary efforts to establish cooperation to consensus achievement. The US-Canada Great lakes joint management went through over 100 years' practice, and is currently still developing and improving itself. With more than 120 years' similar exploration of River Thames, and even over 170 years' progressing of River Rhine, come the successful management modes for the two rivers. And for certainty, they will continue developing and improving.

References

Aad Correljé, Delphine François Tom Verbeke. Integrating Water Management and Principles of Policy: towards an EU Framework[J]. Journal of Cleaner Production,2007,15 (16): 1499 – 1506.

Aljosja Hooijer, Frans Klijn, G. Bas M. Pedroli, et al. Towards Sustainable Flood Risk Management in the Rhine and Meuse River Basins: Synopsis of the Findings of IRMA-SPONGE. River Research and Applications[J]. Special Issue: Towards Sustainable Flood Risk Management in the Rhine and Meuse River Basins,2004, 20 (3): 343 – 357.

Annett Weiß, Milada Matouskova J rg Matschullat. Hydromorphological Assessment within the EU-Water Framework Directive-transboundary Cooperation and Application to Different Water Basins,2008, 603 (1):53 – 72.

Anonymous. Best Practices in River Basin Management Planning[EB. OL]. Guidance Document under the WFD Common Implementation Strategy, accessed on http://forum. europa. eu. int/Public/irc/env/wfd/library,2002.

Ben Page Maria Kaika. The EU Water Framework Directive: part 2. Policy Innovation and the Shifting Choreography of Governance[J]. European Environment,2003, 13 (6): 328 – 343.

Bjørn P. Kaltenborn, Oddgeir Andersen, Christian Nellemann. Second Home Development in the Norwegian Mountains: Is it Outgrowing the Planning Capability[J]. International Journal of Biodiversity Science & Management,2007, 3 (1): 1 – 11.

Bruce Mitchell. Integrated Water Resource Management, Institutional Arrangements, and Land-use Planning. Environment and Planning A. 2005(37): 1335 – 1352.

Cals M. J. R. , Postma R. , Buijse A. D, et al. Habitat Restoration along the River Rhine in the Netherlands: Putting Ideas into Practice. Aquatic Conservation: Marine and Freshwater Ecosystems[J]. Special Issue: River Restoration: The Physical Dimension,1998, 8 (1): 61 – 70.

Christine A. Klein. 2006. The Law of the Lakes: From Protectionism to Sustainability. St. L. Rev. 1259 (2006), available at. http://scholarship.law.ufl.edu/facultypub/5.

Daniel Hering, Angel Borja, Jacob Carstensen, et al. The European Water Framework Directive at the Age of 10: A Critical Review of the Achievements with Recommendations for the Future [J]. Science of the Total Environment,2010, 408 (19): 4007 – 4019.

Einar Jørstad, Ketil Skogen. The Norwegian Red List between Science and Policy[J]. Environmental Science & Policy,2010, 13 (2): 115 – 122.

Emil Dister, Dieter Gomer, Petr Obrdlik, et al. Water Management and Ecological Perspectives of the Upper Rhine's Floodplains[J]. Regulated Rivers: Research & Management, 1990, 5 (1):1 – 15.

Erik Mostert. The European Water Framework Directive and Water Management Research[J]. Physics and Chemistry of the Earth, Parts A/B/C,2003, 28 (12-13): 523 – 527.

Erik Olsen, Harald Gjøsæter, Ingolf Røttingen, et al. The Norwegian Ecosystem-based Management Plan for the Barents Sea[J]. ICES Journal of Marine Science,2007, 64 (4): 599 – 602.

Eva Irene Falleth, Sissel Hovik. Local Government and Nature Conservation in Norway: Decentralisation as a Strategy in Environmental Policy. Local Environment[J]. Special Issue: Nordic Environments,2009, 14 (3): 221 – 231.

Ferreira J. G., Val C. e, Soares C. V., et al. Monitoring of Coastal and Transitional Waters under the E. U. Water Framework Directive[J]. Environmental Monitoring and Assessment, 2007, 135 (1-3): 195 – 216.

Gooch G. D., Stalnache P. Integrated Transboundary Water Management in Theory and Practice [M]. UK IWA Publishing,2006.

Jeremy Carter, Joe Howe. The Water Framework Directive and the Strategic Environmental Assessment Directive: Exploring the Linkages[J]. Environmental Impact Assessment Review,2006, 26 (3): 287 – 300.

Johan Woltjer, Niels Al. Integrating Water Management and Spatial PlanningStrategies Based on the Dutch Experience[J]. Journal of the American Planning Association,2007, 73 (2): 211 – 222.

John Engler. Letter from John Engler, Governor of Michigan, to George Voinovich. Chair of the Council of Great Lakes Governors,1993.

John W Handmer. Flood Hazard Management: British and International Perspectives[M]. CRC Press,1987.

Joseí E. Castro, Maria Kaika, Erik Swyngedouw. London: Structural Continuities and Institutional Change in WaterManagement[J]. European Planning Studies,2003, 11 (3): 283 – 298.

Kliota N., Shmuelia D., Shamirb U. Institutions for Management of Transboundary Waterresources: their Nature, Characteristics and Shortcomings[J]. Water Policy,2001(3):229 – 255.

Koos Wieriks, Anne Schulte-Wülwer-Leidig. Integrated Water Management for the Rhine River Basin, from Pollution Prevention to Ecosystem Improvement[J]. Natural Resources Forum, 1997, 21 (2): 147 – 156.

Kristin Selvik. Biodiversity and Modern Forestry: the Concept of Biodiversity and its Meaning Within Norwegian Forestry Management[J]. Norwegian Journal of Geography,2007, 58 (1): 38 – 42.

Liv Astrid Sverdrup. Norway's Institutional Response to Sustainable Development[J]. Environmental Politics (Special Issue: Sustainable Development in Western Europe: Coming to Terms with Agenda 21),1997, 6 (1): 54 – 82.

Maria Kaika, Ben Page. The EU Water Framework Directive: part 1. European Policy-making and the Changing Topography of Lobbying[J]. European Environment,2003, 13 (6): 314 – 327.

Maria Kaika. The Water Framework Directive: A New Directive for a Changing Social, Political and Economic European Framework. European Planning Studies,2003, 11 (3): 299 –316.

Middelkoop H. , Kwadijk J. C. J. Towards Integrated Assessment of the Implications of Global Change for Water Management-the Rhine Experience[J]. Physics and Chemistry of the Earth, Part B: Hydrology, Oceans and Atmosphere,2001, 26 (7-8): 553 –560.

Nilsson S. , Langaas S. , Hannerz F. International River Basin Districts under the EU Water Framework Directive: Identification and Planned Cooperation[J]. European Water Management Online,2004.

Organisation for Economic Co-operation and Development. Environmental Performance Reviews: United States,2005:245.

Philippe Quevauviller, Panagiotis Balabanis, Christos Fragakis, et al. Science-policy Integration Needs in Support of the Implementation of the EU Water Framework Directive[J]. Environmental Science & Policy,2005, 8 (3): 203 –211.

Pieter Huisman. From One-sided Promotion of Individual Interests to Integrated Water Management in the Rhine Basin[J]. Water Science and Technology,1995, 31 (8): 59 –66.

Rogers P.. International river basins: Pervasive unidirectional externalities. Conference on the economics oftransnational commons[J]. Italy: University of Siena,1992(4): 25 –27.

Sabine E Apitz, Michael Elliott, Michelle Fountain, et al. European Environmental Management: Moving to an Ecosystem Approach[J]. Integrated Environmental Assessment and Management,2006,2 (1): 80 –85.

Sissel Hovik, Morten Edvardsen. Private-public partnership: An exceptional Solution in Nature Conservation in Norway[J]. Local Environment,2006, 11 (4): 361 –372.

Steven E. Chester. Great Lakes Legacy Act, in state of the greatlakes annual report, p5, available at http: // www. deq. state. mi. us/documents/deq-ogl-SOGL03. pdf,2003.

Susanna Nilsson, Sindre Langaas. International River Basin Management under the EU Water Framework Directive: An Assessment of Cooperation and Water Quality in the Baltic Sea Drainage Basin[J]. AMBIO,2006, 35 (6): 304 –311.

Thomas Bernauer. Explaining Success and Failure in International River Management[J]. Aquatic Sciences,2002, 64 (1): 1 –19.

Timothy Moss. The Governance of Land Use in River Basins: Prospects for Overcoming Problems of Institutional Interplay with the EU Water Framework Directive[J]. Land Use Policy,2004, 21 (1): 85 –94.

Treaty Between the United States and Great Britain Relating to Border Waters Between the United States and Canada. 1909. Available at http: // www. ijc. org/rel/agree/water. html.

van Ast J. A. Trends towards Interactive Water Management; Developments in International River Basin Management[J]. Physics and Chemistry of the Earth, Part B: Hydrology, Oceans and Atmosphere,1999, 24 (6): 597 –602.

van der Kleij W. , Dekker R H. , Kersten H. , et al. Water Management of the River Rhine: Past, Present and Future[J]. European Water Pollution Control,1991, 1(1):9 –18.

van Dijk G. M. , Marteijn E. C. L. , Schulte-Wülwer-Leidig A. Ecological Rehabilitation of the River Rhine: Plans, Progress and Perspectives[J]. Regulated Rivers: Research & Management,1995, 11 (3-4): 377 –388.

Willems P. , de Lange W. J. Concept of Technical Support to Science – policy Interfacing with Respect to the Implementation of the European Water Framework directive[J]. Environmental Science & Policy,2007, 10 (5): 464 –473.

Yannis A. Mylopoulos, Elpida G. Kolokytha. Integrated Water Management in Shared Water Resources: The EU Water Framework Directive Implementation in Greece [J]. Physics and Chemistry of the Earth, Parts A/B/C,2007,33 (5): 347 –353.

Constructed Wetlands as Main Alternatives to Conventional Wastewater Treatment Technologies

Tokmajyan H. V. , *Mkrtchyan A. R.* and *Mkrtumyan M. M.*

Yerevan State University of Architecture and Construction, 105/1a Teryan Street, Yerevan, 0009, Republic of Armenia

Abstract: Recently, natural treatment systems are widely used to replace the conventional treatment systems for the treatment of different wastewaters including domestic, industrial, and agricultural wastewaters because of the multiple values and functions of the natural treatment systems. Compared with the traditional wastewater treatment facilities commonly in use, natural treatment systems require lower construction and operational costs providing both economic and ecological benefits. Among those natural treatment systems, constructed wetland is apparently one of the commonly applied system used worldwide for wastewater treatment. The classifications of types of constructed wetland has been given including the Free Water Surface system (FWS) with shallow water depth and the Subsurface Flow System (SFS) with horizontal and vertical types, with water flowing laterally through the beds filling with different media (e. g. sands, gravels), that is described in the article.

Key words: constructed wetland, wastewater, nutrients, macrophytes, effluent, ecological & economic benefits

Nowadays, conventional wastewater treatment and pre-treatment technologies and methods, as well as typical wastewater treatment plants require qualified personnel, high exploitation, maintenance and investment costs that apparently do not provide neither ecological nor economical benefits. This is the main reason that led to the development of alternative methods for wastewater treatment that have broad applicability for almost all the types of wastewater treatment. The natural systems typically require fewer operational personnel, consume less energy, and produce less sludge than the higher-rate systems. Usually these systems are planted with aquatic plants or macrophytes, which rely upon natural microbial, biological, physical and chemical processes to treat wastewater. The aquatic systems include the stabilization ponds, the aquatic systems with floating plants, and the constructed wetland systems. These systems are well suited for small communities and rural areas and small communities because of the low flows of wastewater to be treated and the availability of suitable land area according to the volume of wastewater flow. The treatment systems of constructed wetlands are based on ecological systems found in natural wetlands. A main distinction between constructed wetlands and natural wetlands is the degree of control over natural processes. For instance, a constructed wetland operates with a relatively stable flow of water through the system, in contrast to the highly variable water balance of natural wetlands, mostly due to the effects of variable precipitation. As a result, wetland ecology in constructed wetlands is affected by continuous flooding and concentrations of total suspended solids (TSS), biochemical oxygen demand (BOD), and other wastewater constituents at consistently higher levels than would otherwise occur in nature.

The most important mechanisms to remove aquatic pollutants are listed in Tab. 1 (Brix H., 1993).

As with other natural biological treatment technologies, wetland treatment systems are capable of providing additional benefits. They are generally reliable systems with no anthropogenic energy sources or chemical requirements, a minimum of operational requirements and low costs.

The treatment of wastewater using constructed wetland technology also provides an opportunity to create or restore wetlands for environmental enhancement, such as wildlife habitat, greenbelts, passive recreation associated with ponds, and other environmental amenities.

A necessary requirement to consider previously deciding to apply natural systems for the wastewater treatment of a particular community is to find out if any of the types of natural systems currently applied can be appropriate for this community, in function of the population number, the sources of sewage, the climate and the characteristics of the land. In all cases, it is necessary the characterisation of the wastewater to be treated: characteristics such as flow rate, chemical composition, and their variation over time, are fundamental parameters to be in consideration.

Tab. 1 Removal of pollutants in wetland systems

Wastewater constitent	Removal mechanisms
Pathogens	UV radiation
	Sedimentation / Filtration
	Excretion of antibiotics from roots and macrophytes
Phoshporus	Soil sorption & plant uptake
Nitrogen	Ammonification, nitrification, denitrification, plants uptake, ammonia volatization
Suspended solids	Filtration & sedimentation
Biochemical oxygen demand	Aerobic & anaerobic microbial degradation

Constructed Wetlands (CWs) are engineered systems of wastewater treatment that are designed and constructed to utilize the natural processes that occur in natural wetlands. A constructed wetland consists of a properly-designed basin that contains soils or other selected substrate, water column, and wetland vegetation, as main elements. Other important components that assist in the treatment of the wastewater, such as the communities of microorganisms, develop naturally. The substrate is placed over an impermeable layer constituted by combinations of clays and geotextile liners, in order to protect the subsoil of pollution from wastewater. Many chemical and biological transformations take place within the substrate, which is constituted by soil, sand, gravel, rock, and organic materials such as compost. It serves to provide physical support for plants and microbial populations, and as reactive surface area for complexing ions in order to remove contaminants.

The vegetation represents an important component in both types of systems. The wetland plants stabilize the substrate, slow water velocities allowing suspended materials to settle, filter, absorb and adsorb pollutants, provide support for microorganisms development, and leakage of oxygen from subsurface plant structures creates oxygenated microsites within the substrate. A fundamental characteristic of wetlands as wastewater treatment systems is that their functions are almost solely regulated by microbiota and their metabolism. Microorganisms are major sink for organic carbon and many nutrients, transform a great number of organic and inorganic substances into innocuous or insoluble substances and involve in the recycling of nutrients. They alter contaminant substances to obtain nutrients or energy to carry out their life cycles. The final effectiveness of constructed wetlands managed for wastewater treatment is dependent on developing and maintaining optimal environments for desirable microbial populations (Wetzel R., 1993).

Various types of constructed wetlands differ in their main design characteristics as well as in the processes which are responsible for pollution removal.

CWs for wastewater treatment may be classified according to the life form of the dominating macrophyte, into systems with free-floating, floating leaved, rooted emergent and submerged macrophytes. Further division could be made according to the wetland hydrology (free water surface and subsurface systems) and subsurface flow CWs could be classified according to the flow direction (horizontal and vertical).

A typical Free Water Surface (FWS) CWs with emergent macrophytes is a shallow sealed basin or sequence of basins, containing rooting soil. Dense emergent vegetation covers a significant fraction of the surface, usually more than 50% (Fig. 1).

Besides planted macrophytes, naturally occurring species may be present. Plants are usually not harvested and the litter provides organic carbon necessary for denitrification which may proceed in anaerobic pockets within the litter layer.

Fig. 1   Typical free water surface wetland

FWS CWs are efficient in removal of organics through microbial degradation and settling of colloidal particles. Suspended solids are effectively removed via settling and filtration through the dense vegetation. Nitrogen is removed primarily through nitrification (in water column) and subsequent denitrification (in the litter layer), and ammonia volatilization under higher pH values caused by algal photosynthesis. Phosphorus retention is usually low because of limited contact of water with soil particles which adsorb and/or precipitate phosphorus.

The first type of vegetated submerged wetlands, particularly the Horizontal Flow(HF) CWs consist of gravel or rock beds sealed by an impermeable layer and planted with wetland vegetation.

The wastewater is fed at the inlet and flows through the porous medium under the surface of the bed in a more or less horizontal path until it reaches the outlet zone, where it is collected and discharged. In the filtration beds, pollution is removed by microbial degradation and chemical and physical processes in a network of aerobic, anoxic, anaerobic zones with aerobic zones being restricted to the areas adjacent to roots where oxygen leaks to the substrate. Organic compounds are effectively degraded mainly by microbial degradation under anoxic/anaerobic conditions as the concentration of dissolved oxygen in the filtration beds is very limited. Suspended solids are retained predominantly by filtration and sedimentation and the removal efficiency is usually very high. The major removal mechanism for nitrogen in HF CWs is denitrification. Removal of ammonia is limited due to lack of oxygen in the filtration bed as a consequence of permanent waterlogged conditions. Phosphorus is removed primarily by ligand exchange reactions, where phosphate displaces water or hydroxyls from the surface of iron and aluminum hydrous oxides. Unless special materials are used, removal of phosphorus is usually low in HF CWs. The most important roles of plants in HF CWs are provision of substrate (roots and rhizomes) for the growth of attached bacteria, radial oxygen loss (oxygen diffusion from roots to the rhizosphere), nutrient uptake and insulation of the bed surface in cold and temperate regions. The schematic layout of a constructed wetland with horizontal sub-surface flow is presented in "Fig. 2".

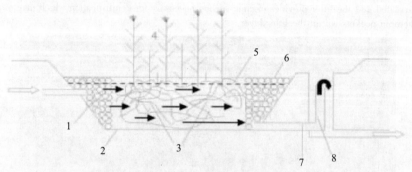

1—inflow distribution zone filled with large stones; 2—impermeable layer;
3— filtration material; 4—vegetation; 5—water level in the bed; 6—outflow collection zone;
7—drainage pipe; 8—outflow structure with water level adjustment

Fig. 2 Schematic layout of a constructed wetland with horizontal subsurface flow

However, the second type—Vertical Flow (VF) CWs did not spread as quickly as HF CWs, probably because of the higher operation and maintenance requirements due to the necessity to pump the wastewater intermittently on the wetland surface. The water is fed in large batches and then the water percolates down through the sand medium. The new batch is fed only after all the water percolates and the bed is free of water. This enables diffusion of oxygen from the air into the bed. As a result, VF CWs are far more aerobic than HF CWs and provide suitable conditions for nitrification. On the other hand, VF CWs do not provide any denitrification. VF CWs are also very effective in removing organics and suspended solids. Removal of phosphorus is low unless media with high sorption capacity is used. The early VF CWs were composed of several stages with beds in the first stage fed in rotation. At present, VF CWs are usually built with one bed and the system is called "compact" VF CWs.

The schematic layout of VFW is presented in "Fig. 3".

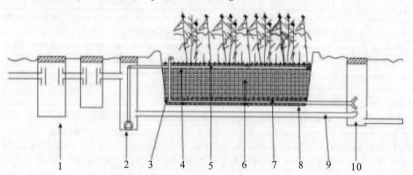

1—sedimentation; 2—pumping well; 3—aeratipn pipe; 4—distribution pipe;
5—distribution layer (wood chips); 6—filtersand; 7—drainage pipes;
8—membrane; 9—recirculation; 10—outlet well

Fig. 3 Layout of a VF constructed wetland system for a single household.

Raw sewage is pre – treated in a sedimentation tank. Settled sewage is pulse – loaded onto the surface of the bed by a level – controlled pump. Treated effluent is collected in a system of drainage pipes, and half of the effluent is recirculated back to the pumping well (or to the sedimentation tank).

Removal of organics is high in all types of constructed wetlands . While in FWS and VF constructed wetlands, the microbial degradation processes are mostly aerobic, in HF constructed wetlands, anoxic and anaerobic processes prevail. The treatment efficiency is similar for FWS and HF CWs, while for VF CWs the percentage efficiency is higher due to higher inflow concentrations. VF constructed wetlands are nearly always used for primary or secondary treatment while FWS are often used for tertiary treatment and HF CWs are often used for treatment of wastewater diluted with stormwater runoff .

Constructed wetlands are found in a wide range of climatological settings, including cold climates, where ice forms on the surface for four to six months of the year. For example, these systems, particularly the submerged type of latter, are found in Canada, North Dakota, Montana, Vermont, Colorado, and other cold – climate areas. Special considerations must be included in the design of these systems for the formation of an ice layer and the effect of cold temperatures on mechanical systems, such as the influent and effluent works. Nitrogen transformation and removal is, however, impaired during very cold periods. Treatment performance for some constituents tends to decrease with colder temperatures, but BOD and TSS removal through flocculation, sedimentation, and other physical mechanisms is less affected. In colder months, the absence of plant cover would allow atmospheric reaeration and solar insolation to occur without the shading and surface covering that plant cover provides during the growing season.

Whatsoever, constructed wetland systems offer an effective and economical self – maintain alternative to conventional treatment systems, and in consequence they appear to have very broad applicability for all wastewater treatment systems in small communities and rural areas.

More details on treatment performance of constructed wetlands for various types of wastewater could be found elsewhere.

References

Knox A S, Nelson E A, Halverson N V, et al. Long-term Performance of a Constructed Wetland for Metal Removal Soil and Sediment Contamination[J]. An International Journal, 2010,19 (6): 667 – 685.

Brix H. Functions of Macrophytes in Constructed Wetlands[J]. Water Science Technology, 1994 (29):71 – 78.

Kadlec R H, R L. Knight J Vymazal, et al. Constructed Wetlands for Pollution Control: Processes, Performance, Design and Operation. Scientific and Technical Report No. 8. International Water Association Specialist Group on Use of Macrophytes in Water Pollution Control[M]. London: IWA Publishing,2000.

Wetzel R G. Constructed Wetlands: Scientific Foundations Are Critical[C] // Constructed Wetlands for Water Quality Improvement, ed. G. A. Moshiri. CRC Press/Lewis Publishers: Boca Raton, FL, USA. CRC/Lewis. 1993.

Kadlec R H. Overview: Surface Flow Constructed Wetlands. In Proceedings of the 4th International Conference Wetland Systems for Water Pollution Control; ICWS Secretariat: Guangzhou, China, 1994:1 – 12.

Vymazal J. Types of Constructed Wetlands for Wastewater Treatment: Their Potential for Nutrient Removal[C] // Transformations of Nutrients in Natural and Constructed Wetlands: Leiden, The Netherlands, ed. . Backhuys Publishers,2001:1 – 93.

Weedon C N. Compact Vertical Flow Reed Beds: Design Rationale and Early Performance. IWA Macrophytes Newsletter,2001(23):12 – 20.

Wittgren H B, Maehlum T. . Wastewater Constructed Wetlands in Cold Climates[C] // R. Haberl, R. Perfler, J. Laber and P. Cooper (eds.), Oxford, UK: Elsevier Science Ltd. ,1996:45 – 53.

Study on Driving Forces of Wetland Change in the Yellow River's Modern Delta

Jiang Zhen[1] ,*Wang Ruiling*[2] and *Wang Xingong*[2]

1. Department of Personnel, YRCC, Zhengzhou, 450003, China
2. Science Research Institute of the Water Resources Protection Bureau, YRCC, Zhengzhou, 450004, China

Abstract:On the basis of the autumn data of remote sensing in 1986, 1996 and 2004, the spatial and temporal changes of wetland landscape are described using the approach of landscape graph spectrum, then the driving forces of wetland change is investigated in the Yellow River's Modern Delta during the past 20 years. The results show that water and sediment resources of the Yellow River, overexploitation of land-use, construction of hydraulic engineering, storm tides and knowledge are the main driving forces of wetland change in the Yellow River's Modern Delta.
Key words:wetland change, driving force, the Yellow River's Modern Delta

The wetland in the Yellow River's Modern Delta (YRMD) is provided with abundant newborn wetland and rare birds. Accordingly, the wetland in the YRMD has become not only one of the international important wetlands, but also the crucial symbol to maintain healthy river. However, in the last 20 years, the wetland ecosystem in the YRMD has suffered tremendous degradation due to the sudden decreased discharge in the river, the increasing human activities and the lack of wetland science etc. Therefore, it is beneficial to the sustainable management of ecological system, if we investigate the driving forces of wetland change in the YRMD.

1　Wetland landscape

This study is focusing on the wetland in close relation to the water coming from the Yellow River in the YRMD. Generally, YRMD starts from Yuwa point, with the Diaokou river in the north and Songchunrong ditch in the south, and was shaped by the man – made changing of the watercourse of the Yellow River entering to the sea in 1976. As we known, the unique wetland ecosystem in the YRMD comes into being by filling – up and frequent fluctuation of terminal tail channel of the Yellow River. The total area of wetland in the YRMD is about 1,570 km^2. Most of the wetland is natural, as lots of seasonal rivers branches off in rivers' mouths to the sea. Especially, many natural wetlands coexist and concentrate in the eastern zone from the Xiaodao river mouth in the south to the Tuhai river mouth in the north. On the contrast, artificial wetland, mainly including paddy field, reservoir and salty field, widely spreads in the Mmidwest of the total wetland in the YRMD. In the sight of spatial distribution, the wetlands in the YRMD vastly centralize in littoral zone, but taper gradually and scatter spatially when shifting into the inlands.

1.1　Hydrological characteristics

The Yellow River is the basic factor to maintain the water system in this region. According to observation by Linjin hydrological station, the mean annual run-off is 41.9 × 10^{10} m^3 in the mouth of the Yellow River, with the discharge difference of 10.6 times. The average discharge is 1,330 m^3/s, ranging from10,400 m^3/s to zero. The Yellow River is also the river with highest sediment concentration in the world. The average annual sedimentriver load is 1.049 × 10^9 t and the average sediment concentration is 25.53 kg/m^3. Recently, with the water flow of the Yellow River, the abrupt flow reduce, the increases of water diversion works and so on, the run-off and the sedimentriver load decreased abruptly by about 60%.

Generally, the wetland ecosystem is situated in the fragile ecological region between land and water. Thus, particular hydrological condition determines that the wetlands are liable to undergo the disturbances by the nature and human activities. Correspondingly, the wetlands are difficult to be restored once destructed in the YRMD.

1.2　Ecological characteristics

As the exceptional habitat of breeding, migrating, and wintering for bird migration in the inland of northeast Asia and around western Pacific Ocean, the unique natural wetlands and abundant biological resources in the YRMD have been one of the most crucial wetlands in the world. Accordingly, in this region, the national nature reserve was established and authorized in 1990 and 1992 respectively. The main aims of the nature reserve focus on the conservation of new-born wetland ecosystem and rare endangered birds.

In the reserve, there are total 393 species (varieties) of plants, mainly including 271 species of angiosperm. Most vegetation is natural with the area of $50,915 \ hm^2$, accounting for 77.9% of the total vegetation. Moreover, there a're 1,466 species of wild animals recorded, consisting of 300 species of terrestrial vertebrates, 583 species of terrestrial invertebrates, 223 species of terrestrial aquatic animals and 418 species of marine aquatic animals. In terms of national priority wildlife, 7 species of birds are listed as the 1st class priority, and 33 species as the 2nd class priority, including Grus japonensis, Circus cyaneus cyaneus etc. In addition, 7 species of birds are in the annex I, 26 species in annex II and 7 species in annex III of convention on international trade in endangered species of wild fauna and flora.

2　Wetland change

On the basis of the autumn data of remote sensing in 1986, 1996 and 2004, the spatial and temporal changes of wetland landscape are investigated in the Yellow River's Modern Delta during the past 20 years, using the approach of landscape graph spectrum as follows.

2.1　During the whole period (1986 ~ 2004)

Landscape pattern of wetlands in the YRMD presents two series of transition in the space-time scale during the whole period. One is the transition of CW to OW (saline lagoon) to SW or MW (reeds) to AW from the coastland to the inland, and the other is the transition of RW to MW (reeds) to SW to AW from the riverbed to the outward.

Landscape pattern of wetlands is unstable in the YRMD. About 66% of the wetlands in the YRMD have changed intensely due to the fragile ecosystems. CW and RW are the two wetland classifications with the larger unchanged areas.

The 1st largest changes of wetland classifications are the process from NW (non – wetland) to OW, AW and CW, and the 2nd largest changes are the process from CW to NW, AW and OW. It is shown that OW and AW have increased rapidly, and that CW is the wetland classification with the largest transferring area.

2.2　During the first temporal sequence (1986 ~ 1996)

The non-wetland has dominantly changed into CW, OW and MW with the area of 655.41 km^2. The increasing wetlands are mainly located in the sand spit of the current watercourse of the Yellow River, on the both sides of previous watercourse of the Yellow River, or between the previous watercourse and current watercourse of the Yellow River.

The CW has obviously changed into OW with the area of 192. 17 km². The increasing OW is mainly located between the previous watercourse and current watercourse of the Yellow River. And the CW near the mouth of previous watercourse has been obviously eroded with the area of 74. 32 km².

2.3　During the second temporal sequence (1996 ~ 2004)

The wetland classifications have dominantly changed into NW with the area of 412. 18 km². The OW has decreased with 189. 20 km² between the previous watercourse and current watercourse of the Yellow River. The CW has reduced by 189. 20 km² near the mouth of previous watercourse and the southern mouth of current watercourse of the Yellow River.

The NW and OW have obviously changed into middle or small sized AW with the area of 120. 88 km². The increasing salty fields accounts for 90% of the increasing AW of middle size. The increasing AW from NW is mainly located on the both sides of previous watercourse and in the south of the current watercourse of the Yellow River. The increasing AW from OW is mainly situated between the previous watercourse and current watercourse of the Yellow River.

The CW has obviously changed into OW with the area of 115. 89 km², and is mainly located on the two sides of previous watercourse and in the south of the current watercourse of the Yellow River.

3　Driving forces

Generally speaking, the driving forces of landscape pattern can be identified into three types, namely a-biotic (physical) factors, biologic factors and artificial factors. The special pattern of wetland landscape has evolved by the fluctuation of watercourse of the Yellow River, the biological succession and intensified human activities.

For all the a-biotic (physical) factors, the hydrological factor directly influences the scope of wetland and its inner ecological process in the YRMD, while the other factors are relatively stable, e. g. climate, soil and topography etc. As special factors, human activities can directly or indirectly change the scope of wetland and its inner ecological process in the YRMD. For the biological factors, the information is lacking and the influences are relatively weak.

Therefore, the discussion of driving factors should focus on the important abiotic (physical) factors and changeful or special artificial factors, in close relation to the hydrologic factors of wetlands.

3.1　Water and sediment resources of the Yellow River

The changes of water and sediment resources of the Yellow River is the essential driving factors for the of space-time pattern of wetland landscape in the YRMD. Because the Yellow River is the basic factor to maintain the water system in this region, and the water and sediment resources are the leading factor to maintain the natural wetland ecosystems in the YRMD.

In general, more water and sediment resources of the Yellow River can guarantee the increase of new-born wetlands in the YRMD. E. g. , the annual new-born wetlands is about 22. 8 km² per year due to more water and sediment resources during the 1976 ~ 1992. However, along with the decrease of water and sediment resources at Linjin station of the Yellow River, the new-born wetlands has been shrunk and even disappeared. E. g. , the coastlands near the mouth of the previous watercourse and the southern mouth of the current watercourse have been obviously disappeared.

Therefore, the decreased water and sediment resources of the Yellow River have resulted in not only the shrinkage of the wetlands, but also the serious unbalance of water system and hydrological ecology in the YRMD. The evolution of wetland landscape has obviously been influenced.

3.2 Overexploitation of land-use

Unreasonable exploitation of land-use is the main driving forces of the evolution of space-time pattern of wetland landscape in the YRMD. Because these kinds of activities have led to important changes of wetland hydrology and its ecosystems, fragmentized the wetlands with the increasing middle or small patches, e. g. coastal wetland, saline lagoon etc. And even the habitats of species of wetland have been damaged or disappeared. In addition, saline lagoon has dramatically increased during the whole period of 1986~2004.

At the same time, with the increasingly unreasonable exploitation of land-use, the salty fields and oil fields etc. have evidently raised, while the natural wetlands have relatively reduced and shrunk. The trends of these changes will surely cause the continuous decrease of natural wetlands and imperil the positive evolution of wetland ecosystems in the YRMD.

3.3 Construction of hydraulic engineering

The construction of hydraulic engineering is an important factor for the changes of wetland hydrology, because this kind of human activities can notably influence the landscape pattern and the hydrological characteristics of wetlands in the YRMD.

Actually, the construction of hydraulic engineering in the upper reaches can change the hydrological cycle and process of wetlands in the lower reaches of the Yellow River. Therefore, the evolution of wetland landscape is unbalanced with the fragmentation and shrinkage of wetlands in the YRMD.

Moreover, the construction of reservoirs leads to more saline lagoon and more fragmentation and shrinkage of wetlands in the YRMD. All these changes have damaged the habitats of species and the biodiversity conservation.

Besides these, dike projects and protection projects against the tide also cause the converse succession, the unbalance of ecosystems and the decrease of biodiversity due to the decrease of a hydrological cycle of wetlands in the YRMD.

3.4 Storm tides

The storm tides can intensify the converse succession of wetland landscape. The YRMD is one of the regions with the most severe storm tides. With the depth of several decades' kilometers, each storm tide has severely damaged and threatened the wetland ecosystems in the YRMD. Moreover, the storm tides can intensify the changes from other wetland classifications to saline lagoon or non-wetlands.

3.5 Knowledge of science, management and law

Because they can mitigate or intensify the pattern changes of wetland landscape to some extents, the knowledge is another important driving force of evolution of wetland landscape by human beings, e. g. the science of wetlands, the management of natural reserve, and the relative laws etc.

Although the exploitation or land – use and the construction of hydraulic engineering have negative effects on the wetlands, those measures can mitigate or restrict the rate and trend of converse succession of wetlands. E. g. ,the strict legislation, the popularization of wetland science, and the improvement of management etc.

In practice, the construction of nature reserve, the implementation of the restoration of wetlands, and the implementation of the unified water regulation since 1999 and the regulation of water and sediment since 2002 have already mitigated the trends of the shrinkage of freshwater wetlands in the YRMD.

Practical Possibility to Manage Radiation Pollution on the Examples of the Region of Krivoy Rog and the Loire River

Sizonenko V.

Institute of General Energy of NAS of Ukraine, 03150, Kiev, Antonovicha 172, Ukraine

Abstract: The paper will show possibilities for managing of possible consequences of accidental events of volley inflow of the muddy mine waters in the Zoltaja River, prognostication of possible consequences of such catastrophic situations. The computer model of the water system of region Zoltje Vody City – Krivoy Rog City has been developed with the new box model of incomplete interfusion with lagging argument. It gave the opportunity for accurate prognostication and short time of preparedness. Examples of prognostication of levels of concentrations of radionuclide (^{238}U) in waters of the Ingulets River, Zoltaja River and Karachunovskoe reservoir depending on the discharges of water of rivers and modes (regulation and catastrophic) of polluted water kick of mine *Nova* are demonstrated. Possibilities of water protection actions are explored at the catastrophic discharge of mine water to the Zoltaja River.

A sharp growth of concentration of tritium in the Loire River Basin is result of interference of the emissions ^{3}H from different NPP stations. Thus there is principle possibility to decrease of maximal concentrations ^{3}H by the change of time of the emissions of the separate stations. At the same time the amplitude and duration of separate emissions remains without modifications. The search of favorable combinations of moments of the ejections can be done with the help of modeling. Results are given.

Key words: box model, water protection, radiation pollution

1 Introduction

There is possibility dynamically to influence on the degree of radio – active contamination of running surface waters by the change of the modes of functioning of water objects and change of the modes of ejections of NPP. Thus positive results can be got not diminishing the amount of the contaminations thrown down in a water. Possibility of the dynamic influencing is supposed by the presence of dynamic adequate model of distribution of radio – active contamination.

The possibilities for managing on the consequences of accidental events of volley inflow of the muddy ^{238}U mine waters in the Zoltaja River (in the region of Krivoy Rog, Ukraine) and some possibilities to manage on the consequences of ^{3}H ejections in the Loire River Basin (France) are presented in the work.

As a dynamic model was used the new box model of partial mixing with lagging argument (UNDBE) which increased accuracy and was developed by author. The model takes into account time of transporting of water masses and intermixing of contamination in some part of volume of the camera at the moment of termination of transporting. The new model is simple, less requiring for full – scale measurements, short time of computation. Short time of calculation gives the opportunity to solve the task of parameters identification. It contributes to increase the accuracy of the model and provides the possibility of its adjustment to a particular water object. The comparison of results of the exercise shows that box model UNDBE gives coincidence with measurements not worse of the more complicated 1D models.

2 Water protection in the region of Krivoy Rog

One of the main problems accumulated for 120-years-old exploitation of the Krivoy Rog metalliferous deposit the necessity of taking up $40 \times 10^6 \sim 45 \times 10^6$ m^3/a mine and annual utilization quarry waters highly mineralized in the volume of $20 \times 10^6 \sim 25 \times 10^6$ m^3/a (Fig. 1).

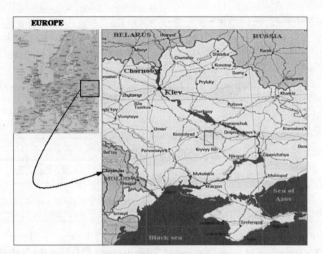

Fig. 1   Krivoy Rog region

Uranium deposit located near-by Zoltje Vody City in the region of Krivoy Rog is one of the largest on territory of Ukraine. One of main mining objects of the city is a mine *Nova*, which works mine of ferrous and uranium ores. Mine water which is pumped out on a surface saturated by uranium, sulfides, chlorides and is used for the technological necessities of mining complex and farther brushes off in tailing dump from where surpluses of water brush off in a Zoltaja River. This in the turn, results in contamination of waters of Ingulets River and Karachunovskoe reservoir-basic source of water-supply of whole region.

The computer model of the water system of region Zoltje Vody City—Krivoy Rog City is developed and the system of prognostication of levels of concentrations of radionuclides in waters depending on the charges of water of rivers and modes of upcast of mine *Nova* waters (regulation and catastrophic) of the Ingulets River, the Zoltaja River and Karachunovskoe reservoir is created. For providing of the operative use a program complex the interface of user which provides the task of initial data and receipt of results of calculations in the graphically formalized kind was developed (see Fig. 2).

Possibilities of water protection actions are explored at the catastrophic discharge of mine water to the Zoltaja River. Protection action is dilution by waters from the storage pool of the Ingulets River. Such measures in many cases enable to access to safe and adequate water and sanitation services during the catastrophic events. The new model gave the opportunity for accurate prognostication and short time of preparedness (less one minute). Time of water transportation from mine *Nova* to storage pool of the Ingulets River more one day. So we have time for modeling and decision-making. Concentrations are shown in three points along the Zoltaja River—the Ingulets River when five days ejection of pollution occurs on the Fig. 2. It is possible to count different scenarios of water discharge from storage pool of the Ingulets River fast and find time and minimal water discharge for decreasing concentration below maximum concentration limit. Resultes are shown on Fig. 3. Water discharge increased from 6 m^3/s to 10. 1 m^3/s.

The program complex has been introduced in the Ukrainian state water – protection organization (Derzvodgosp).

3   Water protection in the Loire River Basin

The research of emissions tritium in the channel of 350 km of the Loire River (France) (Fig. 4) was made in the framework of the IAEA program EMRAS (Environmental Modelling for RAdiation Safety). Information of the in-situ measurements during a half year was used.

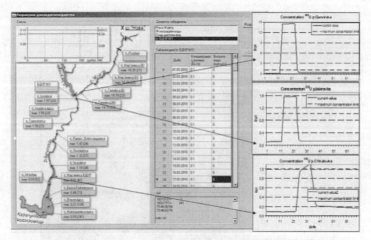

Fig. 2 Interface of user. Escape of mine water (^{238}U) to the Zoltaja River—the Ingulets River – Karachunovskoe reservoir

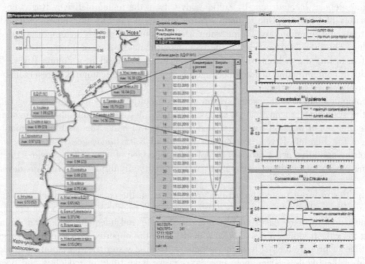

Fig. 3 Escape of mine water (^{238}U) to the Zoltaja River—the Ingulets River—Karachunovskoe reservoir & water protection

A sharp growth of concentration of tritium is result of interference of the emissions ^{3}H from different NPP stations. Thus there is principle possibility to decrease of maximal concentrations ^{3}H by the change of time of the emissions of the separate stations. At the same time the amplitude and duration of separate emissions remains without modifications. The search of favorable combinations of moments of the ejections can be done with the help of modeling.

Example of scenario calculation with the changed moments of some separate ejections of tritium for points Nouan (point after NPP StLaurent) and Angers during half year are resulted on Fig. 5, Fig. 6. The brown line is concentration of ^{3}H before and the orange line after changing moments of ejections. Each scenario calculation for 33 points along the river with an hour time discretization needs 2 min computer time.

A decline of maximal concentration of ^{3}H is 20% ~ 30% for different points along the river.

Radical another way to influence on the size of concentration consists of that, "to collect all of the ejections in one" from different NPP so that the total ejection coincided with the maximal of water discharge (it was strongly dilute).

The scenario calculation "to collect all of the ejections in one" from all NPP in the Loire River Basin for each quarter for points Nouan and Angers are resulted on Fig. 7, Fig. 8. The brown line is concentration of ^{3}H before changing moments of ejections, green line is scenario "to collect all of the ejections in one" and the blue line – water discharge.

In this case we have short high pike of concentration and all another time – practically clean water! It is possible to use alternative sources of water for water supply during short time of high concentration ^{3}H.

A dynamic adequate model of distribution of radio-active contamination gives the possibility to calculate times of ejections from different NPP for obtaining regime "to collect all of the ejections in one".

The shown approaches can be utillized not only in the case of radio-active contaminations.

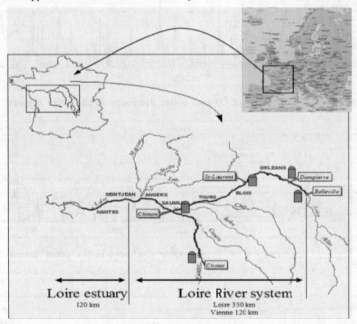

Fig. 4   The Loire River Basin

4   Conclusions

(1) New box model of incomplete interfusion with lagging argument provides the increase of exactness of modeling without considerable complication of mathematical tool due to the account of time of transporting of water masses and more exact account of diffusion.

(2) Program complex with the offered model allows to forecast the transport of contaminations in flowing surface water. It needs less full-scale measurements and short time of computation.

(3) Small time of computation makes possible parametric identification of model and also the work in the mode on line. It gives the possibility of adaptation of model to the object, use in early warning systems and decision support systems.

(4) The regimes of exploitation of HPP and NPP can be used as measures of water protection in the normal emissions of contamination and in conditions of emergency emissions of contamination.

234

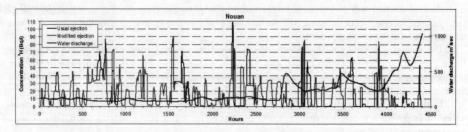

Fig. 5   Concentration of ^{3}H and water discharge in the point Nouan

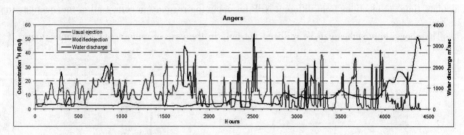

Fig. 6   Concentration of ^{3}H and water discharge in the point Angers

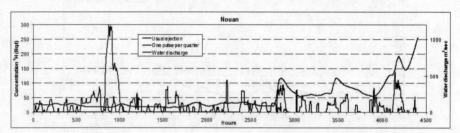

Fig. 7   Concentration of ^{3}H and water discharge in the point Nouan

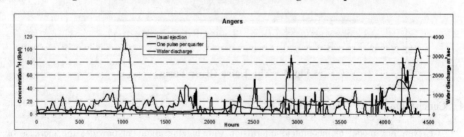

Fig. 8   Concentration of ^{3}H and water discharge in the point Angers

Acknowledgements

The author kindly thanks to the researchers of Ukrainian Hydrometeorological Institute, the Hydrological Forecasting Department of the HydroMet Center in Kiev, Derzvodgosp of Ukrain, the DIREN Centre and EDF (France) for radiological and hydrological data.

References

Sizonenko V P. Increasing Accuracy of the Box Model [C] // Hydroinformatics . 98 – Proceedings of the Third International Conference on Hydroinformatics / Copenhagen / Denmark / 24 – 26 August 1998, V. Babovic & L. C. Larson (ed.), A. A. Balkema, Rotterdam, 1998, 1: 225 – 230.

Sizonenko V P. The Evaluation of Influences of Modes of Operation the Kiev HPP on 90Sr Concentration in the outflow of the Kiev Reservoir [J]. In: The Problems of General Energy, 2000, 2: 58 – 63. (in Russian)

Testing of Models for Predicting the Behaviour of Radionuclides in Freshwater Systems and Coastal Areas / Report of the Aquatic Working Group of EMRAS Theme 1, IAEA, 244p. http: // www – ns. iaea. org/downloads/rw/projects/emras/final – reports/aquatic – tecdoc – final. pdf.

Sizonenko V P. A Case Study for a Numerical Box Model With Lagging Argument in Order to Simulate Transportation of Pollution in Dnieper Reservoirs and Loire River International Conference on Radioactivity in the Environment, 2008, Jun 15 – 20, Bergen, Norway, Abstract Volume, part 2, 225 – 231.

Discusses on Measures of Ensuring Yellow River Water Resources in Yellow River Delta

Tong Yifeng[1] , *Xing Hua*[2] and *Gong Xiaodong*[2]

1. Press and Publication Center, YRCC, Zhengzhou, 450003, China
2. Institute of Yellow River Estuary Research, YRCC, Dongying, 257091, China

Abstract:At present, there are some problems in water sources exploitation and utilization in Yellow River Delta area: conflict between supply and demand of water resources is obvious, efficiency of water utilization is low, water environment is overall bad, Yellow River water diversion amount is more than allocated, water diversion ability of sluices is deficient, water diversion efficiency in irrigation ditches is low, storage capacity of reservoirs is deficient, etc.. So we should carry out following measures to meet the needs of Yellow River water resources in development of Yellow River Delta efficient ecological economic zone: strengthening water resources management and regulation, strengthening saving water, implementing Yellow River water rights transfer, enhancing water environment harnessing, increasing water diversion ability of sluices, improving water diversion efficiency in irrigation ditches, increasing storage capacity of reservoirs.

Key words:water resources development, ensuring, measures, Yellow River Delta

1　Introduction

In 2009, the State Council officially approved "*the Yellow River Delta Efficient Ecological Economic Zone Development Plan*". Taking this as the starting point, the development of the Yellow River Delta area rises for national strategy, become an important part of national regional coordinated development strategy. The Yellow River Delta economic development is inseparable from the Yellow River water resources, which is foundation and guarantee for the development of the national economy in the Yellow River Delta area, and one of the most important constraints to the development of the Yellow River Delta. Since Xiaolangdi Reservoir was built and operated in 1999, water and sediment conditions in the lower reaches of the Yellow River has obviously changed, distribution of annual and interannual runoff has also changed. In the maintenance of no cutoff in the Yellow River, it eased the contradiction between supply and demand of delta region, improved the guarantee rate of water supply, and promoted the healthy development of the national economy in the Yellow River Delta area. But sustainable development of the Yellow River Delta Efficient Ecological Economic Zone will put forward higher requirements on the water resources of the Yellow River. At present, there are still some problems in development and utilization of water resources in the Yellow River Delta, so we should take corresponding measures to solve these problems in water resources management.

2　Problems of development and utilization of water resources in the Yellow River Delta

The main problems existing in the development and utilization of water resources in the Yellow River are: the very conspicuous contradiction of water resources supply and demand, low efficiency of water use, the overall poor situation of water environment, Yellow River Super allocation index phenomenon, more than allocated amount from Yellow River water diversion is, insufficient sluice diversion capacity, low water channel efficiency, insufficient reservoir storage capacity.

2.1　Conspicuous contradiction of water resources supply and demand

The status quo of average annual water supply amount in the Yellow River delta is 3.46×10^9 m^3, required water amount is 4.05×10^{10} m^3, water amount deficient is 0.59×10^9 m^3 with deficient rate of 14.5%, so the contradiction of water resources supply and demand is very conspicuous. Especially irrigation water in the spring is used by about 60% of water used throughout the whole year, but in the same period water from the Yellow River accounts for only about 20% water of the whole years. At the same time, some industrial projects were unable to be built because of the shortage of water resources, which restricted the economic development of the Yellow River Delta Efficient Ecological Economic Zone.

2.2　Low water use efficiency

With conspicuous conflict between water supply and demand, water efficiency of industry and agriculture in the Yellow River Delta region was low and there were existing serious water waste phenomenon.

Industrial water use in the Yellow River Delta mainly has two characteristics: one is huge water consumption. The Yellow River Delta is the oil production area of Shengli Oilfield; oil exploitation needs plenty of water to ensure normal operation of the equipment and water demand of the petroleum chemical industry is also great. The other is the overall low efficiency of industrial water use. Industrial production equipment relatively lagged behind, lacked sufficient water-saving equipment and necessary water procedure.

Agricultural water use waste in the Yellow River Delta area includes the following aspects: firstly, the large water irrigation canal was not complete, no hardening, irrigation through earth canal caused a lot of leakage; secondly, the people attached importance to construction but ignored maintenance, and farmers' water saving consciousness was not strong. Thirdly, the irrigation technology lagged behind, "flooding irrigation" still was used in many places, and water utilization coefficient of agricultural irrigation was only 0.53 in the Yellow River Delta, far lower than the average level of the province.

Because the living water prices were still low, the residents' consciousness to save water was not strong. Popularity rate for water – saving sanitary was not high, resulting in the waste of the living water.

2.3　Poor overall situation of water environment

The Yellow River Delta locates in the lower reaches of the Yellow River with sewage confluence concentration, so large amounts of untreated industrial waste and sewage directly drain into rivers; large amounts of fertilizer, pesticide for agricultural use caused by non – point source pollution; urban sewage treatment rate is low, and urban sewage directly discharged without treatment. The majority of the rivers including the Yellow River are polluted, which makes limited water resources scarcer and causes losses to the production of industry and agriculture. In addition, in coastal areas the groundwater funnel area caused sea water intrusion with the area of 800 km^2. Pollution of river and seawater intrusion not only damages the ecological environment of the area, but also threatens people's health.

2.4　Excess of allocated water index from the Yellow River

Water consumption index from the Yellow River assigned to the Yellow River Delta region by Shandong province was 2.274 × 10^9 m^3, but the annual diverted amount by the Yellow River Delta region was 2.288 × 10^9 m^3 since 1998, which was more than the allocation index. Among them, annual amount diverted by Dongying city was 0.986 × 10^9 m^3, exceeded distribution index of 0.780 × 10^9 m^3. Water diverted from the Yellow River in the year of 2000 ~ 2001 was up to 1.478 × 10^9 m^3, exceeding allocation index by nearly 90%.

2.5　Poor diversion capacity of Yellow River sluice

Since water and sediment regulation in the lower reaches of the Yellow River for the first time in 2002, minimum bankfull discharge has increased from 1,800 m^3/s to 3,810 m^3/s, and average bed elevation at Lijin station has cumulative drop of 1. 8 m. Because river channel incised down caused by water and sediment regulation, water level has dropped at the same discharge and river regime has changed, which resulted in diversion capacity of the culvert and sluice declining. Especially in dry season of spring, some diversion gate does not rely on water; for some sluice on the water, the water level before the sluice is lower than the height of the gate plate, so the sluice can not draw water from the river.

2. 6　Low diversion efficiency of the diversion channel

The Yellow River water becomes the main source of water resources in the Yellow River Delta, but the channel water use coefficient is very low because of relatively backward diversion channel facilities in the previous construction, insufficient channel lining, not perfect canal system supporting facilities and other factors. The national average channel utilization coefficient is around 0. 46, while the coefficient in irrigation area of the Yellow River Delta is only 0. 42.

2. 7　Insufficient storage capacity of the reservoirs

Especially in 1990s, due to the Yellow River often break off, a climax of plain reservoir construction set off in the Yellow River Delta region in order to guarantee the living and production water for the city, oilfield and the local people. But because of the lack of unified planning, economy and technology limitations, the layout of reservoirs was serious unreasonable and there were existing problems of construction quality and reservoir seepage in the plain reservoir. Coupled with the Yellow River seasonal water tends to be relatively stable in the recent years, a great waste of established storage capacity occurred. In 410 built plain reservoirs, the total capacity is up to 1. 189 $\times$ 10^9 m^3, storage capacity is 1. 05 $\times 10^9$ m^3, and we often use average capacity of only about 0. 40 $\times$ 10^9 m^3 with around 40% of storage capacity. On the one hand, a large number of idle storage capacity of the reservoir has been built; on the other hand, water use of irrigation area of the Yellow River downstream, high-pitched zone and brackish water area can not be guaranteed, resulting in unreasonable regional water use.

3　Water resources protection measures in the Yellow River Delta

In the light of the problems existing in the development and utilization of water resources in the Yellow River, we need to strengthen water resource management and scheduling, increase water conservation efforts, implement the Yellow River water right transfer, strengthen water environmental governance, improve the sluice diversion capacity, and channel diversion efficiency, increase water storage capacity of reservoir and so on.

3. 1　Strengthening water resources management and scheduling

3. 1. 1　Determine the reasonable order of water use and supply

According to the characteristics of water use for livelihood, production, ecology in the Yellow River Delta region and their importance, it is very important to identify the scientific and reasonable water use priorities, but also on water right definition of the users. We should set priorities of water users: urban domestic water, domestic water in rural areas, the minimum ecological water, important industrial and mining enterprises (oilfield, power plant), general industry and third sector water, agricultural water use and general ecological water.

Several water sources in the Yellow River Delta area is available for allocation and utilization. In order to rationally use limited water resources and maximize the benefits of water supply, it is very necessary to determine the regional order of water supply sources. Generally we should follow the principle of "first upstream then downstream, first surface water then ground water, first local then outside, first the basin then outside the basin". According to the characteristics of water source in the area, the source of water supply has general order as follows: surface water (water storage projects), surface water (the water diversion projects), unconventional water (sewage treatment water reuse, seawater, brackish water, rain flood etc.), the Yellow River water, the Yangtze River water (water from South to North), shallow groundwater, deep groundwater.

3.1.2   Implementation of scheduling and optimization of multiple sources

The Yellow River Delta region should establish and complete water supply engineering network with rivers and reservoirs links, the integration of urban and rural series, with Jiaodong water route of the South-to-North Water Transfer Project and the Yellow River as the basis, to all levels of canals as a link, to the reservoir, dam for the node. The southern counties (cities, districts) of the Yellow River make full use of diversion water for Qingdao from the Yellow River and Jiaodong water route project, realize co – scheduling of surface water, groundwater, the Yellow River water, the Yangtze River water, unconventional water and other water; the southern counties (cities, districts) of the Yellow River shall use the existing Yellow River Diversion Canal and sluices and dams in the rivers, forming "the vine fruit" engineering system of water supply and the plain reservoir, realizing the joint control of multiple water sources.

3.1.3   Strengthen the unified management of water resources in the Yellow River

The Yellow River Delta is the region of shortage of water resources in Shandong province with contradiction of water supply and demand, so we must implement the most strict water resources management system in the area, and constantly improve and implement the water resource management laws, regulations and policies and measures, delimit three "red lin" of water resources management: first, clear the "red line" for development utilization of water resources, strictly control the total amount of water use; second, clear the "red line" for water function and pollutant carrying limit, strictly control the total amount of sewage into the river; third, is clear the "red line" for water efficiency control, resolutely curb the waste of water.

3.2   Increase water conservation efforts

Agricultural water saving from the following three aspects: The first is to adjust the agricultural structure and crop structure, improve crop layout, cultivation system, farming techniques, and develop drought – tolerant varieties. The second is appropriate to increase the price of water based on considering the local farmers on afford ability. The third is to promote irrigation water – saving measures and technology. Through adopting the above measures, by 2015 popularity rate of field water – saving irrigation achieves 90%, agricultural irrigation water utilization coefficient increases to 0.62.

For industrial water saving, firstly we should enhance the transformation of industrial water, the use of advanced technology, processes and equipment, and increase the water circulation, water repetition rate; secondly, adjust industrial structure, develop water – saving industries, strictly control construction of high water consumption and high pollution industry project. By 2015, repeated utilization rate of industrial water shall reach more than 80%, GDP water consumption of ten thousand Yuan shall reduce to below 55 m^3.

Domestic water saving measures include: strengthening promotion and education of water – saving, strict management and maintenance, research and promotion of various water – saving device, advocated a multiple use of water, sub-quality water etc.

3.3   Implement the Yellow River water right transfer

3.3.1   Internal transfer of water rights in the Yellow River Delta

In order to implement internal water right transfer in the Yellow River Delta, Yellow River Affairs Department at all levels should conscientiously implement the *"Regulations of water fee management licensing and collection of water resources"*. If construction projects need to take the Yellow River water, the applicant shall submit water resources assessment report of construction projects compiled by units with water resources argumentation of construction project quality. For new industrial projects in the scope of the Yellow River Delta, water diversion project from the Yellow River in high-tech Development Zone, new city area, we should implement the transfer of water rights as we can as possible.

3.3.2 Water rights transfer between the Yellow River Delta and other areas of Shandong province

When the interior water right transfer in the Yellow River Delta still can not meet water consumption of the new-built project in the Yellow River Delta, it is necessary to implement water rights transfer between the Yellow River Delta and other areas of Shandong province, so as to supply more water resources for the economic development of the Yellow River Delta. Therefore, we need to build water market of the Yellow River in Shandong province, both parties decide to the quantity and price of water transfer by negotiation.

3.4 Strengthening water environment governance

To strengthen pollution control of the Yellow River and other major rivers, total control of industrial pollutant emission into the rivers; to strictly implement emission permit system for industrial enterprises, and achieve emission standards; to strictly limit the chemical fertilizer, pesticide in agricultural application, control non – point source pollution; to strengthen urban pollution treatment plant construction, improve the urban sewage treatment rate; to improve waterproof engineering, alleviate the seawater intrusion; to make full use of the front flood peak of the Yellow River to clean the pollution in the river, improve the water quality of the Yellow River and the quality of water environment of the delta through cleaning, diluting and self-purification function of the flood.

3.5 Improve diversion capacity of the sluices

For the off – water sluice gates, we need to take measures of construction and renovation of river regulation project to adjust the regime to make them back to the river. For the off – water sluice gates in the dry periods, we can excavate the diversion channel to main flow in the front of gate so as to divert a small flow. For sluice gates which still can not draw water, we only can reform, for example reducing the height of the sluice soleplate, or constructing pumping stations before the gates. Through these measures, we can make the sluice gates divert enough water from the river, even in spring irrigation period.

3.6 Improve the efficiency of the Yellow River diversion channel

Due to difficulty of increasing the supply of water resources in the existing circumstances, it is necessary and feasible to improve channel efficiency of the Yellow River, make more efficient to use water, which is an important part of carrying out efficient agricultural mode.

Increasing the length of channel lining is a critical measure to improve the channel efficiency, so we should determine the reasonable lining structure of canal lining and increase the lining length.

Perfect canal facilities is the key to improve channel efficiency, we should establish a reasonable, scientific channel network from the Yellow River and its facilities, guarantee fundamentally the high efficient use of the Yellow River channel.

3.7 Improve the regulation and storage capacity of the reservoirs

According to the actual situation of the Yellow River Delta, we have two methods of improving the storage capacity of the reservoir: one is the expansion and newly built of plain reservoir; the other is the construction of coastal reservoir. Most existing reservoirs were built in 1980s and 1990s, distribution and quality of which were unable to meet the current requirements. We should carry out scientific planning to repair the potential dangers, to raise a reasonable standard, to expand storage capacity. On the existing basis, combined with the local reality, we can make reasonable planning of new coastal plain reservoirs and plain reservoirs, in order to solve small capacity of the existing reservoirs and to meet the normal water demands.

References

Water Resources Department of Shandong Province. The Comprehensive Planning of Water Resources in Shandong Province[R]. 2008.

Mao Hanying, Zhao Qianjun, Gao Qun. Methods and Modes of Resources Development under Ecological Environment Constraints in the Yellow River Delta[J]. Journal of Natural Resources, 2003 (18): 4.

Oil Produced Water Treatment Using Reed Remediation Techniques—A Case Study from Sudan

Siddig Eissa Ahmed

Faculty of Engineering & Arch, Omdurman Ahlia University, Sudan

Abstract: Produced water is the largest single associated waste product in crude oil production. In the oil fields it is a common phenomenon to find a water layer below an oil layer. Thus, oil production is usually accompanied by water. Worldwide experiences estimated that the total production of water is expected to be in the range of five to ten times that of oil, during the economic lifetime of an oilfield. If this produced water is disposed to the environment directly without treatment may cause environmental pollution. In response to growing pressures on air, water, and land resources, global attention has focused in recent years on finding new ways to sustain and manage the environment. It is a great challenge for oil companies to find a best method to dispose produced water in environmentally friendly and optimum economic conditions. Bioremediation is seen as a promising tool in this endeavor because it can provide an effective approach for managing and preserving the environment.

The objective of this paper is to give highlights on the concept of bioremediation techniques as applied to treat the produced water in Sudan oil fields at Heglig. In the simplest terms, bioremediation is the use of microorganisms (bacteria or fungi) to decompose toxic pollutants into less harmful compounds. Bioremediation is the optimization of biodegradation. Where biodegradation is a natural process by which microbes alter and break down oil into other substances. The resulting products can be carbon dioxide, and water. The optimization can be accomplished by two forms of technology: 1-fertilizing (adding nutrients) or otherwise (e. g using reed plants) and/or 2-seeding (adding microbes). These additions are sometimes necessary to overcome certain environmental factors that may limit or prevent biodegradation.

The paper further discusses the application of this technology to a research/pilot bioremediation project in Heglig-Sudan, which is being implemented with the objectives to biodegrade hydrocarbons from produced water. The project system uses natural plants (reeds – Phragmites Australis) to optimize the biodegradation process. The paper also identifies the potential benefits and gives highlights on the system design and shows some of the very promising results that such a system can naturally biodegrade the hydrocarbons of the produced water. The system operation options, which have been undertaken to obtain optimum results for the biodegradation of the hydrocarbons in the produced water, were discussed. The project is considered as the largest of its kind in the world and it is further hoped that the effects of the project will radiate out beyond Heglig. The system is expected to act as a model approach to oilfield wastewater treatment and will therefore be replicated elsewhere in Sudan as well as in the different parts of the world.

Key words: produced water, reed bed, Phragmites Australis, biodegrading, hydrocarbons

1 Introduction

One of the undesirable by-products of oil production is the generation of produced water. The water oil ratio in some oil wells can be as high as five to ten barrels of water per barrel of oil. Current worldwide oil production is about 0.213×10^6 bbl/d. Somewhere between 1 and 2×10^9 bbl of produced water. In 1993, the U. S. produced 2.5×10^9 barrels of oil and about 26×10^9 bbl of produced water from all sources. This produced water is roughly a trillion gallons; which can fill a football field size tank to a depth of 530 miles. Normally produced water production increases as oilfield ages as can be seen in Fig. 1. In USA prior to 1972 produced water was often discharged to rivers, streams and unlined evaporation ponds-surface water containment. Today 65% of this water is re-injected to maintain formation pressure; 30% is disposed in deep injection wells and 5% of the water is discharged into surface waters such as rivers, natural or artificial lakes, or to the oceans.

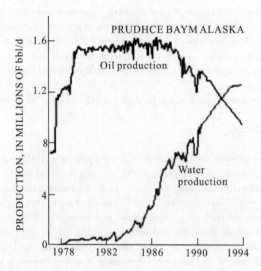

Fig. 1　Water production increases as oilfield ages

Produced water often:

　(1) Is highly saline.

　(2) Contains dispersed and dissolved oil.

　(3) Contains radionuclide.

　(4) Contains treating chemicals.

　(5) Contains heavy metals.

　(6) Contains ammonia.

　(7) Contains hydrogen sulfide.

　(8) Total salinities range from ~ 1,000 mg/L to > 400,000 ppm (median is 50,000 mg/L), where seawater is 35,000 mg/L.

The dissolved organics in produced water include:

　(1) Aliphatic hydrocarbons.

　(2) Aromatic hydrocarbons.

　(3) Phenols.

　(4) Carboxylic acids.

Some of those chemicals come from the reservoirs, and some from the chemicals that are added in the oil processing. With this host of pollutants, disposal of the produced water poses great environmental risks. In offshore activities, the usual practice is to discharge the water into the sea, which is considered to be an environmentally tolerant system, due to its large magnitude. On shore, surface (evaporation ponds) and or subsurface disposal (well-injection) is practiced. These practices have adverse impacts in term of the environment and economy.

　Finding another home for the produced water provides a solution to the produced water disposal problem facing the entire Heglig oil fields operated by the Greater Nile Petroleum Operating Company, Sudan (GNPOC). GNPOC is a consortium composed of CNPC – China; Petronas – Malaysia; ONGC – India; and Sudapet – Sudan. The problem of the produced water disposal in the Heglig field has been stated as:" Disposing of large volume rates of sodic water, contaminated with hydrocarbons and treatment chemicals, into a sensitive and vulnerable environment, at a reasonable cost". In recent years, sustained and intensive research has been carried out on the produced water problem. Investigations into the chemical properties of the water and ways and means to de-toxify it, have resulted in some novel techniques for disposal. One such technique that has attracted GNPOC is bioremediation, where plants are used to remediate the contaminated water. GNPOC through Oceans-esu of UK as consultants, recently managed a project team that performed a pilot/ research project at Heglig oil fields, which are located in the south part of central Sudan. The goal of this project is to treat the hydrocarbons from the produced water and use the treated water for irrigation purposes. This paper gives highlights on the concept of bioremediation techniques as well as the application of this technique to treat the produced water in Sudan oil fields at Heglig.

2 What is bioremediation

Bioremediation is a process related directly to biodegradation. Biodegradation is nature's way of recycling wastes, or breaking down organic matter into nutrients that can be used by other organisms. "Degradation" means decay, and the "bio-" prefix means that the decay is carried out by a huge assortment of bacteria, fungi, insects, worms, and other organisms that eat dead material and recycle it into new forms. For oil biodegradation these new forms can be carbon dioxide, water, and simpler compounds that do not affect the environment.

Bioremediation is the optimization or acceleration of biodegradation. This acceleration can be accomplished by two forms of techniques: 1-fertilizing (adding nutrients) and/ or 2-seeding (adding microbes). These additions are necessary to overcome certain environmental factors that may limit or prevent biodegradation. Knowledge on bioremediation acquired since 1942 allows manipulating environmental factors to enhance natural biodegradation. This knowledge include: Identification of microbes capable of degrading petroleum hydrocarbons; Nutrient requirements of these microbes, such as carbon, nitrogen and phosphorus; Environmental requirements such as oxygen, water and temperature; and metabolic pathways of decomposition for oil fractions. These oil fractions include the aromatics, aliphatic and asphaltenes. Bioremediation can be used for oil spills and has potential to help clean up land oil spills. There are also benefits to bioremediation such as saving money, being ecologically sound, destroying contaminants (not moving them from one place to another) and treating waste on site.

Microbes: Certain enzymes produced by microbes attack hydrocarbons molecules, causing degradation. The degradation of oil relies on having sufficient microbes to degrade the oil through the microbes' metabolic pathways (series of steps by which degradation occurs). Fortunately nature has evolved many microbes to do this job and throughout the world there are over 70 genera of microbes that are known to degrade hydrocarbons (U. S congress,1991 a).

Oxygen: biodegradation is predominantly an oxidation process:" Bacteria enzymes will catalize the insertion of oxygen into the hydrocarbon so that the molecule can subsequently be consumed by cellular metabolism (Bragg et al. , 1991). Because of this, oxygen is one of the most important requirements for the biodegradation of oil.

Metabolic Pathways for Oil Decomposition: Over the last 20 years complex chemical equations have been derived to describe the metabolic pathways in which oil is broken down. "The general outline bioremediation pathways for aliphatic and aromatic hydrocarbons have been formulated and continue to be developed in greater detail with time" (Glaser, Venosa and Opatken, 1991). All of these pathways will result in the oxidation of at least part of the original hydrocarbon molecule.

3　Application of bioremediation at heglig pilot/research project

By using the science of microbiology and chemistry, scientists have been able to recognize the processes and factors that allow biodegradation to take place. This same knowledge also allows trying to increase theses natural processes (bioremediation) by manipulating the environmental factors to achieve results faster than occur naturally. In the following part of the paper the application of bioremediation technique to Heglig oil fields, Sudan will be examined. The key element of the bioremediation used in Heglig oilfields centers around using the common reed (plants) Phragmites australis to enhance the process of biodegradation.

3.1　Project description

The Heglig oil fields are located at the southern part of central Sudan as shown in Fig. 2. The terrain of that area, generally level in the north, rises in the south to the Nuba Mts. The economy is agricultural, with millet as the staple crop and nomadic cattle, goat and sheep herding. The Heglig Oilfield lies in the Savanna belt of the northern hemisphere tropical climate, which is classified as hot and semi arid. The average hot summer and cold winter temperatures are 40 ~ 20 ℃ respectively. The rainy season is from June to October. The average annual rainfall varies from 650 ~ 700 mm.

(1) The bioremediation site at Heglig consists of three distinct areas as follow:

The first area comprises a series of Storage Lagoons (SL), and Reed Bed Lagoons (RB) numbered SL 1 to SL 6 followed by RB 1 to RB 6. The first set of lagoons is signed to be the storage lagoons while the other set of lagoons are converted to horizontal reed beds. The reed beds are planted with Phragmites australis.

(2) The water exiting these lagoons flows under gravity into a balanced distribution canal and then into a much larger ploughed section assigned to forestry and wetland species.

(3) Finally the water exiting into the final stage consisting of two open storage lagoons.

Produced water flows from the oil processing facilities to the two sequential evaporation ponds is pumped along a pipeline to the first receiving lagoon at the present rate of more than 200,000 bbl/d, which is equivalent to about 32,000 m^3/d. The produced water at the end of the pipeline is discharged into a small chamber (delivery basin) within the first lagoon and then cascades over the top of the chamber into the lagoon proper. The flow path for the produced water is from SL1 to SL 6, whiles the flow from storage lagoons into the reed beds; RB 1 to RB 6 can take several combinations of flow pattern. This is because the system is designed to incorporate various flow patterns whether on series or parallel or a combination of such. Transmission of the produced water from one lagoon to the next is by gravity flow through a bottom gate valve. The total capacity of the lagoon system is up to 800,000 m^3 which on the initial daily input discharge as indicated previously; would allow for about 25 d retention time. The total surface area for the lagoons is some 65 km^2. The entire project site covers an area of more than 400 km^2. Fig. 3 and Fig. 4 show respectively the different segments of the bioremediation project, and a longitudinal profile for the bioremediation project including the oil recovery system.

3.2　Heglig produced water quality

Produced water requires treatment for a number of water quality constituents depending on the intended water use. In Sudan, there are no current drinking water regulations covering the use of treated produced water. However, there are standards for discharging the produced water to the environment. Typical water quality characteristics of Heglig oil fields are summarized in Tab. 1.

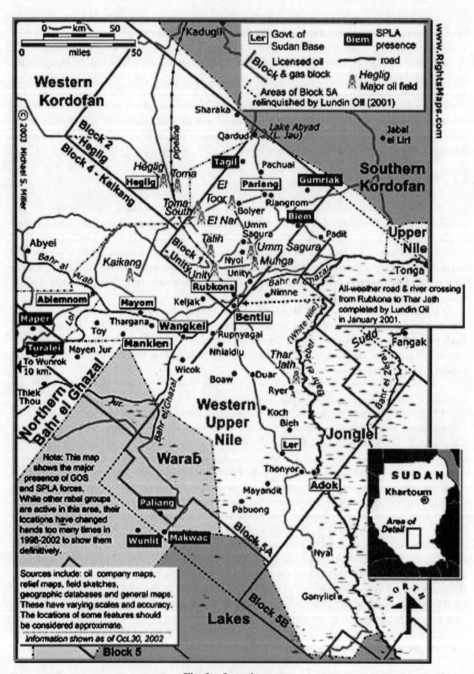

Fig. 2　Location map

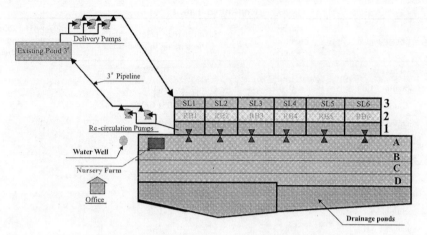

Fig. 3   Schematic diagram of the bioremediation

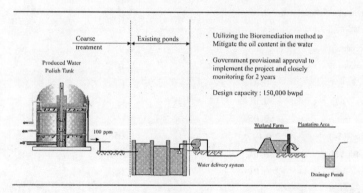

Fig. 4   Longitudinal profile of the bioremediation system in Heglig

Tab. 1   Typical concentration of constituents in produced water, Sudan environmental regulations, and FAO 1985 water standards for irrigation

Constituent	Heglig produced water	Sudan nvironmental regulations (Discharging to fresh water)	FAO 1985 water standard for irrigation
Color	Light brown	Non	
Odor	Mineral oil	Non	
pH	8.2 ~ 9.4	6 ~ 9.0	6.5 ~ 8.4
Total organic carbons	5 ~ 120	—	
Total dissolved solids	1,200 ~ 1,292	1,200	< 2,000
Oil & grease		5.0	
Total suspended solids		30.0	
Biological oxygen demand		30.0	
Chemical oxygen demand		40.0	

Continued Tab. 1

Constituent	Heglig produced water	Sudan nvironmental regulations (Discharging to fresh water)	FAO 1985 water standard for irrigation
Phenols		0.002	
Cadmium (Cd)	< 0.01	0.01	
Chromium	< 0.01	0.05 (Cr VI)	
Copper (Cu)	0.05	0.1	
Lead (Pb)	< 0.05	0.05	
Mercury(Hg)	Not detected	0.001	
Nickel (Ni)	< 0.05	0.01	
Dissolved oxygen		>2	
Sulphides		1.0	
Ammonia (NH_3)	Nil	Nil	
Nitrate (NO_2)	Nil	30	5 ~ 30
Fluoride	3.75	0.5	
Total Phosphates		1.0	
Detergents		0.05	
Cyanides		0.05	
Iron (Fe III)	0.27	1	
Zinc (Zn)	0.01	1	
Tin (Sn)	0.05	—	
Arsenic (As)	0.017	0.05	
Manganese (Mn)	0.01	0.5	
Silver (Ag)	0.01	0.05	
Heavy metals		1	Trace elements
Radioactive materials		Non	
Residual chlorine	41 mg/L	1	< 4 meq/L
Bacterial count		2,500/100 mL	
Boron	0.01		0.7 ~ 3.0
Calcium (Ca)			
Magnesium (Mg)			
Sodium (Na)			

3.3 Hydrocarbons & other Pollutants

As it has been mentioned earlier the intention is to remediate the production water from the Heglig field and to use the remediated or cleaned up water for forestry or agriculture.

The forestry/ agricultural part of the project will need to be managed to minimise the risk factors arising from any pollutants that may occur within the irrigation water. It is apparent that the following pollutant groups have the greatest potential to affect the proposed irrigation water:

(1) Free/dissolved hydrocarbon contamination.

(2) High salinity (dissolved solid) content.

3.3.1　Receptors

The primary receptors for pollutants in irrigation water are as follows:
(1) Soils within the crop production area of the project.
(2) Crops within the system.
Secondary receptors for any pollutants entering the system are the following:
(1) Humans using the crops.
(2) Local water cycle.
(3) Local ecosystems.

3.3.2　Probability matrix for hydrocarbon pollutants in soil / crops

Hydrocarbon contamination has a higher probability of occurrence in the areas of the irrigation that are first in the sequence. These are as follows:
(1) Lagoon areas (proposed bioremediation site).
(2) Areas close to the irrigation channels in the forestry areas.
(3) Areas closest to the "inlet" end of the system.

3.3.3　Probability matrix for soil sodicity or high salinity of irrigation water

Sodicity and high salinity are likely to be closely associated with one another. These both have a higher probability of occurrence in the areas of the irrigation that are sequentially later. This is because the problem will be compounded by evapotranspiration in crop areas, and evaporation from irrigation ditches.

The sequence of risk ought to follow the following pattern:
(1) Areas remote from irrigation ditches.
(2) Areas remote from the "inlet" end of the system.

3.4　Mitigation of pollution occurrences

Having established the likelihood of contaminants reaching the pathway, it is incumbent upon the project design to minimise the risk to the project from such pollution incidents. The following measures are proposed.

(1) Planting of secondary lagoons with common reed Phragmites australis to form a phytoremediation system in advance of the main forestry area. Reed beds are tolerant to moderate levels of hydrocarbon contamination in irrigation water, and are effective at removal of both dissolved and free phase hydrocarbons.

(2) Careful selection of tree species tolerant to saline and sodic conditions, with plants of highest tolerance in the highest risk areas.

(3) Monitoring of key chemical parameters in the system so that the actual chemical conditions are well known and documented.

(4) Potential application of gypsum which can be applied to soils if sodium levels become dangerously high.

3.5　Use of reed beds to optimize microorganisms growth

Particular plant species were highly effective whilst agricultural topsoil with its diverse habitat of microbiology was found to be affective against many types of toxic material. Combining plant species with good top soil allowed continuous flow of water through the plant roots and at the same time allowing natural bacteria to detoxify and purify the water. The plants have three functions as follow:

(1) The very extensive root system creates channels for the water to pass through.

(2) The roots introduce oxygen down into the body of soil and provide an environment where aerobic bacteria can thrive. These organisms are necessary for the breakdown of many types of compounds, in particular the oxidation of ammonia to nitrate, the first step in the biological breakdown of this compound.

(3) The plants themselves take up a certain amount of nutrient from the wastewater.

Normally the effluent treatment system will be formed within a certain area of the site whilst the clean water will be used for irrigation.

Many of the reed bed systems in the UK have used the common reed, Phragmites australis. The root system of the Phragmites is as shown in Fig. 5. The mechanisms for bioremediation through reeds is as shown in Fig. 6.

Fig. 5   Rhyzomes to introduce oxygen into soil

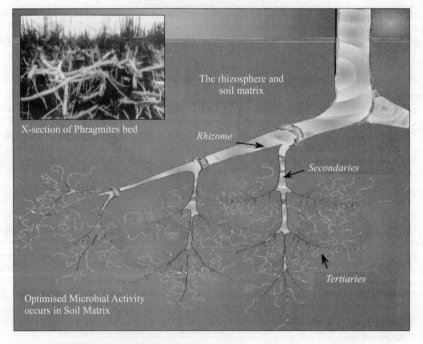

Fig. 6   Optimized microbial activity occurs in soil matrix

4　The bioremediation system operation

4.1　Monitoring program

Generally speaking the monitoring will cover: chemical, microbiological, toxicological and botanical parameters. In order to assess the risks to the forestry and the efficiency of the bioremediation project, 40 monitoring stations are proposed for the system. These points will be monitored for parameters shown in Tab. 1. The monitoring for some of the parameters will be either on daily, weekly, biweekly, monthly, quarterly, and semiannual basis. Bioaccumulation "dissolved organics" such as benzene, toluene, xylene bio-accumulate in plant, soil, and fish as well as heavy metals will be monitored.

4.2　Operating strategy & oil removal

The bioremediation system has been established in about ten months, particularly the treatment system which is consisted of the reed beds. The system used is called the horizontal system where Phragmites have been grown in ridges and allowed to propagate. It is anticipated for the first six months (started on May, 2004) that the focus will be based on the treatment levels of the system as a whole with particular attention paid to the reed beds. This will be based on both the collective treatment of the beds, and also on an individual basis. Several options can be used. However for the mean time only one option is now used and this is as shown in Fig. 7, where the other option shown in Fig. 8 will be used in case of emergency.

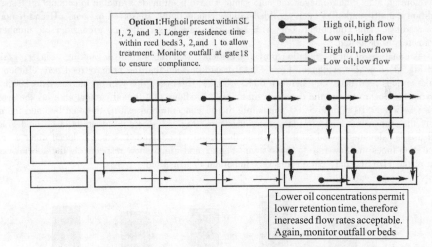

Fig. 7　Treatment pattern of the produced water

5　Hydrocarbon biodegradation

At the moment the analysis of oil in water is carried out by the site laboratories, but the accuracy could not be quoted within 100%. This is due to the equipment and methodology applied, the technique is one more suitable for water in oil analysis; by this it is meant the technique is more accurate the higher the concentration of oil. To rectify this it was recommended that the technique be changed to a more suitable method. Fluorescence spectrophometery was deemed the best available within financial and site operational techniques. The technique is a very detailed and selective one.

252

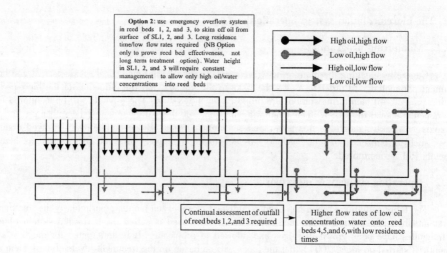

Fig. 8   Treatment pattern of the produced water in case of emergency (high hydrocarbons)

This technique when used correctly can be up to 1,000 more sensitive than UV/Visible techniques. Therefore an M53 Flouro imager and auto sampler have been made available to conduct oil tests. In addition a multi-sonde has been provided to give conductivity, dissolved oxygen, pH and temperature readings. This equipment come with a built in data logger allowing a considerable number of parameters to be tested.

An analysis of oil degradation provides evidence that the system is working. Fig. 9, Fig. 10 and Fig. 11 show this evidence. From Fig. 9 it can be seen that the primary reed beds efficiencies of the various reed beds vary between 60% ~ 80%. However, this can be increased to 100% by varying the flow rates and the detention time. Fig. 10 shows that the oil content entering the system varies between < 10 ppm to > 200 ppm while the oil concentration exiting the reed bed and the outlet of the system is < 1 and in many cases it is zero. Fig. 11 shows the quantity of oil biodegraded and this is calculated based on the equation the quantity biodegraded is the produced water oil concentration times the flow rate to the system. Fig. 12 and Fig. 3 show respectively the special variation of the TDS and pH, parameters very important to irrigation.

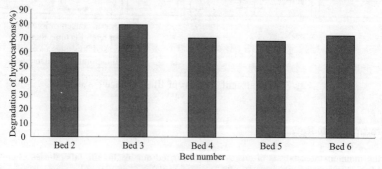

Fig. 9   Degradation of hydrocarbons by bed

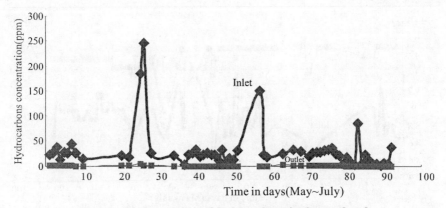

Fig. 10 Biodegradation of hydrocarbons in the produced water(inlet & outlet concentratio)

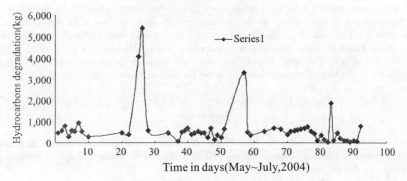

Fig. 11 Hydrocarbon degradation

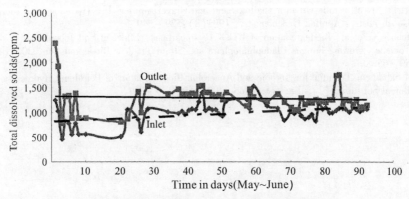

Fig. 12 Spatial variation ot the total dissolved solids

6 Conclusions

This work resulted in a number of promising findings (e. g. the extent of hydrocarbon degrada-

254

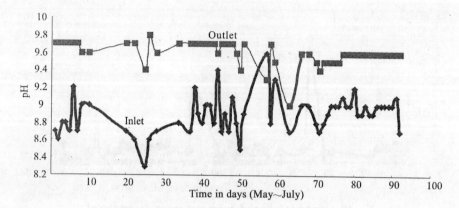

Fig. 13　Spatial variation of produced water pH

tion), which can contribute for the establishment of reed bed bio-treatment to be applied to resolve the problem posed by hydrocarbon in the produced water. The following goals were achieved: high organic matter biodegradation, reduction in the total operational time, and minimization of the generated waste. So far the time spent was on construction and establishment of the bioremediation system (the reed beds). The next phase of the study will focus on detailed monitoring to ensure the compatibility of the Heglig treated produced water for the environment and irrigation.

References

Adriana Ururahy S, Nei Pereira Jr. Oily Sludge Biotreatment[M]. Biotechnology & Ecosystems Department, Petrobas Research Center (CENPES), Rio de Janeiro, Brazil. adrianasoriano@ cenpes. petrobas. com. br

Bernard P C, Bastow M A. Hydrocarbon Generation, Migration, Alteration, Entrapment and Mixing in the Central and Northern Sea[M]. in W. A. England, and A. J. Fleet, eds. , Petroleum Migration, Geological Society, Special Publication, 1991.

Bennett PC, et al. Crude Oil in a Shallow Sand and Gravel Aquifer[J]. Hydrology and Inorganic Geochemistry: Applied Geochemistry, 1993(8):529–549.

Prenafeta F X, et al. Biodegradation of BTEX Hydrocarbons: Effect of Soil Inoculation With the Toluene – growing Fungus Cladophialophora sp. Strain T1[J]. Biodegradation, 2004(15): 59–65.

Jani M Salminen. Potential for Aerobic and Anaerobic Biodegradation of Petroleum Hydrocarbons in Boreal Subsurface[J]. Biodegradation, 2004(15): 29–39.

Exploring for the Setting up the IRBM Action Framework in China
—Taking the Yellow River Xiaohuajian Reaches as an Example

Zeng Xianghui[1] , *Wang Chunyan*[2] , *Ye Yaqi*[2] , *Feng Qiao*[2] , *Wei Jing*[2] and *Niu Hongyan*[2]

1. Department of International Cooperation, Science and Technology, MWR, P. R. China, Beijing, 100053, China
2. China Water International Engineering Consulting Co. ,Ltd. , Beijing, 100044,China

Abstract: In recent years, due to rapid growth of social economy, China is now facing problems of water resources, water disasters, aquatic environment and aquatic ecology associated with river basin. With continuous exploration, people have already recognized the shortage of traditional water management, and start to find the new method to manage the water. Therefore, the integrated river basin management (IRBM) was been paid more and more attention. Chinese government found that the implementation of integrated river basin management is the strategic choice to solve the above four problems, then to protect the health of rivers and lakes, as well as the sustainable development of society and economy. In the process of IRBM in the worldwide, many countries have their own river basin management mode and accumulated a lot of experiences. However, it is a complex project to implement IRBM as it not only requires Long-term exploration, research and practice, but also a solid work foundation to progress steadily. In the meantime, because of the difference of China and EU, EU experiences can not be copied in China directly. To assist in solving current water issues, this paper takes the Xiaohuajian reaches (i. e the river reach from Xiaolangdi to Huayuankou) of the Yellow River as pilot area, adapted to the basin context in conformity with the orientations defined by the Chinese Government and based on the EU experiences, the working process will start from the problems analysis, objective, then to planning and implement, which will construct the integrated river basin management action framework with the aim of exploring the best river basin management model, which can be further carried out on the rest of the Yellow River basin as well as to provide references to other river basins in China.
Key words: integrated river basin management (IRBM), Xiaohuajian reaches of the Yellow River, action framework

1 Introduction

As it is known to all, water is the most essential and important element for human beings and other livings to survive. With the continuous development in our economic society, water arouses an unprecedented contradiction between supply and demand. Under current situations, the problems are classified into two types: on the one hand, the demands of water are increasing and people are paying more and more attention to water environment; on the other hand, the problems are still updating, especially about the mixed pollutant wastewater as well as river basin developments. However, people are not able to deal with such water treatments problems. Thus, aiming to solve the problems mentioned above, the common tendency is to the integrated river basin management (IRBM) after the practices by many countries. China has been facing the global climate problems and challenges since 1990s, including problems about water as well, such as the conclusion:" water excess, water shortage, water contamination, and muddy water". In order to propose a new water resources management strategy to replenish the traditional ideas, IRBM draws unprecedented attention in public. To be honest, since the accompanying problems and contradictions with economy developments in China, the current framework and management process in river basin agent lack delegates of benefits and the relevant essential conversational consulting strategy. Thus, it is also a series of challenges both about the unified management of water resources and integrated management in river basin. Good news comes with the improvements in the unified management of water resources and integrated management in river basin that are brought by governments. Thus, the recommended implement of IRBM is an important way in solving the increasing problems about water resources and environment; meanwhile, it creates advantages in making structures and applying IRBM in china.

The so-called integrated river basin management (IRBM) could be defined as: through the integrated management of combined departments and administrative regions, in order to make best adjustment to the nature, the abundant resources such as water, soil and living beings are developed, utilized and protected comprehensively at river basin scale. Meanwhile, it is a proper way to actualize the maximization of river basin economy and social benefits, as well as the sustainable development in the river basin. Thus, IRBM is the guarantee of water, ecology and sustainable development in economic society. According to aim and principle of IRBM, China needs an integrated framework including law – making, constitution reform, institution – building, scientific support, information share and public participation. Therefore, with the relevant systematic actions, China has the opportunity to transfer the river basin management from traditional method to IRBM.

2 Comparison of water resources management between EU and pilot area

For the past few years, based on the local economic conditions and basic situations, every country has its own features in IRBM and acquires lots of experience. The specific work varies from planning, developing to management. Especially in EU countries, the EU makes several directive documents such as EU Water Framework Directive, EU Groundwater Directive, and EU Flood Directive. All of the performance is encouragement for what the EU members has done about IRBM.

Due to difference of location and national situation between Europe and China, the water resources problems may be different in some aspects. For example, the loss of water and soil in Europe is less than China. However, for the similar water resources problems, the water resources management experiences of WFD can give references to Chinese river basin management. Here the comparison of water resources management between EU and pilot area are shown in Tab. 1.

Tab. 1　Comparisons of water resources management between EU and the pilot area in the Yellow River

	Pilot area in the Yellow River	EU
Water resources management system	Multi – department administration in water resources is serious, lack of coordination between different authorities	The integrated river basin management pattern is adopted
Law construction	The law system for water resources sustainable utilization is not perfect	Manage all the water resources and protect water source by law
Management measures	Administration measure is the main measure, and the supervision system is not perfect	The measures of economy, technology, society, organization and system are adopted integrated
Stakeholders participation	Public participation is not enough	All the water users are encouraged to participate in the water resources management
Water saving and reused	Measures for water saving are not enough with a serious waste problem; meanwhile, the reused rate is low	Various measures for water saving are adopted, and the efficiency of water use is high
Data monitor and share	Monitor is not comprehensive; data quality is poor, and lack of data share	The monitoring data is comprehensive, which is a good base for water quality evaluation

3　Case studies

3.1　Pilot area introduction

The pilot area is located between the Xiaolangdi Reservoir and Huayuankou (hereinafter referred to as "Xiaohuajian"). It is the crossways between midstream and downstream and mainly contain the Xiaohuajian mainstream of the Yellow River, the Luo River and Qin River. The catchment area is 35,883 km². Also, the pilot area are semi-humid region with an average precipitation of 630 mm. As the main tributary of the Yellow River, the Luo River is located on the south bank of the Yellow River, originated from the Lantian County in the Huashan Mountain ranges of Shaanxi Province. The river flows across Shaanxi and Henan Provinces, with the mainstream length of 447 km and the catchment area of 18,881 km². The Yi River is the biggest tributary of the Luo River, with a river course length of 265 km and a catchment area of 6,030 km². The Qin River is on the north bank of the Yellow River, originated from the south ranges of the Taiyue Mountain in Qinyuan County of Shaanxi Province, and flows across Shanxi and Henan Provinces. The river course length is 484 km and the catchment area is 13,532 km².

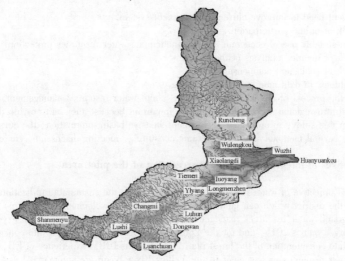

Fig. 1 Water system map in pilot area

The mainstream of the Yellow River at the intake of Xiaohuajian is where the hydrological station of the Xiaolangdi dam is located with an average run-off of 48. 4 × 10⁹ m³, while the outtake of Xiaohuajian locates at the Huayuankou Hydrological Station with an average run-off of 53. 3 × 10⁹ m³. Apart from the by-pass run-off, the average self-producing water flow is 5. 727,97 × 10⁹ m³, of which surface water accounts for 4. 934,15 × 10⁹ m³ and groundwater takes up 3. 28 × 10⁹ m³ (of which water flow without overlaps of surface water is 7. 938,2 × 10⁹ m³). Water resources flow of the Luo River, Qin River and Xiaohuajian mainstream are shown in Tab. 2.

Tab. 2 Water resources flow of Xiaohuajian section

(Unit: 10,000 m³/a)

River name	Surface water	Groundwater	Non-overlap	Total
Luo River	325,894	187,334	28,368	354,262
Qin River	130,324	104,114	32,548	162,872
Mainstream of Xiaohuajian	37,197	31,322	18,466	55,663
Total	493,415	322,770	79,382	572,797

In accordance with the "Division Plan of Water Function Zone in the Yellow River Basin and Northwest Inland River Basin" developed by the Yellow River Conservancy commission (YRCC) in

2002 and the results obtained from water function zone monitoring in 2008, water quality in most water function zones in the area is below the standard, part of the area are even below Level V. The situation now is not promising.

3.2   Water resources management issues outline

Through different ways, such as data collection, specialist consulting and field trip, the existing water resources issues in the pilot area had been summarized. Then through the problem trees analysis method, the inner factors which lead to the water resources issues have been found.

(1) Imperfect laws and regulations in sustainable use of water resources.

(2) Multi-department administration in water resources, lack of coordination between different authorities.

(3) Urgent need to improve current management system.

(4) Lack of public participation.

(5) Non-suitable use of economic means; inadequate water resources protection.

(6) Lack of detailed analysis of the basin.

(7) Lack of ecological compensation.

(8) Problems in data monitoring and sharing.

During the process of analyzing the problems about water resources management, the communication with intra-regional stakeholders are very important because they can provide the actual information. In this study, in order to get the comprehensive basin information, the current problems analysis uses several methods such as telephone consulting, meeting discussion, survey, etc.

3.3   Objectives of integrated water resources plan of the pilot area

The total objective of water resources plan is to achieve the sustainable utilization of the water resources of the district which can guarantee the sustainable development of economy and society. For the integrated water resources management in the pilot area, the present base year is 2008, short-term target year is 2015, and for the Long-term 2030 should be taken into consideration.

Taken into consideration of the latest plan, the previous IRBM experience of EU, objectives of integrated water resources management in the Yellow River Basin are summarized as following:

(1) Flood control and sediment reduction.

(2) Soil and water conservation objectives.

(3) Reduction of emission and improvement of water quality.

(4) Water saving and grey water recycling objectives.

(5) Water shortage relieving objectives.

3.4   Measures involved and selection

In regard to the main issues encountered in the utilization and exploitation of the Yellow River water resources in pilot basin area, the corresponding control measures involved in various planning are summarized as following:

(1) Establish new water resources management system, enhance the integrated allocation of water resources.

(2) Promote the reform of water resources mechanism, improve management efficiency of water agencies.

(3) Establish and improve the legislation system of water resources management.

(4) Public participation in water resources management.

(5) Establish water market, enhance sustainable utilization of water resources.

(6) Detailed analysis of basin.

(7) Establish and improve the research of ecological compensation mechanism on water and soil conservation.

(8) Establish water resources information management system.

FQ On the basis of the cost benefit analysis, the project team concludes the short-term and Long-term measures that could be implemented in mainstream as shown in Tab. 3.

Tab. 3　Measures suggestion

Objectives	Short term	Long term
Flood control	Reinforce embankment	Reinforce embankment, draw up flood risk map, establish alarm system
Water and soil conservation	Construct sediment control dam and silt retention dam	Increase vegetation coverage, return farmland to forest
Water quality improvement	Strengthen the punishment for sewage emission, reduce release of toxic substances	Build wastewater treatment plant, increase wastewater recycling rate, reduce sewage emission
Water shortage alleviation	Reduce water usage per 10_000GDP, Optimize water resources allocation	Reduce water usage for agricultural irrigation, increase industrial water recycling rate
Water – saving and wastewater recycling	Increase water price, encourage water – saving facilities	Increase water price, raise public awareness of water conservation

In this stage, as the planning contains many relevant regulations and every region has its own character, the participations of stakeholders seem so important. Through the consulting and real interaction with stakeholders, the actual problems existing in implementation could be realized. Therefore, it provides the judgment for the refresh optimization in deciding the regional final measures.

3.5　Action Framework (AF) of the Yellow River IRBM in pilot area

Stakeholders take part in the problems analysis, objective, planning and implement in the course of water resources framework of the Yellow River pilot area. The pilot area was selected including the YRCC views from the relevant units and departments. After selecting the pilot area, getting support with YRCC for the data acquisition, a large number of hydrology and water resources reports and data were collected. After the completion of the draft inception report, presentation to the Project Steering Committee meeting and receiving comments and suggestions from the Steering Committee the inception report was finalized.

The procedure of the Yellow River IRBM AF can be described as below:

(1) Inception period: The subject of the analysis (what is analyzed under what conditions) and the objective of the analysis (what are the desired results of the analysis) and constraints (what are the limitations) are specified.

(2) Development period: The analysis and (elements of) solutions to the water resources problems are developed. Major activities are data collection. Individual measures will be developed and screened in this phase. A gradual improvement of the understanding of various characteristics of the water resources system is obtained.

(3) Selection period: The purpose of the selection period is to generate a limited number of promising strategies, important activities in this period are strategy design, impact assessment, evaluation of strategies, sensitivity and scenario analysis and presentation.

Moreover, during this phase public consultations will be needed to inform and involve stakeholders and get their reactions and inputs.

Based on the steps already undertaken in phase 1 and phase 2, the AF has been worked out (Shown in Fig. 2). Based on the experiences in phase 2 is has to adjusted and finalized.

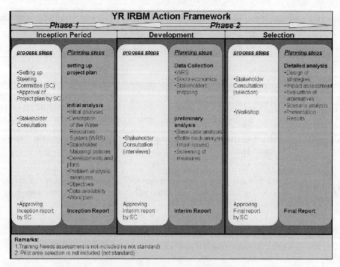

Fig. 2　Action framework of the Yellow River

This framework uses European IRBM experience as basis. In accordance with the governmental planning and with the help of blending the conceptions, the action framework aims to achieve the goal of building an ecological, sustainable, and reasonable water resources management mode. The combination of EU experience and the YR IRBM provides a further reference in developing the river basin area in China.

4　Conclusions and suggestions

The IRBM involves in many departments such as: water, environment protection, national resources, agriculture, forestry, transport, science, and travel. Thus, there would be a demand of participations among all departments and stakeholders in IRBM. As the action of IRBM is a long-period, complex and systematic process, the implement and exploration process is needed, as well as the performance of the pilot work to get more experience. In order to actualize the goal mentioned above, we need to observe the maneuverable principle of unique designing, adjusting measures to local conditions, pilot projects first and impelling step by step. Thus, the development of IRBM can be systematic and gradual. In this study, the Xiaohuajian area is used as the pilot to explore the IRBM in China; also the acquired experience provides the reference for further implementation in larger area.

4.1　Conclusions

(1) The directive suggestions from EU specialists should be emphasized. In this way, they can provide new eyesight of IRBM to make the construction of the IRBM framework effectively.

(2) During the process of understanding the existing problems and solutions in regional water managements, the methods are various. For example, telephone consulting, meeting discussion, face to face communication and survey are all useful methods in acquiring the whole information in basin.

(3) The enthusiasms of stakeholders are the key point in IRBM. In order to make a stronger guarantee in IRBM, the stakeholders' concentrations should be increased and it is better to the reply their suggestions in time in IRBM.

4.2　Suggestions

(1) Because of the economic limitations and the large pilot area, the field research is not

comprehensive. Thus, in the future research processes, in order to get more actual information and public opinions in basin, the selection of pilot areas should be more proper and the frequency of field research should be increased.

(2) Stakeholders are relatively passive during IRBM, in this case, the beneficial mechanism are needed more than ever to encourage the stakeholders participations in IRBM.

(3) In IRBM, the attention, direction, and coordination from governments are significant, especially in the areas with a greater capacity of spanning. The government is able to provide guarantees of strengthening IRBM, comprehensive planning, developing and protection.

References

Yan Heye. The Valid Path to Solve the Water Shortage Problem of Qinhe Area[J]. Water Resource Science and Economy, 2007, 13(2):117 – 118.

Chai Chengguo, Yang Xianghui. The Sustainable Utilization and the Construction of the System of Prevention Evaluation Index in Xiaohuajian Water Source in Yellow River[J]. Water Resource and Water Electricity Technology, 2003,34(3):1 – 2.

Yellow River Conservancy Commission of Ministry of Water Resources. The Flood Control Programing in Yellow River[M]. Zhengzhou: Yellow River Conservancy Press, 2008.

Yellow River Conservancy Commission of Ministry of Water Resources. The Construction and Practice of the Transform System of the Yellow River Water Right[M]. Zhengzhou: Yellow River Conservomcy Press, 2008.

Guo Jianmin, etc. The Yiluo Stream Memory[M]. Beijing: Science and Technology of China Press,2000.

Jiaozuo Yellow River Bureau. Qinhe Memory[M]. Zhengzhou: Yellow River Conservancy Press, 2009.

Wang Bin. The Collected Works Protection and Management Water Environment[M]. Zhengzhou: Yellow River Conservancy Press, 2002.

Yellow River Conservancy Commission of Ministry of Water Resources. The 12th Five – year Plan of Yellow River Basin and other Rivers in Northwest, 2009.

Zhengzhou City Council. Water-saving Society Construction Planning[R].2006.

Luoyang City Council. Water-saving Society Construction Planning[R]. 2009.

Li Qi. Compare of Water Resources Management System in Overseas[J]. Journal of Economics of Water Resources, 1998 (1):62 – 63.

Wan Jun,Zhang Huiying. Basin Management in France[J]. China Water Resources, 2002 (10): 164 – 166.

Liu Zhonggui. Water Resources Protection and Management Summary for Germany, France and Holland[J]. Pearl River, 2000 (3):4 – 6.

Pang Hongjun. Experiences and Significance of Germany Water Resources Management[J]. Shandong Agriculture, 2003 (3):18 – 19.

Martin Griffiths. EU Water Framework Directive Handbook [M]. Beijing: China Water Power Press, 2008.

Tan Wei. "EU Water Framework Directive" and Significance[J]. Law Science Magazine, 2010 (6):118 – 120.

Dong Zheren. Rhine River—disposal Protection and International Cooperation[M]. Beijing:China Water Power Press, 2005.

Wang Weizhong. Urban Water Resources Management and System Exploitation in Overseas City [M]. Beijing:China Water Power Press, 2007.

Andersen L S, and Griffiths M. Potential significance of the EU WFD to China[J]. Express Water Resources & Hydropower Information, 2009, 30(9):85 – 94.

Measurements Beyond River Health: Water
Sustainability Assessment and River Sustainability Index

Wu Huijuan[1], *Richard C. Darton*[1], *Alistair G. L. Borthwick*[2] and *Ni Jinren*[3]

1. Department of Engineering Science, University of Oxford, Parks Road,
Oxford, OX1 3PJ, United Kingdom
2. Department of Civil & Environmental Engineering, University College Cork, Ireland
3. Department of Environmental Engineering, Peking University, The Key Laboratory of Water
and Sediment Sciences, Ministry of Education, Beijing, 100871, China

Abstract: Water is one of our most critical resources: as populations and economies grow, water scarcity and degradation pose a growing threat to socio – economic development and environment protection. It has become more challenging but critical to provide adequate freshwater for the competing needs of human use, whilst at the same time managing the associated Long-term consequences and risks in the environment. The maintenance and restoration of healthy river systems have become important objectives of environment and water resources management.

Integrated assessment is needed in order to understand these challenges and to manage water towards a sustainable future. Tools for river health assessment have been developed to measure the ecological status and natural capital of the river courses. Extending river health measurements to include socioeconomic values imbedded in water resources, this paper presents an indicator-based framework for assessing river sustainability. We discuss how sustainability indicators are applied to assess the specific concerns of water sustainability. In this paper we introduce our approach using the Process Analysis Method of sustainability assessment, which leads to a Composite River Sustainability Index for the Lower Yellow River in China. This methodology presents a systematic way to incorporate all domains of sustainability, and to generate a tailored indicator set.

The paper concludes that a sustainability indicator framework provides a transparent and participatory basis for assessing river sustainability. Such assessments provide the information and overview that are needed for identifying critical issues, and are especially helpful in underpinning policy development.

Key words: sustainable assessment, sustainability index, river health, measurements

1 Introduction

The concept of sustainable development has gained widespread acceptance since it was put on the global agenda by the World Commission on Environment and Development in 1987. The most quoted definition of sustainable development, published in the Brundtland Report (WCED, 1987), refers to "development that meets the needs of the present without compromising the ability of future generations to meet their own needs". It illustrates the dilemma inherent in human development between meeting human needs, in particular the essential needs of the World's poor, and the limitations on the environment's ability to cope with the consequences. The concept of sustainable development recognises the interactions between nature, development, and the meeting of basic human needs. These interactions can be conveniently described in terms of the three pillars of sustainable development, namely, economic development, social development and environmental protection (UNCSD, 2002; N, 2005). Sustainable development envisions a path to a sustainable future in which these interactions are well-balanced.

In the 25 years since the Brundtland Report was published, much effort has been dedicated to applying this concept in practical situations. Quantitative measurements of sustainability are required in order to evaluate to what extent sustainability is being achieved, to track progress towards sustainability, and to provide information and guidelines for development projects (Kates et al., 2001; Kates et al., 2005). As a result, various frameworks and tools have been developed in an effort to obtain integrated measures of sustainability. Sustainable development indicators (SDIs) have been intensively used to improve stakeholder engagement and to guide policy – making (Singh et al. , 2009; Pintér et al. , 2005; Bell and Morse, 2008; Ness et al. , 2007). The indicator set results in metrics which are a collection of carefully chosen measurements that quantify each indicator and cover relevant environmental, economic and human/social impacts. The strength of the SDI set lies in its ability to summarise and focus attention on the essential elements of a complex situation. SDIs are extensively used, for example, in assessing the sustainability of whole countries (UN, 2007; DEFRA, 2010), water resources (Sullivan, 2001; PRI, 2007), companies (GRI, 2006), and manufacturing operations (IChemE, 2002). A key decision is the choice of exactly which indicators to include in a set, and which to omit. This process of selection needs to be both transparent, and to follow a methodology designed to produce an indicator set for a particular purpose (Dalal – Clayton, 2002).

Of the World's natural resources, freshwater is critically important as it sustains the global e-cosystem and human civilization. In the latter part of the 20th Century, awareness dawned that global stocks of water resources are finite (Gorre – Dale, 1992). As population and economies grow, water scarcity and degradation pose a serious and growing threat to socio – economic development and environmental protection. Hence, sustainability of water resources is strategically vital to sustainable development. In this context, the water and development nexus has been highlighted by the work of Falkenmark (1989), Gleick (1996), Rogers (1997, 2008), and others. The importance of water was addressed at the Rio Earth Summit in 1992 and it features in the UN Millennium Development Goals (UNDP, 2000). More recent researches reaffirm the role of water as a fundamental basis of sustainable development. For instance, a group of senior scientists identified freshwater as one out of nine essential planetary systems whose boundaries must be identified and quantified to "ensure a safe operating space for humanity and to prevent unacceptable environmental change" (Rockströ et al. , 2009). Recently, Gleick and Palaniappan (2010) introduced the concept of peak water, referring to the limits of water availability. The term peak ecological water is defined as the point beyond which the total environmental externality costs exceed the total value provided by human use of that water. This concept guides us towards using and managing water in a more sustainable manner. In the recent 6th World Water Forum, the importance of water in enhancing the quality of life and particularly in achieving green economic growth was also addressed (WWC, 2012).

Given that freshwater resources is a fundamental prerequisite to sustainable development, this paper examines how sustainability indicators can be used as a holistic methodology for integrated water assessment. We review previous studies ranging from single metric of water scarcity, to large-scale water sustainability performance assessment at country or basin level. Associated with these case studies, we will discuss a consistent methodology for selecting an indicator set and the applicability of a composite index. Our objectives are to show how to apply SDIs to assess the overall sustainability of the Lower Yellow River, and how a well-chosen indicator set can provide a simple but comprehensive and useful guide to complex issues.

2 Sustainability indicators

The need to assess sustainability has given rise to the development of various approaches and tools, which include indicators, benchmarks, audits, indices, accounting parameters, as well as assessment appraisal and other reporting systems (Bell and Morse, 2008). To better understand the nature of sustainability assessment, Ness et al. (2007) categorised various tools, illustrated in Fig. 1. They used a framework with of the following three categories: ①indicators and indices, which can be further divided into non-integrated and integrated subcategories; ②product-related assessment, which refers to material and/or energy flow assessment and life cycle assessment; and ③integrated assessment, which consists of tools focusing on policy change and project implementation. The framework also employs an overarching category associated with monetary valuation. Though not exhaustive, the framework covers most of the tools that are frequently discussed in literature or used in practice.

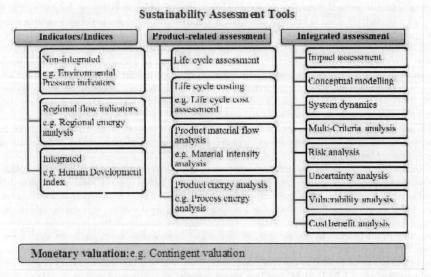

Fig. 1 Categories for sustainability assessment tools (Ness et al. , 2007)

SDIs are increasingly recognised as useful tools, and have been extensively applied to measure different dimensions of sustainability. SDIs are usually identified and adjusted through empirical evidence (Pintér et al. , 2005). They are simple to understand and quantitative. The advantages of SDIs are their ability to summarise complexity and to interpret enormous quantities of data in terms of meaningful information (Godfrey and Todd, 2001; Warhurst, 2002; Singh et al. , 2009). SDIs are often used to visualize longer-term sustainability trends, because they usually have a wide scope, are sensitive to change, and can be continuously calculated (Harger and Meyer, 1996; Ness et al. , 2007). A well-chosen set of SDIs will summarise the performance of a complex system in a way that provides a comprehensive and comprehensible overview. SDIs, which provide early warning information to prevent economic, social and environmental damage, can be used to support decision-making, promote communication, and underpin policies (Berke and Maria, 1999).

In some cases, the individual indicators can be compiled into a composite index. A composite index is constructed according to a theoretical framework or definition (OECD, 2004). Such an approach allows for the creation and evaluation of multi – dimensional concepts which cannot be represented by a single indicator (Singh et al. , 2009; ECD, 2004). Examples such as the well-known Human Development Index (UNDP, 2006) and the Environment Performance Index (Yale, 2002), demonstrate the strength of composite indices in quantifying complex issues. The composite index approach, however, introduces elements of uncertainty. Due to the mechanisms used for including or excluding indicators in the index, and the need to decide how to weight and normalise data, composite indicators have been criticised as subjective (OECD, 2005). Moreover, Saisana and Tarantola (2002) have addressed concerns that composite indices, "if poorly constructed or misinterpreted, may invite simplistic policy conclusions or even send misleading messages". It has therefore been suggested that a combination of uncertainty and sensitivity analyses should be included in order to gauge the robustness of any composite indicator in order to increase transparency and to frame policy – making (Singh et al. , 2009; OECD, 2005). Notwithstanding these difficulties, which must be carefully dealt with, composite indicators can prove to be very useful in practice. A well – structured composite index can be used to interpret a large variety of different data while focusing attention on and simplifying the problem (Atkinson et al. , 1997), facilitate public communication, and further promote accountability (Saisana and Tarantola, 2002).

Many indicators have been developed to assess and monitor water-related vulnerabilities and risks. A widely used water stress or water scarcity indicator was proposed by Falkenmark in 1989.

This simple metric is represented by the total annual runoff available for human use. According to the Falkenmark indicator, a country or region is in the status of "water stress" or "water scarcity" when water supplies fall below $1,700$ m^3 or $1,000$ m^3 per capita per year respectively (Falkenmark, 1989). Gleick (1996) introduced the term "basic water requirements" to describe water used for four basic human needs: drinking water for survival; water for human hygiene; water for sanitation services; and water for certain household needs such as preparing food. Gleick suggested that 50 L water per person per day is the minimum required to meet these basic needs, regardless of an individual's economic, social, or political status. Taking fluctuations in water availability and social adaptive capacity into account, Ohlsson (2000) developed a social resources water stress/scarcity index (Ohlsson, 2000;). Based on the Falkenmark indicator weighted according to the UNDP Human Development Index, the social resources water stress/scarcity index captures the social impacts of water resources, and is claimed to be more useful than earlier indices (Ohlsson, 2000; Brown and Matlock, 2011).

To promote integrated water resources management, a broad consensus has been reached that water should be treated as a social and economic good (Rogers et al., 1997; Gorre – Dale, 1992). To measure the economic value of water resources, Allen (1993;) developed the concept of virtual water to explain how water is embedded in food production and trade (Allan, 1993; Allan, 1998). The virtual water content of a product can be described as the volume of water used through the whole supply chain that leads to an end product. For instance, research into virtual water has found that the production of one kilogram of beef requires $15,000$ L of water (Mekonnen and Hoekstra, 2010). This concept was said to "help people, especially those who live in semi – arid and arid areas, to understand the value of water, and thus opens the door to more productive water use" (SIWI, 2007). Closely linked to the concept of virtual water, the water footprint is a geographically explicit indicator, concerning not only volumes of water use and pollution, but also location (Hoekstra, 2008). Water footprint is a practical term which can be applied to assessment at different scales. Researches show that the water footprint of the average US citizen is $2,840$ m^3 per year, 20% of which is imported from outside the US (Mekonnen and Hoekstra, 2011). The average total global water footprint in the period from 1996 to 2005 was $9,087$ Gm^3/a, of which agricultural production contributed 92% (Hoekstra et al., 2012).

3 Water sustainability assessment

Integrated assessment of water sustainability takes a different approach to the above methods which focus on just one or two critical aspects of the water situation. To be effective, water sustainability assessment must be based on a comprehensive review of all the environmental, social and economic impacts, and provide results which are holistic, well – structured, and easy to understand and use.

Designed as a holistic measurement tool, the Water Poverty Index is a composite index that incorporates quantified estimates of water availability along with socio – economic variables (Sullivan, 2001; Sullivan, 2002; Sullivan et al., 2003; Lawrence et al., 2002). What distinguishes WPI from earlier water resource assessment tools is its attempt to "move away from the conventional, purely deterministic, approaches to water assessment, relying primarily on models and large-scale data" (Sullivan, 2001). WPI aims to provide a simple, transparent, meaningful assessment without compromising the complexity of the issues. With its focus on water-related poverty, WPI examines the extent to which water stress impacts on human populations, especially those which suffer most from inadequate access to water and sanitation. It enables decision-makers at all levels to determine priority needs for interventions in the water sector.

In an effort to address all major water – poverty related concerns, the development of WPI has involved extensive engagement of stakeholders, academic experts, practitioners, and policy-makers worldwide. The following issues were identified by the stakeholders: measures of access; water quality and variability; water for food and other productive purposes; capacity to manage water; environmental aspects; and questions of spatial scale (Sullivan, 2001, Sullivan, 2002). Taking all these into consideration, WPI encompasses five key water-related components, namely, resources, access, capacity, use, and environment. WPI is calculated in a similar way to HDI, and is expressed as

$$WPI = \frac{\sum_{i=1}^{n} w_i X_i}{\sum_{i=1}^{n} w_i} \tag{1}$$

where, WPI is the value for a particular location; X_i refers to the i – th component of WPI; w_i refers to the weighting applied to component X_i; and $n = 5$ is the number of components. In Eq. (1), the five components discussed above are given a standardized score ranging from 0 to 100, where 0 and 100 refer to the worst and the best water poverty situations respectively.

Indicator development is a complex process, requiring widespread consultation. To verify WPI, pilot studies have been carried out at community scale in South Africa, Tanzania and Sri Lanka (Sullivan et al. , 2003). Feedback from the various participants confirmed that the WPI framework was systematic, transparent, and inclusive. According to (Sullivan, 2001), WPI enhances the understanding of complexities inherent in water-poverty issues, and provides a meaningful assessment by integrating physical, socio-economic and environmental aspects. Hence, the WPI framework can be used to identify the community which performs best, or the development of which component could be most beneficial within a community.

Primarily designed for use at community level, the WPI methodology can however be modified and applied at different scales to suit different needs. Further studies have been done to apply WPI at the global-scale, such as the International Water Poverty Index calculated by Lawrence et al. (2002). This index ranks countries, and communities within countries, taking into account both physical and socio-economic factors associated with water scarcity. Fig. 2 presents a zonation map of WPI obtained for 140 countries in 2002 using five grades ranging from very low to very high, adopted by UNEP (2005). The result of this global WPI analysis illustrates the poverty and hydrology hypothesis proposed by Grey and Sadoff (2007). This hypothesis states that many of the world's industrialized countries have an "'easy' hydrologic legacy", which refers to relatively low rainfall variability and easily recovered water resources during their period of development; other nations and societies have remained poor, partly because of their "'difficult' hydrological legacy", where water security can hardly be achieved (desert, flood regions etc.) which has limited their development (Grey and Sadoff, 2007). Brown and Lall (2006) presented a similar finding which shows a "statistically significant relationship between greater rainfall variability and lower per capita GDP".

These studies demonstrate the strength of WPI in providing rapid and holistic assessment, and the potential of WPI as a powerful tool to guide decision – making. With wider application of the WPI, a standardized framework could be set up for the purpose of benchmarking. To implement WPI at different scales, data availability and data collection in a cost – effective manner remain major challenges. In the long run, data collection for WPI assessment can be linked to provincial and/ or national censuses and water monitoring programmes by UN affiliated organisations, for instance the WWAP by UNESCO, (Sullivan et al. , 2003).

4 River sustainability index

The concept of river health evolved from scientific principles, legal mandates, and changing societal values concerning integrated river basin management (Boulton, 1999; Karr, 1999; Norris and Thoms, 1999; Norris and Hawkins, 2000). And the maintenance and restoration of "healthy" river systems have become important objectives of environment and water resources management (Rapport et al. , 1998; Li, 2005). Consequently, guidelines and tools have been developed to assess the ecological condition of river systems, and to monitor river quality. Examples such as the EU Water Framework Directive – River Basin Management Plans (EC, 2000), 'Maintaining the Healthy Life of the Yellow River' Action (Li, 2005), and the River Health Index (Ni and Liu, 2006) all display the influence of this concept.

River health studies have helped focus interest on the concept of river sustainability, while also

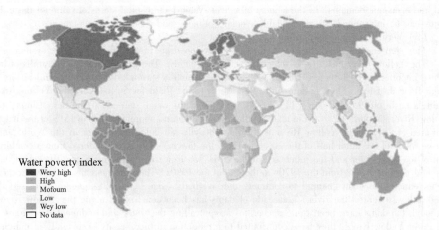

Water poverty index
- Wery high
- High
- Mofoum
- Low
- Wey low
- □ No data

Fig. 2   Water poverty index zonation map in 2002（Ahlenius，2005）

providing a foundation for assessing river sustainability. Distinct from river health, river sustainability comprises not only the natural value of the river course from an ecological perspective, but also the social development and economic activity in the river basin and surrounding areas. River sustainability depends on whether the river system can support the Long-term ecological, economic, and social functions of the river basin as a whole. A broad definition is therefore required of river sustainability in keeping with the principles of sustainability, which should be backed up by quantitative measurements. One way of achieving this is to measure river sustainability by means of a Composite River Sustainability Index (CRSI), following ideas proposed by Ni et al. (Ni and Liu, 2006; Foster et al. , 2008; Ni et al. , 2010).

The CRSI is constructed from a carefully chosen set of indicators. In the present paper, the Process Analysis Method is employed to guide indicator selection. The Process Analytic Method (PAM) was developed by Chee Tahir and Darton (2010) as a formal and transparent methodology by which to generate sustainability indicators. It enables the development of a comprehensive set of sustainability indicators and metrics tailored to a particular river system. Also, by using PAM, the value of a particular indicator can be traced back through the analysis to a particular river function. This is especially helpful in identifying the major sustainability impact generators. The following five tasks are undertaken in applying PAM to a river system:

(1) Overview the river system.

PAM starts with an in-depth review of the river functions, as well as identifying interconnections between river health and the socio – economic state of the river basin.

(2) Select definition of sustainability.

The concept of river sustainability is interpreted in terms of sufficiency, efficiency and fairness perspectives. The sufficiency perspective means the river system should have sufficient runoff of required quality to maintain the ecological health of the river while also supporting social settlements and economic activities within the river basin. The efficiency perspective refers to creating more goods and services while using fewer water resources and creating less waste and pollution. The fairness perspective considers the distribution of benefits and dis-benefits between the competing users of water in the river (which includes the river eco-system itself).

(3) Set system boundary.

The system boundary is determined by two factors: the spatial and temporal scales. Setting the system boundary is very important as it limits the processes to be included in the sustainability framework (Chee Tahir and Darton, 2010).

(4) Set-up sustainability framework.

PAM considers the impact of the river system on the capital residing in the environmental, social and economic domains. In the context of a river system, the capital stores of value for these domains can be interpreted as: river health, social wellbeing and economic development.

(5) Verification and calibration.

This step ensures the indicators and metrics developed are applicable to the river system.

The Yellow River in China is selected for the case study. The Yellow River is culturally of immense importance in China and yet is also known as "China's sorrow" due to its dreadful history of major flood disasters. Flowing across the vast North China Plain for 5,464 km, the Yellow River drains a basin of 795,000 km^2 (Li, 2005). As the main water source for Northern China, the Yellow River supports 12% of the total population of China and supplies water to 15% of the irrigation area of China. The Yellow River is the most heavily silt-laden large river in the World, and during much of the latter half of the 20th Century, the lower Yellow River suffered from a combination of too little water and too much sediment. This was exacerbated by many occurrences of very low and no-flow events from the 1970s to the end of the 1990s. These events caused the water quality to reduce, the main channel to contract, the riverbed to rise, and consequently increased the flood risk. To control the river, a cascade of dams has been constructed along the Yellow River. Although the dams have been quite successful at controlling the water and sediment discharges in the lower Yellow River, they have contributed to a reduction in biodiversity, especially in numbers of certain semi-migratory fish, and are medium-term solutions at best because of reservoir sedimentation. The sustainability of the lower Yellow River is therefore a topic of immense importance, and requires very careful assessment and monitoring. In the present case, a preliminary set of 18 sustainability indicators has been derived for the lower Yellow River after verifying the methodology framework and cconsulting a broad range of stakeholders through the auspices of the Yellow River Conservancy Commission and Peking University. Fig. 3 shows the indicators fitted within the three domains of the CRSI framework. The CRSI is calculated by the method of re-scaled values,

$$CRSI_j = \frac{\sum_{i=1}^{n} w_{i,j} I_{i,j}}{\sum_{i=1}^{n} w_{i,j}} \qquad (2)$$

where, the index CRSI in the given year j is composed of n non-dimensionalized sub-indices $I_{i,j}$ with respective weighting $w_{i,j}$.

Fig. 4 captures the results from early studies on River Health Index for the Yellow River by Ni and Liu (2006), Borthwick (2005). It shows that the general 'health condition' of the lower Yellow River weakened significantly in the latter half of the 20th Century, to the point that it started to fail in its primary ecological functions (Borthwick, 2005). Preliminary analysis of the present CRSI leads to a similar conclusion that the Yellow River can no longer meet the development needs of the current and future generations. In the next phase of our research, we will examine the temporal fluctuations in the CRSI, and explore relations between river behaviour and hydro-climatic trends, population trends by causality, and correlation tests between the index and specific indicators. This could help us better understand the sustainability impacts of different indicators. By comparing different scenarios in terms of CRSI, well-balanced strategies for river rehabilitation and river basin management could be developed.

5 Conclusions

This paper reviews the concept of sustainability and demonstrates how sustainability indicators can be applied to integrated water assessment. Water sustainability assessment tools are discussed, including single metric for measuring a specific concern of water resources management and integrated water sustainability assessment. The study on WPI outlines and measure water issues of a community or a nation following consultation of a broad range of stakeholders. Case studies show that WPI were well-accepted by participants, who found them easy to understand and to use, and that they provided useful information for benchmarking and decision-making.

The new CRSI framework provides a measure beyond river health. By incorporating all do-

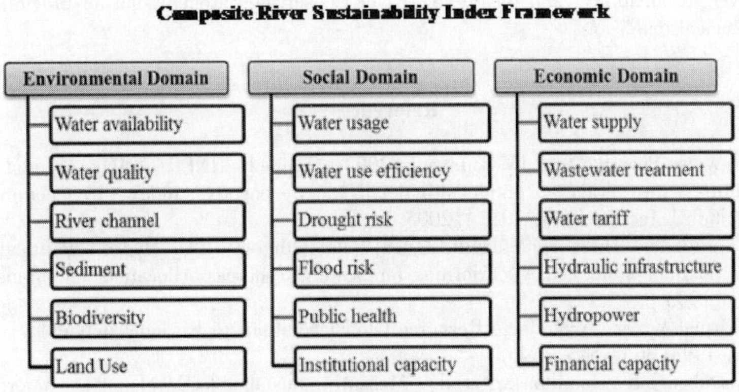

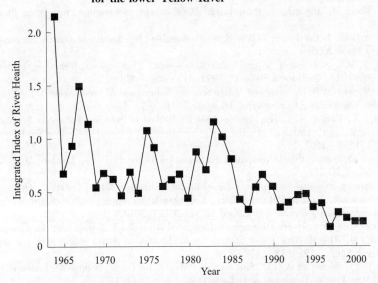

Composite River Sustainability Index Framework
Environmental Domain
Water availability
Water quality
River channel
Sediment
Biodiversity
Land Use
Social Domain
Water usage
Water use efficiency
Drought risk
Flood risk
Public health
Institutional capacity
Economic Domain
Water supply
Wastewater treatment
Water tariff
Hydraulic infrastructure
Hydropower
Financial capacity

Fig. 3 Composite river sustainability index framework for the lower Yellow River

Integrated Index of River Health
0
0.5
1.0
1.5
2.0
1965
1970
1975
1980
1985
1990
1995
2000
Year

Fig. 4 Integrated river health index for the lower Yellow River since 1964 (Ni and Liu, 2006; Borthwick, 2005)

mains of sustainability, CRSI considers not only natural capital, but also social and economic values of river systems. CRSI uses the Process Analysis Method, a systematic approach to set the framework and guide indicator selection. The case study on the lower Yellow River demonstrates CRSI as a holistic methodology for assessing the overall sustainability of the river basin. It also shows that CRSI has the potential to be a very useful tool, in terms of observing trends of river sustainability over time, providing early warning, and underpinning policies for integrated river basin management.

Acknowledgements

We acknowledge financial support from the National Research Foundation and the Economic Development Board (SPORE, COY-15-EWI-RCFSA/N197-1). The authors are particularly grateful to Professor Jinren Ni and his colleagues from Peking University, China for their detailed advice on river sustainability, and the Yellow River Conservancy Commission for providing access to the hydrological data.

References

Ahlenius H. Water Poverty Index by Country in 2002 [Online]. UNEP/GRID – Arendal. Available: http://www.grida.no/graphicslib/detail/water – poverty – index – by – country – in – 2002_d6db [Accessed May 2012]. 2005.

Allan J A. Fortunately There are Substitutes for Water Otherwise Our Hydro – Political Futures Would be Impossible. ODA: Priorities for Water Resources Allocation and Management. London, 1993.

Allan J A. Virtual Water: A Strategic Resource Global Solutions to Regional deficits[J]. Ground Water, 1998(36): 545 – 546.

Atkinson G, Dubourg R, Hamilton K, et al.. Measuring Sustainable Development: Macroeconomics and the Environment[J]. Edward Elgar, Cheltenham, 1997.

Bell S, Morse S. Sustainability Indicators : Measuring the Immeasurable, London, London[M]. Earthscan, 2008.

Berke P, Maria M. Planning for Sustainable Development: Measuring Progress in Plans[R]. 1999.

Borthwick A G L. Is the Lower Yellow River Sustainable[N] Society of Oxford University Engineers News, 2005.

Boulton A J. An Overview of River Health Assessment: Philosophies, Practice, Problems and Prognosis[J]. Freshwater Biology, 1999(41): 469 – 479.

Brown A, Matlock M D. A Review of Water Scarcity Indices and Methodologies[D]. The Sustainability Consortium, University of Arkansas, 2011.

Chee Tahir, A, Darton, R C The Process Analysis Method of Selecting Indicators to Quantify the Sustainability Performance of a Business Operation[J]. Journal of Cleaner Production, 2010 (18):1598 – 1607.

Dalal Clayton D. Sustainable Development Strategies: a Resource Book, London[M]. London : Earthscan, 2002.

Defra. Measuring Progress: Sustainable Development Indicators 2010 [Online]. Department for Environment, Food and Rural Affairs, UK. Available: http://sd.defra.gov.uk/progress/national/annual – review/ [Accessed 12 Jan 2012], 2010.

EC. The EU Water Framework Directive – Integrated River Basin Management for Europe, 2000.

Falkenmark, M. The Massive Water Scarcity Now Threatening Africa – Why isn't it being Addressed Ambio, 1989(18):112 – 118.

Foster J, NI J. , Borthwick A G L. Sustainability Metrics for Large, Sediment – Laden Rivers. M. Sc. MSc Thesis, University of Oxford; 2008.

Godfrey L, Todd, C. Defining Thresholds for Freshwater Sustainability Indicators within the Context of South African Water Resource Management. 2nd WARFA/Waternet Symposium: Integrated Water Resource Management: Theory, Practice, Cases. Cape Town, South Africa. 2001.

Gorre – Dale E. The Dublin Satement on Water and Sstainable Dvelopment[J]. Environmental Conservation, 1992(19): 181.

Grey D, Sadoff C W. Sink or Swim? Water Scurity for Gowth and Dvelopment[J]. Water Policy, 2007 (9): 545 – 571.

Gri. Sustainability Reporting Guidelines, Amsterdam, Global Reporting Initiative. 2006.

Harger J R E, Meyer F M. Definition of Indicators for Environmentally Sustainable Development [J]. Chemosphere, 1996 (33): 1749 – 1775.

Hoekstra A Y, Mekonnen M M, Chapagain A K. Global monthly water scarcity: Blue Water Footprints Versus Blue Water Availability[J]. PLoS ONE, 2012(7).

Hoekstra A Y C, A K. Globalization of Water: Sharing the Planet's Freshwater Resources[M].

Oxford: UK, Blackwell Publishing, 2008.

Icheme. The Sustainability Metrics: Sustainable Development Progress Metrics Recommended for Use in the Process Industries, Rugby, Warwickshire[J]. Institution of Chemical Engineer, 2002.

Karr J R. Defining and Measuring River Health[J]. Freshwater Biology, 1999 (41): 221 – 234.

Kates R W, Clark W C, Corell R, et al. Environment and Development: Sustainability science [J]. Science,2001 (292): 641 – 642.

Kates R W, Parris T M, Leiserowitz A A. What is Sustainable Development? Goals, Indicators, Values, and Practice[J]. Environment,2005(47): 8 – 21.

Lawrence P, Meigh J, Sullivan C. The Water Poverty Index: an International Comparison[J]. EconWPA,2002.

Li Guoying Maintaining the Healthy Life of the Yellow River[M]. Zhengzhou: Yellow River Conservancy Press,2005.

Mekonnen M M, Hoekstra A Y. The Green, Blue and Grey Water Footprint of Farm Animals and Animal Products[J]. UNESCO – IHE, 2010.

Mekonnen M M, Hoekstra A Y. National Water Footprint Accounts: The Green, Blue and Grey Water Footprint of Production and Consumption[J]. UNESCO – IHE, 2011.

Ness B, URBEL – PIIRSALU E, Anderbebc S. L. Categorising Tools for Sustainability Assessment [J]. Ecological Economics, 2007(60): 498 – 508.

Ni J, Liu Y. River Health Diagnosis and Ecological Rehabilitation[J]. China Water Resources, 2006(13).

Ni J, Sun L, Li T, et al. Assessment of Flooding Impacts in Terms of Sustainability in Mainland China[J]. Journal of Environmental Management, 2010(91): 1930 – 1942.

Norris R H, Hawkins C P. Monitoring River Health[J]. Hydrobiologia, 2000(435): 5 – 17.

Norris R H. Thoms M C. What is River Health? [J] Freshwater Biology, 1999(41): 197 – 209.

OECD. The OECD – JRC Handbook on Practices for Developing Composite Indicators. the OECD Committee on Statistics. Paris: OECD,2004.

OECD. Handbook on Constructing Composite Indicators: Methodology and User Guide,2005.

Ohlsson L. Water Conflicts and Social Resource Scarcity. Physics and Chemistry of the Earth, Part B: Hydrology, Oceans and Atmosphere, 2000 (25): 213 – 220.

Pinter L, Hardi P, Bartelmus, et al. Indicators of Sustainable Development: Proposals for a Way Forward. IISD Publications,2005.

Pri. The Canadian Water Sustainability Index Case Study Report. In: MORIN, A. (ed.) Working Paper Series 028. Policy Research Initiative (PRI),2006.

Pri. Canadian Water Sustainability Index, Government of Canada, Policy Reserach Initiative, 2007.

Rapport D J, Costanza R. Mcmichael A J. Assessing Ecosystem Health. Trends in Ecology & Evolution, 1998(3): 397 – 402.

Rockstrom J., STEFFEN W., NOONE K., et al. A Safe Operating Space for Humanity[J]. Nature, 2009(461): 472 475.

Rogers P, Bhatia R, Huber A. Water as a Social and Economic Good: How to Put the Principle into Practice. Global Water Partnership. Stockholm, Sweden: Swedish International Development Cooperation Agency,1997.

Saisana M, Tarantola S. State of – the – art Report on Current Methodologies and Practices for Composite Indicator Development. EUR 20408 EN. Italy: European Commission – JRC, 2002.

Singh R K, Murty H R, Gupta S K, et al. An Overview of Sustainability Assessment Methodologies [J]. Ecological Indicators, 2009(9): 189 212.

SIWI. 2007. Virtual Water Innovator Receives Stockholm Water Prize [Online]. Stockholm: SIWI. Available: http://www.siwi.org/sa/node.asp? node =344 [Accessed April 2012].

Sullivan C. The Potential for Calculating a Meaningful Water Poverty Index[J]. Water International, 2001(26): 471 – 480.

Sullivan C. Calculating a Water Poverty Index[J]. World Development, 2002 (30): 1195 –

1210.

Sullivan C A, Meigh J R, Giacomello A M, et al. The Water Poverty Index: Development and Application at the Community Scale[J]. Natural Resources Forum, 2003(27): 189 – 199.

UN 2005. 2005 World Summit Outcome.

UN 2007. Indicators of Sustainable Development: Guidelines and Methodologies, New York, United Nations.

UNCSD 2002. The Johannesburg Declaration on Sustainable Development: From Our Origins to the Future.

UNDP. 2000. Millennium Development Goals Overview [Online]. Available: http://www.undp.org/content/undp/en/home/mdgoverview.html [Accessed May 2012].

UNDP 2006. Human Development Report 2006 : Beyond Scarcity : Power, Poverty and the Global Water Crisis, New York Basingstoke : Published for the United Nations Development Programme by Palgrave Macmillan

WARHURST, A. Sustainability Indicators and Sustainability Performance Management. Report to the Project: Mining, Minerals and Sustainable Development. Warwick, England: International Institute for Environment and Development, 2002.

Wced. Our Common Future[M]. Oxford : Oxford University Press, 1987.

WWC. Water and Green Growth. Marseille: World Water Council, 2012.

YALE. Environmental Performance Index by Yale University [Online]. Available: http://epi.yale.edu/about [Accessed May 2012].

Philosophy Speculation on System Theory of Maintaining the Healthy Life of the Yellow River

Cheng Yanhong[1] , *Zhai Ran*[2] , *Lou Shujian*[3] and *Jia Meiping*[3]

1. The Yellow River Flood Control and Drought Relief Headquarters Office,
YRCC, Zhengzhou, 450003, China
2. School of Earth Sciences and Engineering, Hohai University,
Nanjing, 211100, China
3. Sanmenxia Multipurpose Project Administration, YRCC , Sanmenxia, 472000, China

Abstract: System theory is the science on studying the general mode, structure and regular pattern of the system. It is also a philosophy theory on studying the ideas and methods of the system. Although the concept of system theory comes from the study on generally things of the nature, it plays the maximum effect on the social sphere and eventually led to the scientific management of modern society. System theory believes that the world is a big system. And the structure of the system determines its function and behavior. The river changes the surface of the earth through the erosion, transportation and stacking interaction, which is constantly changing and replacing. It is an important open system of the survival of humans and much other biology. Due to the uncontrolled development on rivers by human, the rivers of worldwide today are almost facing the crisis of survival. As the most complex refractory river of worldwide, the Yellow River faces the crisis of survival especially seriously. By systematically analyzing the outstanding problems of the Yellow River, the principal contradiction of the Yellow River is less water and more sediment, different sources of water and sediment, and uncoordinated water-sediment relation. The source of healthy life crisis of the Yellow River is that the relationship between human and the Yellow River are demanded. It is the inevitable result of human moving forward along the track of industrial civilization. Therefore, the Yellow River Conservancy Commission (YRCC) formally proposed the new river training concept of maintaining the healthy life of the Yellow River, made it as the ultimate goal of the scientific development and management of the Yellow River, and started to build the corresponding river training systems. In this paper, the basic tenets of Marxist dialectical materialist view is applied to make speculation and interpretation on maintaining the healthy life of the Yellow River from the perspective of the system theory philosophy. Philosophical ontology, epistemology and axiology of maintaining the healthy life of the Yellow River, which is also analyzed. For the current practices of scientific development and management of the Yellow River, the proposal of re-understanding of the Yellow River, in-depth study of the Yellow River and more respect for the Yellow River are studied and proposed.

Key words: system theory, philosophy, speculation, maintain the healthy life of the Yellow River

Although the concept of system theory comes from the study of general things in nature, the field of which palys out most is in the social fields, and eventually leads to the science-oriented modern society management. The system theory is considered that the system is as a structural whole formed through the interaction of multiple elements; the world is a big system; the structure of the system decide the function or behavior of the system; and open system cause formation, evolution and development of the complex self-organizing system of the system .

The river changes the surface of the earth through the erosion, transportation and stacking interaction, which is constantly changing and replacing, it is not only the main part of the magnificent earth, but also the important open systems that human and many lives to survive. The river is flowing in the day and night, nurturing the life and breeding civilization. However, because of the human outrageous development and utilization to the river, the rivers in the world scope almost face the survival crisis. As the most complex and difficult controlling river in the world, the Yellow River-survival crisis is particularly outstanding. By system analysis of problems existing in the Yellow River, it is not difficult to find that " less water and more sediment, different sources of water and sediment, and uncoordinated water-sediment relation", all that are the principal contradiction of the Yellow River, the maintenance of health life of the Yellow River is the ultimate goal of scientific development and management of the Yellow River. This paper apply the basic principle of marxism dialectical materialism view of nature, material view, value, epistemology, dialectics and so on, through system philosophy view, for maintaining healthy life, carrys out the speculative and interpretation, and puts forward some suggestions.

1 The establishment of the concept that the maintenance of health life of the Yellow River

For a long time, the increasing water demand of the economy and society in the Yellow River valley again and again break through the bottom line of the Yellow River life, people fight the river for water, and fight the water for land, the situation is more and more severe, from this, there are produced a series of problems that the water resources shortage, channel shrivel, the flow cutoff, ecological degradation, water pollution and so on, and which in turn constrain the watershed economic and social sustainable development. Based on this, in the beginning of the new year 2004, the YRCC formally proposed the new river training concept of "maintaining the health life of the Yellow River", made it as the ultimate goal of the scientific development and management of the Yellow River, and started to build the corresponding river training systems. The main purposes are below: the first is to improve the people's scientific understanding level of the Yellow River life, observe, analysis, rethink, forecast with the aid of the humanities and social science theory, thereby grasp the harmonious laws that people live together with the Yellow River, and regulate human own social behavior. The second is in order to promote positive practice of comprehensive, coordinated, sustainable scientific outlook on development. The third is in order to make the person's relations with the Yellow River, from a previous relationship to conquer and transformation, turn into a harmony, coexistence symbiotic relationship, warning people in the development and utilization of the Yellow River, only conform to the laws of nature, recognize and maintain the Yellow River's own life value and rights, the Yellow River region economy and society can develop continuously, national culture ability can sustain for generations.

2 The system theory philosophical foundation of maintaining the health life of the Yellow River

The system theory is the knowledge that studies the system of general model, structure and law, and philosophical theory that studying the system idea and system method. According to dialectical materialism, the material world is a unified whole that formed of numerous interrelated, interdependent, mutual restrict, interactions of the objects and processes, it is the philosophy foundation of system universal existence. System theory philosophy includes the ontology, epistemology, value etc. of the system.

2.1 The system theory philosophical ontology of maintaining the health life of the Yellow River

From the philosophy and dialectic point of view, the most common meaning of the life is existence and extinction, any natural body only has the existence and extinction characteristics or process, whether it is the result of the development of self or external influence, undoubtedly have vitality, the Yellow River also is such.

The ontology life of the Yellow River mainly displays in, the first is the water sustainable production and regeneration, which can naturally, regularly, continuously supply water and recycled water; the second is the natural flow capacity, that is the ability of the natural water cycle, which ensure continuous and effective natural flow capacity of the river; the third is the natural growth ability, which can naturally to nourish a certain amount of aquatic organisms and maintain the balance of ecological system; the fourth is the water supply capacity, that is the water resource supply capacity; the fifth is the natural land shaping and landscape ability, that is nurturing and sustaining the watershed wetland, basin, estuary land, and river natural landscape capacity; the sixth is the flood discharge capacity; and the seventh is the pollutant carrying and purification ability.

Every life form on earth, each individual life, or biological species, has its special proccess and rhythm for evolution, development, and the final decay. The core of the Yellow River is water, which is endlessly flowing, the formation, development and evolution of the Yellow River life is a natural process, which obeys the natural law of development, and has tremendous counterforce and normative to the external behavior. The Yellow River is the birthplace and the carrier of all forms of life in the river basin, is an important part of the earth's ecological system, the life law of the Yellow River is not to be transferred by human's will.

2. 2　The system philosophy epistemology of maintaining the health life of the Yellow River

The System philosophy epistemology is the philosophy to explore the nature and structure of human understanding, the relationship between understanding and objective reality, the premise and foundation of understanding, the process and rule of occurrence and development of understanding, the truth standard of understanding and such problems. The cognition activity of the Yellow River healthy life need scientific epistemology for guidance, this includes the dialectical materialist theory of knowledge and some reasonable epistemology conclusion in other schools of philosophy epistemology and effective recognition methodology.

The Yellow River is not only as a natural eco-system, composed of water and issues contained like aquatic animals and plants, microorganism and environmental factors, but also a complete and unified organic whole of the flowing water system including the source, wet land, lakes, and numerous different tributaries and main streams, at the same time it is a open system components of the watercourse system and watershed system. Among the system, rivers and watershed, there are a large amount of exchange of matter and energy, all factors of which effect on the maintaining and health of the Yellow River life. Therefore, the system philosophy epistemology of maintaining the health life of the Yellow River, the first is started from the understanding of the right to live and the right to health of the Yellow River.

2. 2. 1　The right to live of the Yellow River

As the essential character of the existence of the Yellow River life, integrity and continuity are the fundamental prerequisite that the Yellow River realizes its ontology attributes and all the features, but the integrity and continuity of the Yellow River usually through the water, and the dynamic flow to show the vigor of the life. Therefore, the most important guarantee to the survival of the Yellow River is to maintain the basic water amount of the Yellow River, only the water amount is the basic elements to show the Yellow River existence, although different rivers for water amount requirements have differences, there are usually have a basic water amount, this is the characteristics of the rivers that are different from the creeks, and also the necessary conditions of maintaining the river ecological system. Therefore, integrity, continuity and the basic water amount for maintaining these characteristics are the guarantee of the Yellow River survival.

Maintaining the right to live of the Yellow River require us in the use of the Yellow River resources, fully consider not to take the basic water amount of the Yellow River survival, and divide artificial water area, all actions must be in accordance with the ecological law of the Yellow River, and put the respect and safeguard the Yellow River's right to live in an important position in the concept.

2. 2. 2　The right to health of the Yellow River

The right to health of the Yellow River is the higher claim on the basis of the right to live of the Yellow River, and this claim is necessary for the existence and the functions properly performing of the Yellow River. The so-called healthy life of the Yellow River usually means that the ecosystem integrity of the Yellow River is not damaged, the system is in a normal and benchmark state. The health status evaluation of the Yellow River can be determined by the flow capacity, water quality, estuarine wetland health level, biodiversity and the water supply satisfaction level to both sides and other indicators. The human understanding in the field of health life of the Yellow River usually means that the Yellow River ecological functions and services in the economic functions can be in normal condition, but both of them are closely and often contradictory. One side, human will try to strengthen the economic function of the Yellow River, but on the other hand, large-scale human economic activities damage the ecosystem health of the Yellow River, make the economic function of the Yellow River weakening. The idea of the Yellow River healthy life needs us to recognize that the ecological function is the support force of the economic function, and requires us to pay attention to the ecological function of the Yellow River.

2.2.3 Maintaining the right to life and health of the Yellow River

From Marxist dialectical point of view to know, neither have the main body only right and have no obligation, nor the main body only obligations without the rights, any rights and obligations are symmetrical.

The river carries on obligation for human, inevitable have own right. As the right owner, it has the right to preserve their own survival benefits, challenge the action that infringing on its interests. When the human damage to its survival interests, it has the right to respond. Giving the Yellow River the basic rights of life and health are also provided us responsibilities and obligations of the Yellow River, this means that the Yellow River is no longer just the resource for us to develop and use, instead, need to give necessary respect, like other lives.

Therefore, maintaining the health life of the Yellow River means that: ① we must content the ecological water demand of the Yellow River first, economic activity can only carry out in accordance with the Yellow River ecological law, the economic water can not occupy the ecological water; ② the large pollutant in economic activities can not directly discharged into the Yellow River, lead to the water quality continues to decline, to make all kinds of ecological system to the survival of the environmental constantly tend to worse. In the natural and open ecological system of the Yellow River, can not only emphasize needs of human, but also from the Yellow River own requirement, coordinate the contradiction between the human and the river. In short, maintaining the right to the health and life of the Yellow River is not only requirements of the basic water for the Yellow River self-maintenance ability, also requires a clean water quality, the stable riverway, the health river ecosystem and so on.

2.3 The systematology philosophy axiology of maintaining the health life of the Yellow River

Marx philosophy emphasizes the value is attributes of things to meet human's certain need, human is the subject and reference frame of the value evaluation. Value has its objective existence form, and also the subjective report form. Value in many fields have specific morphology, such as social value, individual value, economic value, legal value and so on. Depart from the human and human evaluation, anything is of no value.

From object and task of the philosophy, systems philosophy theory of value in the system of philosophy has the important position, which is manifested in the value guidance function and value pursuit of the mission of Marx's philosophy, it reveals the value foundation and the function of Marx's philosophy, strengthen the practice of the philosophical guidance, enhance the modernity and reality force of the philosophy, enrich the philosophical basic connotation.

On the value of the Yellow River, at present, usually divided into: instrumental value and intrinsic value. The Yellow River has an external tool value, such as the Yellow River economic value, aesthetic value, history and cultural value and so on, will no doubt get the consistent admitting. At present, more and more scholars think, the confirmation of the Yellow River intrinsic value is related to whether the Yellow River has their own rights. Therefore, from the basic requirement of containing the healthy life of the Yellow River, through the philosophy theory of value, the understanding of instrumental and intrinsic value of the Yellow River, can guide us in scientific management development and the management practice of the Yellow River, make the Yellow River better benefit the Chinese nation.

3   The philosophy introspection of health life crisis of the Yellow River

The carrying capacity of each river to natural and economic social system is limited, the load of river life keep sustainable development only in the range of its carrying capacity. The formation, development and evolution of the Yellow River health life is a natural process, has its own law of development, and has huge reaction and normative to the outside world action. Therefore, once the crisis of the Yellow River health life happpens, the other systems relied on the Yellow River will lose the existing basis. The history tragedy of Lop Nor in Tarim river basin and Loulan ancient city, are the shocking examples.

3.1   The main expression of health life crisis of the Yellow River

Nearly twenty years, with the rapid expansion in population increase and the development of economy and society, the Yellow River bears pressure increases more, using the downstream flow cutoff for the mark, the Yellow River healthy life crisis mainly shown is in the following respects:

Firstly, the downstream riverbed is significant atrophy, flood discharge ability drops quickly. Because for a long time people "contend for the land with river", "contend for water with river" (excessive diversion, heavily crowed out river ecological water use), do not follow the Yellow River own development rule, the long-term sustained small flow in downstream channel, resulted in the main river channel deposited serious and shrunk endlessly, the transverse gradient increasing, the situation of the "secondary suspended river channel" is ceaseless aggravate, reduce the flood discharge capacity of the river greatly.

Secondly, the upper and middle of channel shrinkage is expanding, under the same flow, the water level is rising, the suspended river "on ground" only in the downsteam has appeared before, now does in the upper and middle of river channels. the reality of the river continued to shrink in the upper and middle reaches, has already showed the fragility of the Yellow River's life .

Thirdly, the Yellow River water resources can not make ends meet, the sustainable support ability are faced with serious crisis. At present, the Yellow River accounted for 2% of the river runoff, supports the development of 12% of the total national population and 15% of the farmland. The utilization rate of the water resources in river basin has reached 70% , far exceeding the internationally recognized warning line of 40% .

Fourthly, the extremely fragile ecological environment of the Loess plateau area has not been fundamentally changed.

3.2   The philosophy cognition and introspection of health life crisis of the Yellow River

The human beings use and transform the nature since existing on the earth, maintain the survival needs and social material civilization construction through obtaining the natural material. For a long time the effect on nature is limited, the disturbance that the nature suffered is in the limits of recovery. Since enter the industrial civilization, the demands of water greatly increase, humans began to develop and use the rivers, transform the rivers blindly, and has been inconsistent with the river long time.

Because the human ignore the health of the Yellow River's life for a long time, the development and demand of the Yellow River has far exceeded its capacity, causing the first flow cutoff of the Yellow River in 1972, after that, there are 21 flow cutoffs in 26 years, in 1997 the flow cutoff were up to 226 d, the river reaches were 704 km, the water use were all emergency in urban and rural, industry and agriculture.

At present, the ecological crisis and the river crisis in global scope forced us to reflect the relationship between human and nature from the angle of philosophy. For a long time, people on the ecological crisis and the crisis awareness are limited to the technical and economic level. By the 1960s and 1970s, the environmental protection movement are gradual risen, searching for the reason of environmental crisis from the western social culture has formed a strong trend. Some scholars see that behind the technology and economy, there are more deep-seated reason resulted in the destruction of environment and ecology, led to the ecological crisis and the river crisis. More philosophers began to call to abandon the philosophy of human domination of natural, and to build the natural ecological philosophy of respect; abandon anthropocentrism, establish the value direction of harmonious development of human and nature.

The author thinks about that the ecological crisis to the philosophical level is to deepen understanding of the ecological crisis, it break that the ecological crisis is just the improper economic and technical implementation, is partial, occasionally superficial understanding, turn to establish a comprehensive new ecological view of nature and values with harmonious relationship between human and nature. Just in this reflection trend, people on the river crisis have a more clear understanding and reflection.

The source of the generant crisis of the healthy life of the Yellow River is the relationship destroyed between human and the Yellow River, which is the inevitable result that human development forward along the industrial civilization track. Because the value pointer of the industrial civilization is the narrow anthropocentrism, and as plunder and conquer the nature is its core value, then, the river crisis is inevitable. The conquer and control of the river by industrial civilization leads to a global rivers crisis, directly threat to the continuation and development of human civilization. Humans have only put on river crisis to criticism of civilization height, can truly find solutions to overcome the current increasingly serious crisis of the rivers. This not only needs to adjust human's political and economic system, also need to adjust the human values, establish the basic attitude of respect for the river.

4 Some suggestion for maintaining the health life of the Yellow River

Marx and Engels have a basic idea that the relationship between human and the natural environment is subject to the interpersonal relationship. Therefore, the interpersonal harmony is the premise of the harmony between the human and nature. The contemporary social ecology also emphasizes the relationship between human and nature in social relations in investigation, emphasize the priority of the social reform to establish healthy ecological. For the reality, it is obviously that interpersonal disharmony cause the disharmony between human and natural. For example, people destroy the natural environment and healthy life of rivers as an excuse of economy development. However, many times what we lack is not the development, but the balanced, harmonious development.

The human developing history has repeatedly shown, the rivers health status directly affects the economic development and people's lives, the river's destiny directly mapping of the human economic society vicissitudes. Therefore, in view of the current practice requirements of the Yellow River scientific management and development, combined with the system theory of speculative philosophy of maintaining healthy life of the Yellow River, for maintaining the healthy life of the Yellow River propose the following rationalizations.

4.1 Anew understanding of the Yellow River

Anew understanding of the Yellow River, on the basis of original understanding, advance with the times, without old idea, thinking to look at the new situation, deepen the particularity of understanding of the Yellow River, understanding the status and importance of the Yellow River in China water. Since ancient times the water-sediment relationship of the Yellow River is not balance, the imbalance before the Republic of China is covered by overflowing hazards all over. Since the people manage the Yellow River, the Yellow River is taming and peace all year round, but the crisis lurked in it, the riverbed in lower reaches of the Yellow River is taller $2 \sim 4$ m than that in 1950, approximately 1.0×10^{10} t sediment deposited in river channel. It is mainly caused by the unbalanced water-sediment relationship, realized this characteristic, work combined with the Yellow River reality, we must learn from the previous the river training experience, persist in developing Marx's doctrine to guide practice training, advance firmly the water-sediment regulation, realize the balanced water-sediment relationship of the Yellow River, on this basis to maintain the healthy life of the Yellow River.

4. 2　Further study the Yellow River

The objective of understanding the Yellow River is to solve the Yellow River's problems, which need to rely on the scientific method. The Yellow River now has many problems need us to further study. For example, the "Aug. 2004" flood, the Xiaolangdi reservoir discharge is about $2,500 \text{ m}^3/\text{s}$, the water adding from Xiaolangdi reservoir to Huayuankou interval is not too much, but the flow in Huayuankou section is $4,000 \text{ m}^3/\text{s}$ when the flood progressed, why appears such a "hump phenomenon" Is it the high concentration of sand flow in the process of evolution undergone some changes? Many similar such problems appear in 30 years, what is its mechanism? How is the relation with the change of the riverbed? And with the sediment particle size distribution? So far, it is not still revealing the changing rule. Such questions are too many.

The further study of the Yellow River is very important for the promotion of the harmonious development of economic society, therefore, we must research from the fundamental interests of the people; combining the economic, social, ecological development request with the Yellow River's problems; explore boldly, innovate continuously, practice bravely, study for the continuous new occurrence in the development and management of the Yellow River; advance firmly the scientific development and management, modernization of the Yellow River, advance with the times to study. Only mastering the Yellow River rules, establish perfect theoretical and operation system, guide better practice of maintaining healthy life of the Yellow River.

4. 3　More respect for the Yellow River

The proposal to maintain healthy life of the Yellow River by the Yellow River Conservancy Commission is from the respect point of the Yellow River. Before the 20th Century, the river training goal has always been stifling all the floods, so far, it can be said there is not much problem to control the flood, and now, before the Republic of China, the situation that "two dike breaches in three years and one river course change in a hundred years" unlikely appears, however, simply controlling flood take many other problems, controlling the flood is not meaning to manage the river. In 1958, a large flood of $22,000 \text{ m}^3/\text{s}$ occurred in Huayuankou station, the flow capacity in downstream main channel is $8,000 \text{ m}^3/\text{s}$; in 1982, a flood of $15,300 \text{ m}^3/\text{s}$ occurred in Huayuankou station, the flow capacity in downstream main channel is $6,000 \text{ m}^3/\text{s}$; in 1996, a flood of $7,600 \text{ m}^3/\text{s}$ occurred in Huayuankou station, the flow capacity in downstream main channel is $3,000 \text{ m}^3/\text{s}$; in the 21th Century, the Yellow River did not have a more than $4,000 \text{ m}^3/\text{s}$ of the flood, and in 2001, the flow capacity of the main channel in downstream Huayuankou station is $1,800 \text{ m}^3/\text{s}$. If we do not take decisive measures to manage, shape the Yellow River's main river channel, if the Yellow River without a main channel, with wandering and "secondly suspended river" development, once the flood occurred, it will have a direct impact on levee, even a burst and diversion risk.

Respect for the Yellow River must know correctly all the problems of the Yellow River, treat these problems dialectically. Wang Shucheng, the former Minister of MWR, said the Yellow River management needs not only scientists, but also philosophers. We need comprehensively considering and analysis, there must be flood in river, flood is equivalent to human's pulse, no pulse, and there is no life. The Holland experts firstly proposed that flood control turns to flood management, the domestic also gradually change, in order to achieve flood resources. Now people have realized deeply that human is not the master of the nature, but the friends; human is not the destroyer of the nature, but the protector.

The YRCC need to focus on the top-level design, with ensuring the safety of flood and ice control as the centre, plan the development of the whole Yellow River, economic and social development, human and nature harmonious development, realize comprehensive, harmonious, sustainable development of the river basin economy and society. At work, do not those merciless things, timely handle all appeals, and try to maintain the legitimate rights and interests of all the parties. To strengthen the support ability of the Yellow River for economic and social development, improve the livelihood of the people; solve the most intense, pressing needs of people, and the problem related to people's long-term interests; in flood control and drought relief, ensure the safety of the flood and ice control, pay attention to flood control and drainage work in city; in the exploitation and utilization of water resources, meet the people's need, protect the legitimate rights and interests of the masses affected; meet the urgent demand of the economic and social development, and maintain the healthy life of the Yellow River.

References

Wang Huimin. The System Theory and Method of River Basin Sustainable Development [M]. Nanjing: Hohai University Press, 2000.

Mao Zedong. On Practice and On Contradictions [C] Works of Mao Zedong. Party School of the CPC Central Committee Press.

Marxist-Leninist Works. Reprint in 2010. Party School of the CPC Central Committee Press.

E Jingping. The Economic Society and the Flood and Drought Disaster [J]. China Flood Control and Drought Relief, 2006.

Li Jianhua, Fu Li. Modern System Science and Management [M]. Beijing: Science and Technology Press, 1996.

Tian Yingkui. The Study of Scientific Outlook on Development, Serial Number is 2011 ~ 2042, the Branch Campus of Party School of the CPC Central Committee Draft.

Li Guoying. Speculations and Practice of Yellow River Management [M]. Beijing: China Water Conservancy and Hydropower Press; Zhengzhou: Yellow River Conservancy Press, 2003.

Shang Hongqi. Maintaining the Healthy Life of Rivers [C] // Proceedings of the 2ed National Yellow River Conference. Zhengzhou: Yellow River Conservancy Press, 2005.

YRCC. The Value and Ethics of Rivers, YRCC [M]. Zhengzhou: Yellow River Water Conservancy Press, 2007.

Hou Quanliang. Ecological Civilization and River Ethics [M]. Zhengzhou: Yellow River Conservancy Press, 2007.

Cheng Xiaotao, Wu Yucheng, et al. Research on New Concepts of Flood Management and Flood Control Security System [M]. Beijing: China Water Conservancy and Hydropower Press, 2004.

Zhang Daojun, Zhu Maiyun, et al. On Sustainable Development of River Basin Ecological Environment [M]. Zhengzhou: Yellow River Conservancy Press, 2001.

Problems and Countermeasures of Hydraulic Construction of Floodplain Areas in the Lower Yellow River

Su Yunqi, *Hou Xinxin* and *Wang Ziying*

Yellow River Institute of Hydraulic Research, Key Laboratory of Yellow River
Sediment Research, MWR, Zhengzhou, 450003, China

Abstract: As a part of flood discharge channel and a settlement of people living along the river, floodplain areas in the lower Yellow River faces complex management issues, as well as the contradiction between how to make people rich and the Yellow River management. So it results in the poor living environment and backward economic development of floodplain areas, and is inconsistence with the construction of new socialist countryside. Therefore, much attention should be paid to how to strengthen hydraulic construction, how to change the poverty, and how to promote economic development of floodplain areas in the lower Yellow River. Based on analyzing the characteristics and problems of floodplain areas, the paper puts forward some countermeasures for strengthening hydraulic construction and promoting economic development of floodplain areas.

Key words: floodplain areas, hydraulic construction, economic development, the lower Yellow River

1 The basic situation of floodplain areas in the lower Yellow River

There are more than 120 floodplain areas in the lower Yellow River (from Mengjin County to Kenli County). The total area is $3,544.45 \text{ km}^2$, accounting for 83.6% of the river area, in which the area of cultivated land is 223 km^2. And it involves in 43 counties of 15 cities in Henan Province and Shandong Province. Owing to historical reasons, the resident population in floodplain areas of Yellow River is about 1.895 million.

The lower Yellow River is a suspended river, and has broken many times in history which resulted in a heavy disaster to people. Under the leadership of the Chinese Communist Party, the Yellow River disaster has been controlled effectively. It achieves the great feat that the river hasn't broken for more than 60 years, and the living standard of the people along the river improves constantly. The floodplain areas of the Yellow River is the place that the people in floodplain areas depend on for production and living; and as an important part of the Yellow River channel, playing a role in draining flood, detention of flood and sand in flood season. The floodplain areas in the lower Yellow River especially in Henan province has the wide channel, shallow riverbed and unstable main channel which result in the frequent occurred phenomena of overbank and flood along the dyke. Because of the influence of the Yellow River flood, the producing and living conditions are rough and the economic development is relatively backward.

For a long time, the hydraulic construction is relatively backward. In 1950s and 1960s, the government invested to construct some hydraulic projects. But because of insufficient fund, incompletion and bad quality of matching projects and lacking maintenance, many projects have been used more than durable years and damaged heavily. Therefore, most farmland can't be irrigated effectively. During drought, although close to the Yellow River, the farmland can't be irrigated.

Because of the underdeveloped traffic, information block, weak basic industry and lacking investment, the floodplain areas has poor economic development. It is hard to solve the problem about development radically by unstable economy with farming. The research result of floodplain treatment in the lower Yellow River in 2008 showed that the floodplain areas in the lower Yellow River were still one of the poorest areas.

2　The characteristics of floodplain areas in the lower Yellow River

The floodplain areas in the lower Yellow River have characteristics differing from Huang – Huai – Hai Plain. The existence of production dykes makes the lower channel become a secondary suspended river, and results in the flood overbank frequently. After flood overbank, the ground elevation increases, and the canals silt. In some floodplain areas, flood forms new cross flow and oblique flow. It changes original beach landform, and leads the original projects to damage or out of action. In order to build water escape platform, some villagers dig in their farmland or around the villages, make pits and ponds everywhere. Because of incomplete water conservancy facilities and the poor soil owing to flood overbank, farmland irrigation guarantee rate is low. Most income of the people in floodplain areas is used to heighten the building platform. The extra expenditure aggravates the people's burden.

3　The implementing situation of hydraulic construction of floodplain areas

In order to improve agricultural production condition and promote economic development in floodplain areas, approval by Ministry of Water Resources and national agriculture comprehensive development office, Yellow River Conservancy Commission (YRCC) implemented the first, second and third stage of hydraulic construction of floodplain areas in the lower Yellow River during 1988 to 1997 continuously, which were divided into 4 kinds: irrigation, drainage, silt beach to improve soil and production road.

The third stage of hydraulic construction of floodplain areas in the lower Yellow River made agricultural production condition in floodplain areas promote preliminarily, and yield improve significantly. Owning to large population, wide area and large project quantity, water conservancy infrastructure construction in floodplain areas was mainly about key projects, but matching projects incompletely. So the whole benefit of projects couldn't play full. With operation for many years, projects were damaged and aged heavily, and the condition of irrigation and drainage was still bad.

During hydraulic construction of floodplain areas, depending on scientific and technological progress to develop agriculture and water conservancy, YRCC promoted scientific and technological demonstrations and achievements actively. Based on investigation deeply and according to the characteristics of floodplain areas, it took "water-saving, energy-saving, land-saving, preventing water-caused damage, reducing the cost of projects and irrigation, improving the grain yield" as a goal, combined scientific research, demonstration and achievement extension, accomplished comprehensive planning and construction of 12 science and technology demonstration districts successively, and implemented more than 80 scientific and technological projects in total. The work of science and technology, demonstration and achievement extension improved the scientific and technological level, project quality and investment benefit of hydraulic construction of floodplain areas, promoted concentrated development, and guided the work of floodplain areas. In the districts of science and technology demonstration and extension, the farmland could be irrigated in drought and drained in flood; the basic agricultural production condition was improved radically; and the original medium-low yield farmland was strengthened to high yield farmland. Science and technology, demonstration and achievement extension also promoted hydraulic construction of the whole floodplain areas.

4　The problem of hydraulic construction in floodplain areas

Although the hydraulic construction of floodplain areas in the lower Yellow River has made great achievements, there is still a big gap with the goal of "comprehensive well-off society construction" and "construction of socialist new countryside" in new period. With the current social and economic development, many new problems need to be solved, especially the problems with agricultural production and hydraulic construction of current rural social and economic development.

4.1　The problem of national policy

More than 30 years ago, the state implemented preferential policy of abolishing agricultural tax for floodplain areas in the lower Yellow River. But the state abolished agricultural tax all over the country since 2006. It means the preferential policy for floodplain areas loses its significance. Especially accumulation labor and compulsory labor which hydraulic construction of floodplain areas depend on was cancelled after tax reform, the healthy development of hydraulic construction of floodplain areas has a new problem.

Moreover, the state enacted implementing scheme that to give the farmers subsidy directly according to cultivated area. Because this scheme was calculated by former tax cultivated land, the floodplain areas in the Yellow River were eliminated. According to "tentative compensation methods for expropriation of flood detention basin" promulgated by the State Council in 2000, the floodplain areas in the Yellow River were not included in the list of flood detention basin. It still were not included in "the revised list of national flood detention basin" issued by Ministry of Water Resource in 2010. It means the floodplain areas cannot obtain national economic compensation.

Therefore, in the case of bad natural conditions and lack of national support, the living standard of floodplain areas is worse. The paper wrote by Zhang Hui, the Associate Researcher of Henan Academy of Agricultural Sciences, proved the poverty of floodplain areas: by the end of 2008, there was 1.035,5 million people living in floodplain areas in the Yellow River, among which the number of people with annual net income per capita of less than 1,196 Yuan was 278,000, that with annual net income per capita of less than 1,500 Yuan was 363,000, and the poverty rate reached to 35%. The reason of the poverty of people in floodplain areas is instable income and no-guarantee. It needs to strengthen hydraulic construction of floodplain areas and implement preferential poverty alleviation policy to guarantee the people's living and change the poverty as soon as possible.

4.2　The problem of production dikes

The State Council promulgated the policy in 1974 that "the floodplain areas in the Yellow River should break production dikes, build water escape platform, and the grain with a quarter can feed people for the whole year". Because of some floodplain areas with poor soil, "the grain with a quarter to feed people for the whole year" is hard to meet the basic need of the development of floodplain areas. Owing to the frequent flood, it is hard to develop other industries except for agriculture. And due to low education level, the migrant labors are mostly as physical labors, lack of employment skills, and work unsteadily. All the factors lead in the people of floodplain areas with lower income, poverty and relying on agricultural economy highly.

The production dikes are built for protecting cultivated land and building, which limit the flood in main channel without overbank and deposit naturally. The sediment can only deposit between production dikes. It makes the lower channel become secondary suspended river with high channel, low beach and lower beside dikes, and makes the capacity of flood drainage decrease and the threat of flood increase. Without production dikes, flood is supposed to submerge the floodplain areas. Therefore, the production dikes are the lifeline of the people in some floodplain areas, and need to repair to keep safe. But for the demands of flood control and a long period of peace and order of the Yellow River, the production dikes shouldn't be kept. So the production dikes fall into vicious circle that "destroy, repair, re-destroy and re-repair".

4.3　The problem of industrial structure adjustment

During the period of planned economy, the main aim of agriculture was to produce grain. With development of economic society, agricultural production is being marketization, the pursuit of which is not only the grain yield but also economic benefit. Agricultural industrial structure is adjusted from wheat and autumn grain in double-cropping system to combination of multi-variety, multi-system and multi-structure. Therefore, on the one hand, how to adjust economic structure of floodplain areas is worth studying because floodplain areas cannot adjust industrial structure of other areas indiscriminately; on another hand, farmlands with different industrial structures have different needs of water conservancy facilities and different benefit distribution, and then have different demands of hydraulic construction in floodplain areas.

4.4 The problem of the property of hydraulic projects

As a kind of infrastructure, hydraulic projects are public welfare. The beneficiaries are not only farmers and agriculture, and also all the people and units in various trades.

Most of hydraulic projects in floodplain areas were built in 1950s and 1960s, and have bad quality of matching projects. Although it had built and repaired some projects in 1990s, the problems with weak foundation, incomplete matching projects and low irrigation efficiency still exist. Through more than ten years' operation, many projects damage and age heavily. Nowadays, hydraulic projects as a kind of infrastructure cannot meet the needs of the rapid development of agricultural economy. Therefore, the construction of hydraulic projects in floodplain areas has become a restrictive factor of enrichment. It needs the state and local government to invest to perfect the construction of hydraulic projects in floodplain areas, and increase the living standard of people in floodplain areas.

4.5 The problem of investment psychology

The unique natural environment of floodplain areas brings up the unique psychological characteristics of the people in floodplain areas, which is disaster environmental psychology of conditioned reflex. It means most people maintain vigilance the flooded possibility of the Yellow River at any time. The psychology is good for preventing flood, but bad for making and implementing the plan of becoming rich and the enthusiasm of production input owing to afraid of flood. Considering about the threat of flood, the local and out-of-town people are unwilling to invest. So the economic development of floodplain areas is difficult. In order to help the people to overcome this psychology, the government departments need to consider and meet the needs of the people in floodplain areas, and construct new environment with harmony between human and nature.

5 Countermeasures of economic development in floodplain areas

With the use of Xiaolangdi Reservoir and the gradual perfection of control works of the lower Yellow River, the ability of artificial control on flood and river regime is greatly enhanced, the probability of flood overflow is greatly reduced, and the condition to develop hydraulic construction of floodplain areas is mature. Considering the characteristic and the policy adjustment of floodplain areas, it should develop hydraulic construction to improve the ability of flood control, improve poverty, promote economic development, and then to ensure safety of flood passing. The main countermeasures of economic development are as follows.

5.1 Classified management of floodplain areas

Through investigation, floodplain areas in the lower Yellow River are divided into four districts by main function: flood passing district, flood detention basin district, cultivatable district and livable district. Based on districts, floodplain areas should be planned overall and manage individually, combined river regulation, production safety and economic development to realize common prosperity.

5.2 Strengthen infrastructure construction

Considering the people's needs with flood control, infrastructure construction is strengthened by newly-construction, reconstruction and repair. It should increase the investment strength in infrastructure construction in policy, especially water conservancy projects construction. It is of vital important to strengthen the investment of agriculture and tertiary industry, perfect agricultural development and infrastructure conditions, increase the production and living standard, and realize economic sustainable development of floodplain areas in the lower Yellow River.

5.3　Strengthen scientific research of hydraulic construction in floodplain areas

Hydraulic construction has lots of problems because floodplain areas in the lower Yellow River are large. Therefore it should strengthen scientific research of hydraulic construction in floodplain areas. On the basis of the characteristics of floodplain areas, it should be important to select suitable technology and measures to introduce, demonstrate and extend, to give full play to the guild and demonstration of science and technology, to realize prosperity by science and technology, and to offer a technological support to hydraulic construction.

5.4　Technological training

It is important that hydraulic construction should be suitable for the characteristics of floodplain areas. But how to mange and use is more important. According to the characteristic of the people in floodplain areas such as low cultural level and weak technology force, it is necessary to develop training work of water conservancy construction and management, and make the people to master the methods of projects construction and maintenance knowledge, in order to prolong service life of water conservancy projects and produce more social and economic benefits.

5.5　Practical technology introduction as priority

According to geographical position, landform and water conservancy projects status of different floodplain areas, on the premise that flood passing is unaffected, take developing agricultural production and realizing maximum benefit as target, introduce, demonstrate and extend practical technology suitable for local water conservancy, agriculture, animal husbandry and processing industry, and impel the people's enrichment.

5.6　Industrial structure adjustment of floodplain areas

Based on concrete conditions of floodplain areas, it is necessary to change present single economic mode, adjust industrial structure with comprehensive development mainly and market as a guild, and broaden production field. Then it is important to develop grain production steadily, increase the proportion of forestry, animal husbandry and fishery, and increase the commodity rate of agricultural products. By means of standardized dike construction in the Yellow River, develop ecological tourism resources, and seek new economic growth point for floodplain areas. Develop green agriculture of floodplain areas, drive base construction non-pollution agricultural productions, and promote economic development.

5.7　Developing socialized service system

The socialized service system and professional service organization of floodplain areas is still weak at present. For example, the services of water conservancy, agricultural technology, feed and processing are not good. It needs related departments to support and explore economic situation of developing the integration of agriculture, industry and commerce and the coordinated process of production and marketing, extend the management mode that companies drive bases and bases drive farmers, strengthen the production service, and promote industrial structure adjustment of floodplain areas.

5.8   Speeding up policy research of floodplain areas

Under the situation of constructing the human-oriented and harmonious society, new policy of floodplain areas needs to be reconsidered and researched to care for the people's vital interests, improve the conditions of production and living, and solve the conflict between economic development and regulation of the lower Yellow River properly. One of important subjects is to make policy according with interests of the people in floodplain areas. And it is also a guarantee for implementation of the Yellow River management strategies.

Research on the Key Period and Quantity
of Ejina Oasis's Ecological Water Requirement in the Heihe River

Yang Lifeng[1] , *Si Jianhua*[2] , *Hou Hongyu*[1] and *Zhang Yue*[3]

1. Yellow River Engineering Consulting Co. , Ltd. , Zhengzhou, 450000,China
2. Cold and Arid Regions Research Institute of the Chinese Academy of Sciences,
Lanzhou, 730000, China;
3. North China University of Water Resources and Electric Power, Zhengzhou, 450003, China

Abstract: Through analyzing remote sensing materials of the Ejina oasis in different times, the composition and scale of the ecological system are determined, and meanwhile the characteristic and ecological function of different ecological system are analyzed in this paper. Based on that, the law and quantity of Ejina oasis's natural ecological water requirement are put forward. Furthermore, the relationship between natural ecology and groundwater level, and the effects on natural ecology brought by the groundwater's change, are analyzed further, and the conclusion that the key periods of natural ecological water requirement within the year are April and August is proposed in this paper. In short, the research results in this paper provide scientific basis for ecological water requirements of different ecological recovery targets and water resource's reasonable allocation, and determining the scale of Huangzangsi temple reservoir in the upstream.

Key words: Heihe River, ecological, water demand, research

1 Introduction

Heihe River basin is an area where its local water resources are extremely lacking. Its water resources can't meet the need of its social and economic development and the ecological environment balance (Yan Dapeng, Wang Li, 2011). Vegetation is the subject of ecology and environment in the arid region who directly determines the environment quality (Li Weihong et al. , 2008). Water is the most important factor to maintain the ecosystem of the arid areas of sustainable development. Aired areas water is an oasis otherwise desert (Jia Baoquan et al. , 1998). In order to protect the natural vegetation of the desert region, we need to study the rational ecological water requirement. Scholars have conducted a lot of research on arid zone ecological water demand. Wang Fang et al. (1999) generalize that the ecological water requirement is the maintenance of ecosystem stability and natural ecological protection and artificial ecological construction consumed the amount of water. Liu Changming et al. (2011) proposed the method of how to estimate the balance of water and heat, salt and water balance, water and sediment balance and regional water balance in the calculation of the ecological environment of water;Cui Shubin(2011) generalize that the ecological water demand should be referring to a specific region within the ecosystem water demand; Dong Zengchuan (2001) generalize that the eco-environmental water demand is water ecosystems to maintain ecological and environmental function necessary for the water consumption; Wang Genxun generalize that the ecological water of Heihe River basin water demand for ecological water demand of artificial oasis natural oasis; Zhao Wenzhi et al. (2001) proposed arid vegetation ecological water demand can be divided into critical ecological water demand, the optimal ecological water requirement and saturated ecological water demand on the basis of the study in the arid areas of vegetation water; Xia Jun generalize that the vegetation ecological environment water demand in arid areas is to ensure that the plants normal healthy growth and simultaneously suppress the desertification and alkalization, the required minimum quantity of water and even the development of desertification. In short, different scholars have different understanding of the concept of vegetation ecological water demand; due to different scholars have different research purposes, different research methods and different subjects. However, in the process of ecological water demand study, scholars have less consideration of ecological water demand of the key period. Heihe River basin, for example, the role and significance of the Heihe River water to downstream natural oasis lies not only in the full assurance of water volume, especially in the replenishment time selection (Li Junqing, 2009), if the replenishment miss the key period of the downstream oasis ecological water demand, will directly affect the normal growth of the oasis vegetation. To maintain Ejina oasis health, this is first that stability and sustainable development of ecological water demand of the key period and water demand. We need study the Ejina oasis eco-critical water demand of its water demand; it is of great significance to ensure the smooth realization of the Heihe River governance objectives.

2 Ejina oasis ecological structures

In order to obtain the Ejina oasis ecological structures, we select two image data of the 1980s (1987) and 2008, Ejina oasis land use is divided into 10 first classes, 16 secondary classes, in accordance with the national land use of the two classification criteria. Secondary class divide into three levels according to the vegetation coverage, high coverage (>75%), cover (15% ~75%) and low coverage (5% ~15%).

1987, the total area of Ejina oasis about 3,630 km^2, the proportion of arbor, shrubs, herbaceous plants, arable land were 12.1%, 26.9%, 60.3%, 0.7% of the total oasis area. Arbor area is 438 km^2, of which: the area of coverage >75% is 133 km^2, the area of coverage between 5% to 15% is 144 km^2; shrub area is 977 km^2, of which: the area of coverage >75% is 311 km^2, the area of coverage between 15% ~75% is 666 km^2; Herbaceous area is 2,189 km^2, of which: the area of coverage >75% is 468 km^2, the area of coverage between 15% ~75% is 1,721 km^2; the cultivated area is 25 km^2. Non-oasis area is 56,305 km^2, of which: the settlements area is 12.2 km^2, the water area is 216 km^2.

2008, the total area of Ejina oasis is about 3,445 km^2, the proportion of arbor, shrubs, herbaceous plants, arable land were 11.1%, 26.5%, 61.0%, 1.3%. The arbor area is 383 km^2, of which: the area of coverage >75% is 126 km^2, the area of coverage between 15% to 75% is 152 km^2; the shrub area is 913.48 km^2, of which: the area of coverage >75% is 303 km^2, the area of coverage between 15% ~75% is 610 km^2; The herbaceous area is 2,103 km^2, of which: the area of coverage >75% is 516 km^2, the area of coverage between 15% ~75% is 1,586 km^2; the cultivated area is 45 km^2. Non-oasis area is 56,490 km^2, of which: the settlements area is 17.4 km^2, the water area is 218.4 km^2.

3 The ecological water demand of Ejina oasis

3.1 Calculation method

Using averiyanov formula (Guo Bin et al., 2010)

$$E = a(1 - H/H \max)^b \times E_0 \tag{1}$$

where, E is phreatic evaporation, mm; H is groundwater depth, m; H_{max} is the limit of groundwater depth, m; E_0 is water surface evaporation, mm; a, b are plant-related undetermined coefficients.

We get to the phreatic evaporation of different groundwater depths through the calculation of the evapotranspiration model; ecological water demand of the vegetation ecology is the product of the vegetation ecosystem area and phreatic evaporation in the groundwater depth. Namely:

$$W = E \times A \times 10^{-3} \tag{2}$$

where, W is vegetation ecological water demand, m^3; A is the calculate area of the oasis (including vegetation transpiration, bare soil evaporation), m^2; E is phreatic evaporation, mm.

Determination of the natural vegetation ecological water: May to October, Euphratica, Elaeagnus, Tamarix plant transpiration water consumption were measured by stem flow instrument, low sparse herbaceous plant individual transpiration was measured by LI-6400, analysis of changes of herbaceous plant individual water in the downstream of Heihe River determine the key period of typical water consumption of individual plants.

Groundwater depth determination: long-term observations of groundwater depth from Ejinaqi Water Authority, periodic groundwater depth survey point survey data from two surveys of Alashan desert eco-hydrology research and experiment station of Cold and Arid Regions Environmental and Engineering Research Institute, Chinese Academy of Sciences in 1999 and 2008. According to the different spatial location, 8 representative long-term observation wells were chosen on East River region; according to the different spatial location, 6 representative long-term observation wells were chosen on West River region. The two regions totally have 15 long-term observation wells, the longest data series from 1987 to 2008.

3.2 The calculated results

Based on the above calculation method, calculated natural ecological water demand of Ejina oasis, the results are shown in Tab. 1.

Tab. 1 Ejina oasis monthly ecological water demand in different periods.

Year	Monthly ecological water demand ($\times 10^8$ m^3)												
	1	2	3	4	5	6	7	8	9	10	11	12	Total
1987	0.094,4	0.167,4	0.346,5	0.786,7	1.046,5	1.135,8	1.319,3	1.068,3	0.780,3	0.434,6	0.190,1	0.9	7.46
2008	0.070,9	0.137,6	0.349	0.542,3	0.747,5	0.838,2	0.878,2	0.751,1	0.566,8	0.330,2	0.177	0.079,2	5.467,9

It can be seen from the table, Ejina oasis natural ecological water demand is 5.47×10^8 m^3 in 2008, which plant transpiration water and phreatic evaporation water is 4.04×10^8 m^3, the waters of ecological water demand is 1.43×10^8 m^3; Ejina oasis natural ecological water demand is 7.46×10^8 m^3 in 1987, which plant transpiration water and phreatic evaporation water is 5.68×10^8 m^3, the waters of ecological water demand is 1.78×10^8 m^3. From the water demand during the year, the maximum water demand in July, followed by June, the minimum water demand in January.

4 Study on the key periods of water in the natural ecological year

Tree, shrub and herbaceous plant are the dominant vegetation of Ejina oasis. Among those oasis vegetation, tree accounting for 12.1%, shrub accounting for 26.9%, herbaceous plant accounting for 61.0%. Using the conservation of isotopic mass method, the Cold and Arid Regions Environmental and Engineering Research Institute, China Academy of Science, has analyzed the contribution function of water source for the riparian forest plants. The results show that different water source has different contribution on the oasis vegetation. The contribution to the trees by groundwater was 93%, and 90% to the shrub. While,97% demand for the herbaceous plant was supplied by the soil water within 80 ~ 100 cm. In another word, groundwater was mostly used to maintain the normal growth and development of trees and shrubs in the downstream of Heihe River, and herbaceous plants maintain most of the surface water.

The analysis results of the phenological characteristics in Ejina oasis show, the buds of Trees and shrubs generally in late March - early April buds began to germinate, while the germination period of the herbaceous plants is mainly occurred in early April; According to the mean monthly process of change chart of groundwater level in the long-time observation wells, we can see that after April each year, the water level is decreased continuously, 8 ~ 9 month dropped to the lowest point, the groundwater depth is the maximum month up to August. Although the spring groundwater table relatively shallow, but the herb root is short, usually only 40 ~ 80 cm, groundwater above the depth of 1.5 m is not able to meet the needs of its normal growth and development. The herbaceous plants accounted for more than 60% of the total oasis area, the study shows that spring is period of the herb seeds germinate, sprout reproduction and multi-tillering (branching). If the spring irrigation is conducted timely before the herbs turn green or turn green, the reviving rate can be increased by 20%. At the same time, it allows the number of species of herbaceous communities, community level, community diversity and dominant species and its dominance to be increased. Therefore, timely irrigation in the spring, which is the most critical for the growth and development of downstream herbs, can effectively improve the herb germination self-renewal and rejuvenation.

According to the analysis on the relationship between the phenological rhythm of populus euphratica and flow of rivers, the period of flowering and seed dispersal period have an extremely close relationship with the river. Populus euphratica flowering from April 8 to May 10, the seed dispersal period from July 28 to August 23, those two most important breeding in time corresponding to the Ejina River spring flood (March and April) and autumn floods (July and August). A large number of populus euphratica seed spread in August, August is also the flood period of Heihe under natural conditions, therefore, with seed germination conditions, sexual reproduction can be successfully completed. At present, the Heihe annual upstream discharged water is not fixed, but not occurred at the twice flood period of Ejina River. Heihe watershed in value not only to fully meet the water demand of populus euphratica community, the choice of the watershed time should also be consistent with the two flood season. Populus euphratica seed germination is very sensitive to water conditions. In the natural environment, during the period that the spread of populus euphratica seed, only the place of a suitable water conditions created by the flood, seeds can germinate and seedlings to form. Therefore, the water condition is the dominant and limiting factor of populus euphratica seed germination. Populus euphratica seeds concentrated spread in August. During this time, the lower limit of soil moisture water content of not less than 13%. Populus euphratica intrazonal vegetation attributes determine its survival, reproduction and distribution changes which must be deeply dependent on the river, and must remain consistent with the river's flood period. Also shows that only during the flood period of the river, the environment conditions can meet the living demand of the populus euphratica seed security germination (ie, seed germination and seedling formation). Therefore, the only special condition of populus euphratica sexual reproduction is a certain intensity of river flooding process which consistent with the seed dispersal period. In other words, it requires enough moisture in August.

The populus euphratica clone population pattern and clone growth configurations play a decisive role in populus species update and population stability. The populus euphratica clone (Root suckers buds) is closely related to groundwater depth and silt accumulated thickness. Root suckers sprout buds at the time of its annual late growth (late July to early September). Groundwater level in the 1.5 ~ 4 m range, showed a relatively higher potential of clonal growth, an average of about $0.19/m^2$, shallow groundwater levels (< 1.5 m) or deep (> 4.0 m), the number of ramets compared decreases, an average of $0.03 /m^2$ and $0.09/m^2$. It shows that for the successful completion of the asexual reproduction of populus euphratica, the groundwater level in August must remained in the 1.5 ~ 4 m. Populus euphratica populations updates in a living environment under the conditions of soil silt thickness of 60 ~ 100 cm, it will have a strong stability. It shows that the more number of flooding times, the greater the thickness of the accumulated silt in the soil profile. This feature requires that each year should be at least twice a flood, and corresponds to the populus euphratica flowering and seed dispersal period (ie April and August).

During the growing period, the water consumption process of different plant as follows: the maximum water consumption month of populus euphratica and dwarf shrub sophora alopecuroides and alhagi sparsifolia and karelinia caspica is August; the maximum water consumption period of herb reed is from August to September, while the herb Splendens is August. From the plant community scale, we can see that he maximum water consumption period of populus euphratica forests and shrub tamarix chinensis and herb reed is from June to August. The changes of groundwater level during the year shows that, after April, the groundwater level continued to decline, it down to the lowest point in August and September, the highest number of the maximum groundwater level appears in the August. During this period the vegetation transpiration strongest, and it also the most serious water shortage period. Therefore, from the water consumption process of plant physiological, another critical season is August.

According to the above analysis on the research results, in the natural ecological years, Ejina Oasis critical water requirement period is April and August.

5 Conclusions and recommendations

Using remote sensing interpretation and field surveys, determine the scale and changes of Ejina Oasis in the 1980s (1987) and the present year (2008). In 1987, the scale of Ejina oasis is 3,630 km², the scale of it reduce to 3,445 km² in 2008. During this period, the scale of the oasis reduce about 185 km², among which, the trees reduce 55 km², shrubbery area decreased by 64 km² and grassland decrease 102 km². Tree and shrub and herbaceous plant are the dominant vegetation of Ejina oasis.

The shrub and herb accounted for 87.6% of the total area of the oasis. In April, the groundwater level can meet the bud of trees and shrubs, and the germination of herbaceous. But part of the regional the groundwater level is deeper with lower soil moisture content, so it is impossible to achieve the conditions necessary for herb germination. If there is no flood or irrigation to create a suitable water conditions, its seeds can't germinate, and it is impossible for seedling formation. Ejina River history under natural conditions, the formation of water seasonal changes during the year, April to June and July to September flow, respectively, accounting for about 21.3% and 56.2% of the annual flow, which is from spring snow melt and summer precipitation in concentrated form, and higher in April and August, respectively, accounted for 11.3% and 23.5% of the annual flow (Feng Shengwu, 1988; Wang Genxu, et al., 1998; Li Sen, 2004).

Studies have shown that the key periods of water demand are April and August. Transport the required amount of water in ecologically key period of protection, reasonable regulation of the library of water and river runoff, and protect the water demand in April and August, in April to ensure that at least 0.80×10^8 m³ lose water, in August to ensure that at least 1.08×10^8 m³ lose water to ensure downstream oasis vegetation physiological water requirement will not dry; taking into account the pre-dry and the next month water consumption and; taking into account the pre-dry and the subsequent month of water consumption and, if possible, with water as much as possible to increase the critical months of the water configuration.

Since July 2000, the implementation of the Heihe River water regulation on populus protection, rejuvenation update played a good effect, but subject to the constraints of the conditions of the lack of engineering measures, and each year the water time is greatly arbitrary. Heihe watershed in value not only to fully meet the water demand of populus euphratica community, especially in the choice of the time. Otherwise, the main watershed of the meaning will be lost. Also the role in restoration and maintenance of populus euphratica community living environment will be greatly reduced. Therefore, during the watershed implementation process, should follow the ecological and biological characteristics, particularly the survival of populus euphratica community property and distribution of intrazonal, must also take into account the guarantee of water and watershed time of selection, especially the watershed time should coincide with the key period of the ecological water demand.

References

Yan Dapeng, Wang Li. The Ecological Water Regulation Program of Heihe River[J]. Yellow River, 2011, 33(1):54 –55.

Gong Jiadong, Dong Guangrong, Li Sen. Degeneration of Physical Environment and its Control in Ejina Oasis at the Lower Reaches of Heihe River[J]. Journal of Desert Research, 1998, 18 (1):44 –50.

Li Junqing, Lu Qi, Chu Jianmin, et al. Research on the Populus Euphratica in Ejina Oasis[M]. Beijing: Science Press, 2009.

Zhang Wuwen, Shi Shengsheng. Study on the Relation between Groundwater Dynamics and Vegetation Degeneration in Ejina Oasis[J]. Journal of Glaciolgy and Geocryology, 2002, 24(4): 421 –425.

Guo Bin, Wang Xinping, Li Ying, et al. Prediction on Ecological Water Demand in the Mainstream of the Tarim River Based on Ecological Restoration[J]. Progress in Geography, 2010,

29(9):1121 –1128.

Jia Baoquan, Xu Yingqin. The Conception of the Eco-environmental Water Demand and its Classification in Arid Land—Taking Xinjiang as an Example[J]. Arid Land Geography, 1998, 21 (2): 8 – 12.

Wang Fang. Research on the Ecological Water Requirement in Arid and Semi-arid Land[D]. Beijing: China Water Conservancy and Hydropower Research, 1999.

Liu Changming, Chen Yongqin, Yu Jingjie, et al. New Technologies and New Methods of the 21st Century Chinese Hydrological Scientific Research[M]. Beijing: Science Press, 2001.

Cui Shubin. A Research on Issues of Water Demand Foreco-logical Environment[J]. China Water Resources, 2001 (8):71 – 74.

Dong Zengchuan, Liu Ling. Study on the Water Resources Allocation of West China[J]. Water Resources and Hydropower Engineering, 2001(3):1 – 4.

Wang Genxu, Cheng Guodong. Water Demand of Eco-system and Estimate Method Inarid Inland River Basins[J]. Journal of Desert Research, 2002, 22(2):129 – 134.

Zhao Wenzhi, Cheng Guodong. Review of Several Problems of Ecological Hydrological Processes in Arid Land [J]. Chinese Science Bulletin, 2001, 46(22):1851 – 1857.

Xia Jun, Zheng Dongyan, Liu Qinge. Study on Evaluation of Eco-water Demand in Northwest China[J]. Hydrology, 2002 (5):12 – 17.

Wang Genxu, Cheng Guodong. Changes of Hydrology and Ecological Environment during Late 50 Years in Heihe River Basin[J]. Journal of Desert Research, 1998,18(3):233 – 238.

Li Sen, Li Fan, Sun Wu, et al. Modern Desertification Process in Ejina Oasis and its Dynamic Mechanism[J]. Scientia Geographica Sinica, 2004, 24(1):61 – 67.

Discussion on Harnessing Measures of the Xiaobeiganliu Reaches of the Yellow River

Zheng Shixun, *Fan Yongqiang* and *Jia Feng*

Yellow River Shanxi Bureau, YRCC, Yuncheng, 044000, China

Abstract: In recent years the silting in the Xiaobeiganliu reaches of the Yellow River has been serious, causing new problems to river flood control, engineering layout and works defense. Based on the practice of the Yellow River harnessing, this paper analyzes the problems existing in the harnessing of the Xiaobeiganliu reaches of the Yellow River in six aspects, namely, the channel flood discharge capacity has declined, the river pattern can not be effectively controlled, the phenomenon of serious danger caused by mere small flood has occurred frequently, the floodplain land suffered from serious salinization, the construction standard of flood control roads were too low and the construction of flood control non – engineering measures has lagged behind. It is put forward such harnessing measures as: carrying out water and sediment regulation; carrying out research on the law of "bottom tearing" scouring; speeding up construction of river harnessing projects; building village protective projects; raising construction standard of floodplain roads; making preparation earlier for flood control emergency rescue; and strengthen construction of flood control non-engineering measures.

Key words: harnessing of the Xiaobeiganliu reaches of the Yellow River, existing problem, harnessing measure

The Xiaobeiganliu reaches of the Yellow River refer to the reaches of the Yellow River trunk stream between Yumenkou and Tongguan. In history, the reaches were frequently stricken by flood disasters, and collapse of banks and beaches often occurred, endangering the life and property of the people along the banks. Since the beginning of the Yellow River control by the people, great success has been achieved in the harnessing of the reaches, providing an important guarantee for social stability and economic development. In recent years, the reaches have been seriously silted, and the channel has been wandering frequently, bring new problems to river flood control, engineering layout and works defense, and also having some influence on social stability and steady economic development along the Yellow River. It is essential to analyze the problems existing in the harnessing of the Xiaobeiganliu reachesof the Yellow River, propose comprehensive harnessing measures and advance the harnessing of Xiaobeiganliu of the Yellow River better and faster to provide greater guarantee for society and economy along the Yellow River.

1 Current situation of the Xiaobeiganliu reaches harnessing of the Yellow River

The Xiaobeiganliu reaches are the reaches between Yumenkou and Tongguan of the Yellow River trunk stream, which are part of the middle Yellow River. The river bed of the reaches is wide and shallow, the flow is disordered, central bars and braided channels have formed, the truck stream keeps wandering, being a typical accumulation wandering channel. In history, the reaches were frequently stricken by flood disasters, and collapse of banks and shores frequently occurred, endangering the life and property of the people along the banks. The harnessing of the reaches began in 1968, and up to the end of 2011, 36 river harnessing works altogether with total length of 147. 49 m and 1,107 groins had been built on both banks. The construction of these works has preliminarily straightened out the river pattern of the Xiaobeiganliu reaches, prevented continued collapse and recession of high banks and shores, protected villages, facilities for living and production, cultural relics and historic sites along the river, providing an important guarantee for local economic development and social stability.

2　Main problems existing in harnessing of the Xiaobeiganliu reaches

In recent years, the Xiaobeiganliu reaches of the Yellow River have suffered from serious silt-ation, the river had no main channel, and the main stream has been wandering, bringing new prob-lems to river flood control, engineering layout and works defense.

2.1　River siltation is serious and flood discharge capacity has declined

Over the past thirty-odd years since 1977, no large-scale "bottom tearing" scouring has oc-curred in the Xiaobeiganliu reaches, the river bed has been continuously silted and lifted, the channel has continuously shrunk and its flood discharge capacity has declined year after year. Es-pecially, in the more than ten years from 1986 to now, the water entering the Xiaobeiganliu reaches has drastically decreased, which has further worsened the imbalance of incoming water and sedi-ment. The channel has been lifted 6 cm a year on average, the main channel has shrunk, so the bankfull discharge has dropped from round 5,000 m^3/s to about 3,000 m^3/s, and in some reaches to less than 2,000 m^3/s.

2.2　Existing works have been separated from the flow and the river pattern can not be ef-fectively controlled

In recent years, due to serious river siltation and constant wandering of the channel, many ex-isting works have been separated from the flow and can not give their benefits into play. According to the river pattern survey data before the flood period of 2011, only 26% of works on both banks touched the flow (Wang Junkuan, Fan Yongqiang, Feng Quanmin et al., 2011). In the reaches where no works have been built such as Zhichuan of Hancheng, the upper end of Yunlin works of Heyang, Chayukou, Lu'an of Dali, Tonghekou of Tongguan, Xifan of Wanrong, Shundi to Hanji-azhuang of Yongji and Fenghuangzui of Ruicheng, the main flow washed banks and floodplains, di-rectly endangering the safety of villages, roads, bridges and electric pumping stations along the riv-er. According to incomplete statistics, in recent ten years the collapse of banks and floodplain was 52.76 km long in total, the collapse width was usually 100 ~ 1,000 m, and about 91.1 thousand mu of floodplain land was destroyed due to collapse. For example, the bank of Tongguan reaches of Tongguan collapsed for 3.2 km, and the bend top of the bank was less than 10 m from State Road 310, about 50 m from Lianyungang-Horgos Highway, having already endangered the safety of the three important communication lines: State Road 310, Lianyungang-Horgos Highway and South Tong-Pu Railway; in 2009, the bank of Shundi-Hanjiazhuang reaches of Yongji collapsed for about 6 km, so that 4.8 × 10^4 mu of farmland in Zuncun floodplain was inundated, the development fruit in the floodplain was ruined and more than ten villages such as Xixiang were threatened by the flood.

2.3　Adverse river pattern frequently appeared and serious danger caused by mere small flood constantly occurred

Due to river bed lifting with siltation, the flood discharge capacity of the main channel de-creased, which weakened the river harnessing works' function of stabilizing river pattern, increased the occurrence chance of "transverse river", "slanting river" and "flood discharge along dykes", and increased the possibility of burst in works, seriously threatening the safety of the works on the Xiaobeiganliu reaches of the. Yellow River. For example, in the late August, 2008, the discharge at Longmen Hydrological Station of the Yellow River was always between 700 m^3/s and 1,700 m^3/s, the whole of the Nanxie Control Works of Hancheng touched the flow, the main flow formed a slant-ing river in front of Dams No. 1 ~ 4, directly lashed the connected dam and the dam head of the works, causing danger of foundation stone missing, slope stone collapse and foundation earth expo-sure to occur in succession at Dams No. 1 ~ 4, with accumulative danger length of 242 m and dan-ger number of 19 dam-times, losing 3,504 m^3 of stone, 970 m^3 of earth and 0.725,5 × 10^6 Yuan of investment; from 2008 to 2009, the maximum discharge was only 1,150 m^3/s at Longmeng station, the Dams No. 16 ~ 22 of Xiaoshizui Works encountered danger of foundation stone collapse in succession, with accumulative danger length of 576 m and danger number of 24 dam-times, losing 3,494 m^3 of stone and 0.516,3 × 10^6 Yuan.

2.4 Front – back elevation difference of works has increased and floodplain land salinization is serious

Due to constant uplifting of the river bed, the elevation difference between the side facing the water (front) and the opposite side (back) has increased, now the front – back elevation difference is mostly larger than 1.50 m, with maximum of 3.35 m, forming the trend towards suspended river. In Zuncuntan, Xifantan and Lianbotan on the left bank, the front – back elevation difference is about 1 ~ 2 m, and the swamp is nearly 10 × 10^4 mu; in Zhaoyitan and Ximintan on the right bank, the elevation difference is around 2 ~ 3 m, causing rising of ground water level in the floodplain area and increasingly serious swamping and salinization. Among the 20 × 10^4 mu of arable floodplain land, 109 thousand mu has become swamp or saline-alkali land, of which nearly 10 thousand mu is open – water area, some areas have become desert, some can only grow suaeda salsa and large area of floodplain land has become barren (Xue Xuanshi, 2009).

2.5 Construction standard of flood control roads is too low

The flood control roads on both banks of the Xiaobeiganliu of the Yellow River are mostly earth roads or sidewalks level with the floodplain, and are uneven. Some roads close to high banks are easily clogged due to landslides; some roads level with the floodplain are of low construction standard, they are in poor condition and will become muddy during rainy seasons; once flood flows over bank, the roads will be submerged and become impassable. The roads along the Yellow River whose construction was funded by local governments are mostly of low construction standard, they are level with or a little higher than the floodplain; after flood overbank, they are liable to be flooded, blocking the traffic. In October 2009, the maximum discharge at Longmen station on the trunk stream of the Yellow River was only 1,600 m^3/s, yet flood overbank occurred in Wanrong reaches, a section of a road about 4 km long along the river was flooded and could not be used normally.

2.6 Construction of flood control non-engineering measures lagged behind

At present, the flood control non-engineering measures of Xiaobeiganliu are not imperfect in flood control teams, thinking and construction of information technology, and can not adapt to the requirement of fighting large floods and dealing with serious dangers. In construction of flood control teams, due to lack of fund, mobile rescue teams have not been set up, the number and quality of masses protective teams are difficult to guarantee; in thinking, since the runoff of the Xiaobeiganliu of the Yellow River has always been small in recent years, lack of vigilance and lucky mind are widespread among leaders and masses along the Yellow River; in construction of information technology, information on water regime, works situation and danger is mainly collected and transmitted manually, which is inaccurate and slow, often causing opportunity for rescue to be missed, and the management departments can not connect to the flood control network of upper levels, so they can not get information on flood control in time.

3 Discussion on harnessing measures of Xiaobeiganliu Reaches of the Yellow River

3.1 Continue to carry out water and sediment regulation to reduce silting in Xiaobeiganliu Reaches of the Yellow River

For five years running since 2006, YRCC has carried out the experiment of utilizing and optimizing spring flood process to scour Tongguan river bed to lower its elevation, accumulatively washing away 0.154×10^8 t of sediment deposited in the Xiaobeiganliu channel of the Yellow River and lowering Tongguan elevation by 0.56 m (Hou Suzhen, Tian Yong, Lin Xiuzhi et al. , 2011) , which has effectively improved the siltation pattern of Xiaobeiganliu of the Yellow River and deepened the understanding of sediment law. In recent years, the water and sediment regulation of the lower Yellow River has extended to the reservoir on the upper reaches, and the flood discharged from the upper reservoirs also has had good effect on the scouring of the Xiaobeiganliu reaches. According to statistics, during the period from 2005 to 2011, a total of 1.30×10^8 t of sediment was washed away and the channel was washed 0.15 m on average (Yang Ren, Li Jiwei, Jia Xinping et al. , 2001) . Therefore, using and optimizing flood from upper reaches to carry out water and sediment regulation is an effective measure to improve siltation pattern of Xiaobeiganliu of the Yellow River, improve the flood discharge capacity of the channel and increase bankfull discharge.

3.2 Study law of bottom tearing and make use of the scouring effect of bottom tearing to reduce river siltation

The "bottom tearing" occurring in the Xiaobeiganliu reaches of the Yellow River has great scouring effect on the river bed, it can scour the river bed 2 ~ 9 m deeper within a short period, with maximum scouring length of about 130 km. After "bottom tearing", the channel flood discharge capacity increased, the channel became orderly and the flood stage went down, which were all conducive to flood control safety. In recent years, there have been few flood processes of large discharge and high sediment content, the condition of "bottom tearing" has been difficult to form, and large – scale "bottom tearing" has never occurred since 1977. To study the law of "bottom tearing" and mold manually "bottom tearing" process by means of water and sediment regulation with use of reservoirs on the upper reaches will greatly reduce river siltation and increase flood discharge capacity.

3.3 Speed up construction of river harnessing works and perfect flood control works system

After years of harnessing, the flood control works system of Xiaobeiganliu reaches of the Yellow River has begun to take shape, and it has played an important role in control of the Yellow River trunk stream and protection of banks and villages and has produced remarkable benefit in flood control and disaster reduction. However, the river regime of Xiaobeiganliu reaches of the Yellow River has not been effectively controlled. The main channel has been wandering frequently, in some reaches without defense there are large spaces, banks have collapsed from time to time, overbank flood has occurred now and then, and disasters such as flood disasters are still serious, directly endangering the life and property of the masses along the river. Therefore, it is essential to further carry out construction of river harnessing works.

3.4 Construct necessary village protection works to protect the safety of villages along the river

During the spring flood period of 2009, large area of overbank flood occurred in Zuncuntan of Yongji, the overland water was about 0.2 ~ 0.4 m deep and the overbank flood lasted about one day; and the overland flow directly rushed to Xixiang Village etc. and threatened their safety. The safety problem aroused great attention of the Yellow River Flood Control and Drought Relief Headquarters and provincial, city and county leaders of the Shanxi Province, who immediately organized rescue. To protect villages from flood attacks, we should construct necessary protective works around villages. The protective works should be able to protect against floods once in 20 years at Longmen Hydrometric Station on the Yellow River, their crest should be 6 ~ 8 m wide, ant they should be protected with riprap stones.

3.5 Raise construction standard of floodplain roads to provide condition for flood control emergency handling

Construction of floodplain roads should go through argumentation, and is up to certain flood control standard, especially the effect of overbank flood that has higher chance of occurrence should be taken into consideration, so that the roads may be used for a longer time. In construction of floodplain roads, the requirement of flood control danger handling should also be considered so that, in case of large flood, the masses can retreat out and materials for emergency handling can be transported in.

3.6 Make preparations in advance for flood control and emergency handling

According to survey, on both banks of Xiaobeiganliu of the Yellow River about 100 thousand residents live below the historical highest flood level, and there is about 1,000 thousand mu of usable floodplain lands on both banks, which are important agricultural production bases of Shanxi and Shaanxi Provinces, where the task of flood control is very heavy. Before flood period, the management units should make preparations for flood control and emergency handling in aspects of thinking, responsibility system, technology, materials, teams and facilities. In thinking, strengthen publicity to make the leaders and masses along the Yellow River overcome carelessness and lucky mind; in responsibility system, allocate specific tasks to particular individuals in forms of meetings and responsibility certificates, etc. ; in technology, constantly revise flood control plans and engineering emergency handling plans, and make sure that the plans are scientific and practical; in materials, prepare adequate materials for flood control, and store some reserve materials; in teams, have sufficient personnel and carry out necessary training and exercise; in facilities, mainly get mechanical equipment for emergency handling ready, so that when danger occurs, they can have sufficient emergency handling capacity and can control the danger within a short period of time to ensure perfect safety.

3.7 Strengthen construction of flood control non – engineering measures to guarantee flood control safety as far as possible

Strengthen water law enforcement, and make the cadres and the masses along the river be aware of the basic situation of Xiaobeiganliu reaches of the Yellow River and carry out production in accordance with the basic law of river evolution. Take the requirement of flood control into consideration before constructing infrastructure such as drilling wells and erecting power lines in the channel, prohibit constructing structures or planting long-stalked crops within the channel, prohibit constructing cofferdams, which may hinder normal flood discharge. Set up mobile flood control rescue teams to improve emergency handling capacity and level in flood control. Strengthen construction of flood control information technology. Set up an internal LAN to achieve communication with the Yellow River Flood Control and Drought Relief Headquarters and local flood control headquarters, carry out through monitoring of works and realize automatic collection of data of first-line water regime and engineering condition.

References

Wang Junkuan, Fan Yongqiang, Feng Quanmin, et al. Report on River Regime Survey of the Xiaobeiganliu Reaches of the Yellow River Prior to 2011 Flood Period [R]. Zhengzhou: Flood Control Office, YRCC, 2011.

Xue Xuanshi. Recent Channel Scour-and-Fill of the Xiaobeiganliu of the Yellow River and its Effect. http://www. docin. com/p-20924622. html,2009.

Hou Suzhen, Tian Yong, Lin Xiuzhi, et al. Analysis of Experiment Result of Optimizing Spring Flood to Scour and Lower Tongguan Elevation [R]. Zhengzhou: People's Yellow River, Issue, 2011.

Yang Ren, Li Jiwei, Jia Xinping, et al. Feasibility Study Report of Recent Harnessing Project of the Yumenkou—Tongguan Reaches of the Yellow River [R]. Zhengzhou: Yellow River Engineering Consulting Co. , Ltd. , 2001.

Study on River Health Based on Natural-social Coupled Function

Li Yinghong and *Yu Dahuai*

School of Marxism, Hohai University, Nanjing, 210098, China

Abstract: The development and protection of the river should be based on the optimum unity between river value and human needs. The value of the river must be recognized and the relying on each other between river value and human needs should also be stressed. Through the exposition of the relationship between river and human civilization, river health is evaluated by the nature-society compound ecosystem. The dominant position of river value can be established. The ultimate goal of managing the river is to maintain the healthy life of river, which is in order to build the river health evaluation system based on environmental management.

Key words: functional coupling, river health, life, ethics

River health and sustainable utilization of water resources, affected by climate change and human activity, is the key issue for human survival and development. The concept of river health derived from the 1980s when western countries started to protect the ecosystem of river. Both in Europe and North America, river conservation actions were carried out in many countries by modifying the development of water law and environmental law to strengthen the river's environmental assessment and ecological protection. The concept of river health emerged accordingly. At present, more attention is paid to protect river ecosystem and maintain the healthy life of rivers. The standard of river health is mostly to be analyzed from the ecological point of view. Biological health of the river, or watershed ecosystem health is more concerned. For example, people focus on water chemistry, physical and biological integrity of river health in the United States. In Australia, healthy river is defined as the adaption to local environment, social and economic characteristics and to support for river ecosystems, economic activities and social functions. In South Africa, retention of water is mostly studied to maintain the healthy life of rivers. In China, water conservancy researchers are also putting forward proposals for the concept of the maintenance of river health. In the Second International Yellow River Forum (October 19, 2005), the concept of healthy life of rivers was firstly proposed. Liu Changming (2005) pointed out that the healthy life of rivers must be based on the healthy water cycle. Dong Zheren (2005) believed that the key of protecting river health is to maintain the biodiversity of river ecosystems. Wang Zhaoyin (2005) argued that river health consisted of water security, sediment balance, river stability, river use and water ecology. Hou Quanliang & Li Xiaoqiang (2007) tried to relate the concept of humans and other organisms with river and set up the river health assessment system. River health at home and abroad is mostly studied from the viewpoint of biophysical ecology. The evaluation index of river health mostly reflects the river ecosystem health indicators. From a broader research vision, river health and sustainable human development should be linked from the comprehensive society-economic, human and natural health perspective.

There are many rivers in China. The total length of rivers in China is 420,000 km. The number of those rivers whose basin is greater than 100 km^2 is about 50,000. These rivers are very important for the economic development and protection of people's life. In recent years, the exploitation of China's rivers has caused the ecosystem deterioration and heavy watershed pollution with floods, drought and water shortage, soil erosion and other issues, and thereby endangers the river's economic and social value. Under the influence of global climate change and human activities, how to maintain river health and use water resources sustainably are the important problem that people must understand and investigate.

1 Ecological crisis and the existence filemma of rivers

Rivers were the birthplace of human civilization. Human beings can not live without rivers, and they influence on each other. Rivers, as an organic whole, include the water, watercourse, embankments, the animals and plants growing in water, as well as the people living nearby the river. This could be called the rivers' community. Humans, other biological and natural objects are all the members of the community. Rivers and its biodiversity are coexistent, constituting a coupling between ecological environment and life. For a long time, humans settled around the rivers, and the rivers were maximal utilized and reformed by the ways of building dam for generating, developing water resources, flood control, irrigation and city water supply and so on. On one hand, the rivers provide food, agriculture, industry, and water for humans. On the other hand, they have the business, transportation, entertainment, and many other service functions. In general, rivers are the foundation of human health. Changes of the river not only change the nature, but also affect the human society. The status and the future of human survival and development are increasingly affected by the river changes. Healthy river ecosystems are closely related to human health, while the survival and health of the river largely depend on human behavior.

The river health standards must be dynamic. In different time or areas, the standards actually reflect the human values produced under a corresponding background. For a long time, the relationship between the man and the river had historical variability, mainly including three stages:

In the period of agricultural civilization, the river could keep its natural condition. In the absence of human intervention, each river was experiencing the process of life evolution for growth, development and decline. The river not only had the natural value, but also had the social value. Although the river has already been developed and utilized by human, people were still in awe of the river. This, to a certain extent, has kept the river ecological in balance.

In the period of industrial civilization, the demand of water allocated by the construction of water conservancy, soil and water conservation and inter-basin water transfer project, which had been beyond the capacity of which river could provide, was increased sharply. A large number of industrial and domestic wastewater were drained into the river, which resulted in drying up rivers, dry lakes, deterioration of water quality and the loss of biodiversity. This situation directly resulted in an unprecedented crisis of the river ecosystem. This subsequently brought a purposeful river treatment. Since industrial revolution, the epistemology divided the world into subjective and objective parts which are opposite. According to this epistemology, only human beings could be called subjects and possessed intrinsic value. Any others did not have subjectivity and intrinsic value except for human beings. This exactly and profoundly accounted for the phenomenon that the relationship between people and rivers became worse.

In the period ecological civilization, the cumulative effects of water conservancy and hydropower projects were gradually revealed. Global warming has brought many changes to the rivers situation. The rivers face more serious ecological environment problems and challenges. This forced people to seek more reasonable ways to develop. Some ecologist suggested that: "Provide the living space to the river, reconstruct the life network of the river". arousing reflections about'' to reposition the relationship between the man and the river' in the international community.

For each river, the affordability of natural and social systems are limited. Only in the range of its capacity, the rivers' life can continue. However, if the rivers' life systems were crisis, it might cause the structure and functions of ecosystem disorder and break the balance of the ecosystem. What is worse, other species living in the river will lose the basis for survival. As a result, the economic and social developments and even human life will be gradually failure. At present, the rivers are in poor health in China, mainly in several aspects such as the aggradations of river channels and their shrink, as well as the degradation of river ecosystems. According to the report on the environmental condition of China in 2010, it shows that the pollution in the seven key river systems including Yangtze River, the Yellow River, Pearl River, Songhua River, Huaihe River, Haihe River and Liaohe River was mild. Fig. 1 illustrates the percentage of water quality grades of the seven key river systems in 2010. It can be seen that in the 409 national level controlling sections in 204 rivers, the sections with Grade I ~ III water quality accounted for 59.9% ; Grade IV ~ V accounted for 26.0% , and worse than Grade V accounted for 16.4%.

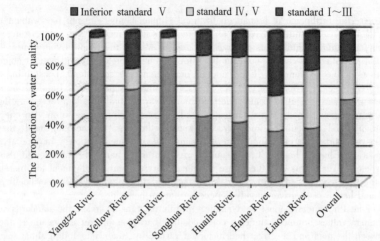

Fig. 1   Percentage of water quality grades of the seven key river systems in 2010

The water quality was good for Yangtze River and Pearl River, followed by Songhua River and Huaihe River. Huaihe River and Liaohe River had medium level pollution, while Haihe River suffered from heavy pollution. Based on the standards of water environment ecosystem health, the water environment ecological system of the main rivers was in an unhealthy state. The water pollution not only has become the biggest threat on the river health, but also has brought threat to the safety of drinking water in some cities. According to incomplete statistics, about 3.6 billion people did not have clean water to drink. In addition, the waste water discharge amount increased from 315 × 10^6 t in 1980 to 63.1 × 10^9 t in 2002. It was reported that more than half of the seven major river systems were polluted. 90% of water was heavily polluted in city. 50% of urban water did not meet the drinking water standards, and 40% of the water was undrinkable.

The deterioration of river water circulation and ecological system has already warned about the issues about serious health and how people view sustainable use of water resources. Human's dependence on rivers, is not only reflected in the civilization creation, vicissitudes, and the various influence on material and spirit world of human, but also the importance of remaining the integrity of biological chain, protecting human living conditions and keeping balance of ecological system. Therefore, maintenance of the rivers' life is no other than the maintenance of human life. To make rivers have the ability to provide sustainable support for human economic and social development, a healthy body should be maintained for rivers. Environmental philosopher Ralston said: "In any community, the power without ethical restrictions is vulgar and destructive." Today, rivers' crisis is just the embodiment of this destructive. To the water environment in 21st century, the demand for water continue to rise, and future situation of the water environment in China is "very serious". The water crisis is caused by the severe water shortage. Resources and water pollution not only damage the ecological balance, but also resulte in the further aggravation of water imbalance between supply and demand. Moreover, it will also cause huge economic losses. According to statistics, due to the shortage of water, the yearly economic loss has been up to 120 × 10^9 Yuan in industry. The annual economic losses caused by water pollution are equivalent to 6% of national financial revenue in that year. So, water crisis not only troubles national economy and people's livelihood seriously, but also becomes the importantly restrictive factors of the social and economic development in our country.

2　Content of coupling nature-social function of river health

A healthy river should meet the conditions as follows: first, it should be continuous runoff, and should have the ability to coordinate water-sand as well as keeping a good water quality; second, it should keep various biological chains in ecosystem function well to achieve a benign cycle of the ecosystem as well as to meet the demand of human and other creatures for survival and development. The concept "Health" is generally applied to humans and animals as a series of medical physiological and chemical indicators can be used to show the health. World Health Organization points out: health is a kind of physical, psychological and social well being, not merely the state of the absence of disease and weakness. "River health" is a concept of social property. Each river owns its different characteristics and social backgrounds, the choice for the types and the quantities of health indicators will be different, so different countries and regions have different standards for river health. Therefore, we define the meaning of river health as follows: On the premise of maintaining river natural, ecological and social services functions as a relatively balance, river can basically achieve the normal and good function for water, material and energy cycles, keep a certain level for ecological and social services functions, meet sustainable development needs of human society, and form a benign circle of balancing human development and protection of rivers. Connotation of river health reflects the conflict between human social-economic development and natural environment protection, and mirrors the human values under social backgrounds.

Coupling is a physics concept. It is a phenomenon that two or more systems or movement patterns influence each other so as to be together through a variety of interactions. Ecosystem is organic coupling. Healthy river should be that the natural function and social function fit each other on the basis of interaction, thereby make each part effectively play their respective functions, and achieve coupling for each subsystem function. River health evaluation should include: ecological indicators, physical and chemical indicators, socio-economic indicators and indicators of human health, namely when a river healthy is evaluated, it should not only give full consideration to the need of maintaining natural functions, but also give full consideration to the balance of river's natural function and social function. Finally, river health evaluation should be changed according the natural and social conditions.

2.1　Natural function of rivers

The so-called natural function of rivers is that rivers play the role in the evolution and development of the earth and nature. Natural functions of rivers are as follows: taking part in earth's water cycle, such as the hydrological functions, transfer function of matter and energy, function of shaping the topography, maintaining cycle of river ecosystems and species balance and so on. As a natural ecosystem, the river is a dynamic organic part, coexisting with the water, the animals, the plants, the microorganisms and the environment, which constitute a mutually coupled ecosystem. Healthy river should keep a strong ability to carry, store, and self purify. Only the river itself maintains a healthy body condition, can it have the ability of supporting economic and sustainable social development of human society?

2.2　Social function of rivers

River ecosystem is capable of regulating climate, improving the ecological environment, safeguarding biodiversity, and completing certain social services and many other functions. Healthy river should not only maintain the significance of ecological, but also emphasis on the functioning of social services. This function/role can be divided into two aspects. One is in the physical level, including a variety of flood control science and technologies formed on the process of flood control activities conducted by human beings, water conservancy project and the resort in convenience of life as well as social and economic benefits, et al. The other is in the psychological level, including literature and art, aesthetics, ethical, philosophical, social customs, leisure and entertainment and so on. Under the condition that rivers have a certain natural structure. meet reasonable ecological needs, and provide a more favorable social services to meet the needs of sustainable development during the corresponding period of human society, that is to say, healthy river is that it can maintain the river's natural, ecological functions and social service functions to achieve a balanced state. The social function of the river is the support of river systems for human social and economic systems, which is the significance of the human maintaining river health.

As a complex river ecosystem combined by society, economic and nature, it should not only be ecologically reasonable, but also econcmically viable, reasonable structure and good operating functions, and can provide ecological services for nature and human needs. In the ecosystem of river and its basin, whether the river functions couple plays a decisive role. The river system of functional coupling must have strong adaptive mechanisms to maintain the system operation, adapt to the external environment, and have the ability to using external environment to promote its evolution and achieve integrated cooperation between all internal functional systems and all functional systems (Fig. 2):

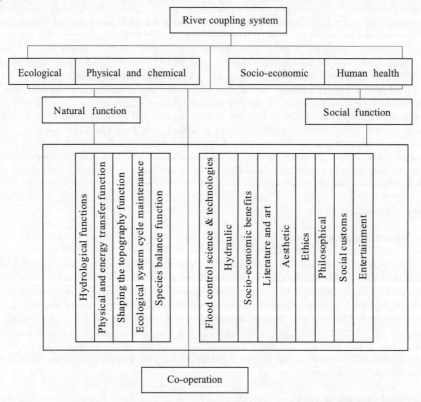

Fig. 2 Coupling relationship between natural-social function of healthy river

In the coupling theory, each factor forming coupling system constitutes the integral part of the coupling system. Any development or change of each factor may lead to changes of other factors and even the changes of the coupling system. Flow, current, flooding, wetlands, water quality constitute the rivers' ecosystem but coupled to each other as life forms. Under the condition of basic balance of natural and social functions, healthy rivers should include: a smooth and stable riverbed, good water quality, sustainable river ecosystem and the proper amount of surface runoff, that is to say, the basic standard for healthy river is the river has both normal natural function and social function. "Coupling" relationship between various functions of the river is not the simple sum, but comprehensive reflection of interaction, interdependence, mutual coordination among subsystems.

3　Establishing a new relationship between human beings and rivers

Under a new historical condition, it is necessary to maintain a relative balance between rivers' social function and natural function, which relates to people's value orientation at different backgrounds. The content of healthy river based on natural-social function coupling calls for people's rethinking about their value orientation , re-positioning the relationship between human and rivers, which gives moral concern and major status for rivers. Recognition for the relationship between human and river as well as human and nature promotes the concept of rivers' healthy life drawing.

3.1　Core values formation of rivers' healthy life

Starting from the river ecosystem integrity, the river's life is regarded as an organic whole. With a vision based on modern science the river's value ignored by human. is evaluated On the concept of river's value, which is that rivers are taken into the scope of moral a new type ethical relationship between the river and the human is formed. Ethics of the river asks people to give necessary respects and moral concerns to rivers, and make the mind-set change on the issue of dealing with relationship between human and rivers, which means treating rivers as lives and forming the core value of rivers' healthy Life. As Li Guoying said, "Rivers' Life" has two meaning as follows: On one hand, rivers support the existence and development of human; On the other hand, rivers can be regarded as organic lives. He also pointed out that: "rivers have lives as human beings do." Rivers' healthy life and rivers' ethics provides a new angle of recognizing rivers, which is a watershed for rivers' control theory.

3.2　Establishing mechanisms for monitoring and evaluating rivers' health

The assessment our country's rivers' healthy life should be based on China's national conditions. The choice of health index for rivers should concern both uniformity and difference in rivers' health. The standards for evaluating river health should be established according to specific rivers, and build different health lives index or structures. It is necessary to reasonably allocate water resources, maintain rivers' ecosystem, establish healthy index for rivers and put them into the integrated planning for watershed or region water resources into the exploitation/development or management of water resources.

On one hand, people should keep a reasonable degree on requirements for the development and utilization of rivers to ensure the sustainable use of water resources; on the other hand, people should protect and recover river ecological system to ensure their health. River health is related to water resources, environment, judicature and administration, electric power, agriculture, industry, traffic, fishery, and other departments. Public participation and social supervision should be strengthened.

According to interest coordination and unified planning, local governments' performance evaluation index system should contain achievements of river basin protection, and set up the mechanism of rewards and punishments for local governments.

It is necessary to build a perfect emergency response and emergency system for the river blanking and pollution prevention, and dissolve the water crisis as soon as possible as well as to develop green economy and environmental protection industry. Combining the government's function and the market mechanism, connecting environmental protection and economic benefit, putting the environmental protection into the system of the economic development and managing it as an industry will make economic entities benefit from pollution control and protection of environment, which is in line with its goals of maximum profit. Therefore environmental protection becomes a conscious act for people. Production, consumption and environment protection then get a harmony. Only coordinating every department well, can we utilize the river with maximize efficiency while maintaining river health.

3.3 Ethical discussion of rivers stakeholders

Designing and construction of river engineering change the original structure of resource allocation and proportion, which makes the distribution problem of social interests apparent. How can we make a balance between utilization of water resource and river protection? Establishing a consultation mechanism, achieving a health standards' consensus among the river developers, the guardian and the social public will relief the conflict of interest between water resources exploitation and environment protection. The ultimate solutions for contradiction between people and rivers can not merely rely on "dialogue" between the river and the natural agent, but also the "dialogue" and "negotiation" based on the platform of the interest between the stakeholders and others, and between these stakeholders and the society. These stakeholders are as follows: river basin management institutions, departments of water supply, water electricity, agricultural, shipping, environment protection, forestry and local residents. People in the same river basin should put aside their interests' differences and give up self interests' standard theory to establish the mutual river moral rules and coordinate each stakeholder's interest in society.

Reaching consensus based on justice principles in the interaction framework can make us evaluate legitimacy and rationality of river development activities from the angle of value and ethic. Although the interest subjects could not reach to the complete consensus, the mode of dialogue/negotiation provides us with a more effective way to solve the conflict between human and river nature. This deeply conversion of thinking mode may radically improve the some crisis including the disaster of rivers ecosystem.

Starting from our country's actual situation, rivers' health should be reached under the conditions of reasonable natural structure. Namely maintaining a balance among natural function, ecological function and social service function of rivers. Based on Marx's view that the unity of people and the nature is "people realize the naturalism and the nature realize the humanitarian", environmental philosophy holds that the conflict between man and nature is really solved by the ontology. The river is a life community with individual dignity. Any natural noumenon is with double life of human material and cultural. The river should be endowed life forms beyond narrow science vision. It is necessary to give essential care to river health with a unique humanistic feeling to keep social and cultural atmosphere of river health. Coordinating the relationship between the natural function and social function of the river within its threshold is the self-consciousness of humans. Giving river the meaning of life, changing the value orientation, and setting up the main body status for value of the rivers are the essential way to develop the relation between human and rivers.

References

Boulton A J. An Overview of River Health Assessment: Philosophies, Practice, Problems and Prognosis[J]. Freshwater Biology, 1999(41):469 – 479.

Karr J R. Defining and Measuring River Health[J]. Freshwater Biology, 1999, 41(2):221 – 234.

Krasnicki T, Pinto R, Read M. Australia Wide Assessment of River Health: Tasmanian Bioassessment Report (TAS Final Report) [R]. Canberra: Department of Primary Industries, Water and Environment, 2002.

Liu Xiaoyan. Scientific Issues on Ideas of River Health[J]. Yellow River, 2008:10.

Rolston H. Environmental Ethics[M]. Translater Yang Tongjin. Bei jing: China Social Sciences Press, 2000.

Huang Shaofu, Zhao Han. Studying on Vendor Selection with Analytic Hierarchy Process and Random Disposal of Data Envelopment Analysis[J]. Journal of Anhui University of Science and Technology: (Natural Science Version).

Wang Wei, Li Chuanqi. Study on Healthy Life of Rivers [J]. Yellow River, 2005(7).

Hou Quanliang, Li Xiaoqiang. Discussion on River Health Life[M]. Zhengzhou: Yellow River Conservancy Press, 2007.

Li Guoying. Keep Healthy Life of Rivers[J]. China Water Conservancy, 2005(21).

Li Guoying. Keep Healthy Life of Rivers—a Case Study of the Yellow River[J]. Yellow River, 2005(11).

Ye Ping. Discussion on River's Natural Life [M]. Zhengzhou: Yellow River Conservancy Press, 2009.

Chao Mengqing. On the Reflection and Reconstruction of Ecological Ethics Noumenon[J]. Morality and Civilization, 2007(3).

Establishment on Ecology Healthy Target and Typical Rivers Appraisal of the Haihe River Basin

Hou Siyan[1] , *Shi Wei*[1] , *Wang Liming*[1] and *Lin Chao*[2]

1. Research Institute of Water Resources Protection Science of Haihe River Resources Committee, Tianjin, 300181, China
2. Haihe River Basin Water Resources Protection Bureau, Tianjin, 300181, China

Abstract: To establish a river health indicator system special for the Haihe River Basin and evaluate the health conditions of Haihe river basin objectively and comprehensively, the author starts from the river ecological health and holds the opinion that firstly, the river health means that a river can play both its natural functions and social functions and secondly, the evaluation of the health conditions of a river requires a normal condition as the benchmark which shall be regarded as the standard of comparison and health assessment. Besides, as for the indicator selection, the indicators about natural functions have been given preference after the analysis of the advantages and disadvantages of the river health indicators in foreign countries and the review of the experience in river health evaluation in China and based on the excessive exploitation of the social functions of the plain tracts of the Haihe River which causes severe ecological environment problem that the natural functions decline. Yongding River and Weiyunhe River are typical examples for the evaluation of river ecological health conditions. The comprehensive index of the river reach of Yongding River from Sanjiaduan to Dongzhou Bridge is 35.5 and the overall assessment is unhealthy. That of the river reach from Dongzhou Bridge to Qujiadian is 43.6 and the overall assessment is subhealthy. The serious problems of Yongding River is that the water volume decreases sharply and the river channel cuts off. The comprehensive index of the river reach of Weiyunhe River from Xuwanchuang to Sinvshi is 22.5 and the overall assessment is unhealthy. The serious problems are that the water is severely polluted and due to the pollution, the organisms in the water are close to extinction.

Key words: connotation, benchmark, health indicators, the Haihe River Basin, typical rivers

1 The connotation of river ecological health

River health is a new concept suggested together with the term, ecosystem health. From the viewpoint of river control, the suggestion of river health has practical significance. It can promote the communication between scientific communities and the public and arouse the concern of the society in the influence imposed by people on rivers.

There is no uniform definition of river health. Different experts may have different understanding. There is a popular and general opinion in foreign countries that a healthy river ecosystem cannot only maintain the structure and functions of the ecological system but also include human and social values. The concept of health covers both ecological integrity and human values.

As people's understanding of ecological health goes deeper, the definition of ecological health becomes clear. If a river is ecologically healthy, it means that the river ecosystem cannot only maintain chemical, biological and physical integrity but also keep the river being able to provide services for the human society. Liu Changming et al thinks that what the river health reflects is how people recognize and confirm the functions of rivers, so in principle a healthy river must have both ideal social functions and natural functions. Therefore, the sign of a healthy river should be the performance of the natural functions under the prerequisite that both natural and social functions do work.

2 The benchmark of river health

River health is a relative concept. A benchmark may be required to judge whether a river is healthy or not. The assessment of health conditions must be based on comparison.

For example, the "Victoria River Health Strategy" of Austria regards the rivers before European settlers as the benchmark. However the European countries ecologically healthy rivers have main ecological features and functions of the rivers before WWW II. According to the analysis and comparison of the information on water volume of the Haihe River Basin in the 1960s when Haihe River Basin has not been affected by any water conservancy project, the water volume presents distinct characteristics. Some rivers in the history are ephemeral streams. Take Weiyunhe River and Yongding River in the 1960s for example, the channel flow is in accordance with rainfall and the peak flow appears during June and August. The natural tends are obvious. See Fig. 1 So the design and calculation of the water volume of Haihe River Basin takes the 1960s as the reference point.

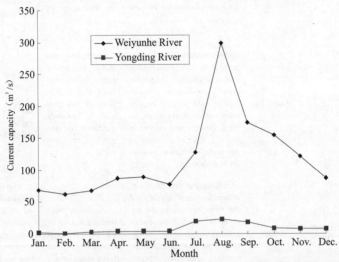

Fig. 1 Annual variations of typical rivers during certain historic period

In recent years many river health benchmark methods have been developed. The following are the four main ones. The first is that the river ecosystem can be recovered completely to an ideal condition. For this method is unrealistic, it is rarely adopted. The second is the sample reach comparative method. It means to regard the unaffected or slightly affected reach as reference. The more the monitored reach deviates from the sample reach, the worse the health condition is. The third one is to set up a principle which is based on water quality indicators. Every water quality indicator must have a range which can be modified according to the local conditions. The fourth is an aggregative model of river conditions. An aggregative river health model shall be set up to describe the physical, chemical and ecological conditions of the river basin.

3 Application of river health evaluation indicators in China and abroad

Many studies have been done on the indicator system and the evaluation method of the river health conditions. Lots of special indicator systems and evaluation methods have been developed, such as the Austrian River Assessment System and the Index of Stream Condition. Britain monitors the river health condition through River Habitat Survey. The Department of Water Affairs and Forestry of South Africa launches the River Health Programme in 1994 and provides a hierarchical framework which can be widely applied to river biological monitoring. The above methods are all developed according to the local features and demands so they may all have some disadvantages. See Tab. 1 for the main content, advantages and disadvantages.

Tab. 1　Main river health assessment methods in the world

Assessment Method	Proposer	Country	Brief introduction	Advantages	Disadvantages
AUSRIVAS	Simpson Norris (1994)	Australia	Modify PIVPACS Approach when collecting and analyzing the assessment data according to the characteristics of rivers in Australia so that the model can be widely applied to assessing the health conditions of rivers in Australia	Be able to predict the biomass which should live in the river theoretically; the result can be grasped by managers in an easy way	This method only takes big invertebrate into consideration and does not connect water quality with habitat degradation and biological conditions
ISC	Ladson (1999)	Australia	Construct a indicator system consisting of river hydrology, morphological characteristics, riparian zone conditions, water quality and aquatic organisms; score every river for its indicators and reference points; the total score is the comprehensive assessment index	Integrate the main characteristic factors about the conditions of the river and provide guidance for scientific management	Lack in the reflections on the changes of individual index; the selection of reference river reach is subjective
RHS	Raven (1997)	Britain	Assess the natural characteristics and qualify of the river habitats by the indicators such as background information, riverway data, sediment features, vegetation type, bank erosion and riparian zone conditions	Connect well the habitat indicators, fluvial morphology and river organisms	The internal connection of the selected indexes and organisms is not clear; some data from assessment are qualitative so it is hard to carry on mathematical statistics
RHP	Rowntree (1994)	South African	Evaluate the healthy conditions of rivers from seven indicators which are invertebrate, fishes, bank vegetation, habitat completeness, water quality, hydrology and shape	Apply biotic community indexes to show the influence of river system on external interferences	In practical application, it is difficult to obtain some indicators

Although the concept of river ecosystem health has been introduced into China for many years, the rive health assessment begins to be regarded as an effective way to inspect the river ecosystem in recent years. The large river basin administrations now have all carried out relevant work.

Changjiang Water Resources Committee suggests the healthy Changjiang indicator system which consists of the general objective layer, the system layer, the condition layer and the indicator layer. This is the first "healthy" river quantitative indicator system in China that can adopt numerical expressions. Yellow River Conservancy Commission in recent years has also put forward the governing idea of "keeping Yellow River healthy". Liu Xiaoyan suggests nine factors such as low-limit flow, bankfull discharge, wetland area and water quality classification as the indicative factors according to the specific conditions of Yellow River. The Pearl River health assessment indicator system is planned to consist of the group layer, the property layer, the classification layer and the indicator layer, including twenty-six indicators such as the river bank and bed stability, the surface area rate, the connectivity with the surrounding natural ecology, the fish habitat and the fishway condition.

It can be seen that the existing river health indicator systems in China are also established according to the special characteristics of different rivers. For example, Changjiang River Basin has the indicator of navigable depth reliability in view of its navigation expectation; Yellow River Basin should guarantee a smooth and safe water-sand channel because of its incredible sediment-carrying capability; The Pearl River Basin focuses on the indicators such as excessive salt content and control of water loss and soil erosion.

4 Establishment of Haihe River health indicator system

Now a general opinion holds that the sign of a healthy river should be the performance of the natural functions under the prerequisite that both natural and social functions do work. However, due to the excessive development of the social functions, the natural functions of Haihe River are declining. The unbalance due to the inclination of the natural functions to the social functions further causes the current consequence that "all of the rivers become dry and all of the water is polluted". Thus a series of ecological problems appears which severely restricts the sustainable development of social economy. So the realization of the social service values must also depend on the healthy river ecosystem. As for the natural functions the following indicators are selected according to the problems of Haihe River, including water volume, water quality, biomass and connectivity. As for the social functions, only some representative flood-prevention indicators are used which can also been included in the river level. The indicators such as the water resources development ratio and the landscape construction are abandoned. The purpose is to restore the natural functions of the damaged rivers to some extent.

All of the sub-indicators should be assessed and calculated through detailed indicators. The water volume index is assessed through relative dry-up length, relative dry-up days, relative cut-off days and average annual flow deviation rate. The water quality index is expressed by the water pollution index. The biotic index is calculated through the biological diversity and the vegetation coverage. The connectivity index is calculated through the number of the gate dams per hundred kilometers. The flood prevention standard index is worked out by the comparison of the current flood control capacity and the flood prevention standard. The study works out the comprehensive assessment index based on the weighted values, including five grades: "healthy, almost healthy, subhealthy, unhealthy and to be collapsed". Sufficient water volume, excellent water quality and sound habitats are the highest signs indicating a river is ecologically healthy. See Tab. 2 for the comprehensive river ecosystem health assessment standard of Haihe River Basin.

Tab. 2　Comprehensive river ecosystem health assessment standard of Haihe River Basin

Overall performance	Assessment Standard	Sub-indicators				
		Water volume index	Water quality index	Biotic index	Connectivity index	Flood prevention standard index
To be collapsed	0 ~ 20	Dry throughout the year	Polluted severely	Extinct	Very poor	More than three levels lower than the standard
Unhealthy	20 ~ 40	There is water seasonally	Polluted moderately	Single	Poor	Three levels lower than the standard
Subhealthy	40 · 60	The water volume is general; seasonal dry-up	Polluted slightly	General in diversity	Medium	Two levels lower than the standard
Almost healthy	60 ~ 80	The water volume is general; intermittent cut-off	Acceptable	Rich in diversity	Good	One level lower than the standard
Healthy	80 ~ 100	The water volume is sufficient and there is water throughout the year	Clean	Very rich n diversity	Excellent	Almost satisfy the standard

5 Assessment methods

According to the weight proportional relationships among water quality, water volume and habitat in the research articles in China and abroad and the research on allocating the weight of sub-indicators by means of expert consultation, the analytic hierarchy process is adopted to determine the weight of the indicators. At last the consistency check has been done to eliminate errors due to any subjective judgment. The total weight of all detailed indicators is recorded as W_i. And the weight of the detailed indicators respectively in the sub-indicators is recorded as c_i. When the average annual flow deviation rate% in water volume is lower than -90%, it is thought to be lower than the acceptable minimum flow (refer to Turner Method) which may exert great influence on the ecological environment. The water volume index becomes the dominating index of the ecosystem and the weight increases to 0.46. Meanwhile, the weight of the water quality index decreases to 0.25. In addition, the weight of the biotic index is 0.14, the weight of the flood prevention standard index is 0.1 and the weight of the connectivity index is 0.05.

Assessment of sub-indicators:

$$P_i = \sum_{i=1}^{n} r_i \times c_i$$

By this equation, we can work out the assessment of the biotic index, the water volume index, the water quality index, the connectivity index and the flood prevention standard index.

Comprehensive assessment: the comprehensive index of the river ecosystem health assessment can be worked out according to the corresponding weight and quantized value.

$$P_i = \sum_{i=1}^{n} r_i \times W_i$$

where, r_i is numerical values of the indicators; W_i is total weight of the detailed indicators; W_i is the respective weight of the detailed indicators in the sub-indicators; P_i is the result of indicator assessment.

6 Assessment of typical rivers

6.1 Yongding River

According to data summary and site survey, as for Yongding River, the problems, for example, the water volume decreases sharply and the rivers cut off or dry up, become more and more serious. The reach from Lugou Bridge to Qujiadian dries up for 365 d and the dry-up length is 146.8 km. So the water volume index should be the decisive index for the overall performance of the ecological health. The weight should also be adjusted accordingly. Through the comparison with the cut-off and dry-up conditions in the 1960s, we can figure out the relative dry-up length, the relative dry-up days and the relative cut-off days and further the water volume index.

The water pollution is not serious in Yongding River when there is water in the river reach. The basic requirements can be satisfied. When the river is dry, the water quality index is 50.

There are two flood gates on Yongding River. The function of Sanjiadian River Sluice is to supply water and that of Lugou Bridge River Sluice is to control flood. Both of them are in Beijing. According to the flood prevention plan of Haihe River Basin, Yongding River flood prevention design standard is once in one hundred years. The existing flood prevention system can satisfy the standard practically.

According to the basic information about the river in all aspects, we can work out the assessment result of the ecological health conditions of Yongding River. See Tab. 3 for details.

6.2 Weiyunhe River

According to investigation, the water in Weiyunhe is seriously polluted. The chemical oxygen demand in Guantao exceeds 200 mg/L. The aquatic life is close to distinction. The vegetation coverage in flood land is about 60%. By the comparison with the cut-off and dry-up condition of Weiyunhe River in the 1960s – 114 cut-off days, 12 dry-up days and 39 km dry-up length, we can work out the relative dry-up length, the relative dry-up days, the relative cut-off days and further the water volume index.

Tab. 3   The assessment result of the ecological health conditions of Yongding River

River Reach	Overall performance		Sub-indicators										
			Biotic index		Water volume index		Water qualify index		Connectivity index		Flood prevention standard index	
	Score	Assessment	Score	Assessment	Score	Assessment	Score	Assessment	Score	Assessment	Score	Assessment
Sanjiadia-Municipal Boundary	35.5	Un-healthy	38.5	Single in biological species	14.3	Dry up through-out the year	50.0	Slight-ly pol-luted	20	Bad	100	Practi-cally satisfy the stand-ards
Hebei Boundary-Dongzhou Bridge	35.5	Un-healthy	38.5	Single in biological species	14.3	Dry up through-out the year	50.0	Slight-ly pol-luted	20	Bad	100	Practi-cally satisfy the stand-ards
Dongzhou Bridge-Qujiadian	43.6	Sub-healthy	58.5	General in biologi-cal diver-sity	14.3	Dry up through-out the year	71.2	Ac-cept-able	20	Bad	100	Practi-cally satisfy the stand-ards

According to investigation, there are two flood gates on Weiyunhe River. The role of Zhuguantun Check Sluice is for irrigation and shipping and that of Sinvshi Check Sluice is for flood control. Both of them are located in Wucheng County, Shandong. According to the flood prevention plan of Weiyunhe River, Weiyunhe River flood prevention design standard is once in 50 years. The existing waterways together with the flood storage and retention areas can only handle the flood once in 10 years.

According to the basic information about the river in all aspects, we can work out the assessment result of the ecological health conditions of Weiyunhe River. See Tab. 4 for details.

Tab. 4   The assessment result of the ecological health conditions of Weiyunhe River

River Reach	Overall performance		Sub-indicators										
			Biotic index		Water volume index		Water qualify index		Connectivity index		Flood prevention standard index	
	Score	Assessment	Score	Assessment	Score	Assessment	Score	Assessment	Score	Assessment	Score	Assessment
Xuwan-cang Sin-vshi	22.5	Un-healthy	20.0	Single in biological species	53.9	Moderate water volume and sea-sonal dry-up	0	Se-verely pollu-ted	70	Good	40	Two levels lower than the stand-ard

References

Meyer J L. Stream Health: Incorporating the Human Dimension to Advance Stream Ecology[J]. Journal of North American Benthological Society,1997(16):439 – 447.

Liu Changming, Liu Xiaoyan. Healthy River: Essence and Indicators[J]. Acta Geographica Sinica, 2008,63(7):683 – 692. (in Chinese)

Dong Zheren. River Health Connotation[J]. China Water Resources,2005,(4):15 – 18. (in Chinese)

Rapport D J, Costanza R, McMichael A J. Assessing Ecosystem Health[J]. Tree, 1998 ,13 (10):397 –402.

Ladson A R,White L J. An Index of Stream Condition: Reference Manual[M]. Melbourn:Department of Nature Resource and Enviroment,1999.

Ladson A R, White L J, Doolan J A, et al. Development and Testing of an Index of Stream Condition for Waterway Management in Australia[J]. Freshwater Biology,1999(41):453 – 468.

Parsons M, Thoms M, Norris R. Australian River Assessment System: Review of Physical River Assessment Methods—A Biological Perspective [C] //. Monitoring River Heath Initiative Technical Report no 21, Commonwealth of Australia and University of Canberra, Canberra. 2002.

Brizga S, Finlays B. River management: the Australasian experience Chischester[M]. New York : John Wiley & Sons, 2000.

Cai Qihua. Safeguard Health Yangtze River, Promote Harmonious Relation of Human and Water[J] . Yangtze River,2005,36(3):1 – 3. (in Chinese)

Liu Xiaoyan. Construction of Healthy Indicators System in Yellow River [J]. China Water Resources,2005(21):28 – 32. (in Chinese)

Lin Mulong,Li Xiangyang,et al. Probe into the Index System for Evaluating the Health of the Rivers in the Pearl River Basin [EB/OL] .2004 – 04 – 05. Pearl River Water Resources Network. 2004 – 04 – 05. (in Chinese)

Zhao Huanchen , Xu Shubai, He Jinsheng. Hierarchy Analysis Method a Simple New Decision Method [M]. Beijing: Science Press,1986. (in Chinese)

Applying Index of Biotic Integrity Based on Fish Assemblages to Evaluate the Ecosystem Health of Weihe River

Wu Wei and *Xu Zongxue*

College of Water Sciences, Beijing Normal University, Beijing, 100875, China

Abstract: In Weihe River Basin, ecosystem is deteriorating gradually under dual pressures from rapid growth of population and development of economics, which is important to identify the biological situation and take measures to restore damaged ecosystem. Biological integrity in Weihe River was assessed based on the 45 sample sites using a fish Index of Biotic Integrity (IBI) in this study. In this investigation, thirty – nine fish species were collected in Weihe River, including twenty – four fish species from *Cyprinidae*, nine fish species from *Cobitidae*, one fish species from *Siluridae*, one fish species from *Bagridae*, 3 fish species from *Gobiidae*. *Cyprinidae* and *Cobitidae* were the main fish community structure in Weihe River. All sites were categorized as reference and impaired sites by TWINSPAN and canonical correspondence analysis (CCA). Nine metrics sensitive to environmental changes were determined to differentiate reference and impaired sites by statistical methods, and scoring them by stepped method, and five classes were defined to represent the biological integrity in the Wei River. Results showed that only 10 sites were classified as excellent, 9 sites as good and 9 sites as fair, while the other 9 sites and 14 sites were classified as poor and very poor, respectively, in which more than half of the sites were subjected to poor or very poor. Biological integrity in the main stem was better than that in its tributaries—the Jing River and Beiluo River. In summary, the results obtained above indicated that biological integrity in the Wei River was intensely destroyed by human activities, e. g. reservoirs, water diversion projects, urbanization, and industrial wastewater effluent. The fish index of biotic integrity developed in this study to evaluate biological health was proved to be effective, and it will be a great reference to watershed management and ecosystem restoration in Weihe River.

Key words: fish index of biotic integrity, environmental factors, canonical correspondence analysis, qualitative habitat evaluation, community structure

1 Introduction

The IBI has been widely used for aquatic integrity assessments across the world since it was developed by Karr(1981) in the United States. Several types of organisms have been proposed as biological indicators of river health. Indices of biotic integrity based on algae (Hill et al., 2000), benthic macroinvertebrates (Weisberg et al., 1997; Baptista et al., 2007) and fish (Bozzetti & Schulz, 2004; Tejerina – Garro et al., 2006) was the most frequently used in river and stream assessments. Fish assemblages have been regarded as an effective biological indicator of environmental quality and anthropogenic stress in aquatic ecosystems, not only because of its position at the top of the food chain, but also due to their sensitivity to subtle environmental changes (Karr, 1981; Simon & Lyons, 1995).

Fish index of biotic integrity is considered an acceptable tool for evaluating biological health and has been widely used in the United States (Dauwalter & Jackson, 2004; Tejerina – Garro et al., 2006), Europe (Breine et al., 2004; Joy & Death, 2004), Brazil (Bozzetti & Schulz, 2004; Casatti et al., 2009) and other countries (Kleynhans, 1999; Qadir & Malik, 2009). The original IBI method was developed on the basis of biological condition in the midwestern United States and modified by different researchers according to the characteristics of different basins. Oberdorff and Hughes (1992) modified some of the metrics originally used for the Seine River Basin, which successfully extended the applicability of IBI to a different region.

In contrast, there is little application to the rivers in China, more similar research in Liaohe

River (Pei et al. , 2010; Song et al. , 2010) and the Yangtze River (Zhu & Chang, 2008; Liu et al. ,2010), nearly no research in Weihe River. The most recent investigation of the fish assemblages in Weihe River Basin was in the 1980s. Over the past 30 years, biological conditions have worsened significantly due to intensive human disturbance. Unfortunately, no existing publications address the biological conditions and changes in biotic integrity in the study area. The objectives of this study were as follows: ①develop the index of biotic integrity for fish assemblage in Weihe River; and ②analyze the spatial variation of the IBI and biological degradation in this basin.

2 Material and methods

2.1 Study area

Weihe River is the largest tributary of the Yellow River, which originates in the Niaoshu Mountain and flows through Gansu, Ningxia and Shaanxi Provinces. Weihe River has a total length of 818 km with a catchment's area of 1.35×10^4 km^2. The Jing River and Beiluo River are the first and second largest tributaries of the Weihe River Basin. Precipitation in the flood season, from July to October, accounts for approximately 60% ~ 70% of the annual precipitation, leading to greater runoff in the flood season. There are many reservoirs and water diversion projects in the whole basin, and streamflow in the river decreases greatly, especially in the non – flood seasons, while flood occurs frequently in the flood season. Weihe River is heavily polluted, most of them are in extremely bad condition. Also, ecosystem function is deteriorating dramatically, characterized by declining fish richness, shrinking wetlands and decreasing coverage of forest vegetation.

2.2 Data collection

In the present survey, 45 sites were sampled across the whole basin in October 2011, including 23 sites in the main stem, 13 sites in the Jing River and 9 sites in the Beiluo River (Fig. 1). Fishes were collected by electrofishing for 30 min during stable flow periods. Although this fish – collection method may not yield a complete inventory of all species, it did result in sample sizes large enough to assess the biological situation of the basin (Oberdorff & Hughes, 1992).

Water temperature (W. tem), pH, conductivity (Cond), salinity (Sali), saturation (Satu), alkalinity (Alk), total dissolved solids (TDS), dissolved oxygen (DO), COD_{Mn}, water hardness (W. hard), TN, TP, suspended solids (SS), and SiO_3 were measured in the investigation. Water velocity and depth were measured using a current meter, and sediment concentration was measured using a set of sediment samples sieved using screens with different apertures of 16 mm, 8 mm, 4 mm, 2 mm and 1 mm.

2.3 Data analysis

2.3.1 Selection of reference sites
According to similar research (Reynoldson et al. , 1997), reference sites with the least – disturbed environments were determined based on a qualitative habitat evaluation index and water quality. TWINSPAN (Hill, 1979) and CCA was applied to selection of reference sites.

2.3.2 Selection of IBI metrics
The metrics developed by Karr (1981), including twelve fish community parameters, can be grouped into two sets: species composition and richness and ecological factors. Considering the present situation of Weihe River ecosystem and the application of IBI (Karr, 1991; Oberdorff & Hughes, 1992; Simon & Lyons, 1995; Pei et al. , 2010), the parameters for this study were selected from twenty-four candidate metrics (Tab. 1). These metrics are categorized into five groups: species composition and richness; trophic composition; tolerance; reproductive guilds; abundance and individual condition.

Twenty-four parameters are included as the candidate metrics of IBI, and a subset of metrics

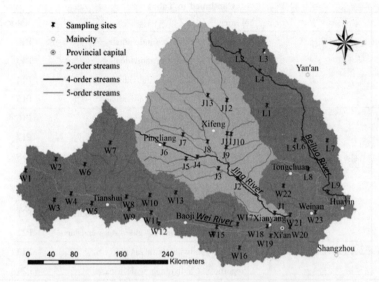

Fig. 1   Location of sample sites in the Weihe River

from these twenty-four candidate metrics will be selected using statistical analysis (Wang et al. , 2005; Baptista et al. , 2007). Firstly, metrics were eliminated if the values for that metric at 90% of the total sites are zero. Secondly, the sensitivity of each metric will be judged according to the degree of interquartile overlap in box-and-whisker plots in Spss 16.0, metrics exhibiting no overlap or little overlap in interquartile range will be retained and confirmed by Mann-Whitney U-test (Barbour et al. , 1999). Finally, the Pearson's correlation is calculated for the retained metrics. For two metrics with an obvious correlation (| R | >0.8), one will be retained to simplify the index because one metric sufficiently represents the pattern of information . The retained metrics constitute the primary index of biotic integrity.

Tab. 1   Candidate metrics used in assessment of fish communities

Category	Metrics	Abbreviation
Species composition and richness	Number of species	P1
	Total biomass	P2
	Shannon – Wiener index	P3
	Number of Siluriformes species	P4
	Number of Cypriniformes species	P5
	Number of Perciformes species	P6
	Number of Cyprinidae species	P7
	Number of Cobitidae species	P8
	Proportion of pelagic species	P9
	Proportion of benthic species	P10
	Proportion of species in the middle and low tiers	P11

Continued to Tab. 1

Category	Metrics	Abbreviation
Trophic composition	Proportion of carnivore individuals	P12
	Proportion of omnivore individuals	P13
	Proportion of herbivore individuals	P14
	Proportion of insectivore individuals	P15
Tolerance	Proportion of sensitive species	P16
	Proportion of tolerant species	P17
Reproductive guild	Proportion of species with pelagic eggs	P18
	Proportion of species with demersal eggs	P19
	Proportion of species with viscid eggs	P20
	Proportion of species with special spawning behaviors	P21
	Proportion of species that guard eggs and larve	P22
Abundance and individual condition	Number of individuals in sample	P23
	Proportion of individuals with diseases	P24

2.3.3　Calculation of the IBI

The stepped scoring method is modified based on the scoring system of Karr, which is scored as 1, 3, 5 for each metric on the basis of comparisons with the distribution of metric values at reference sites (Casatti et al. , 2009). Calculation of metric scoring thresholds is based on the distribution of values at reference sites, with the lower threshold established at the 25th percentile and the upper threshold at the 75th percentile (Schleiger, 2000).

The integrity classes are classified into five groups, i. e. , excellent, good, fair, poor and very poor. The 25th percentile of reference sites is calculated to separate the excellent and good condition (Klemm et al. , 2003), and other three thresholds are obtained by quartering 25th percentile value, with each quarter as very poor, poor, fair and good (Zhang et al. , 2007).

3　Result analysis

3.1　Characteristic of fish community structure and water environmental factor analysis

Thirty – nine fish species were collected in Weihe River, including twenty – four fish species from Cyprinidae, nine fish species from Cobitidae, one fish species from Siluridae, one fish species from Bagridae, 3 fish species from Gobiidae. Cyprinidae and Cobitidae were the main fish community structure in Weihe River. Contrast with investigation in 1980s (Xu & Li, 1984), fish species decreased largely, whereas with similar community structure.

The environmental factors were divided into two categories: water quality factors and hydrology/geomorphology factors. The primary environmental factors were identified based on factor analysis and Pearson's correlation test, respectively. The result of the analysis on the water quality of the sample sites is summarized in Tab. 2 and Tab. 3. The first four components explain 70.95% of all of the water quality factors. Satu, DO, Alk and COD_{Mn} contributed the most to the first component; Cond, Sali and W. hard were the major contributors to the second component; and SS was the biggest contributing factor for the third component. Thus, Satu, DO, Alk, COD_{Mn}, Cond, Sali, W. hard and SS were the primary factors that could best represent all of the other factors. Pearson's correlation results (Tab. 3) indicated a positive correlation between Cond and Sali, between Cond and W. hard, between Sali and W. hard, and between Satu and DO. Cond and DO were critical to fish life and thus were retained as the primary environmental factors. Hydrological and geomorphological factors were similarly analyzed. In summary, the primary environment factors influencing the fish assemblages were DO, Cond, Alk, COD_{Mn}, SS, Altt, Flux and velocity.

According to the CCA analysis (Fig. 2), the fish assemblages were significantly influenced by altitude, DO, Alk, COD_{Mn} and velocity. Velocity was the most important factor impacting fish community structure, because it was important for fish spawn and growth.

Tab. 2   Factor analysis results of water quality factors

Component	Initial eigenvalues			Water quality factors	Component			
	Total	% of ariance	Cumulative%		1	2	3	4
1	*4.133*	29.525	*29.525*	W. tem	−0.100	−0.395	−0.003	*0.765*
2	*2.618*	18.703	*48.227*	pH	−0.178	−0.195	−0.041	−0.485
3	*2.066*	14.758	*62.985*	Cond	0.208	*0.915*	0.002	−0.002
4	*1.115*	7.962	*70.948*	Sali	0.132	*0.966*	0.006	0.017
5	0.985	7.036	77.983	Satu	*−0.763*	−0.008	0.115	−0.097
6	0.956	6.831	84.814	DO	*−0.830*	−0.038	0.071	−0.451
7	0.618	4.414	89.228	SS	−0.082	0.304	*0.828*	0.194
8	0.497	3.549	92.778	TDS	0.568	0.045	0.649	−0.014
9	0.392	2.802	95.580	TN	0.470	−0.038	0.380	0.074
10	0.246	1.754	97.334	TP	0.018	−0.209	0.649	−0.388
11	0.160	1.145	98.479	Alk	*0.790*	0.361	0.177	−0.178
12	0.141	1.007	99.486	W. hard	−0.066	*0.970*	0.055	0.042
13	0.057	0.404	99.890	COD_{Mn}	*0.781*	0.087	0.048	0.043
14	0.015	0.110	100.000	SiO_3	0.396	0.133	−0.429	0.489

Tab. 3   Pearson correlation coefficient of water quality factors

	Cond	Sali	Satu	DO	SS	Alk	W. hard	COD_{Mn}
Cond	1							
Sali	*0.929*	1						
Satu	−0.272	−0.096	1					
DO	−0.200	−0.190	*0.722*	1				
SS	0.223	0.260	0.121	0.047	1			
Alk	0.450	0.417	−0.427	−0.532	0.187	1		
W. hard	*0.857*	*0.920*	0.050	−0.017	0.333	0.292	1	
COD_{Mn}	0.238	0.200	−0.436	−0.613	0.083	0.641	0.031	1

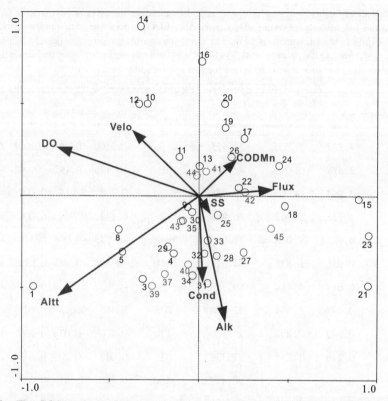

Fig. 2 The CCA analysis of the primary water environmental factors and sample sites

3.2 Calculation of IBI

TWINSPAN was applied to cluster the sample sites based on the five primary water quality factors, and the results indicated that sample sites were fall into one cluster, which were low conductivity, alkalinity, suspended solid, COD_{Mn}, and high DO implying a good water environment. QHEI, reflecting hydrology and geomorphology situation, was utilized to identify the sites with less human disturbance. The sample sites, W1, W5, W10, W12, W13, W16, were chose as the reference sites; these sites are mostly located in the upstream tributaries in remote, thickly forested mountains with less human disturbance.

Metrics that are sensitive to environmental changes were measured to assess the biological integrity of the river. First, the percentage of zero values across all sites was calculated for each metric to determine which metrics would be used to distinguish reference sites and impaired sites. For metrics P4, P18 and P19, the percentages of zero values were greater than 90%, causing them to be eliminated from the candidate metrics. Box-and-whisker plots were created for other twenty-one metrics to determine the amount of overlap of the interquartile range. Nine metrics (Fig. 3) with no or little overlap were selected, then tested to reveal the existing differences between reference sites and impaired sites using the Mann-Whitney U-test. Pearson's correlation results in Tab. 4 showed that there was an obvious correlation between P1 and P5 (| R | >0.8), which meant that only one of those two metrics was necessary to represent the condition of biotic integrity. P1, which was widely used in IBI metrics, was retained as the metric of IBI in this study. According to the above analysis, eight metrics were selected as the primary index of biotic integrity: P1, P2, P8, P14,

P16, P20, P21 and P24. Metrics scoring criterion was present in Tab. 5, and the integrity classes (Tab. 6) were categorized as five groups, i. e., excellent, good, fair, poor and very poor.

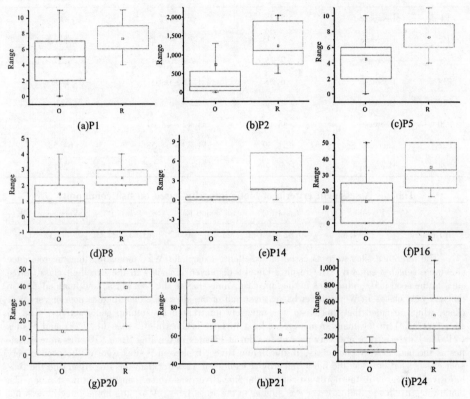

R: the reference sites   O: the impaired sites

Fig. 3   Box – and – Whisker plots of nine candidate metrics sensitive to environmental changes

Tab. 4   The correlation coefficient between nine candidate metrics

	P1	P2	P5	P8	P14	P16	P20	P21	P24
P1	1								
P2	0.224	1							
P5	0.983	0.222	1						
P8	0.370	−0.117	0.443	1					
P14	0.396	−0.024	0.332	0.029	1				
P16	0.541	0.167	0.533	0.043	0.279	1			
P20	0.586	−0.021	0.610	0.248	0.324	0.641	1		
P21	0.300	0.114	0.303	0.352	0.007	0.107	−0.111	1	
P24	0.441	0.176	0.474	0.230	−0.024	0.426	0.465	0.023	1

Tab. 5　Metrics scoring criterion for the stepped scoring method

	Response to stress	5	3	1
P1	Decrease	>8.75	6～8.75	<6
P2	Decrease	>1,705	837.5～1705	<837.5
P8	Decrease	>2.75	2～2.75	<2
P14	Decrease	>6.03	1.58～6.03	<1.58
P16	Decrease	>46.88	25～46.88	<25
P20	Decrease	>48.61	35.42～48.61	<35.42
P21	Increase	<51.39	51.39～64.58	>64.58
P24	Decrease	>560.75	291～560.75	<291

Tab. 6　Assessment criteria of biotic integrity based on fish community

	Excellent (E)	Good (G)	Fair (F)	Poor (P)	Very poor (VP)
IBI criteria	>54.42	40.82～54.42	27.21～40.82	13.61～27.21	<13.61

The reference sites were assessed as excellent, except for W5, indicating that the reference sites were selected suitably. The result (Fig. 4) displayed ten sites in the excellent class, three sites in the good class, nine sites in the fair class, nine sites in the poor class, and fourteen sites in the very poor class. In Weihe River basin, over half of the sites were classified as poor or very poor class, which revealed that this basin was intensely interrupted by anthropogenic activity. The 16 sites (69.6%) in the main stem were at least in fair class, while 2 sites (8.7%) and 5 sites (21.7%) were in poor and very poor class. But at 13 sites in the Jing River, 10 sites were classifies as the poor or very poor class, in the Beiluo River, 6 sites in 9 sites were in the poor or very poor class. That is to say the biotic integrity of main stem is better than the Jing River and the Beiluo River. Integrity in the northeast zone of the Jing River was worse than the southwest zone. The biotic integrity in the Beiluo River was similar to the Jing River. Also, the biotic integrity was not good, but it showed the east region of the Beiluo River showed worse integrity.

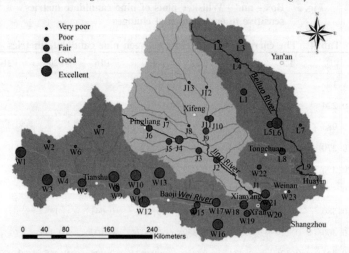

Fig. 4　Biotic integrity classes of the stepped scoring method in Weihe River

4 Discussions

According to Fig. 4, more than half of the sample sites were classified as poor or very poor, indicating that the biological integrity of Weihe River was damaged. According to this investigation, 37 fish species were sampled in the main stem, 14 fishes in the Jing River, and 20 species in the northern Luohe River. Compared with last fish sample (Xu & Li, 1984), fish fauna was similar, whereas fish species decreased a lot. There was worse fish biotic integrity in the northern region of Weihe River basin for their location in the loess plateau. Large cities such as Tianshui, Baoji, Xi'an and Weinan caused significant disturbances in the river, such as domestic sewage from the cities, industrial wastewater and surface runoff from agriculture. These factors negatively influenced the structure and composition of fish assemblages in streams passing through cities and adversely affected the habitats that fish communities live in, which was analogous to observations made by Qadir and Malik (2009). To control floods in the river, as well as irrigation for crops in the non – flood season, many dams and reservoirs were constructed in Weihe River Basin, including Fengjiashan Reservoir, Majiazui Reservoir and Baojixia Irrigation Project, which altered the flow regime in the river, impacting fish spawning and decreasing fish abundance. Studies by Marchetti and Moyle (2001) demonstrated that hydrologic variability between years and seasons indicated that flow regime had a large effect on the fish assemblages.

The coefficient of similarity (Zhang, 1998) was an indicator describing the similarity of two regions based on the species. The coefficient of similarity of the Jing and northern Luohe River was 82.35%, indicating that fish community assemblages between these two basins were similar. In contrast, there were not high similarities between Weihe River and the Jing and northern Luohe Rivers (54.9%, 63.15%). In summary, the integrity of the whole Weihe River was not good because of severe human disturbance.

The assessment was conducted based on only one fish sample, conducted in October, 2011, when there was a high water velocity and flow in the river and better water quality compared with non – flood season. Some fishes in the upstream reaches migrated downstream along with the flood water released from the dam. Therefore, extra fish may have been sampled, and the data may thus reflect this source of error. Additionally, an assessment of biological integrity based only on the fish community is incomplete. Due to differential responses observed between fish and invertebrates, Siligato and Böhmer (2001) recommended that an accurate and comprehensive ecological assessment of streams should be based on both fish and invertebrate assemblages. However, according to Pyron (2008), a single sample per year is adequate for characterizing the fish assemblages of a large river such as the Wabash River. Therefore, a single sample period could reflect basic fish assemblage structure, indicating the biological integrity of Weihe River. Based on an independent t – test, there was a remarkable difference in biological integrity between the reference sites and impaired sites in the present work ($P < 0.05$). This indicates that the IBI method was suitable for evaluating the biotic integrity of Weihe River.

A correlation analysis was performed on the final scores for the IBI and the primary environmental factors. The IBI (Tab. 7) were positively correlated with DO and QHEI, and negatively correlated with Alk and Cond. Environmental factors such as Cond, DO, Alk, the primary chemical factors defined before, were correlation with IBI scores, while SS and COD seemed no clearly relationship with IBI scores. QHEI was the rapid assessment metric for habitat quality, which reflects the degree of disturbance due to anthropogenic activity in the basin. The correlation between the QHEI and IBI scores meant that the IBI method could describe the biotic integrity of Weihe River to some extent. Streams with high physical habitat quality possess more fish species than streams with low physical habitat quality. Rathert et al. (1999) applied regression tree analysis and multiple linear regression analysis, and found that the species richness of native freshwater fish is positively correlated with low annual temperature extremes, warm summer air temperatures, low spatial variability in temperatures, high stream density, historical and present connectivity to great river basins, and high introduced fish species richness.

Tab. 7 Correlation between the IBI scores and the primary environmental factors

	IBI	Altt	DO	Alk	Velo	COD	Flux	SS	Cond	QHEI
IBI	1									
Altt	0.045	1								
DO	*0.442* **	0.196	1							
Alk	*−0.339* *	*0.447* **	*−0.532* **	1						
Velo	0.150	−0.024	*0.311* *	−0.261	1					
COD	−0.254	0.201	*−0.611* **	*0.641* **	−0.266	1				
Flux	0.077	*−0.336* *	−0.135	−0.150	0.247	−0.119	1			
SS	−0.169	0.157	0.047	0.187	0.088	0.083	*0.520* **	1		
Cond	*−0.427* **	0.483 **	−0.196	*0.450* **	−0.218	0.238	−0.232	0.223	1	
QHEI	*0.365* *	0.007	*0.374* *	−0.191	*0.382* **	−0.153	0.164	−0.044	*−0.427* **	1

Note: * *. Correlation is significant at the 0.01 level.
   *. Correlation is significant at the 0.05 level.

5 Conclusions

The initial attempt of fish index of biotic integrity in Weihe River Basin has shown that:
(1) Biological integrity in the main stem was better than the Jing River and northern Luohe River, as a whole, Weihe River was interrupted intensely by human activities like dam construction, industrial wastewater without treatment and urbanization, which had negative impact on fish assemblage and diversity.
(2) The CCA analysis showed that Altitude, DO, Alk. , COD_{Mn} and velocity were the primary environmental factors, which were more harmful to fish history.
(3) This research demonstrated that IBI was an effective and important tool for biological integrity assessment. This research provides an important reference and scientific basis for the management of water resources and the restoration of ecosystem in Weihe River Basin.

Acknowledgements
This study is supported by the Sino – Swiss Science and Technology Cooperation Project (2009DFA22980), the Ministry of Science and Technology, P. R. China, and the Major Special S&T Project on Water Pollution Control and Management (2009ZX07012 – 002 – 003), the Ministry of Environmental Projection, P. R. China. Meanwhile, we thanked colleagues at the Dalian Ocean University for the help in data analysis and the identification of samples. The authors also appreciated the sample team for their assistance with the field work.

References

Baptista D F, Buss D F, Egler M, et al. A Multimetric Index based on Benthic Macroinvertebrates for Evaluation of Atlantic Forest Streams at Rio de Janeiro State, Brazil[J]. Hydrobiologia, 2007,575(1): 83 –94.

Barbour M T, Gerritsen J, B. Snyder & J. Stribling. Rapid Bioassessment Protocols for Use in Streams and Wadeable Rivers[J]. USEPA, Washington,1999.

Bozzetti, M. & U. H. Schulz. An Index of Biotic Integrity based on Fish Assemblages for Subtropical Streams in Southern Brazil[J]. Hydrobiologia 529(1): 133 – 144.

Breine J, Simoens I, Goethals P, et al. Belpaire. A Fish – based Index of Biotic Integrity for Up-
    stream Brooks in Flanders (Belgium). Hydrobiologia, 2004, 522(1): 133 – 148.

Casatti L, Ferreira C P, Langeani F. A Fish – based Biotic Integrity Index for Assessment of Low-
    land Streams in Southeastern Brazil[J]. Hydrobiologia, 2009, 623(1): 173 – 189.

Dauwalter D C, Jackson J R. A Provisional Fish Index of Biotic Integrity for Assessing Ouachita
    Mountains Streams in Arkansas, USA[J]. Environmental Monitoring and Assessment, 2004,
    91(1): 27 – 57.

Hill B, Herlihy A, Kaufmann P, et al. Use of Periphyton Assemblage Data as an Index of Biotic
    Integrity[J]. Journal of the North American Benthological Society, 2000, 19(1): 50 – 67.

Hill M O. TWINSPAN: A FORTRAN Program for Arranging Multivariate Data in an Ordered Two
    – way Table by Classification of Individual and Attributes[M]. Cornell University Press,
    1979.

Joy M K, Death R G. Application of the Index of Biotic Integrity Methodology to New Zealand
    Freshwater Fish Communities[J]. Environmental Management, 2004,34(3): 415 – 428.

Karr J R. Assessment of Biotic Integrity Using Fish Communities[J]. Fisheries, 1981,6(6): 21
    – 27.

Karr J R. Biological Integrity: a Long – neglected Aspect of Water Resource Management[J]. Ec-
    ological Applications, 1991,1(1): 66 – 84.

Klemm D J, Blocksom K A, Fulk F A, et al. Development and Evaluation of a Macroinvertebrate
    Biotic Integrity Index (MBII) for Regionally Assessing Mid – Atlantic Highlands Streams[J].
    Environmental Management, 2003,31(5): 656 – 669.

Kleynhans C. The Development of a Fish Index to Assess the Biological Integrity of South African
    Rivers[J]. WATER SA – PRETORIA – 25, 1999:265 – 278.

Liu M, Chen D, Duan X, et al. Assessment of Ecosystem Health of Upper and Middle Yangtze
    River Using Fish – index of Biotic Integrity[J]. Journal of Yangtze River Scientific Research
    Institute, 2010,27(2): 1 – 7.

Marchetti M P, Moyle P B. Effects of Flow Regime on Fish Assemblages in a Regulated California
    Stream[J]. Ecological Applications, 2001,11(2): 530 – 539.

Oberdorff T, Hughes R M. Modification of an Index of Biotic Integrity Based on Fish Assemblages
    to Characterize Rivers of the Seine Basin, France[J]. Hydrobiologia, 1992,228(2): 117 –
    130.

Pei X, Niu C, Gao X, et al. The Ecological Health Assessment of Liao River Basin, China, Based
    on Biotic Integrity Index of Fish[J]. Shengtai Xuebao/Acta Ecologica Sinica, 2010,30(21):
    5736 – 5746.

Pyron M, Lauer T E, LeBlanc D, et al. Temporal and Spatial Variation in an Index of Biological
    Integrity for the Middle Wabash River, Indiana[J]. Hydrobiologia, 2008,600(1): 205 –
    214.

Qadir A, Malik R N. Assessment of an Index of Biological Integrity (IBI) to Quantify the Quality
    of Two Tributaries of River Chenab, Sialkot, Pakistan. Hydrobiologia, 2009,621(1): 127 –
    153.

Rathert D, White, D, Sifneos J, et al. Environmental Correlates of Species Richness for Native
    Freshwater Fish in Oregon, USA[J]. Journal of Biogeography, 1999,26(2): 257 – 273.

Reynoldson T, Norris R, Resh V, et al. The Reference Condition: a Comparison of Multimetric
    and Multivariate Approaches to Assess Water – quality Impairment Using Benthic Macroinver-
    tebrates[J]. Journal of the North American Benthological Society, 1997:833 – 852.

Schleiger S L. Use of an Index of Biotic Integrity to Detect Effects of Land Uses on Stream Fish
    Communities in West – central Georgia[J]. Transactions of the American Fisheries Society,
    2009,129(5): 1118 – 1133.

Siligato S, Hmer J B. Using Indicators of Fish Health at Multiple Levels of Biological Organization
    to Assess Effects of Stream Pollution in Southwest Germany[J]. Journal of Aquatic Ecosystem
    Stress and Recovery (Formerly Journal of Aquatic Ecosystem Health), 2001,8(3): 371 –
    386.

Simon T P, Lyons J. Application of the Index of Biotic Integrity to Evaluate Water Resource Integrity in Freshwater Ecosystems [C] //. Biological assessment and criteria: tools for water resource planning and decision making. Boca Raton, Florida, Lewis Publishers, 1995: 245 – 262.

Song Z, Wang W, Jiang Z, et al. An Assessment of Ecosystem Health in Taizi River Basin Using F – IBI[J]. Journal of Dalian Ocean University, 2010,25(6): 480 –487.

Tejerina – Garro F L, de Mérona B, Oberdorff T. A Fish – based Index of Large River Quality for French Guiana (South America): Method and Preliminary Results[J]. Aquatic Living Resources, 2006,19(01): 31 –46.

Wang B, Yang L, Benjin H. A Preliminary Study on the Assessment of Stream Ecosystem Health in South of Anhui Province Using Benthic – Index of Biotic Integrtiy[J]. Acta Ecologica Sinica, 2005,25(6): 1481 –1490.

Weisberg S B, Ranasinghe J A, Dauer D M, et al. An Estuarine Benthic Index of Biotic Integrity (B – IBI) for Chesapeake Bay[J]. Estuaries and Coasts, 1997,20(1): 149 –158.

Xu T, Li Z. Studies on Fishes Fauna of the Weihe River[J]. Journal of Xinxiang Normal College, 1984(4): 73 –78.

Zhang Y. Coefficient of Similarity – An Important Parameter in Floristic Geography[J]. Geographical Research, 1998,17(4): 429 –434.

Zhang Y, Xu C, Ma X, et al. Biotic Integrity Index and Criteria of Benthic Organizms in Liao River Basin[J]. Acta Scientiae Circumstantiae, 2007,27(6): 919 –927.

Zhu D, Chang J. Annual Variations of Biotic Integrity in the Upper Yangtze River Using an Adapted Index of Biotic Integrity (IBI)[J]. Ecological Indicators, 2008,8(5): 564 –572.

Environmental Impact Study on Comprehensive Planning along Heihe Watershed

Du Yuan[1] ,*Dong Hongxia*[1] ,*Yu Jun*[2] ,*Gao Xuejun*[1] and *Dang Yonghong*[1]

1. Yellow River Engineering Consulting Co. , Ltd. ,Zhengzhou, 450003, China
2. Henan Water and Power Engineering Consulting Co. , Ltd. ,Zhengzhou, 450008, China

Abstract: Located in the northwest inland area of China, the Heihe River Basin is a part of resource water – deficient area where the water resource is hard to meet the need of local economic development and ecological balance. The comprehensive planning for the Heihe River Basin aims to achieve the win – win result of ecological environment protection and economic and social development by basing on ecological construction and environment protection, centering on scientific management, appropriate allocation, efficient utilization and effective protection of water resources and comprehensively considering the requirement for ecologic environment construction and regional economic and social development over a long period of time. In view of the influence of the implementation of the comprehensive planning for the Heihe River basin on the water resource allocation, water environment, ecological environment, land resources, social environment, and so on, the analogy analysis method, expert consultation method and so on are utilized to conduct the influence analysis of various environmental factors involved with the engineering measures in the comprehensive planning for the basin. From the point of view of environmental protection, the engineering layout, scale, method, environmental feasibility of development and rationality of planned environmental goal are demonstrated and analyzed and from the point of view of ecological environment protection, the suggestions on optimization and adjustment are put forward so as to provide the basis of further making a decision on the comprehensive planning for the Heihe River basin.

Key words: Heihe River Comprehensive Planning for Basin, environmental impact

1 General

The Heihe River, the second largest inland river of China, originates from the middle of the Qilian Mountain and flows through a total of 14 counties (city, district and Qi) of Qinghai, Gansu, and Inner Mongolia. The Heihe River basin stretches from the Dahuangshan Mountain in the territory of Shandan County in the east, borders on the Shiyang River Basin in the east, is Bounded in the west by the Heishan Mountain in the territory of Jiayuguan, neighbors the Shule River basin in the west, and reaches the Sino – Mongolian Boundary in the north, with a total land area of 143,000 km[2]. Due to its deeply landlocked location, dry climate, a small quantity of precipitation and a great quantity of evaporation, the Heihe River basin lacks in water resources, severe in ecological environment, imbalanced in water resource utilization, thus causing water for ecological use to be squeezed to occupy by water for economical use ceaselessly.

The purpose of the comprehensive planning for the Heihe River basin is to resolve the aggravated and outstanding water affairs conflict of increasingly severe ecological environment system of the Heihe River basin, to further strengthen the ecological environment protection of the Heihe River basins, to promote the sustainable socio – economic development in the basin. The comprehensive planning for the Heihe River basin includes: water conservation planning, water resource development planning, ecosystem rehabilitation, soil and water conservation planning, cascade planning, flood control and river regulation planning, water resource protection planning, comprehensive management planning for basin. The comprehensive planning for the Heihe River basin is detailed as follows: agricultural water conservation, industrial water conservation, urban life water conservation; water resource allocation, urban water supply safety planning, rural portable water safety planning; ecological rehabilitation project, ecological migration, fodder grass base construction, water and soil conservation project; regulation and storage project planning, cascade hydropower station development planning; revetment, embankment, channel dredging, reservoir (gate) danger elimination and reinforcement; surface water resource protection, underground resource protection; perfection of system and mechanism, establishment and perfection of legal systems, establishment of water quantity dispatching management information system, reinforcement of management capability, etc.

2 Environmental evaluation procedures and methods of comprehensive planning for the Heihe River Basin

2.1 Evaluation procedures

In accordance with the Regulations for Environmental Impact Evaluation of River Basin Planning, this environmental impact evaluation workflow is illustrated in Fig. 1.

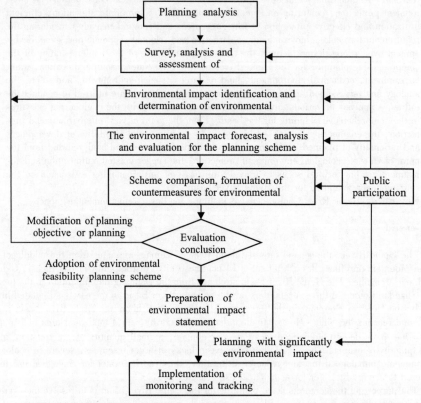

Fig. 1 Environmental evaluation procedures of comprehensive planning for the Heihe River Basin

2.2 Evaluation method

According to the characteristics of the planning scheme, this environmental impact evaluation is made using the following methods, as outlined in Tab. 1.

Tab. 1　Method for environmental impact evaluation of Heihe River Basin planning

Evaluation links	Planning evaluation method
Planning analysis	Expert consultation method
Existing environmental survey analysis	Data collection method, remote sensing interpretation method and field investigation method
Environmental impact identification and evaluating indicator	Listing method and analogy analysis method
Environmental impact forecast, analysis and evaluation	Statistical analysis method, analogy analysis method and expert and public consultation method
Scheme comparison and countermeasures on environmental protection	Comprehensive analysis method and analogy analysis method
Public participation	Actual survey and statistical analysis method

3　Environmental impact identification and evaluating indicator of comprehensive planning for the Heihe River Basin

According to the engineering layout, scale, implementation sequence, existing environment, environmental problems of the Heihe River Basin, on the basis of the environmental identification result and the environmental protection objective, the analysis and determination has been made for the environmental impact assessment system and the indicator system which consists of water resources, aquatic environment, ecological environment, land resources and social environment. The concrete structure is outlined in Tab. 2.

Tab. 2　Environmental protection objective and evaluating indicator of comprehensive planning for the Heihe River Basin

Theme of environment	Environmental protection objective	Evaluating indicator
Water resources	1. Protect the surface water resources of river basins and promote the sustainable utilization of water resources; 2. Protect groundwater resources, and keep balance of supply and drainage of groundwater resources; 3. Improve the utilization efficiency of water resources and overall development of service basins; 4. Help relieve the conflicts of water for production, life and ecology to ensure the sufficient ecological water consumption in the middle and lower reaches; 5. See that a certain quantity of water will flow into the eastern Juyanhai Lake	1. Water resources availability of basins ($\times 10^8$ m^3) 2. Surface water resource development and availability (%) 3. Groundwater resource exploitation ratio (%) 4. Water deficient ratio (%) 5. Probability of water for production, life and ecology 6. Water use efficiency (%) 7. Quantity of water flowing into the lower reaches ($\times 10^8$ m^3)
Water environment	1. Maintain and protect the environment functions of rivers (lakes and reservoirs) and ground water; 2. Restore and improve the status of low – temperature water for project; 3. Realize the goal of discharging waste water up to standard	1. Water quality compliance rate in water function areas (%) 2. Status of restoration of drained low – temperature water 3. Waste water drainage rate up to standard

Continued Tab. 2

Theme of environment	Environmental protection objective	Evaluating indicator
Ecological environment	1. Protect the watershed ecosystem function and keep the balance of ecology; 2. Protect the biodiversity of basins; 3. Protect ecological sensitivity and ecologic environment of fragile zones; 4. Enable the ecological environment of river basins to be restored to the level in the 1980s; 5. Prevent water and soil erosion of basin	1. Area of woodland and grass-land (km^2) 2. Area of restoration area of forest and grassland ($\times 10^4$ mu) 3. Area of returning land for farming to forestry and grassland ($\times 10^4$ mu) 4. Ecologic water requirements of Ejina Oasis on the lower reaches of the river ($\times 10^8$ m^3) 5. Scale of oasis on the lower reaches of the river (km^2) 6. Quantity of water flowing into the eastern Juyanhai Lake ($\times 10^8$ m^3) 7. Construction of soil and water conservation forest ($\times 10^4$ mu)
Land resources	1. Conduct rational exploitation and protection of land resources; 2. Prevent desertification of lands and deterioration of lands	1. Harnessing area of saline – alkali soils ($\times 10^4$ mu) 2. Restoration of the ecosystem (km^2)
Social environment	1. Promote the sustainable socio – economic development of basins; 2. Guarantee flood control safety; 3. Guarantee water supply safety; 4. Improve the urban, domestic and agricultural water supply conditions; 5. Improve inhabited environment of people; 6. Promote the water – saving social construction; 7. Protect environmental sensitive spots	1. Flood control standard 2. Installed capacity and annual generating capacity ($\times 10^4$ kW and $\times 10^8$ kW $\cdot$ h) 3. Irrigated area ($\times 10^4$ mu) 4. Water-supply quantity $\times 10^8$ m^3 and reliability (%) 5. Number of ecological migration ($\times 10^8$ persons)

4   Environmental impact forecast and evaluation of comprehensive planning for the Heihe River Basin

Firstly, carry out strength analysis of the planning development in combination with the planning scheme, then predict and evaluate the impact of relevant environmental elements from the layout, scale and development order of planning scheme in combination with planning type and characteristics.

4. 1   Impact on hydrology and water resources

The planned development of cascade hydropower stations and the regulation storage reservoir will level the runoff process of the downstream reaches, and have a certain influence on the hydrological regime, that is, reduce the drainage water quantity during high-water period, and increase the drainage water quantity during low-water period. Also, the water stretching, water depth increasing and velocity slowing will result in the obvious change in fluvial morphology, longitudinal continuity and lateral continuity.

Allocation of water resources will ensure that downstream ecological water demand will accelerate the water-saving transformation to improve the water supply guarantee rate, promote the adjustment of industrial structure and agricultural planting structure.

4. 2　Impact on Water Environment

4. 2. 1　Impact on water quality

After implementation of the planned measures on control scheme of total pollutants discharged into rivers and water pollution prevention, it is predicated that the urban sewage treatment rate will be up to 80% in 2020, 90% in 2030; the compliance of the water function area in the sub water system will be annually increased to 100% from existing 78. 6%, and the surface water quality will be improved on an overall basis. After the implementation of the water source protection project and the groundwater protection project so planned, the water quality of the water head area will be effectively protected; the groundwater pollution problems in local areas can be solved effectively; and the safety of water for living of residents along the riverbanks and that of water for industrial and agricultural production will be ensured.

4. 2. 2　Impact on water temperature

The construction of the hydropower stations and regulating reservoirs may stratify water temperature of the reservoirs. The drainage of low-temperature water of the reservoirs may have an adverse impact on the growing, fattening and wintering of downstream aquatics and fish in particular. The said impacts mainly find expression in the change in composition of fish fauna, the delay in fish breeding season, slow of fish egg incubation and fish growth, etc. Besides, the low-temperature may have an impact on the irrigation of agricultural lands to a certain extent.

4. 2. 3　Groundwater level impact analysis

The implementation of water-saving modification work in middle-reach irrigation area will reduce the supplement groundwater from water leakage in farmland, which might affect the groundwater water level. Therefore, the scheme of water diversion shall be strictly implemented and the super-standard exploitation of groundwater shall be strictly forbidden.

4. 3　Ecological environmental impact

4. 3. 1　Terrestrially ecological impact analysis

The reservoir filling and inundating involved in the development of cascade hydropower stations and the key regulation storage project may cause the permanent losses to the vegetations in the reservoir area and reduce the vegetation area and habitats of terrestrial animals in the area. The reservoir regulating operation may change the hydrological regime of the downstream reach of the dam site, cause a certain reducing reaches and have an impact on the growth of vegetations on both banks. With water storage of reservoirs, improvement of microclimate of reservoir area and rise of groundwater level around reservoirs, the provision of more sufficient water supply for the vegetations around reservoirs and the increase of coverage of vegetations around reservoirs can increase the plant diversity and animal diversity. And the reservoirs will not have an obviously adverse impact on the composition of terrestrial animal fauna, population structure, magnitude of stocks and rare species. The ecological project and water-and-soil conservation project can improve the ecological environment, inhibit the trend of ecological environment deterioration, obviously increase the species and coverage of vegetations in the area, and significantly improve the vegetation productivity.

4. 3. 2　Hydrophytically ecological impact analysis

The development of the cascade hydropower stations and the key regulation storage project may obstruct interflow of the river ecosystems, make the river ecosystem to be divided into discontinuous environmental units, and have an influence on survivals and propagation of semi-migratory fish. Worse still, the fish habitat fragmentation may result in the metapopulation of different sizes, the

gene exchange and genetic diversity of populations may be affected to a different extent, and survival forces of species. The impound water area and slack water area formed by reservoir filling will be
beneficial to the population growth of widespread fish, but narrow the agreeable inhabits for the
growth and propagation of the special fish adapting to the rapids environment.

4.3.3 Downstream wetland impact analysis

After the implementation of the planned project, the quantity of water flowing into the eastern
Juyanhai Lake can reach 0.5×10^8 m^3 when the quantity of water flowing from the Zhengyixia Reservoir is 9.5×10^8 m^3, the 0.1×10^8 m^3 water can be drained down to the Weilu Beihaizi. In the
meanwhile, the artificially complementary planting and seeding as well as the fenced enclosures will
be arranged such that the groundwater level can rise and the wetland rehabilitation can be promoted
significantly.

4.4 Impact on land resources

The flood-control project, regulating reservoirs and cascade hydropower station construction,
reclamation of saline-alkali soils, ecological construction, soil and water conservation project, etc.
will have an impact on the land uses.

The flood-control project, regulating reservoirs, and cascade hydropower station project land
and reservoir inundation will reduce the grasslands, woodland and cultivated areas, and have an
adverse impact on land resources. The reclamation of saline-alkali soils in the irrigation areas will
be able to change the low-yield field into the high-yield field, make for the agricultural production
and have an advantageous influence on the land resources. The planting and seeding as planned in
the ecological construction and water-and-soil conservation project can increase the coverage rate of
woodlands and grasslands in the area, improve the ecological environment and have a helpful influence on the land resources.

4.5 Impact on the social environment

4.5.1 Impact on affected residents

The flood-control project, cascade hydropower project, regulating reservoir project, and etc.
so planned involve the issues of inundation and resettlement of affected residents. The occupancy of
cultivated land resources may have an adverse impact on the local land use and agricultural productions. In case of improper resettlement of affected residents, the relatively serious social problems
may arise.

4.5.2 Impact on social environmental of the project area

The planning implementation will ensure water utilization in the irrigation area of the basin;
enhance the irrigation water utilization efficiency in the irrigation area; promote the sustainable development of the economy in the project area, improve water shortage of the basin; gradually establish the integrated system of development, utilization, allocation, economization and protection of
water resources in the middle-west part of the Heihe River; enhance the water resources utilization
efficiency; realize the appropriate allocation and efficient utilization of water resources; guarantee
the flood-control safety of the basin, avoid the property losses and ecological disasters caused by
floods and realize the moderate and efficient socio-economic development of the basin.

4.5.3 Impact on environmentally sensitive spots

(1) Impact on the national nature reserve—Qilian Mountain, Gansu.

The construction and operation of the Huangzangsi Reservoir, Taolaixia Reservoir, Shancheng-
he Reservoir and Shihuiyao Reservoir in Minle County, Daciyao Reservoir and Suyoukou Reservoir
in Ganzhou District, Baishiya Shimen Reservoir in Shandan County, Fenglehe Second-cascade Hydropower Station planned in the middle-west sub-water system, and Guanshan River Hydropower
Station will have an adverse impact on the nature reserve. The construction and inundation will, to

a different extent, cause the losses to the vegetations of the reserves. After reservoir filing, the meadow development in the marshlands and wetlands around the reservoir will increase the botanical diversity and the animal diversity.

(2) Impact on the national nature reserve—Heihe Wetland in Zhangye, Gansu. The fenced enclosure is planned for the national nature reserve-Heihe Wetland in Zhangye, Gansu with an area of 615,000 m, which can reduce the man-made interferences and influences and help to restore the ecological environment of the wetland. The flood-control project on the middle reaches of the Heihe River is in the experiment area of the Heihe Wetland Reserve in Zhangye so its implementation will have no influence on the water supply of the wetland. The flood-control project is of linear distribution so the project area will have a slight influence on the reserve. However, the construction may disturb protected animals. Huangzangsi Reservoir replaces plain reservoirs in the middle reaches, of which 8 reservoirs involve the Heihe Wetland Reserve in Zhangye. Due to the abandonment of plain reservoirs, the water surface area of the Heihe Wetland Reserve has been reduced by 6.37 km^2. About 88% of the inundated area of the Zhengyixia Reservoir is located in the Heihe Wetland Reserve, and the construction and operation of the reservoir will have an adverse impact on the nature reserve.

(3) Impact on the national nature reserve – Huyanglin Scenic Spot at Ejina Qi, Inner Mongolia.

No project is planned in the reserve. The establishment of the windbreak forests outside the Huyanglin Reserve and the implementation of the planning will increase of the quantity of water flowing into the lower reaches of the river and will of positive influence to the reserve.

(4) Impact on geological parks, forest parks and scenic spots.

The implementation of planning can guarantee that plenty of water will flow into the Huyanglin Scenic Spot and the eastern Juyanhai Scenic Spot in the Alxa Desert Global Geopark China, good to ecological restoration. According to the planning, the shrubberies such as red willows, Haloxylon ammodendrons and Russian olives and herbal plants such as fenugreeks, which cover an area of 10.8 km^2, will be replanted around the eastern Juyanhai Lake, and will increase the area of the woodlands and grasslands. Fenced enclosure will be planned for the eastern Juyanhai Lake and the woodlands and grasslands around, which can support the self-restorability of natural vegetations, invigorate and naturally regenerate the vegetations around the Juyanhai Lake, of great benefit to the Juyanhai Scenic Spot in the Alxa Desert Global Geopark China.

The planning is not made in the Huyanglin National Forest Park in Ejina, Alashan Meng, Inner Mongolia, and the Huyanglin Scenic Spot in Ejina. However, thanks to the location of the Huyanglin National Forest Park on the lower reach of the Heihe River, the implementation of the planning scheme can ensure that sufficient water will flow into the Huyanglin National Forest Park and the Huyanglin Scenic Spot, beneficial to the ecological environment protection.

(5) Impact on cultural relic.

According to the preliminary investigation, there are 7 cultural relics in the inundated area of the Zhengyixia Reservoir, including 5 historical and cultural relics under state protection (the relics of Tiancheng Great Wall in Ming Dynasty, the relics of Shanzuidun Beacon Tower, the relics of Huowang Temple, the relics of Xiaohuangdun Beacon Tower in Han Dynasty, and the relics of the Great Wall in Houzhuang) and 2 historical and cultural relics under county protection (the relics of Xiangshan Temple and the relics of Tiancheng Castle). During the project implementation, the competent department in charge of cultural relics should bring forward the concrete opinions on protection.

5　Heihe River Basin comrrehensive planning scheme rationality demonstration

The environment rationality demonstration of planning scheme mainly aims to analyze the environment rationality of goal and development orientation, the environment rationality of development scale and planning layout, the accessibility of environmental protection goal and evaluation index, the sustainable development influence of river basins, the sustainable development influence of river basin economy, as well as the comprehensive influence on coordinated and sustainable development strategy of economy, society and ecological environment.

5. 1 Planning layout rationality analysis

(1) Huangzangsi Reservoir can effectively regulate the water resource, supply the water to midstream and downstream in due time, substitute for the 19 plain reservoirs in midstream irrigation area, reduce the evaporation and leakage loss, improve the water diversion condition in midstream irrigation area and promote the water-saving reformation in irrigation area. Seen from the water diversion in midstream and downstream irrigation area and the water regulation of Heihe River, the construction of Huangzangsi Reservoir will play an important role in the aspects of water resource development, rational arrangement and regulation management of Heihe River. Located in the boundary between midstream and downstream of Heihe River, Zhengyixia Reservoir can effectively regulate the water resource, supply the water to midstream and downstream in reasonable way, improve the water delivery efficiency, substitute for the 10 plain reservoirs in innovation irrigation area, reduce the evaporation and leakage loss in plain reservoir, and effectively save the water resource. The joint regulation of Huangzangsi Reservoir and Zhengyixia Reservoir in upstream and midstream can realize the unified management of water resource of Heihe River. The water discharge from upstream reservoir can rationally distribute the water yield to users under Zhengyixia Valley through the regulation of Zhengyixia Reservoir, so as to ensure the ecological water supply in Erjina Oasis. Therefore, the layout of Heihe main stream storage regulation project is very rational based on the Long-term benefit of whole river basin.

(2) At present, the main segments and channels of Heihe River show the disadvantages of low water delivery efficiency, wide and shallow stream channel, dispersed stream channel and large evaporation and leakage loss. According to the unified management and regulation requirement of water resource of Heihe River, it is necessary to take the engineering measure, improve the water delivery efficiency and promote the water regulation effect of river channel, increase the effective water yield of Erjina Oasis, and recover and rescue the ecological environment in downstream. Therefore, main segment water delivery project, stream channel regulation project and flood control project are urgent and rational.

(3) There are 31 plain reservoirs in the midstream and downstream innovation irrigation area of Heihe River, with gross available storage of $0.735,8 \times 10^8$ m^3. The evaporation and leakage loss of reservoir takes up $30\% \sim 40\%$ of gross storage capacity. There are 2,300 km main canal, 2,290 km branch canal and 5,910 km lateral canal in the midstream and downstream innovation irrigation area of Heihe River, with lining ratio of 45%, 40% and 20% respectively. The matching condition is not suitable for the water shortage degree in the irrigation area of river basin, and the utilization factor of irrigation water is only 0.52. Therefore, water-saving reformation project should focus on the water diversion entrance integration, canal system regulation, abandoned plain reservoir utilization, canal system lining, farmland matching construction and water-saving technology development, in order to effectively solve the various problems in irrigation area. The interaction of different projects can change the water resource waste in irrigation area fundamentally. This layout is rational.

(4) Relevant control measure shall be taken according to the current ecological environment in upstream, midstream and downstream, in order to recover the ecological environment to the level of 1980s. Upstream shall focus on the ecological restoration and soil control; midstream shall focus on the farmland shelter-forest system update, grain for green wetland restoration protection and oasis periphery protection; downstream shall focus on the oasis restoration and ecological irrigation to promote the ecological restoration. This specific layout can rationally and effectively solve the ecological deterioration in different areas.

(5) In Heihe River Basin, the water quality of upstream is good, which reaches the grade-Ⅱ requirement; the water quality of midstream is becoming poor due to the influence of industrial sewage, domestic sewage and agricultural waste water, which reaches the grade-Ⅲ requirement during flood season and the grade-Ⅳ requirement during non-flood season; the water quality of Mayinghe River is seriously polluted, which reaches the grade-Ⅴ requirement. For the development and utilization of water resource, we shall reduce the water resource waste and prevent the water pollution,

which are complementary to each other. Therefore, it is necessary to take the water resource protection measure, protect the portable water resource and build the sewage treatment plant. This layout is effective and rational.

5. 2　Ecological environment rationality analysis of planning scheme

The water-saving reformation project of irrigation area can reduce the ecological water demand in agriculture and improve the ecological situation in river basin. The unified water resource management and regulation project system can ensure the ecological water yield in downstream. The implementation of ecological restoration and soil control planning can increase the area and coverage of ecological vegetation in river basin, such as enclosed forestation, grain for green, grassland degeneration control, artificial irrigation and artificial afforestation. Meanwhile, such planning scheme has obvious effect on the adjustment of river basin industrial structure, the increase of ecological water proportion, the change of ecological concept and the local socio-economic progress. Storage regulation project and stepped power station construction will cause the negative influence to aquatic organism, so relevant unit is authorized to research on this issue and propose the effective relief measure during planning process, aiming to control the negative influence of aquatic organism within acceptable range. Through the effective water pollution control system, the industrial sewage discharge qualification rate in river basin and the urban domestic sewage discharge qualification rate are 100% and 80% respectively, which means the realization of water quality goal in water function area and surface water resource protection. This planning scheme is rational from the view of ecological environment.

5. 3　Socio-economic rationality analysis of planning scheme

The implementation of flood control planning can improve the flood control ability of river basin, protect the safety of human life and property and avoid the threat of flood to town, industry, traffic, production facility and domestic installation. It not only plays an important role in promoting the stable development of local economy and improving the life quality of resident, but also has significant effect in accelerating the national defense development.

The utilization factor of irrigation water in river basin will be increased to 0. 55 from 0. 52 at present; irrigation requirement will be reduced by 1.68×10^8 m^3; the repeated utilization factor of industrial water will be improved to 74. 9% from 52. 3% in base year; the water yield of every ten thousand yuan of industrial added value will be reduced to 50 m^3 from 112 m^3 in base year; the comprehensive leak rate of urban water supply network will be reduced to 10. 1% from 19. 2%; the water yield of every ten thousand yuan of GDP will be reduced to 127 m^3 from 354 m^3 in base year; the water-saving society with high water utilization efficiency is built initially. Water resource development will provide the clean energy for the local socio-economic development, which can save about 25.8×10^4 t coal and provide the clean energy for the socio-economic development in river basin. The construction of upstream and midstream main segment storage regulation project will solve the "bottleneck" drought caused by uncoordinated water supply and use process in river basin during May and June of each year fundamentally, ensure the water yield of irrigation area during irrigation peak time and keep the sustainable development of agriculture in irrigation area. The implementation of urban and rural drinking water safety planning can solve the rural drinking water safety problem and ensure the urban water supply safety. Therefore, this planning scheme is rational in the socio-economic view.

5. 4　Adjustment of planning scheme

The partial content of preliminary scheme refers to the core area and buffer area of nature reserve, which is contrary to the principle of prohibiting development area control in Nature reserve Regulation of People's Republic of China and Division of National Development Priority Zones. Additionally, project construction will cause the serious influence to the ecological environment in nature reserve. Therefore, this part of planning content is adjusted during planning compilation.

6 Heihe River Basin comprehensive planning strategy measure

The minimum preventive and remedy strategy and measure shall be taken according to negative influence, which can slow down the environmental influence. In planning scheme, requirement shall be proposed to important construction project in the aspects of environment admission, ecological protection, pollution control and environmental management.

6. 1 Environmental management

To realize the planning goal, it is necessary to take the environmental management measure during planning implementation and operation. During planning implementation, we shall strengthen the environmental management, supervise the environmental protection facility and thus reduce the environmental and ecological damage of project construction. During planning operation, we shall pay attention to the environmental management, find the environmental problem timely, analyze the environmental influence of project operation according to the environmental monitoring result during operation, take proper measure to reduce the negative influence, and propose the comment or suggestion in the view of environmental protection.

6. 2 Water environmental protection

Take the water resource protection measure during planning implementation; demarcate the water resource reserve; assist the government department in developing the water resource protection regulation according to the relevant national water resource protection laws and regulations; intensify the water-saving measure; fully promote the construction of water-saving society; accelerate the construction of sewage treatment plant; reduce the pollution of Heihe River; strengthen the propaganda work; improve the water resource protection awareness of entire people; enhance the law enforcement force; reduce the occurring of emergency water pollution and illegal pollution discharge; strengthen the water quality monitoring and supervision; implement the water environment monitoring and main facility construction; improve the automation level of water quality monitoring; satisfy the demand of water resource protection and management; strengthen the construction sewage discharge management during planning implementation.

6. 3 Ecological protection

Strengthen the construction management during project construction; minimize the damage of temporary land on vegetation; plant the proper vegetation in the basin of abandoned reservoir artificially; strengthen the ecological protection propaganda in river basin; improve the ecological protection awareness of entire people in river basin; reduce the damage to ecological vegetation; adjust the industrial structure; change the traditional grazing method in grazing district; decrease the overload rate of grassland; promote the restoration of ecological vegetation; strictly restrain the wasteland reclaim in river basin; supervise the implementation of ecological protection measure such as grazing prohibition, enclosed forestation and grain for green; strictly manage the distribution of water resource; and prevent the ecological water is misused as production water.

7 Conclusions

According to the stipulation and requirement of national environmental impact assessment, the ecological environment impact possibly produced by implementation on comprehensive planning along Heihe River watershed is predicted and evaluated. The rationality of planning scheme is analyzed from such aspects as planning layout, ecological environment and economic society to put forward the countermeasures to prevent or relieve the unhealthy influence on ecological environment so as to provided the support for the comprehensive planning layout and scheme along Heihe River watershed and provide the appraisal on environment influence assessment for comprehensive planning layout along continental river watershed in the northwest.

The comprehensive planning along Heihe River watershed aims to improve the ecological environment along the watershed and prompt the economic development along the watershed. After the planning scheme is implemented, the demand and supply contradiction for water resources in the midstream and downstream of water system in the east along Heihe River watershed will be relieved to improve the current status of water quality along the watershed and change the situation that the agricultural irrigation water inside the watershed occupies the ecological environment water for a long term, the consumption of water for ecological environment of the water system in the east will be increase 0.71×10^8 m^3, the consumption of water for ecological environment of the water in the Midwest will be increase 0.61×10^8 m^3, the water volume entering into Ejin Banner will increase 0.58×10^8 m^3. The increase of ecological environment water will stabilize the scale of oasis and a certain lake area downstream to improve the natural rehabilitation capability of ecological environment along watershed and effectively prevent the shrinkage and degradation trend of forest and grassland inside the watershed as well as provide the reliable guarantee for sustainable development of economic society along watershed.

References

Zou Jiaxiang. Technical Manual of Environmental Impact Assessment: Hydropower Engineering [M]. Beijing: China Environmental Science Press, 2009.

Gu Hongbin, Yu Weiqi Cui Lei. Preliminary Research on Environmental Impact Assessment of Hydropower Planning of China[J]. Hydropower Station Design, 2007.

Li Xiangyun. Environmental Impact Assessment for Hydropower Development and Planning[J]. Environment and Sustainable Development, 2011.

Li Ying, Jiang Guzheng. Function of Strategic Environmental Impact Assessment in Water Resources Development Planning of Basin[J]. Yangtze River, 2010.

Using Virtual Water Strategy in the Treatment of the Yellow River

Yang Yisong and *Su Yunqi*

Yellow River Institute of Hydraulic Research, Zhengzhou, 450003, China

Abstract: Virtual water strategy is that water shortage countries or regions use the commodity attribute of water by importing water-intensive products from water – rich countries and regions to meet the needs of local water resources and food security needs. Water resources shortage of the Yellow River Basin not only affects the social , economic development in river basin and the basin ecological, but also affects the comprehensive management of the Yellow River and the national eco-security. The water resources shortage will be more seriously with the development of social and the population increasing. Water resources shortage has become the main factor restricting economic development and environment restoration in the Yellow River Basin, which makes the management of the Yellow River more difficult. With virtual water strategy in the Yellow River Basin, it can make the most use of water resources by the virtual water trade and can use the water resources of other basins without cross-basin water transfer project, thereby it might alleviate the water shortage crisis and promote the sustainable development and improve environment gradually and provide an adequate of water resources and ensure enough time for the management of the Yellow River. It is a complex and systematic project to use the virtual water strategy to manage the Yellow River. It needs the cooperation of all water administrative departments and local governments in the basin and river basin agency and central water administrative departments and other relevant departments to establish a virtual water trading platform and the management control system. In the meantime, it also needs to research on the ecological environment of the Loess Plateau and vegetation restoration, adjust the water resources management to provide enough water as far as possible to restore vegetation of the Loess Plateau. So the radical cure of the Yellow River may be done.

Key words: water resource, virtual water, virtual water strategy, Yellow River Basin, the management of the Yellow River

The Yellow River is famous not only because she is the cradle of Chinese nation, but also because its sediments top the world. She is the important water sources for the northwesten China and the north China and the threat to eco-security sources in North China. The water resource in the Yellow River has the features of large inter annual variations and long period of continuous low flow. As a result, the water resources shortage is more serious in the dry period and low water period. With the development of society and population increasing, the water resources shortage be further intensified and the water resources security situation will be even more severe in the Yellow River basin, which not only affects the social and economic development but also affects the management of the Yellow River and the food security and the eco-security of China. In this paper, it discuses how to apply virtual water strategy to alleviate the pressure from water resources shortage and to provide opportunities to make the radical cure for a sound Yellow River.

1   The water resources in the Yellow River

The water resources security in China is very seriously. The total water resources of China is $28,124 \times 10^8$ m^3, 6% of the global water resources. It is the fourth of the world after Brazil, Russia and Canada. But the per capita is the 110th in the world, only the 1/4 of the world average, and 1/5 of the United States', 1/ 50 of the Canada 's. China is one of the 13 poorest countries in terms of the per capita water resources. The multi – year average annual runoff of the Yellow River Basin is 534.8×10^8 m^3, it is only 2% of the country's and the per capita is only 473 m^3 or 23%

of the country's, which should supply water for 15% of the arable land area and 12% of China's population. At the same time, it must transport the sediment to sea for it contains so much sediments which further reduces the amount of water usable for economic and social development. Despite the shortage of water resources in the Yellow River, nearly 1/3 gross water must supply to outside the basin part area and cities. In 1990s, the average natural runoff of the Yellow River is only 437×10^8 m³. The measured volume is only 119×10^8 m³ in Lijin Section, and the actual consumption of runoff in the upstream is already up to 318×10^8 m³, which is about 73% of the total natural runoff and already exceeded the carrying capacity of the Yellow River. The amount of groundwater that 93×10^8 m³ was exploited in 1980, now has increased to about 140×10^8 m³ and it has exceeded the exploitable volume in some areas.

The water resources shortage is about 66.04×10^8 m³ every year in the Yellow River Basin at present. It is predicted that it would be up to 75.32×10^8 m³ in 2020 s. The shortage of water resources is about 26.57×10^8 m³ even if the western route of South – to – North Water Transfer Project and the Hanjiang – to – Weihe River Project are in operation (Xue Songgui, 2011) (the details in Tab. 1). The trans-basin water diversion project can alleviate The shortage of water resources in the Yellow River Basin, but it will likely be anhydrous for diversion during the function shortage of China, and it will be dangerous for the Yellow River in which the ecological environment and social economic development and eco – security were be affected seriously. To deal with the shortage of water resources in the Yellow River Basin, we had put forward some new measures (Yang Yisong, Wang Peiping, in Haihong, 2010; Yang Yisong, Bian Yanli, 2011). In this paper, it is only discussing the idea to alleviate water resources shortage by using virtual water strategy in the Yellow River Basin.

Tab. 1   The supply and demands on water resources in the Yellow River in different periods[*]

(Unit: $\times 10^8$ m³)

Planning	The demands	The supply				The shortage	The ratio of water shortage (%)
		Surface water	Ground – water	The others	The total		
Base year	485.79	304.82	113.22	1.72	419.75	66.04	13.6
In 2020s	521.13	309.68	123.70	12.43	445.81	75.32	14.5
In 2020s with the Hanjiang – to – Weihe River Project	521.13	321.57	123.70	12.43	457.70	63.43	12.2
In 2030s	547.33	297.33	125.58	20.36	443.18	104.16	19.0
In 2030s with the Hanjiang – to – Weihe River Project and the western route of South – to – North Water Transfer Project	547.33	375.12	125.28	20.36	520.76	26.57	4.9

Note: * it is from Xue Songgui, 2011.

2   The concept of virtual water and virtual water strategy

The concept of virtual water is made by Tony Allan in 1993 (Tony Allan, 1993). And it is the water embedded in key water – intensive commodities such as wheat. Tony Allan (1993, 1994, 1997, 1998, 1999) believed that virtual water might be a tool to deal with the problem of water shortage. Producing goods and services generally requires water. The water used in the production process of an agricultural or industrial product is called the 'virtual water' contained in the product. For producing 1 kg of grain we need for instance 1,000 ~ 2,000 kg of water, equivalent to 1 ~ 2 m³. Producing livestock products generally requires even more water per kilogram of product. For producing 1 kg of cheese we need for instance 5,000 ~ 5,500 kg of water and for 1 kg of beef we need in average 16,000 kg of water (Chapagain and Hoekstra, 2003). Virtual water has also been called 'embedded water' or 'exogenous water', the latter referring to the fact that import of virtual water into a country means using water that is exogenous to the importing country. Exogenous water is thus to be added to a country's 'indigenous water'. Compared to the real water resources, its convenient transportation characteristics make trade became a useful tool to alleviate the shortage of water resources (Haddadin, 2003).

Virtual water strategy is the strategy which can input water resources by virtual water trade, namely the regions where water is scarce could achieve their water security by importing water – intensive products from those whose water is abundant. The trade between countries and regions is actually import or export of water resources. The purchase of virtual water quantity is equivalent to annual runoff of the Nile by import on food subsidies every year in the Middle East.

3 Discussing the virtual water strategy in the Yellow River

3.1 Feasibility analysis

The Yellow River is about 5,464 km, which is originated from Bayankala mountain in Qinghai province and flows through Qinghai, Sichuan, Gansu, Ningxia, Inner Mongolia, Shaanxi, Shanxi, Henan, Shandong. And it goes into Bohai, in Kenli County of Shandong province. The Yellow River Basin is about 79.5×10^4 km^2. There are several different natural conditions in the Yellow River Basin. As a result, it is different to consume the quantity of water resources to produce similar products in different regions within the watershed, which provides a basis for the implementation of virtual water strategy within the basin. It needs about 2.6 m^3 water to produce 1 kg grain in Gansu province(Xu Zhongmin, Long Aihua, Zhang Zhiqiang,2003), but it is only 1.3 m^3(Li Sujuan, Qian Guoling, Li Shanshan,2009) in Shandong province. So the demand of water resources can range from 5.2 m^3 to 2.6 m^3 to produce 2 kg grain in Gansu and Shandong respectively, and it can save about 1.3 m^3/kg when producing the 2 kg grain only in Shandong province. If it imports 1,000 t grain production from Shandon into Gansu instead of producing it in Gansu, the water saved is about 1.3×10^6 m^3 in the Yellow River Basin. Similarly, it can import goods from the other basins, this is equivalent of inter – basin water transfer function in terms of effectiveness and can ease the pressure of water shortage in the Yellow River Basin.

3.2 The effects of the management of the Yellow River

One of key issues in the Yellow River management is how to coordinate the water resources and sediments of the Yellow River, namely how to use the limited water resources to ensure sustainable economic and social development, to protect and recover the vegetation of the Loess Plateau in the middle reaches of the Yellow to make the sediments into the Yellow River reduce gradually and wash the sediment in the lower reaches of the Yellow River into the sea, so that the Yellow River can be safe for a long period. But with the development and the population increasing, water will be in shortage as a resource, and water resources shortage will increase year by year in the Yellow River Basin(Xue Songgui,2011). Without water resources from outside, the ecological water will be reducing further and the ecological environment of the Loess Plateau will worsen further. It is impossible to reduce the sediments flowing into the Yellow River. The flood will be inevitable in the Yellow River after the Xiaolangdi Dam reservoir lost the function of intercepting sediment.

It can alleviate water resources shortage in the Yellow River Basin and gain time and chance to manage the Yellow River with the virtual water strategy. By the virtual water strategy, it can carry out the adjustment of industrial structure and promote the virtual water trade in the basin on one hand, on the other hand, it can make the most use of the limited water resources to meet a greater economic, social and ecological benefits by importing some water – intensive products from outsides. As mentioned above, it can import grain production to supply the upper and middle reaches to meet the demands of increasing population instead of producing it in the upper and middle reaches of the Yellow River, which may save much more water resources in the basin. For example, if it imports 2×10^6 t grain production to supply the upper and middle reaches from Shandong province, the saved water resources may be about 26×10^8 m^3, which is more than the amount of the Hanjiang – to – Weihe projection. If the grain is imported from outside the basin, the saved water resources may be about 52×10^8 m^3, which is approximately equivalent to 1/3 of the water amount of the west route of South – to – North Water Diversion Project's. Therefore, with the virtual water strategy, it can alleviate largely the pressure of water resources shortage and reduce the ecological water for economy and slow down effectively the deterioration of ecological environment. And it can provide very good chance to harness the yellow River.

3.3　The Relationship between the virtual water strategy and the South – to – North Water Transfer Project

Many of the scholars who study on virtual water exaggerated the function of virtual water. And some think it can solve fundamentally the water resources shortage of our country by the virtual water strategy and the South – to – North Water Transfer Project is not necessary to constructure. Meanwhile, some people in water resources management has ignored the virtual water strategy in mitigating water shortage pressure and indulged in the South – to – North Water Transfer Projection. It has many success model to alleviate the shortage of water resources pressure by the virtual water strategy in world (Allan,19971998; Hoekstra A Y,2003), but this does not prove it is totally fit for the Yellow River Basin to solve the water resources shortage and veto the South – to – North Water Transfer Projection.

The distribution of the water resources and land source is extremely lopsided in China. There lives about 7 hundred million populations in the South, where it has only 1/3 farmland but 80.9% of the water resources and only 30% grain production of the country. And there lives about 6 hundred million population in the North, where it has 2/3 farmland and 70% grain production but only 19.1% of the water resources of the country. As a result, it is impossible to make the region self-sufficient in grain in the South, not to mention the supply to the North. And the waste water often happens in the South. If the South depends on imports form the international grain market for the Long-term without the extra food supplies produced in the North, the domestic food shortages will be appear and the starvation will be happened once the grain of international market declines sharply. Then the food security and social stability will be declined in China. Therefore, the fundamental method to solve the shortage of water resources is through the implementation of water diversion project from south to north, which can give full play to its social benefit, economic benefit and ecological benefit with water resources in China.

Although the virtual water strategy can not replace the role of South – to – North Water Transfer Project in the Yellow River Basin, the function of virtual water strategy can not be ignored. It can be used for the optimization of industrial structure, improving the use efficiency of water resources, and alleviating the shortage of water resources pressure. Especially, if virtual water strategy is used scientifically and rationally before the diversion water transfer projects operate, it can well ease the pressure of water resources shortage, maintain social and economic stability and development and slow down the continued deterioration of the ecological environment in the Yellow River Basin. And once the water diversion projects operate, it can make the limited water resources play a greater role and it is possible to make Yellow River safe for ever by the proper measures.

4　Conclusions and suggestions

It is good for alleviating water resources shortage pressure, the development of society and economy and maintenance of ecological environment in the Yellow River Basin by the application of virtual water strategy. It also provides an opportunity to harness the Yellow River very well. But the virtual water strategy could not replace the role of South – to – North Water Transfer project because of the uneven distribution of water and land resource in China. Only by giving full play to the role of virtual water strategy and with the function of South – to – North Water Transfer Project, it is possible to solve the problem of water resource s shortage in the Yellow River Basin and maintain the sustainable development of economy and society, safeguard national food security and ecological safety, and provide enough water resources for harnessing the Yellow River.

References

Xue Songgui. Summary of Yellow River Basin Integrated Water Resources plan[J]. China Water Resource, 2011(23):108 – 111.

Allan J A. Fortunately There are Substitutes for Water Otherwise Our Hydro – Political Futures Would be Impossible[C] // Overseas Development Administration. Prioritized for Water Resources Allocation and Management. London, United Kingdom: ODA, 1993:13 – 26.

Allan J A. Overall Perspectives on Countries and Regions[C] // Rogers P, Lydon P. Water in the Arab World: Perspectives and Prognoses. Massachusetts: Harvard University Press, Cambridge, 1994:65 – 100.

Allan J A. Virtual Water: Long Term Solution for Water Short Middle Eastern Economies[R]. The 1997 British Association Festival of Science, University of Leeds, 1997.

Allan J A. Virtual Water: Strategic Resource[J]. Ground Water, 1998, 36(4):545.

Chapagain A K, Hoekstra A Y. 'Virtual Water Trade: a Quantification of Virtual Water Flows Between Nations in Relation to International Trade of Livestock and Livestock Products [R]. IHE Delft, The Netherlands, 2003.

Haddadin M J. Exogenous Water: a Conduit to Globalization of Water Resources [R]. IHE Delft, The Netherlands, 2003.

Hoekstraa A. Y. (editor). Virtual Water Trade: Proceedings of the International Expert Meeting on Virtual Water trade[R]. IHE Delft, The Netherlands, 2003.

Li Sujuan, Qian Guoling, Li Shanshan. Research on the Virtual Water Structure Construction of Agricultural Products in Shandong Province[J]. China Business(Economic Theory), 2009 (6):79 – 80.

Turton A R, Moodley S, Goldblatt M, et al. An Analysis of the Role of Virtual Water in Southern Africa in Meeting Water Scarcity: an Applied Research and Capacity Building Project[R]. Johannesburg: Group for Environmental Monitoring(GEM) and IUCN(NETCAB), 2000:2 – 8.

Xu Zhongmin, Long Aihua, Zhang Zhiqiang. Virtual Water Consumption Calculation and Analysis of Gansu Province in 2000[J]. Acta Geographica Sinica, 2003, 58(6):861 – 869.

Yang Yisong, Wang Peiping, Yu Haihong. Ecological Thinking About the Harnessing of Yellow River by Using Three Kinds of Water Resources[C] // Proceedings of the 4th International Yellow River Forum on Ecological Civilization and River Ethics. Zhengzhou: Yellow River Conservancy Press, 2010.

Yang Yisong, Bian Yanli. Building a Harmonious Relationship Between Water Resource and the Environment on the Loess Plateau: How to Restore Its Vegetation[C] // 2011 International Symposium on Water Resource and Environmental Protection (ISWREP2011), 2011:2834 – 2837.

The Assessment Techniques and Application of Watershed Management Strategies Based on Complex System Theory

Ni Hongzhen, *Wang Tangxian*, *Liu Jinhua* and *Zhao Jing*

China Institute of Water Resources and Hydropower Research, Beijing, 100038, China

Abstract: The theory based on the human-nature coupled system (Coupled Human – Nature, System, referred to as CHN) that uses a holistic point of view for the action mechanism and characteristics of the large complex system of atypical water resources, by the interaction of the variables to each subsystem proceed multi – process simulation, description and information transfer , in which a set of simulation model are built which can have a clear and dynamic diagnosis of water resources and in favour of making quick administrative decision of coordinated growth between environmental economy and water resources to meet research needs of the modern water resource planning and water resources complex problem and meet the water resources urgent requirement of our watershed and regional adaptive management. The paper simulates and analyzes the effect of all possible water resource management regulation project and scenarios such as the Yellow River Basin water conservation, pollution control (reuse), water transfer programs to the basin's economic and social development and ecological environment protection. It can provide decision support for adaptable and dynamic water resources management mechanism of the Yellow River Basin to response to changing natural risks and human development needs and scientific revision of water resources development planning, to improve the efficiency and level of water resources scientific management under the new situation.

Key words: CHN, water resources complex systems, multi-process scenario simulation, planning revision

Natural evolution and Long-term human disturbance, water resources and economic, social, ecological, environmental constitute an intricately large systems which is extremely complex and interactive changes, in which water is one of the most sensitive controlling factor. The randomness and complexity of the system change will continue to increase with time, so the complex system global base on the human-natural coupling (Coupled Human – Nature, the System, referred to as CHN) are proposed according to the mechanism and characteristics of the water resources complex system, to build a set of simulation model which can have a clear and dynamic diagnosis of water resources and in favour of making quick administrative decision of coordinated growth between environmental economy and water resources in order to meet research needs of the modern water resource planning and the water resources urgent requirement of our watershed and regional adaptive management. It can better identify the problem and objective, scientific decision-making, implement the most stringent water resources management system, and improve the efficiency and level of water resources scientific management under the new situation.

The paper proposes the future water resources management possible scenarios of the Yellow River Basin based on objective fact and water resources planning, and set the boundary of the model. Using the model to simulate and valuate various possible future scenarios of Yellow River Basin can provide decision support for the Yellow River Basin to enhance water demand management, strengthen water-saving and pollution control, carry out the revision of water resources planning, and improve water use efficiency and effectiveness.

1 Theoretical framework of water resources complex systems

CHN refers to the overall system of humans and the natural interaction, the complex mode

(Patterns) and processes (Processes) in these systems as a single-discipline institute which can't be found and described, so the tight integration social and natural sciences become philosophical and methodological basis of the CHN system research.

From CHN perspective to understand system mechanism of humans adapt to a changing environment, and as a basis for design infrastructure and management system that is reliable (reliable), the elastic (resilient) and anti-jamming (robust), it can guide the human individual and social behavior to adapt to the changing natural environment, so as to achieve the purpose to improve efficiency and operability of water resources management system .

The water system is a very typical coupling system of human and the natural. In the system, there have a complex interactions between human subsystems (such as the daily life, industrial and agricultural production, urbanization construction and other economic and social development activities) and natural subsystems (such as climate, rainfall, runoff, interaction between surface and groundwater, ecological environment evolution, etc.). Both interactive produce engineering facilities and management subsystem, and any single discipline theory and method can not completely describe the complexity of the system. Holism methods must be used to research from the basic characteristics of the human-natural coupling system.

CNN, based on the systems perspective, integratedly applies theoretical methods such as macro / measurement / resources / industrial economics, hydrology and water resources, water environmental studies, complex adaptive system, water resource planning and rational allocation of, and uses modern modeling and analysis techniques such as modern statistics, macroeconomic forecasting, input-output analysis, population and urbanization projections analysis, analysis and forecast of water-saving quota, the rational allocation of water resources, water balance analysis, river basin water pollution parameters estimate, multi-objective group decision making, water resources integrated scheduling, complex adaptive systems analysis, on the basis of full understanding of the development ecological environment and economic and social and water resources supply-demand relationship and mutual influence. CNN from the basin water resources system involved in human subsystem (such as the daily life of human, industrial and agricultural production, urbanization construction and other economic and social development activities) and natural subsystems (such as climate, rainfall, runoff, and surface groundwater interaction, ecological environment evolution, etc.), which choose a small size basic continuous function equations that have adequate timing section, regional spatial scales, to subdivided space cells, water cycle node and the role of subject and object elements. It not only contain operational process of the economic and social and the evolution of ecological environment in the single individual of the human and natural subsystems and using variable to proceed multi-process direct and dynamic description in the processes of water resources circulation and use , but also contain dynamic coupling simulation of complex interactions and feedback between the various subsystems, the overall model who can have a complete description of the coordinated development between water resources and environmental economy of water resource system complexity based on a holistic point of view shown in Fig. 1.

Model features include: prediction function (including population and urbanization forecast, macroeconomic trends, prediction of water demand function, prediction function of changes in water quality), analog functions (including development process simulation of the national economic, various process of water demand, process of use (consumption) water changes, balance process of water supply and demand and water quality change processes), optimize the coordination function (optimize coordination between the national economic structure and speed development, optimization functions of industrial structure, coordination functions between the multi-objective) and decision-making functions (function of sensitivity analysis and strategic choice analysis). The model has a brief description as follow Tab. 1.

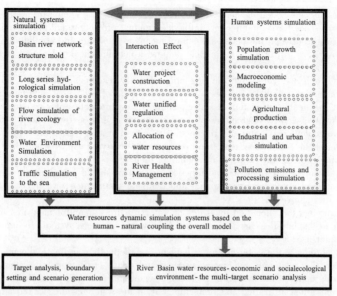

Fig. 1   Simulation theory and methodology of the water resources system based on the human-natural coupling

Tab. 1   Model systems profile of the Yellow River Water Resources in coordination with the environmental and economic development

The name of the model	The main function	Remarks
Population projection model	Development forecast of total population and urban and rural development	Operate independently or embedded in the overall model
Assessment Model of water resources use result	According to the history and current status of water and economic statistics to estimates efficiency and effectiveness of water	Operate independently
Macroeconomic models	Forecasts development of economy, irrigation area, grain yield with input-output analysis, expanded reproduction theory, agricultural production function	Operate independently or embedded in the overall model
prediction model of water demand and water-saving	Forecast water demand and water saving, water-saving investment demand with macroeconomic model	Operate independently or embedded in the overall model
Prediction model of water pollution load emission and regulation	Forecast pollutants emissions, reducing the investments amount of pollution treatment with macroeconomic model	Operate independently or embedded in the overall model
Model of Water supply-demand analysis	Simulate adjustment calculation of water supply-demand of for each partition and node	Operate independently or embedded in the overall model
Model of multi-objective and group decision	Multi-objective coordination and solution, multiple decision-makers consultation	embedded in the overall model
The overall model of coordinated development between water resources and environmental and economic	The whole optimization and simulation of water systems, economic systems, agricultural systems, environmental systems and the ecosystem	In combination with other models, build the overall model and run the calculations

Water Resources – Social economic-ecological environment is a composite system consisting of three subsystems. They have interactions, influence, mutual promotion and mutual restraint, therefore. There needs to take all the necessary control means to intervene and regulate the complex system of water resources, in order to ultimately achieve a coordinated development of the composite system.

The overall objective of system the promotion of water resources sustainable utilization, sustainable economic and social development and virtuous cycle of ecological environment, guarantee renewable and sustainable utilization of water resources, and promote water resources use efficiency, and improve the security degree of ecological environment and environmental, with minimal resources costs to achieve maximum effectiveness. The core of system regulation is to improve the coordination level of the complex system, improve the overall carrying capacity and the level of resistance to risk, the content and elements of regulation include:

First, the composition elements and functions of system should be adjusted, making the balance, optimization of water resources spatial and temporal distribution, water eco – environmental carrying capacity improve;

Second, through the allocation measures of water resources, control measures of economic – social system, to regulate the carrying capacity of the water resources system, to improve efficiency and level of water use;

Third, according to the water cycle mechanism, synthetically coordinate allocation of water resources between the economic and social systems and eco – environmental system relations, to change the water status of the system of economic, social and ecological environment;

Fourth, the adjust utilization and protection way of the water resources, and adjust the water resources allocation between different regions and different water industry , makes the water cycle and the distribution relationship and carrying capacity chance toward to conducive renewable and sustainable use.

In system, the economic subsystem is the foundation of the entire system, and water resources subsystem is a link between the various subsystems. Economic development can not only cause environmental pollution, but also promote environmental governance; not only increase the water demand of all walks of life, but also provide the economic foundation for the development of water supply. Water resources as a short board of regional economic development has become the key natural resources of the target competition. Competition to for capital and water resources in each subsystem make the entire system develop in direction that is in favor of their own interests. For a long time to maintain system coordination and balanced development, allocation of funding and resources in subsystem must be with the development of the system, do a reasonable adjustment, therefore, the model should have a dynamic mechanism be closely linked for each module, especially water and macroeconomic module.

2 Program settings and parameters determination

Temporal and spatial relations of model, the relationship set between the model system. Planning period: the status level of year 2005, the forecast year is 2010, 2020 and 2030. The analog of Yellow River water regulation is: 1999 to 2007.

The margin settings of model. Include: watershed indicators, unified scheduling requirements, the available water of the local water resources in Yellow River, holding dirt capacity of water environment, configuration program outside the basin water, reservoir operation rules.

Calibration and settings of model parameter. Include: physical water balanced parameter calibration, water-saving quota and sewage quota under different scenarios, scenario set of water investment parameters, scenario set of the macroeconomic control parameters.

Model validation. The focus is to check balance between Yellow River control node measured runoff and simulate water from model on the past 10 years, and the efficiency of water use and water supply quantity.

Setting the scene of coordinated development program. Forecast scenarios program that have 36 sets are set from the three aspects of water conservation, pollution control and water transfer program.

Scenario analysis intended to set the scene from the three aspects of water conservation, pollution control and water transfer program, in which, saving have three sets of scenarios (ordinary water-saving, and strengthen water-saving, super water-saving), pollution control and reuse have three scenarios (general governance sewage, and strengthen governance sewage, super governance sewage), water transfer have four sets of scenarios (no water diversion, water diversion scheme1, 2, 3) by this combination of 36 sets of scene programs. are shown in Tab. 2.

Tab. 2   Coding scheme of scenarios

Water-saving	Coding of water-saving	Governance sewage	Coding of governance sewage	Water diversion	Coding of Water diversion	Coding scheme of scenarios
Ordinary water-saving	0	General governance sewage	0	No water diversion	0	Coding of water-saving + Coding of governance sewage + Coding of Water diversion
Strengthen water-saving	1	Strengthen governance sewage	1	Water diversion scheme1	1	Such as;000 is ordinary water-saving, general pollution-control and no water diversion, 111 is strengthen water-saving, strengthen governance sewage, and water diversion scheme1 ⋯
Super water-saving	2	Super governance sewage	2	Water diversion scheme2	2	
				Water diversion scheme3	3	

Scenarios and parameter water-saving. Water-saving mode is divided into three types, and to be characterized by the different water quota.

General water-saving mode. Model that remain the existing water-saving input on the basis of status water-saving level and water-saving measures, and consider the water quotas and water consumption trends since the 1980s, so water-saving mode are determined .

Strengthen water-saving mode. Mainly on the basis of the general water-saving, water-saving investment are further increased, to enhanced water demand management, inhibit the excessive growth of water demand, further improve water use efficiency and water saving level. The water-saving mode is identified after meeting the ecological and environmental water basic protection needs. The overall characteristics of this mode is the implementation stricter water-saving measures, more stringent efforts to adjust industrial structure, increase water-saving investment.

Super water-saving. The water supply can't meet the reasonable requirements of the economic and social development, so the coercive measures are readjust the industrial structure, even at the cost of being forced to shut turn and stop some enterprises in many areas. The most stringent water – saving system is implemented to do everything possible to reduce the water quota of unit output or product, the economic and social was coercive style development.

Effect of water – saving can be reflected from the water quota, so the water quota is used to represent the above three models.

Water diversion scenarios and their characterization parameters. Water diversion mode is divided into three types, the rate of sewage treatment, the rates of reuse after sewage treatment are to represented.

General governance sewage. Model that remain the existing governance sewage input on the basis of status governance sewage level and governance sewage construction, and ensure that sewage into the rivers don't increase as a constraints.

Strengthen governance sewage. Model based on general governance sewage further increase governance sewage efforts and reuse level for meet restrain of the dirt holding capacity and amount of sewage into the river. The overall characteristics of the mode are the implementation of strict pollution control measures and assessment requirements, and increase investment in water governance sewage efforts.

Super governance sewage. Model based on super governance sewage further increase governance sewage efforts, in addition, to meet the objectives and requirements of the water environment, should continue to improve the quality of the water environment, and the focus is to increase its reuse, increase water supply capacity of the Yellow River.

Mode of governance sewage are represented by the rate of sewage treatment and the reuse rate of sewage, different scenarios of sewage treatment rate and reuse rate are shown in Tab. 3.

Tab. 3   Scenario set of governance sewage indicators in the Yellow River basin

Indicators	General governance Sewage		Strengthen governance Sewage		Super governance Sewage	
	2020	2030	2020	2030	2020	2030
Sewage treatment rate	55%	70%	65%	85%	80%	95%
Reuse rate	20%	30%	30%	35%	35%	40%

The paper conducts prediction and calibration about emission intensity of COD and ammonia nitrogen from non-agricultural industries and urban residents in Yellow River Basin according to relevant planning.

Configuration program of water transfer outside the basin. According to the receiving area design of the Western Route 1 of South Water to North, the project is provided by the Yellow River Engineering firms planning, under 80×10^{10} m^3 scale of water diversion. Configuration schemes are shown in Tab. 4.

Tab. 4   The transferred water configuration program of project 1 of west line
$$(\text{Unit}: \times 10^9 \text{ m}^3)$$

River inside and outside	Departments or provinces		Configuration of water		
			Plan 1	Plan 2	Plan 3
River outside	Departments	Key cities	24.3	24.3	24.3
		Energy base	17.9	17.9	17.9
		Eco – irrigation area of hei shanxia	0.9	3.9	8.9
		Shiyang River	2	4	4
		Subtotal	45	50	55
	Provinces	Qinghai	5	5	5
		Gansu	10	12	12
		Ningxia	9.6	11.6	15.3
		Neimenggu	14.2	14.7	15.2
		Shanxi	4.2	4.7	5.5
		Shaanxi	2	2	2
		Subtotal	45	50	55
River inside			35	30	25
Total			80	80	80

Considering the water volume of 1×10^9 m^3 in 2020 and 1.5×10^9 m^3 in 2030 of Chinese Jinan Wei project, Water Configuration program in the outer Basin of Yellow River Basin water are shown in Tab. 5.

Because the model use the monthly scheduling method to simulate runoff, water use, water consumption monthly of watershed, the monthly runoff process of project 1 of south water to north and Jinan Wei project is the model of one of the boundary conditions in which different runoff processes will produce different consequences under the conditions of the same total water resources. The model use the monthly runoff of project 1 of south water to north as the input, project 1 add 8×10^{10} m³ of total water resources. Its monthly flow process is shown in Tab. 6.

Tab. 5　Water configuration program of Chinese Jinan Wei project + south water to north project (Unit: ×10⁹ m³)

Water configuration program	year	Qing hai	Gansu	Ningxia	Nei menggu	Shaanxi	Shanxi	Henan	Shan dong	Subtotal of River outside	River inside	Total
Plan 1	2020					10	0	0	0	53	35	88
	2030	5	8	9.6	14.2	19.2	2	0	0	58	35	93
Plan 2	2020					14.7	2	0	0	56	30	86
	2030	5	8	11.6	14.7	19.7	2	0	0	61	30	91
Plan 3	2020	5	8	15.3	15.2	15.5	2	0	0	61	25	86
	2030	5	8	15.3	15.2	20.5	2	0	0	66	25	91

Note: Shiyang River water distribution is not in the Yellow River Basin, so the quantity of water configuration in Gansu province of the Yellow River Basin is 8×10^8 m³.

Tab. 6　The monthly runoff process of project 1 of south water to north

(Unit: m³/s)

Month	June	July	August	September	October	November	December	January	February	March	April	May
Average flow	321.4	325.2	316	318	309.6	262.4	242.4	234	224.6	212.8	0	277.4

The 56 older series processes of monthly inflow of Chinese Jinan Wei project are provided by Yellow River Water Conservancy and Hydropower Survey and Design. Years of average monthly runoff process are shown in Tab. 7.

Tab. 7　The monthly runoff process recommended by Chinese Jinan Wei project

(Unit: ×10⁸ m³)

	January	February	March	April	May	June	July	August	September	October	November	December
1×10^9	0.591,2	0.414,0	0.612,9	0.863,0	0.877,6	0.739,8	1.062,8	1.003,7	1.026,9	1.029,3	1.014,3	0.764,3
1.5×10^9	0.847,8	0.601,1	0.887,8	1.271,0	1.322,0	1.114,1	1.646,8	1.545,3	1.611,6	1.583,6	1.480,6	1.088,3

The setting of the reservoir scheduling rules. Reservoir operation rules impact regulation and distribution of basin water, the reservoir scheduling rules need to set up in the model. This model considers nine reservoirs which take up water supply, power generation, flood control and so on. They are very complex.

In order to reflect the actual reservoir scheduling rules, and avoid making the model too complex to solve difficult, the paper make simplification as follows:

(1) According to the monthly data analysis of reservoir inflow, outflow and storage capacity since 1999 to 2007, each reservoir monthly minimum outflow can be get, as the flow limit of the reservoirs. This rule actually contains to generate electricity, ice, and many other considerations;

(2) The water level of the reservoir in the flood season shall not exceed flood limit level, the water level in non-flood season shall not exceed the maximum water level, at the same time water level shall not be less than the dead water level. This rule includes the consideration of flood control and water supply scheduling.

3 Effect analysis and program recommended

Optimization and calculation of model can get the optimal result of the economic and social development and ecological environment state under each program scenario, can get the result of optimal allocation of water resources in areas and departments under the various scenarios. According to the results of the analysis, researcher can compare the possibility of future changes, as well as decision makers' preferences, decision-makers according to the actual situation, choose the adaptive program of water resources management in the context of change, revise and adjust relevant plan to adapt to the changing situation. According to economic development and possibilities of investment constraint, studies suggest that recommended scenarios is 110 that is coordinated development between water resources and economic and social based on carrying capacity of water resources in the Yellow River. The model use technical ideas of "demand decided by supply" which is the constraints of the available water supply, keeping the intensity of water-saving, governance sewage and reuse and water transfer and distribution outer basin program as a key factor. The model simulate optimization results of different water transfer schemes under the premise of meeting the economic, food, and water quality objectives and requirements , to provide decision-making basis.

According to the principles of the possible and feasible combination, indicators predict results of 10 sets of the most likely combination of scenarios are shown in Tab. 8.

Tab. 8　Macroeconomic effects analysis of Scenario program in Yellow River Basin

Program code		000	001	110	111	112	113	220	211	212	213
Water – saving situations		General	General	Strengt hen	Strengt hen	Strengt hen	Strengt hen	Super	Super	Super	Super
Situations of governance sewage		General	General	Strengt hen	Strengt hen	Strengt hen	Strengt hen	Super	Strengt hen	Strengt hen	Strengt hen
Water diversion Situation		No water diversion	Plan1	No water diversion	Plan 1	Plan 2	Plan 3	No water diversion	Plan 1	Plan 2	Plan 3
GDP ($\times 10^8$ Yuan)	2005	14,306	14,306	14,306	14,306	14,306	14,306	14,306	14,306	14,306	14,306
	2020	39,924	42,943	45,246	46,348	46,348	46,348	48,360	48,731	48,731	48,731
	2030	70,083	85,334	86,203	97,282	99,117	100,833	97,418	102,813	104,183	105,415
Developing speed	2006 ~ 2020	7.08%	7.60%	7.98%	8.15%	8.15%	8.15%	8.46%	8.51%	8.51%	8.51%
	2021 ~ 2030	5.79%	7.11%	6.66%	7.70%	7.90%	8.08%	7.25%	7.75%	7.89%	8.02%
	2006 ~ 2030	6.56%	7.40%	7.45%	7.97%	8.05%	8.12%	7.98%	8.21%	8.27%	8.32%
The energy industry accounted for GDP. (%)	2005	13.8	13.8	13.8	13.8	13.8	13.8	13.8	13.8	13.8	13.8
	2020	14.0	14.3	14.5	14.5	14.5	14.5	14.2	14.7	14.7	14.8
	2030	13.7	14.8	14.4	14.9	15.0	15.1	14.0	14.9	15.0	15.1
Total area of Irrigation ($\times 10^4$ mu)	2005	8,868	8,868	8,868	8,868	8,868	8,868	8,868	8,868	8,868	8,868
	2020	9,118	9,405	9,309	9,669	9,669	9,669	9,502	9,826	9,821	9,818
	2030	9,425	10,735	9,830	11,258	11,326	11,416	10,091	11,584	11,651	11,802
Total Food output ($\times 10^4$ t)	2005	4,377	4,377	4,377	4,377	4,377	4,377	4,377	4,377	4,377	4,377
	2020	4,853	4,991	4,945	5,117	5,117	5,115	5,120	5,226	5,223	5,221
	2030	4,952	5,599	5,176	5,866	5,902	5,952	5,419	6,118	6,153	6,233

Continued to Tab. 8

Program code		000	001	110	111	112	113	220	211	212	213
The amount of COD reduction ($\times 10^4$ t)	2020	70	72	87	87	87	88	111	91	91	91
	2030	100	109	134	142	142	143	170	152	150	150
Water – saving investment ($\times 10^8$ Yuan)	2020	306	316	463	471	471	471	680	674	675	675
	2030	528	572	885	945	946	950	1,356	1,389	1,388	1,384
Water saving ($\times 10^{11}$ m^3)	2020	39	40	51	51	51	51	60	59	59	59
	2030	56	59	78	81	81	81	94	95	95	95
Water – supply investment ($\times 10^8$ Yuan)	2020	448	653	472	677	677	677	550	678	678	678
	2030	906	2678	969	2,754	2,844	2,996	1,095	2,753	2,843	2,990
The amount of water supply ($\times 10^{11}$ m^3)	2005	430	430	430	430	430	430	430	430	430	430
	2020	441	452	443	453	453	453	448	454	454	454
	2030	454	513	457	518	521	526	465	518	521	526
Water consumption ($\times 10^{11}$ m^3)	2005	239	239	239	239	239	239	239	239	239	239
	2020	271	277	272	279	279	279	274	279	279	279
	2030	282	317	286	320	321	324	288	320	321	324
Investment of governance sewage ($\times 10^8$ Yuan)	2020	455	490	557	726	586	595	592	25	592	624
	2030	675	745	829	998	939	926	930	70	923	923
Standard treatment plant (Block)	2020	88	94	109	114	116	116	140	113	119	115
	2030	127	141	158	177	175	177	190	175	174	174

3. 1 Analysis of the effect based on the water resources carrying capacity

Comparative analysis can be found by the simulation results of the program:

(1) The insufficient of total amount of Yellow River water resources constrain on the development of the economic society of the Yellow River Basin.

(2) In the Yellow River Basin the shortage of the water resources restrict the economy development, water conservation has obvious promotion and protection on the Yellow River Basin in upgrading of industrial structure and the sustained rapid development of economic and social.

(3) Sewage treatment can not only meet the objectives and requirements of water environment governance, but also be conducive to promoting economic development, and the investment of sewage treatment and reuse has a higher macro – economy effect.

The analysis of economic effect. Considering the utilization of the Yellow River water resources, water saving and the possibility of sewage reuse, carrying capacity of water resources, based on the development request of water supply economy, under the recommended 110 project scenarios, GDP is expected to reach $8.620\ 3 \times 10^{12}$ Yuan on the Yellow River Basin in 2030, and the economic development speed is 7.45% within a calculated period of 25 years. The adjustment of industrial structure is obvious, the proportion of three industries adjusted from the 9.8 : 49.8 : 40.3 in 2005 to the 2.2 : 46.6 : 51.2 in 2030. Energy industry accounted for the proportion of GDP increased from 13.8% in 2005 to 14.5% in 2020 and 14.4% in 2030. The whole basin irrigation area is 9.83×10^7 mu in 2030, added about 9.62×10^6 mu in 2005; in 2030 grain output will reach to 5.176×10^7 t, per capita output of grain reach 396 kg, be slightly higher than the 391 kg status. Within the 25 – year period to 2030, including water supply (mainly recycled water), water conservation and governance sewage investment reach 2.683×10^{13} Yuan, average annual water – saving investment is 1.07×10^{10} Yuan. Up to 2030, the whole basin-wide need new construction of 158 sewage treatment plants with the daily processing capacity of 100,000 t.

The utilization analysis of water resources. In 110 program scenarios, the expected total amount of water supply and water demand in the whole basin is 4.57×10^{10} m^3, more about 2.8×10^9 m^3 than 2005. Within adding about 2.8×10^9 m^3 water supply, other water supply which is mainly water reuse increased by 2×10^9 m^3. Composition from the water demand in 2030, compared with 2005, agricultural water demand is reducing 2.4×10^9 m^3, urban and rural residents increased by 2.4×10^9 m^3, nonagricultural industries increased by 2.4×10^9 m^3, and eco-an increase by 0.3×10^9 m^3. It should be notedthat the water demand of basin-wide energy industry increased from 1.75×10^9 m^3 in 2005 to 2.78×10^9 m^3 in 2030, adding 1.03×10^9 m^3.

In order to meet to strengthen water conservation objectives and requirements, 25 – year period from 2006 to 2030, the basin-wide water-saving investment is 8.85×10^{10} Yuan to achieve water savings of 7.8×10^9 m^3.

From the water consumption of the whole basin, total water consumption is 2.52×10^{10} m^3 of river outside in 2005, it is predicted that it will reach 2.86×10^{10} m^3, an increase of 3.4×10^9 m^3 in 2030. From waste water emissions, basin-wide industrial and urban waste water discharge of 4.2×10^{10} m^3 in 2005, it is predicted that it will reach 7.1×10^{10} m^3 in 2030.

In 2030, in order to meet the requirements of the control amount of COD into the river , the basin-wide need reduce 1.34×10^6 t COD, corresponding to need to build 158 sewage treatment plants with the daily processing capacity of 1×10^7 t standard.

3.2 The effect analysis of outer basin water transfer for the Yellow River Basin Development

Using amorous scene "Yes – no" contrastive analysis method, the result are shows as follows:

Outside the basin water transfer have a huge promotion for the Yellow River Basin economic and social sustainable development. According to model calculations, the implementation of outside the basin water transfer, even in the river with water minimum scenario, GDP is expected to reach 8.53×10^{12} Yuan, more 1.53×10^{12} Yuan than no outer basin water transfer scenarios, especially after the implementation of water diversion, the whole basin economic growth rate is 7.1% from 2021 to 2030, increased by 1.3% from 5.8% than the non-diversion. Even in the situation of strengthening water conservation and governance sewage, the whole river basin GDP of plan 1 increase by 13% than no water diversion scenarios in 2030, increased GDP reached as high as 1.1 Yuan. Comprehensive comparison of all other programs shows that the benefit of implementation of water transfer to macro – economic is enormous.

Outside the basin water transfer is important safeguard mechanism to build "energy basin". Studies have shown, the development potential of energy industry is limited only supported by water conservation and local water use, and unsustainable after 2020. In a variety of water conditions, proportion of the energy industry can be maintained to continue to rise. Such as 2030, after the implementation of West Route of South – to – North, energy industry occupies GDP proportion will achieve 14.9% ~ 15.1%, which improved obviously than without water diversion conditions.

Water transfer can effectively promote the grain production capacity of the Yellow River basin. The principle of water transfer is the protection of water requirement for the energy and the urban development, but by the water exchange and water recycling, it can effectively increase the amount of water used for agriculture. According to the model calculations, even the water channel with the minimum water plan 1 by 2030, compared with 2005, under the 001,111 and 211 three projects, the irrigated area will be added 1.867×10^7 mu, 2.39×10^7 mu, and 2.716×10^7 mu. Grain yields will be added to 1.222×10^7 t and 1.489×10^7 t, and 1.742×10^7 t.

The outer basin water transfer plays an important role in safeguarding the health of the Yellow

River. According to model calculations, without no water transfer case, due to economic and social development, increasing water demand will further squeeze the ecological water. The ecological water of the Yellow River Basin may also increase by 3×10^9 m^3 to 3.5×10^9 m^3. With the outer basin water transfer, water resources consumption of the Yellow River Basin increases by 6.4×10^9 m^3 to 7.1×10^9 m^3 than the status quo. The ecological water demand in Yellow River Basin will increase by 2×10^9 m^3 to 2.7×10^9 m^3. It can be seen, water conservation, governance sewage, reuse and other measures can ease contradictions between the development of economic and social and water. But it can reduce the water use of the Yellow River ecological environment endanger the construction of health Yellow River; outside the basin water transfer can not only increase the water demand of economic and social development, but also the real increase water consumption of ecological environment in the Yellow River. That can be considered that outside the basin water transfer is the core measures to maintain Long-term stability and health in the Yellow River.

About the final optimization, decision-makers can base on feasibility economic development and carrying capacity of water resources to make scientific judgment. The model only provides an effective analytical tool and reference.

4   Conclusions

The simulation model of complex water resources system based on Coupled Human – Nature System (CHN) theory can detailedly and effectively portray the complex relationship of water resources, economy, society and ecological environment, each subsystem and their interactions, and it is an effective technology tools for simulating water resources planning and management program revision and adjustment effect under changing environment.

By setting up the scenario of possible water management programs, the model of water resources-economy society-environment economy coordinated development can deal with the future of the changing in environment economic risk. The simulation analysis of model found that water resources strategy which is water-saving priority, sewage control – oriented, multi – channel source, strict management is an important guarantee for the sustainable economic and social development in the Yellow River Basin. Outside the basin water transfer has huge macro – economic and ecological environmental effects on the Yellow River Basin.

References

Wang Tangxian, Wang Hao, Ni Hongzhen, et al. Model and its Application of Coordinated Development between Water Resources and Environment – economic [M]. Beijing: China Water Power Press, 2011.

Zhao Jianshi, Wang Zhongjing, Weng Wenbin. Theory and Model of Water Resources Complex Adaptive Allocation System[J]. Journal of Geographical Sciences, 2003, 13(1): 112 – 122.

Zhao Jianshi, Wang Guangqian, Weng Wenbin. The Research on Integrated Hydrologic-Economic-Environment-Institutional Yellow River Basin Model on Theory of Water Resources Complex Adaptive System, Proceedings of 1st International Yellow River Forum on River Basin Management[M]. Zhengzhou: Yellow River Conservancy Press, 2003.

Zhao Jianshi, Wang Zhongjing, Weng Wenbin. Study on the Whole Model of Water System [J]. Science in China (Series E), 2004(34)(supplement extra edition I):60 – 73.

China Institute of Water Resources and Hydropower Research. Study on Harmonized Water Development and National Economy [R]. 2005.

Yellow River Conservancy Commission. Yellow River Water Resources Planning Report [R]. 2008.

Study of Flood Erosion Efficiency and its Adjustment in the Lower Yellow River during the Storage Periods of the Sanmenxia Reservoir and the Xiaolangdi Reservoir

Li Xiaoping[1] , *Lv XiuHuan*[2] and *Li Wenxue*[3]

1. Institute of Hydraulic Research, Yellow River Conservancy Commission,
Zhengzhou, 450003, China
2. Department of International Cooperation, Science and Technology, Yellow River
Conservancy Commission, Zhengzhou, 450003, China
3. Yellow River Engineering Consulting Co. , Ltd. , Zhengzhou, 450003, China

Abstract: This paper analyzes erosion efficiencies of floods in the Lower Yellow River during the storage periods of the Sanmenxia Reservoir and the Xiaolangdi Reservoir. The results indicate that erosion efficiency of total load at the downstream of the dam increases with the increase of average flood discharge if discharge is less than 4,000 m³/s and maintains a constant value of 20 kg/m³ when discharge is greater than 4,000 m³/s. By investigating the erosion efficiency of grouped particles, we find that the decreasing extension of erosion efficiency of fine particles is higher than the increasing extension of erosion efficiency of coarse one, which is the crucial reason for erosion efficiency of total load holding at constant when discharge is greater than 4,000 m³/s. Supply of fine particles from bed materials is the major factor that affects erosion efficiencies of total load. It is suggested that the average flood discharge during the storage period of the Xiaolangdi Reservoir should be controlled at about 4,000 m³/s to obtain the highest erosion efficiency, thus to make good use of the limited water resources and improve channel drainage capacity of the Lower Yellow River.

Key words: erosion efficiency, storage period, total load, size fraction, flood of lower sediment concentration, supply of bed material, the lower Yellow River

1 Introduction

In Oct. 1999, the Xiaolangdi Reservoir began to store water. During the preliminary storage period, regimes of flow and sediment load entering the Lower Yellow River changed significantly. Water discharged from the reservoir was basically clear, and the Lower Yellow River underwent a state of erosion and adjustment. In the past, numerous scholars had conducted remarkable studies on flow carrying capacity and bed material variation for low sediment concentration flow. However, studies about erosion efficiency and the factors influencing it for low sediment concentration flow created by regulation of large reservoirs in the master stem of the Yellow River are rarely seen. How to regulate regimes of flow and sediment load by reservoir to intensify erosion effect, to increase flood discharging capacity, and thus to reduce threaten brought by the so called 'double phased suspended river' at downstream are tough problems need to be solved urgently. This study focuses on erosion efficiency and its variation characteristics, investigates factors that affect it, and finally proposes flow discharges for high erosion efficiency.

2 Data

Data were selected from floods occurred during the storage periods of the Sanmenxia Reservoir and the Xiaolangdi Reservoir. Only floods of clear water or low sediment concentration flow ($Q/S < 0.015$) formed by reservoir density flow were considered. During storage period, sediments discharged from reservoir by density flow were very fine, and their influence on erosion efficiency could be neglected.

In the Lower Yellow River, sediments are usually classified as follows:

Fine particle if $d < 0.025$ mm, median particle if 0.025 mm $< d < 0.05$ mm, coarse particle if $d > 0.05$ mm, in which d is sediment diameter.

In this study, 41 flood events were selected. Erosion had occurred in the Lower Yellow River for all the selected floods. Among of them, 27 flood events happened during the storage period of the Sanmenxia Reservoir, 14 happened during the storage period of the Xiaolangdi Reservoir, and sediment concentration was less than 15 kg/m³ in 38 flood events. The threshold values of flow and sediment for these floods are listed in Tab. 1.

Tab. 1 The threshold values of flow and sediment of floods

Item	Duration (d)	Runoff ($\times 10^9$ m³)	Sediment ($\times 10^6$ t)	Average flow rate (m³/s)	Average sediment concentration (kg/m³)	Erosion ($\times 10^6$ t)	Erosion efficiency (kg/m³)
Max.	2.8	0.39	0	649	0	0.8	1.7
Min.	43	7.01	86.4	5707	30.6	120.11	22.0
Total	617.8	122.32	614.3	2292	5.0	1770	14.4

3 Analysis of erosion efficiency

The erosion efficiency is here defined as sediments scoured from river bed and transported to the Lijin station by 1 m³ of water during flood. It is expressed as η_e with the unit of kg/m³, and usually computed by

$$\eta_e = \frac{\Delta W_s}{W} \tag{1}$$

where W_s is the amount of erosion; W is the amount of water (sum of the water at the Xiaolangdi station, Heishiguan station and Wuzhi station).

Considering the variables of runoff, flow rate, sediment concentration, water diversion, flood duration and their relationships, by some transformation one can get

$$\eta_e = S_i - \alpha S_0 - \beta S_d \tag{2}$$

in which S_i is average sediment concentration of the Xiaolangdi, Heishiguan and Wuzhi Stations; S_0 is average sediment concentration at the Lijin Station; and S_d is average sediment concentration for diverted water.

3.1 Erosion efficiency of total load

Based on statistic analysis of flood dada, a relationship between erosion efficiency and average flood discharge is established (see Fig. 1). Fig. 1 shows that erosion efficiency increases with the increase of average flood discharge and reaches its maximum (about 20 kg/m³) at a discharge about 4,000 m³/s. For discharge greater than 4,000 m³/s, erosion efficiency maintains at a constant value basically. Data obtained both from the storage period of the Sanmenxia Reservoir and the storage period of the Xiaolangdi Reservoir are well blended in Fig. 1, which means that this relationship is of universal for the Lower Yellow River. It should be noticed that for a given period of time the amount of sediment scoured from riverbed was large even though erosion efficiency keeps at a constant value for discharge greater than 4,000 m³/s. Considering the facts that sediment concentration for flow released from reservoir was low and particles were fine, thus erosion efficiency depended mainly on flood discharge and the supply of bed material.

354

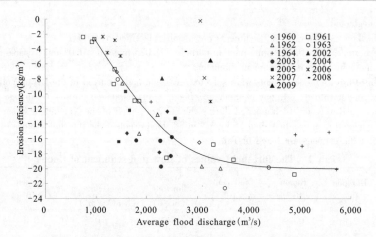

Fig. 1 Relationship between erosion efficiency and average flood discharge for total load in the Lower Yellow River during storage period

3. 2 Erosion efficiency for grouped particles

Fig. 2 illustrates the relationships of erosion efficiency with grouped particles. For coarse particles ($d > 0.05$ mm) , erosion efficiency is almost zero except for large flood discharge. For medium size particles, erosion efficiency increases with the increase of flood discharge and reaches its maximum value of 5 kg/m³ at a discharge about 2,200 m³/s. Afterward, erosion efficiency maintains at constant value. For fine particles, erosion efficiency also increases with the increase of flood discharge and reaches a maximum value about 12 kg/m³ at a discharge of 3,000 m³/s. However, for discharge greater 3,000 m³/s, erosion efficiency decreases with the increase of discharge. Erosion efficiency for total load does not increase when flood discharge is greater than 4,000 m³/s is mainly due to the decrease of erosion efficiency of fine particles.

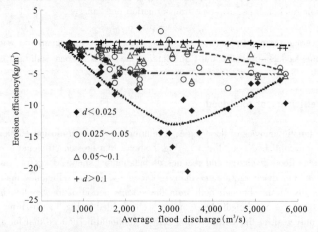

Fig. 2 Relationships between erosion efficiencies and average flood discharges for total grouped particles

4　Analysis of factors influencing erosion efficiency

4.1　Adjustment mechanism of erosion efficiency

Eq. (2) indicates that erosion efficiency reflects the variation of average sediment concentration in the lower reach. For flow of low sediment concentration during reservoir storage period, sediment concentration at the Lijin station is approximately equal to the sediment carrying capacity, that is $\eta_e = S_*$. According to Zhang,

$$S_* = K\left[\frac{v^3}{gR\omega}\right]^m \tag{3}$$

in which v is flow velocity, R is hydraulic radius and equals to h for wide and shallow channel, ω is particle fall velocity. Particle diameters of suspended load in the Lower Yellow River are usually less than 0.1 mm and they falls in a viscous state. Thus, the Navier-Stokes equation is applicable,

$$\omega = \frac{1}{18}\frac{\gamma_s - \gamma}{\gamma}g\frac{d^2}{v} \tag{4}$$

Considering Manning's equation.

$$v = \frac{1}{n}h^{2/3}J^{1/2} \tag{5}$$

and the continuity equation.

$$Q = Av \tag{6}$$

By substituting Eq. (5) into Eq. (6), one can get

$$h = \left(\frac{nQ}{BJ^{1/2}}\right)^{3/5} \tag{7}$$

By substituting Eq. (4), Eq. (5) and Eq. (7) into Eq. (3), one can get

$$S_* = K\left(10.91\frac{J^{6/5}vQ^{3/5}}{g^2n^{12/5}d^2B^{3/5}}\right)^m \tag{8}$$

in which K and m are obtained from field data analysis.

In Eq. (8), J can be considered as constant during a flood event, the viscosity coefficient v can also be treated as constant during flood season. Thus, sediment carrying capacity (S_*) is mainly affected by flow rate (Q), roughness (n), diameter of suspended particle (d), and channel width (B). Among these factors, influences of n and d are most significant.

4.2　Analysis on factors affecting erosion efficiency

4.2.1　Variation of hydraulic parameter (v^3/h)

Field dada analysis indicates that flow velocity and water depth in the Lower Yellow River increased with the increase of flow rate. This characteristic failed if flood overflowed channel bank. To illustrate the influence of flow rate on sediment carrying capacity, plots of hydraulic parameter against flow rate are given in Fig. 3 (Huayuankou station) and Fig. 4 (Gaocun station). It can be seen that hydraulic parameters increased with the increases of flow discharge at both stations for floods observed in the storage periods of the Sanmenxia Reservoir and the Xiaolangdi Reservoir. Consequently, sediment carrying capacity increased with the increase of flow rate.

Fig. 3 and Fig. 4 also show that at the later phase of storage period, value of hydraulic parameter was smaller than it was in the earlier phase (especially for Gaocun station). Due to continuously erosion, channel became wide at the later phase of storage period and channel roughness became larger owing to armoring process, which resulted in the decrease of hydraulic parameter. Therefore, erosion efficiency at the later phase of storage period was smaller than it was at the earlier phase.

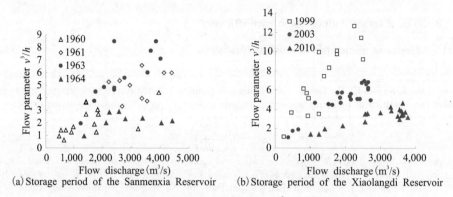

Fig. 3 **Relationship between hydraulic parameter (v^3/h) and flood discharge at the Huayuankou station during storage period**

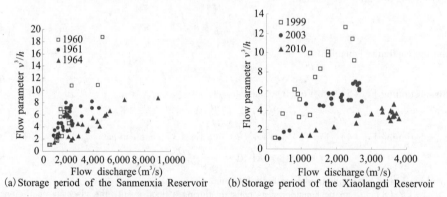

Fig. 4 **Relationship between hydraulic parameter (v^3/h) and flood discharge at the Gaocun station during storage period**

4.2.2 Variation of particle size

Before the construction of the Sanmenxia Reservoir, about 80% of the bed materials were coarse particles and 10% were medium size particles . Due to flooding, contents of fine and medium size particles in the flood plain were, about 42% and 21% respectively, higher than they were in the main channel. At the end of storage period, bed materials in the channel of the Lower Yellow River became coarser (Fig. 5). From 1986, the Lower Yellow River had undergone a series of low flow year, channel had aggradated and bed materials had become finer. Samples taken in 1999 showed that the contents of fine, medium and coarse sediments in bed materials accounted for 23% , 31% , and 46% respectively. Fig. 6 indicates the armoring effects of channel bed after the Xiaolangdi Reservoir impounding water. For instance, the medium diameter of bed material at Gaocun station increased from 0. 045 mm in 1999 to 0. 125 mm in 2010. With the increase of duration of clear water released from the Xiaolangdi Reservoir, most of the fine and medium size particles in bed materials in the Lower Yellow River were scoured and carried away. The diameter of bed material was basically greater than 0. 05 mm.

Field data shows that erosions happened in the whole reach of the Lower Yellow River during the storage periods of the Sanmenxia Reservoir and the Xiaolangdi Reservoir. During the storage period of the Sanmenxia Reservoir, river training works were imperfect, channel widening and deepening occurred together (Fig. 7(a)). From Sep. 1960 to Oct. 1964, 2. 32 $\times 10^9$ t of sediment were eroded. Among which, 578 $\times 10^6$ t were scoured from the main channel and 174. 2 $\times 10^6$ t were

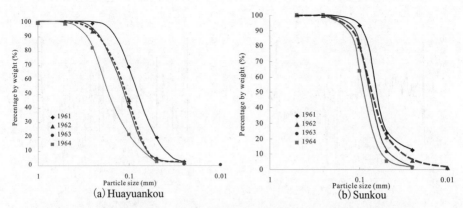

Fig. 5 Size distributions of bed materials for typical channel sections in the Lower Yellow River during storage period of the Sanmenxia Reservoir

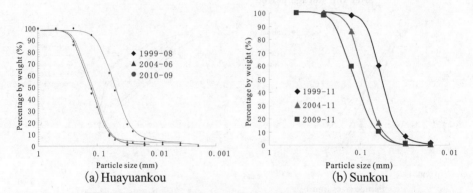

Fig. 6 Size distributions of bed materials for typical channel sections in the Lower Yellow River during storage period of the Xiaolangdi Reservoir

scoured from the flood plain. From 1986, river training works has been perfected gradually, thus bank collapse was not significant and erosion in the Lower Yellow River happened mainly in downward direction during the storage period of the Xiaolangdi Reservoir (Fig. 7(b)). The total mount of sediment eroded from the channel of the whole lower reach was measured at 1.076×10^9 t. The complicated channel response to flow regime for this period was given by Ref. [2].

During storage period, suspended loads were mainly supplied from riverbed. Mean diameter of suspended particles varied simultaneously with the mean diameter of bed material. For flood of different magnitude, erosion capability increased with the increase of flood discharge, so did the sediment carrying capacity. Bed materials of fine and medium size would be suspended and carried away first. Bed materials of coarse size, which was the major contents of bed material, could hardly be carried away due to inefficient sediment carrying capacity. Therefore, the armoring process of riverbed was quick for large flood (Fig. 8). With the development of armoring process, less and less bed materials could be suspended and transported and thus erosion efficiency of total load could not be increased further. Therefore, erosion efficiency of total load depended mainly on the supplies of fine and medium size particles from bed materials.

358

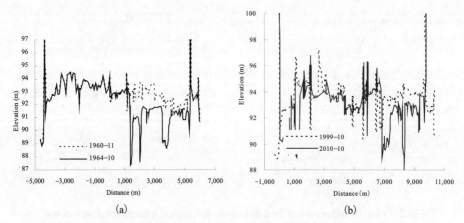

Fig. 7　Channel configurations at the Huayuankou section before and
after the storage period

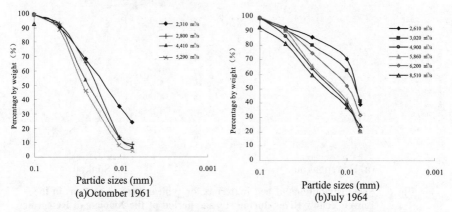

Fig. 8　Size variation of suspended material with flood discharge at the Lijin
station during storage period

4.2.3　Variation of roughness

Roughness represents the resistance of channel boundary to flow. It is usually expressed by Manning's coefficient n. In the Lower Yellow River, roughness is very large for small flow discharge. With the increase of flow rate, roughness decreases gradually and reaches its minimum value at certain flow rate. Afterward, roughness may increase slowly or maintain a constant value. Some studies show that the minimum value of roughness at the Lower Yellow River happens for flow rate between 1,000 m^3/s to 1,500 m^3/s.

During storage period, the larger the flood was, the more intensify the erosion and the quicker the armoring process (represents by equivalent roughness Δ) was. However, water depth (h) also increased with riverbed erosion, relative roughness (Δ/h) did not change much. Thus, roughness did not vary significantly with the increase of discharge during flood.

In the late phase of storage period, channel had been widened and armoring process had been finished. For a same magnitude flood, water depth was small and equivalent roughness was large. The relative roughness was large, so did the roughness. Therefore, erosion efficiency for flood of same magnitude was small during the later phase of storage period when compared with what it was in the earlier phase.

5 Remarks and suggestions

Analyses of erosion efficiencies during the storage periods of the Sanmenxia Reservoir and the Xiaolangdi Reservoir indicate that: erosion efficiency of total load increases with the increase of flood discharge and reaches its maximum value ($20 \ kg/m^3$) at the discharge about $4,000 \ m^3/s$, and afterward, erosion efficiency maintains at a constant value ($20 \ kg/m^3$) basically. By analyzing erosion efficiencies of grouped particles, we find that the decreasing extension of the erosion efficiency of fine particle is greater than the increase extension of the erosion efficiency of coarse particles.

Main factors affecting erosion efficiency include average flood discharge, size of suspended particle, configuration of channel cross-section, roughness, and etc. Among them, size of suspended particle is the dominant factor. During storage period, size of suspended particle is closely related to the size of bed material. With the increase of flow rate, fine and medium size particles are less available in bed material, thus erosion efficiency does not increase further with the increase of discharge. Analysis indicates that the contents of fine and medium size particles in bed material are essentials to erosion efficiency of total load.

Based on analysis given in this paper, It is suggested that the average flood discharge during the storage period of the Xiaolangdi Reservoir should be controlled at about $4,000 \ m^3/s$ to obtain the highest erosion efficiency, thus to make good use of the limited water resources and improve channel drainage capacity of the Lower Yellow River.

References

Fei Xiangjun. Study on Sediment Carrying Capacity of Low Sediment Concentration Flow in the Lower Yellow River [J]. Yellow River, 2004, 25(5):16 – 18.

Xu Jiongxin. Complicated Response of Channel Adjustment During Erosion Period of Clear Flow in the Lower Yellow River [J]. Advance in Water Science, 2001, 12(3):291 –299.

Liu Jinmei, Wang Shiqiang, Wang Guangoian. Preliminary Analysis on Unbalanced Sediment Transport in Long Distance of Alluvial River [J]. Journal of Hydraulic Engineering, 2002 (2):47 –53.

Zhang Ruijin. River Sediment Dynamics [M]. Beijing: China Water and Power Press, 1998;184.

Zhao Ye'an, Zhao Wenhao, et al. Basic Laws of Channel Evolution in the Lower Yellow River [M]. Zhengzhou: Yellow River Conservancy Press, 1998;16.

Mai Qiaowei, Shen Hongxin, et al. Study on Sediment and Channel Characteristics of the Lower Yellow River [R]. Yellow River Institute of Hydraulic Research, 1959;23.

Su Yunqi, Li Yong, Shen Guanqing, et al. Variation of Roughness and its Influence on Flood Control in the Lower Yellow River in Recent Years [R]. Yellow River Institute of Hydraulic Research, 1999;30.

Zhao Ye'an, Yao Weiyi, Liu Hailing, et al. Propotype Experiments for River Training in the Reach of Wenmengtan Resetllment for the Xiaolangdi Reservoir [R]. Yellow River Institute of Hydraulic Research, 1998;240.

Extreme Flow Changes in the Delta of West River and North River, China

Kong Lan[1], *Gao Zhengyi*[1], *Chen Junxian*[1], *Hu Lianghe*[1] and *Chen Xiaohong*[2,3]

1. China Water Resources Pearl River Planning Surveying and Designing Co. , Ltd. ,
Guangzhou, 510610, China
2. Center for Water Resources and Environment, SunYat-sen University, Guangzhou, 510275, China
3. Key Laboratory of Water cycle and Water Security in Southern China of Guangdong
Higher Education Institutes, SunYat-sen University, Guangzhou, 510275, China

Abstract: In this study, we analyze the long extreme flow series of two hydrological stations of the Delta of West River and North River by using statistical techniques. The results indicate that: ①the trend of the extreme flow is complicatedly affected by both natural and human activities. The trend of the extreme flow is somewhat similar between Sanshui and Makou hydrological stations. The annual minimum flow has an obviously downward trend while the annual maximum flow has an upward trend in the delta of West River and North River during 1959 ~ 2009; ②there are different changing characteristics of extreme flow in different time intervals, both the annual minimum flow and the annual maximum flow arrive to the extreme values in 2000s, the extreme flow has an upward trend from 1959 to 1980 at < 90% significance level, the trends of annual minimum flow at Sanshui and Makou hydrological stations are significant at > 95% level during 1959 ~ 1992, 1993 ~ 2009 and 1959 ~ 2009, the annual maximum flow has an upward trend at both stations from 1959 to 2009; ③Mann-Kendall analysis of extreme flow series shows that the changes of the annual minimum flow are greater than the annual maximum flow. This posed a new challenge for policy-making aiming to enhance human mitigation to water hazards and water resource management. The results of this study will be of great significance in water resources management and better human mitigation of the natural hazards in the delta of West River and North River.

Key words:extreme flow, trends, the Mann-Kendall trend test method, the Delta of West River and North River

1 Introduction

The delta of West River and North River has one of the most complicated deltaic drainage systems in the world with approximate one hundred channels. The delta is heavily populated though contribute much to the socioeconomic development of human society. The hydrological characteristics of the delta are complicatedly affected by both natural and human activities (Hou WD, et al. , 2004; Zhang et al. , 2009). Accordingly, the delta receive increasing concerns from hydrologists, policymakers, ecologists especially during last decades (Chen et al. , 2004; Zhang et al. , 2006; Lu et al. , 2008; Wang et al. , 2009; Kong et al. , 2010). The West River and the North River that flow through Guangdong Province are two major river systems, and their flood is a well-known problem. With the rapid development of social economy, the issue of water resources shortage has been becoming more and more seriously in dry season. In particular, the contradiction between the water supply and demand of dry season is a prominent problem because of increasing salinity since 1990s. The authors analyze the change characteristics of the annual maximum and minimum flow time series under the influence of natural factors and human activities, which has referential significance to guide the planning and sustainable utilization of water resources in the Delta of West River and North River.

2 Data and methods

Sanshui and Makou hydrological stations with long-term annual minimum flow and annual maximum flow data in the delta of West River and North River, China, were analyzed for this research (Tab. 1). The hydrologic data before 1989 are extracted from the Hydrological Year Book (published by the Hydrological Bureau of the Ministry of Water Resources of China) and those after 1989 are provided by the Water Bureau of Guangdong Province. The quality of the hydrological data was firmly controlled before release.

Tab. 1 Hydrological stations analyzed in the paper with annual extreme flow

Station	River	Longitude	Latitude	Date used(years)
Sanshui station	The North River	112°50′E	23°10′N	1959 ~ 2009
Makou station	The West River	112°48′E	23°07′N	1959 ~ 2009

In this paper, the main analysis method is the Mann-Kendall trend test method (M-K) (Mann, 1945; Kendall, 1975; Libiseller, 2002; Zhang et al., 2006; Kimpel et al., 2007). The catastrophe analysis of climatic changes was widely used in the research on climatic evolution in the 1970s. However, in this study, the M-K method has been used to explore the changing features of annual minimum and maximum flow series in the Delta of West River and North River, China.

The Mann-Kendall test can be used to assess the trends of hydrological or climatic elements time series. Assume the time series under study is $x_t(x_1, x_2, x_3, \cdots, x_n)$, which is stable, The definition of the test parameter S is as follows:

$$S = \sum_{i=1}^{n-1} \sum_{j=i+1}^{n} \text{sign}(x_j - x_i) \tag{1}$$

The $\text{sign}(x_j - x_i)$ is as follows:

$$\text{sign}(x_j - x_i) = \begin{cases} 1 & (x_j\text{-}x_i) > 0 \\ 0 & (x_j\text{-}x_i) = 0 \\ -1 & (x_j\text{-}x_i) < 0 \end{cases} \tag{2}$$

When $n \geq 10$ and the S is approximate normal distribution, the statistical test value Z is calculated as:

$$Z = \begin{cases} (s-1)/\sqrt{n(n-1)(2n+5)/18} & s > 0 \\ 0 & s = 0 \\ (s+1)/\sqrt{n(n-1)(2n+5)/18} & s < 0 \end{cases} \tag{3}$$

If $Z > 0$, it shows that the series has an upward trend, if $Z < 0$, the series has a downward trend, and if $Z = 0$, the series has no changing trend. When the absolute value of Z is greater than or equal 1.28, 1.64, 2.32, 90%, 95%, 99% confidence levels are accepted.

The Mann-Kendall test can show detailed changing trends of different periods in the analyzed time series and does not require the data to be normally distributed. Assume the time series under study is $x_1, x_2, x_3, \cdots, x_n$, and mi denotes the cumulative total of samples so that x_i. $x_j(1 \leq j \leq i)$, where n is the number of the samples. The definition of the statistical parameter dk is as follows:

$$d_K = \sum_{i=1}^{k} m_i \quad (k = 2, \cdots, n) \quad \text{with} \quad m_i = \begin{cases} +1 & x_i > x_j \\ 0 & x_i \leq x_j \end{cases} \quad (j = 1,2,\cdots,i) \tag{4}$$

Under the null hypothesis of no trend, the statistic d_k is distributed as a normal distribution with the expected value of $E[d_k]$ and the variance $var[d_k]$ as follows:

$$\begin{cases} E[d_k] = \dfrac{n(n-1)}{4} \\ Var[d_k] = \dfrac{n(n-1)(2n+5)}{72} \end{cases} \tag{5}$$

Under the assumption above, the definition of statistical index of UF_k is as follows:

$$UF_k = \frac{[\,d_k - E[\,d_k\,]\,]}{\sqrt{var[\,d_k\,]}} \qquad (k = 2, \cdots, n) \qquad (6)$$

UF_k follows the standard normal distribution. In a two-sided test for trend, the null hypothesis is rejected at the significance level of α. The UF curve shows the changing trend of a time series. The time series will be in downward trend if $UF < 0$ and vice versa. If the UF value is greater than the critical values, then this upward or downward trend is at $>95\%$ significance level.

3 Results and discussions

3.1 Trends of annual minimum flow

Fig. 1 shows that the annual minimum flow has an obviously downward trend at Sanshui and Makou hydrological stations during 1959 ~ 2009. The coefficient of variation is 0.802 at Sanshui hydrological station and it is 0.565 at Makou hydrological stations, which indicate that the annual minimum flow has significant changes. It can be seen from the UF curves of Fig. 2 that the annual minimum flow has decreased dramatically since 1990s at $>95\%$ significance level. The analysis shows that at Sanshui hydrological station, during 1950 ~ 1982, the annual minimum flow has a upward trend, which in 1962 ~ 1975 is significant at the 95% level, and at Makou hydrological stations, there is a downward trend in 1960s and an upward trend during 1971 ~ 1986 (Fig. 2).

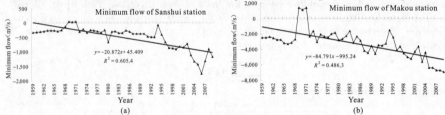

Fig. 1   The processes of the annual minimum flow

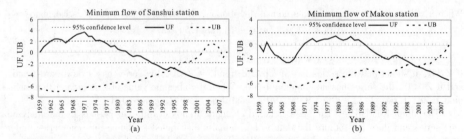

Fig. 2   The M-K trends of the annual minimum flow

3.2 Trends of annual maximum flow

Fig. 3 demonstrates that the annual maximum flow has an upward trend at Sanshui and Makou hydrological stations during 1959 ~ 2009. The coefficient of variation is 0.360 at Sanshui hydrological station and it is 0.285 at Makou hydrological stations, which indicate that the changes of the annual maximum flow are not as significant as the ones of the annual minimum flow. One can see from Fig. 4 that at Sanshui hydrological station, during 1968 ~ 1984 and 1997 ~ 2009, the annual maximum flow has an upward trend, which in 2006 ~ 2009 is significant at the 95% level, and has a downtrend in 1985 ~ 1996. At Makou hydrological station, the trend of the annual maximum flow is somewhat similar to the one of Sanshui hydrological station, but the former is not significant at the 95% level (Fig. 4).

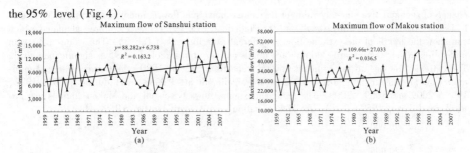

Fig. 3   The processes of the annual maximum flow

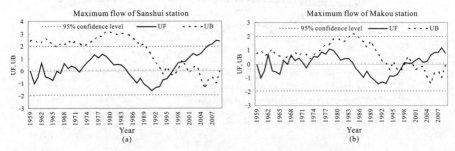

Fig. 4   The M-K trends of the annual maximum flow

3.3   Changes of different time interval

Tab. 2 shows that there are obvious changes of extreme flow in different age. The minimum flow has a downward trend, while the maximum flow has an upward trend. In 2000s, both the annual minimum and maximum flow arrive to the extreme values.

Tab. 2   The changes of average extreme flow in different age

(Unit: m³/s)

Time interval	Minimum flow		Maximum flow	
	Sanshui station	Makou station	Sanshui station	Makou station
1960 ~ 1969	−259	−2,800	7,882	28,922
1970 ~ 1979	−212	−1,499	8,814	30,780
1980 ~ 1989	−362	−2,984	6,826	25,580
1990 ~ 1999	−560	−3,597	10,525	31,690
2000 ~ 2009	−1142	−5,613	11,260	32,840
1959 ~ 2009	−497	−3,200	9,033	29,884

The increasing trend of the curve of accumulative anomaly value stands for the positive anomaly and vice versa. The curve can directly show long-term trends and sustainable change of series. Fig. 5 implies similar changing properties in terms of the curves of accumulative anomaly value at Sanshui and Makou stations, i. e. the annual minimum flow has increasing trends during 1959 to early 1990s and decreasing trends during early 1990s to 2009; the annual maximum flow has decreasing trends during 1959 to early 1990s and increasing trends during early 1990s to 2009; the change of accumulative anomaly value of the annual maximum flow is unobvious during 1959 to early 1980s. Based on the above analysis, the statistical M-K test values (Z) are calculated which can show the trends of the extreme flow of different time intervals (Tab. 3). The values (Z) with under line indicts the upward or downward trends are very remarkable at >95% significance level. The extreme flow has an insignificant upward trend from 1959 to 1980, the trends of annual minimum flow at Sanshui and Makou stations are significant at >95% level during 1959 ~ 1992, 1993 ~ 2009 and 1959 ~ 2009. The annual maximum flow at Sanshui and Makou stations has upward trend during 1959 ~ 2009, but Makou station's isn't significant at >95% level.

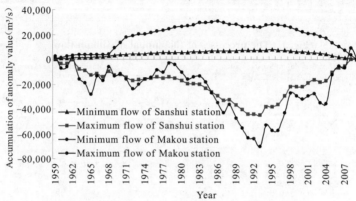

Fig. 5   The curves of the accumulative anomaly value

Tab. 3   The statistical M-K test values (Z) of extreme flow

Time interval	Minimum flow of Sanshui station	Minimum flow of Makou station	Maximum flow of Sanshui station	Maximum flow of Makou station
1959 ~ 1980	0.45	1.02	0.85	0.56
1959 ~ 1992	-2.96	-2.05	-1.33	-1.31
1993 ~ 2009	-3.50	-4.24	0.37	4.12
1959 ~ 2009	-6.43	-5.48	2.40	0.76

4   Conclusions

In the past decades, altered extreme flow processes across the Delta of West River and North River were observed. The direct consequences of these hydrological alterations are higher risk of flood inundation in flood seasons and more difficult human withdrawal of fresh water resource due to more frequent salinity intrusion in the dry seasons at the estuary. We obtained some interesting and important conclusions based on the aforementioned analysis:

(1) The trend of the extreme flow is very complicate that is affected by both natural and human activities which are including the changes of precipitation and human-induced deepening of river channels. The annual minimum flow has an obviously downward trend and the annual maximum flow has an upward trend in the Delta of West River and North River during 1959 ~ 2009.

(2) The changing characteristics of extreme flow are different in the different time intervals. In 2000s, both the annual minimum and maximum flow arrive to the extreme values. The extreme flow has an insignificant upward trend from 1959 to 1980, the trends of annual minimum flow at Sanshui and Makou stations are significant at >95% level during 1959 ~ 1992, 1993 ~ 2009 and 1959 ~ 2009. The annual maximum flow at Sanshui and Makou stations has an upward trend during 1959 ~ 2009. Therefore, the changes of annual minimum flow are greater than those of annual max-

imum flow.

References

Chen X H, Chen Y Q. Hydrological Change and its Causes in the River Network of the Pearl River Delta[J]. Acta Geographica Sinica, 2002, 57 (4):429 – 436. (in Chinese)

Chen X H, Zhang L, Shi Z. Study on Spatial Variability of Water Levels in River net of Pearl River Delta[J]. Journal of Hydraulic Engineering, 2004(10):36 – 42. (in Chinese)

Hou W D, Chen X H, Jiang T, et al. Temporal Change of Flow Distribution in River Network of the Delta of West River and North River[J]. Acta Scientiarum Naturalium Universitatis Sunyatseni, 2004(43)(supplment):204 – 207. (in Chinese)

Kendall M G. Rank Correlation Methods[M]. Griffin: London, UK, 1975.

Kimpel I, Mandabu D, Yamanaka, et al. Precipitation in N. Cpell between 1987 and 1996[J]. International Journal of Climatology, 2007, 15(2):245 – 256.

Kong L, Chen X H, Zhang Q, et al. Impacts of Rising Sea Level on Water Level Changes Along the Pearl River Estuary, China[J]. Ecology and Environmental Sciences, 2010,19(2): 390 – 393. (in Chinese)

Libiseller C. A Program for the Computation of Multivariate and Partial Mann-Kendall Test[D]. Sweden:University of Linkö Ping, 2002.

Lu Y J, Jia L W, Mo S P, et al. Changes of Low Water Level in the Pearl River Delta Network [M]. Beijing: China Water and Powr Press, 2008. (in Chinese)

Mann H B. Nonparametric Tests Against Trend[J]. Econometrica , 1945(13):245 – 259.

Wang L N, Chen X H, Li Y A, et al. Rules of Runoff Variation of Xijiang River Basin[J]. Journal of China Hydrology, 2009, 29(4):22 – 25. (in Chinese)

Zhang Q, Xu C Y, Becker S, et al. Sediment and Runoff Changes in the Yangtze River Basin during Past 50 Years[J]. Journal of Hydrology, 2006(331):511 – 523.

Zhang Q, Xu C Y, Chen Y Q, et al. Abrupt Behaviors of the Streamflow of the Pearl River Basin and Implications for Hydrological Alterations across the Pearl River Delta, China[J]. Journal of Hydrology 2009:274 – 283.

Zhang Q, Xu C Y, Chen Y Q, et al. Spatial Assessment of Hydrologic Alteration across the Pearl River Delta, China, and Possible Underlying Causes[J]. Hydrology. Process,2009 (23): 1565 – 1574.

Study of the Modern Yellow River Delta Wetland Landscape Fragmentation and its Driving Forces

Wang Shiyan , *Mao Zhanpo* , *Liu Jinke* and *Liu Chang*

China Institute of Water Resources and Hydropower Research, Beijing, 100038, China

Abstract: In this paper, the wetland landscape features of the Modern Yellow River Delta have been interpreted from the remote sensing images of the years of 1976, 1986, 2000 and 2008. Based on the landscape ecological theory, the wetland landscape fragmentation and the driving forces are analyzed. The results show that: in general, the maximum patch area, the mean patch size and the mean proximity index of wetlands are continuously decreased. The index of wetland patch density is increased. These results indicate that the wetland landscape in the Modern Yellow River Delta is keeping fragmentizing during recent 32 years. However, the fragmentation process is very complex and the constructed wetland types and the natural wetland ones in the Modern Yellow River Delta area show different changing trends. In the long run, the natural factors of river runoff and sediment load of the Yellow River are the key factors. While in a short time, anthropogenic factors such as agricultural reclamation, coastal aquiculture and oilfield development are the dominating factors of estuarine wetland landscape fragmentation.

Key words: the Modern Yellow River Delta, wetlands, landscape fragmentation, driving forces

Landscape pattern fragmentation is refers to a process from simple to complex tendency which caused by disturbance from natural and anthropogenic factors, that is landscape from a single, homogeneous and continuous and as a whole tends to be complexity, heterogeneous and discontinuous patch mosaic. A deeper degree of landscape pattern fragmentation will lead to an increase in distance between patches of different landscape types, low connectivity and block of regional substances and energy exchange. The landscape pattern fragmentation which caused by human activities is one of main reasons of biology diversity loss .

Estuary wetland is one of important types of wetlands, it is affected directly by both the upstream runoff of rivers and seawater erosion of river mouth, the special hydrology condition determines that the wetland ecosystems are easier to suffer disturbance from nature and human activities, and the ecological conditions are easy to be destroyed. With the development of social economy and population increase, wetland ecosystems in the estuary area are in degradation under the influence of human activities. Nowadays, more than 20% of the estuarine wetland vegetation has ceased to exist, and the rest are disappearing rapidly. As the most complete, most extensive and the youngest estuarine wetlands in China's warm temperature, the Yellow River Delta wetlands were damaged at a large extent and the wetlands present a degradation tendency because of sharp reduction of water and sediment resources, river channelization from diversion dikes' construction, and agricultural development and urbanization of the river deltas, and so on. At present, the international wetland academic circles pay great attention to wetland degradation and that wetland rehabilitation, wetland degradation and recovery, conservation and reasonable utilization had been regarded as one of the most important issues in previous international wetland conferences. Furthermore, it has been found in the course of the study that the Modern Yellow River Delta wetland degradation process is accompanied by significant fragmentation phenomenon. Landscape pattern fragmentation is not only the representation of wetland degradation, but also one of degradation driving factors. So clarify the fragmentation situation of the Yellow River Delta wetland and changing trends in a long time scale, both have certain positive significance to the study of the degradation characteristics of the Yellow River delta and the wetland recover policy establishment.

In recent years, there are lots of researches which used RS and GIS technology on the Yellow River Delta, however, most of these researches are only pay attention to the landscape area, wetland patch density in a certain period, and also only have a brief description and analysis to the landscape pattern fragmentation phenomenon, that long time series, long periods, comprehensive and in-depth study on the Yellow River Delta wetland landscape pattern fragmentation is very few, the wetland fragmentation driving forces study is also less.

In this paper, the Modern Yellow River Delta is as a research unit, whose wetland landscape fragmentation has been studied based on 1976, 1986, 2000 and 2008 multi-source remote sensing images, which reflecting the changes of the modern Yellow River Delta wetland landscape with a larger spatial and temporal scale. The paper is aiming to clarify the fragmentation degree of the Yellow River Delta, and the changing situation during 1976 to 2008, and to explore the driving forces of wetland landscape fragmentation. The study is helpful for the identification of wetland degradation mechanism and to provide a scientific basis for degraded estuarine wetland recovery and reconstruction.

1 Study area

The Yellow River Delta lies between Bohai Bay and Laizhou Bay; it was a piece of land surface delta formed by the Yellow River sediment deposition, sea reclamation since 1855 when the Yellow River changed its course into Bohai sea. As one of the three biggest deltas in China, the Yellow River Delta is the world's fastest-growing deltas. Chronologically, The Yellow River Delta can be divided into Ancient, Modern and the Modern Yellow River Delta. The Yellow River Delta concerned in the paper refers to the Modern Yellow River Delta (Fig. 1), that is the estuarine area formed from Yuwa of Kenli County, north from Tiaohe bay and south to Songchunrong ditch mouth. The study area is of the monsoon climate of warm-temperature zone with the distinct four seasons and climate demarcation line of cold and hot, dry and humid. It's drought and windy in spring, hot and rainy in the summer, cool and clear in the autumn, and cold, dry and snowless in the winter, with distinct inland climate characteristics. Regional mean annual precipitation is 613.6 mm, the average annual temperature is 12.8 ℃.

In the long term interaction among the river and sea and land, the Yellow River Delta formed one integrated and unique regional ecosystem, also a typical estuarine wetland ecosystem. The Yellow River Delta wetland is the most youngest estuary wetland in China, it has the most extensive and integrated new-born estuary wetland ecosystem in China's warm temperature zone, and it is an important transfer station, wintering area and breeding area of the west pacific birds migration in the northeast Asia continent, which is an important region for species protection, wild birds migration and estuary ecological succession. The Yellow River Delta wetland ecosystems are also a unique region which possesses an important ecological protection value among the three biggest River deltas involving the Yangtze River Delta, the Pearl River Delta and the Yellow River Delta. Ecological balance of the region's wetland ecosystems has been an important sign of the Yellow River health.

2 Material and methods

2.1 Data sources

The study uses four remote sensing data, respectively, for 1976 of Landsat MSS images, 1986, 2000, 2008 of Landsat TM / ETM images. Through the field investigation, GPS positioning for images interpretation, and combined with materials collecting such as regional topographic maps, soil maps, vegetation maps and other related socio-economy data, thematic interpretation of those remote sensing images have been fulfilled and also interpretation accuracy has been confirmed by GPS points collected in the field.

2.2 Wetland classification

The Modern Yellow River Delta has a wide variety of wetland types and landscape types. It is necessary to build a classification system before the image interpretation and wetland change analysis. Based on Ramsar wetland classification system, China Wetland Inventory Outline, and also

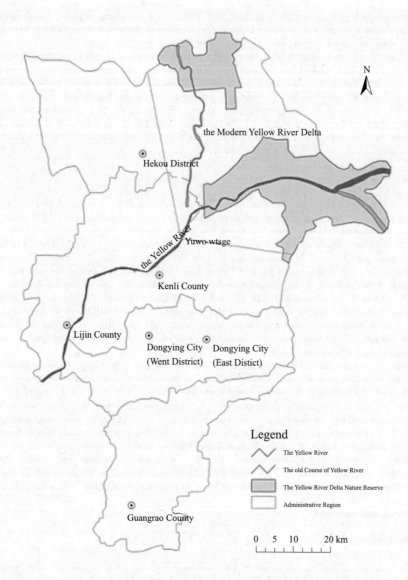

N

the Modern Yellow River Delta

Hekou District

the Yellow River

Yuwo-wtsge

Kenli County

Lijin County

Dongying City
(Went District)

Dongying City
(East Distict)

Legend

The Yellow River

The old Course of Yellow River

The Yellow River Delta Nature Reserve

Administrative Region

Guangrao County

0   5   10   20 km

Fig. 1   The location of the Modern Yellow River Delta

with reference to some relative literatures, the Modern Yellow Delta wetland can be divided into three levels in accordance with the ecological characteristics. The total wetlands are divided into two major categories of natural wetlands and artificial wetlands, firstly. In natural wetlands, and then divided into the type of coastal wetland, river wetland, marsh wetland and so on. Artificial wetlands can be divided into reservoir pond wetland, ditch wetland and so on. Next the third level division of wetlands has been conducted. Tab. 1 shows the final wetland classification system and its distribution of the Modern Yellow River Delta.

Tab. 1 Wetland types and distribution of the Modern Yellow River Delta

Level I	Level II	Level III	Distribution
Natural wetlands	Coastal wetland	Sub – littoral flat bench	Mainly distributes in these coastal shallow waters with water depth is no more than 6 m when low tide
		Middle tidal flat bench	Tide dip zones between average high water line and the low water line
	River wetland	Channel Wetland	The Yellow River channel and other river channels
		Floodplain	Seasonal water areas mainly distributed on the both sides of the Yellow River channel
	Marsh wetland	Marsh wetland	Floodplains, intertidal zone, supralittoral zone and inland saline area
	Meadow wetland	Reed meadow	Mainly distributed in the floodplain, coastal estuary and the intertidal zone
	Others	Saline wetland	Scattered in the delta area
Artificial wetlands	Reservoir pond wetland	Reservoir pond	Lowland areas of the Delta
	Upland farmland and lowland paddy wetland	Paddy fields	Distributed in arable land area of the water sources and irrigation facilities
	Ditch wetland	Ditch wetland	Agricultural ditches and drains, mainly distributed in agricultural areas
	Saltern field	Salt fields and aquiculture ponds	Mainly distribute at coast region and nearly estuary area

2.3 Data processing

The remote sensing images are been processed with 3S spatial information technology, including images bands combination, images fusion, spatial information enhancement, projection and transformation, statistical analysis of spatial information. In the course of images interpretation, through the established wetland classification system, and typical wetland remote sensing thematic interpretation signs built in the field GPS, the four remote sensing thematic images are been visually interpreted. And a confusion matrix has been calculated to analysis the interpretation accuracy. The general interpretation accuracy can reach 96% which shows that the result is acceptable. Fig. 2 is the wetland distribution map of the modern Yellow River Delta of 1976, 1986, 2000 and 2008.

2.4 Study methods

With the development of landscape ecology, a lot of mathematic methods were introduced into the researches. Quantitative analysis approaches which reflect landscape pattern fragmentation are more and more, according to their contents, these approaches can be divided into three parts: landscape patches fragmentation analysis, landscape heterogeneity analysis, and landscape elements interaction fragmentation analysis. In this paper, those common indicators including patch area, patch density, average patch area, and average adjacent degree index are used to analyze the wetland landscape fragmentation of the Modern Yellow River Delta.

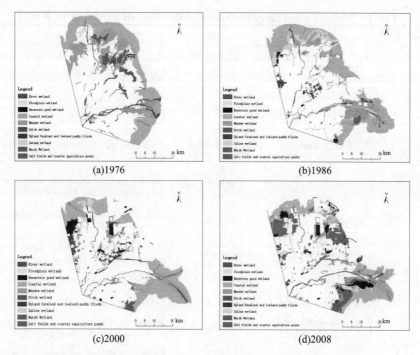

(a)1976   (b)1986

(c)2000   (d)2008

Fig. 2 Wetland distribution of the Modern Yellow River Delta

2.4.1 Patch density (PD)

The ratio of amounts of some type patches to landscape area represents the degree which the landscape matrix divided by this type of patch, i. e. the patch density of this landscape component in the whole landscape (namely porosity). This indicator is composed of landscape patch density and landscape elements patch density. Landscape patch density refers to the unit area patch amount which including all the heterogeneous landscape elements in the landscape. Landscape elements patch density refers to the unit area patch amount of certain kind of landscape element in the landscape. Its formula is:

$$P_{PD} = \sum n_i /A \tag{1}$$

where, P_{PD} is patch density value; $\sum n_i$ is the patch amount of research area or the patch amount of some kind of landscape element; A is research area or a certain kind of landscape patch area.

The higher the value of P_{PD}, the higher the degree of landscape type boundary fragmented, also indicating that the higher the degree of this type of landscape or landscape elements. Conversely, the well – preserved landscape types have relatively higher connectivity and lesser fragmentation degree.

2.4.2 Landscape mean proximity degree index (MPI)

A type of landscape pattern area divided by the square of the nearest distance from its similar type of patch, adds to all patches, and divided by the total number of this type of patches, and then we can get the MPI of this type of landscape. In landscape level, MPI equals to average proximity degree index of all landscape patches, it can measure the proximity degree between patches, and manifest the fragmentation degree of the landscape. A small value of MPI indicates that the dispersion degree between patches but high fragmentation degree. A high MPI value indicates that the proximity degree between patches is high, patch connectivity is well, and fragmentation degree is low.

$$M_{MPI} = \sum_{i=1}^{n} \left(\frac{a_i}{h_i} \right) /n \tag{2}$$

where, M_{MPI} is average proximity degree index; n is the landscape patch count; a_i is patch area of landscape i; h_i is the nearest distance from a patch to the same type patch.

3 Wetland landscape fragmentation analysis

3.1 Change of landscape patch area

The fragmentation of patch area can be reflected by the change of maximum patch area and average patch area. The modern Yellow River Delta Wetland landscape and their patch area change of the Modern Yellow River Delta can be seen in Tab. 2.

Tab. 2 Changes in wetland area of the Modern Yellow River Delta

(Unit: hm²)

Year	Artificial wetlands	Natural wetlands	Maximum patch area	Average patch area
1976	524.79	126,401.04	10,895.07	968.9
1986	8,763.93	102,160.35	10,528.23	396.16
2000	18,682.65	79,843.14	8,764.62	351.94
2008	38,277.99	66,645.81	8,287.93	231.64

As shown in Tab. 2, it can be seen from 1976 to 2008, the area of artificial wetlands in the Modern Yellow River Delta is increasing continually, while that the area of natural wetlands, wetland maximum patch area and average patch area are obviously decreasing. During the period of 1976 to 1986, artificial wetlands area increase by 37,753.20 hm², while natural wetlands area decrease by 59,755.23 hm², this change of wetland area is concerned with urban construction of Dongying City, Shengli Oilfield development and construction, and regional agricultural cultivation of wetlands, which shows that with the intensifying of human activities, artificial wetlands area in the Modern Yellow River Delta increases ceaselessly, and wetland landscape pattern affected by human disturbance is more obvious. During the period of 1976 to 2008, the maximum patch area decreases 2,607.14 hm², decreases by 24%, in which, in the yeas of 1986 to 2000 decreased the most, decreases by 1,763.61 hm². In more than 30 years, average patch area of wetland changes in the larger, a total reduction of 737.26 hm², 76% reduction. This shows that wetland landscape patch amount of the Modern Yellow River Delta is increasing and average patch area is reducing, namely the wetland landscape fragmentation phenomenon is increasingly evident.

3.2 Change of landscape patch density

Patch density reflects the relationship between patch quantities and patch areas, expressed per unit area on a number of patches. Larger values indicate that the higher the degree of landscape fragmentation. In order to more fully reflect the wetland fragmentation of the study area, artificial wetlands and natural wetlands are calculated for patch density respectively, and the calculation results and overall patch density of the total wetlands are compared (Tab. 3).

Tab. 3 Patch density of wetlands in the Modern Yellow River Delta

Type	Patch density(individual/100 hm²)			
	1976	1986	2000	2008
Artificial wetlands	0.190,6	1.483,4	0.856,4	0.603,5
Natural wetlands	0.102,8	0.146,8	0.150,3	0.334,6
Total wetlands	0.103,2	0.252,4	0.284,1	0.431,7

As shown above, the total wetlands patch density and natural wetlands patch density of the Modern Yellow River Delta show an increasing trend, of which during 1976 ~ 1986, 1986 ~ 2000 and 2000 ~ 2008, the total wetlands patch density increase respectively 145%, 13% and 52%, as for the patch density of the natural wetlands increases by 42%, 2.4% and 123% respectively. Artificial wetlands are controlled by the human factors, the patch density change trend and the former two are quite different, of which between 1976 and 1986, patch density of artificial wetlands increases by 678%, while appears to reduce from 1986 to 2000 and from 2000 to 2008, and decreases by 42% and 30% respectively. Thus, affected by many factors, the Yellow River Delta wetland landscape fragmentation shows different characteristics. In the 1970s, the Modern Yellow River Delta disturbed by human activities was less, basically no artificial wetland. At the same time, natural wetlands were also less affected by human disturbances, the patch density was small, and fragmentation was not obvious. Until 1980s, with the development of agriculture and industry, especially the oilfield development, regional economic construction, and agriculture rice planting widely, the artificial wetlands area of reservoirs, ponds, upland farmland and lowland paddy, and culture ponds in the Modern Yellow River Delta increased significantly, which resulted to the increase of patch density of artificial wetlands and their landscape fragmentation. Meanwhile, with the increase of artificial wetlands and human development, natural wetlands were more severely damaged, patch density increased and fragmentation deepened. After 1990s, along with the coastal salt industry and the rapid development of aquiculture, artificial wetlands such as salt filed, shrimp ponds and fish ponds appear to coastal direction spatial gathering changes, and the landscape fragmentation of artificial wetlands is reduced. In contrast to the artificial wetlands, natural wetlands, the patch density increases significantly, increase the degree of fragmentation.

3.3 Change of mean landscape proximity index

Landscape mean proximity index (*MPI*) can reflect landscape connectivity and adjacent degree, can comprehensively reflect the landscape fragmentation degree. According to above quation, the *MPI* of wetlands of the Modern Yellow River Delta has been calculated (Tab. 4).

Tab. 4   The Mean Proximity index (*MPI*) of wetland landscapes in the Modern Yellow River Delta

Type	MPI			
	1976	1986	2000	2008
Artificial wetlands	0	49.96	69.60	199.89
Natural wetlands	890.62	622.04	378.74	185.05
Total wetlands	617.29	500.27	208.62	219.75

As shown in Tab. 4, artificial wetlands mean proximity index increased continually in more than 30 years. The main reasons are that since 1980s the constructions of reservoirs, ponds, ditches, paddy fields and salt fields in the Modern Yellow River Delta was ceaseless aggravating, especially the coastal beach development was increasingly strengthened. The artificial wetlands present sectors – connecting distribution, the wetland landscapes connectivity is increased, and thereby the *MPI* shows a significant increasing tendency. While in 2008, the *MPI* value of natural wetlands is only 20% of that in 1976, about 80% fewer than that in 1976. This is mainly because of human disturbances including reduced incoming water of the Yellow River, regional oilfield exploration, urban infrastructure construction and agricultural development, which has caused natural wetland fragmentation. The *MPI* of total wetlands also shows a decreasing change, the wetland *MPI* value in 2008 decreases by 397.5, about 64% compared with that in 1976, which explains regional wetland connectivity decrease overall, the discretization degree among wetlands increases, and wetland landscape fragmentation is generally intensifying.

4   Driving forces analysis of wetland landscape fragmentation

Due to reduction of water incoming of the Yellow River and serious interference of human activities, regional wetlands in the Yellow River Delta have presented a shrunken sign. Biodiversity in the Delta region and the primary productivity of terrestrial ecosystems and coastal ecosystems suffered a serious damage, that lead to the estuary wetland ecosystems shrink, and wetlands have disappeared. Therefore, the main driving forces of the Yellow River Delta wetland degradation, shrinking and landscape fragmentation come from two aspects: natural driving factors and human driving factors, of that, natural factors mainly refer to climate change, runoff and sediment variation in the Yellow River; human factors mainly include farmland reclamation, coastal aquiculture and oilfield development.

4.1　Natural driving forces

4.1.1　Climate factors

Climate factors which have an effect on the Modern Yellow River Delta landscape fragmentation are mainly air temperature and precipitation. According to related researches, during the period of 1954 to 2002, regional temperature and precipitation of the Yellow River Delta changes only slightly. Air temperature has a trend to rise and the average annual temperature increases by 0.012 ℃ during year 1986 to 2001. While annual average precipitation shows a weak decreasing trend, the average annual reduction of 1.41 mm. Meanwhile, Reed marsh, meadow wetlands and coastal wetlands area and air temperature show negative correlation. Water is a key factor to wetlands existence and precipitation has an important positive effect on the development and succession of regional wetland ecosystems. Therefore, the wetland landscape fragmentation of the Modern Yellow River Delta has a certain relationship with regional precipitation and air temperature, but not obvious. The less rainfall, higher temperature, evaporation amount is larger; and wetlands will be shrank and degraded due to lack of water.

4.1.2　The Yellow River runoff and sediment yield

The Yellow River siltation epeirogenesis is the fundamental driving force of wetland growth and decline of the Yellow River Delta, and the deposition and development of the Yellow River terminal channel have a decisive function to the wetland formation and maintenance of the Yellow River Delta. The Yellow River coming water and sediment changes, not only affect the wetland sedimentation and erosion, but also affect the wetland area maintenance, evolution of wetland landscape patterns, wetland ecosystems balance and overall wetland ecological function.

According to relevant studies, the landscape patterns change of main wetlands types (reed marsh wetland, meadow wetland, and coastal wetland) in the Yellow River Delta has a positive correlation with the Yellow River incoming runoff and sediment, with the increase of river runoff and sediment the wetland ecological space distribution will be expanded obviously. According to Li, when annual average runoff below 1.50×10^8 m^3, annual average sediment yield below 4×10^8 t, and wetland landscape area decreases significantly with runoff and sediment reduction. And among the three factors, runoff and sediment yield make a decisive function. It is thus clear that the change of the Yellow River runoff and sediment yield has an enormous influence to wetland landscape of the Yellow River Delta, and also it is one of the decisive factors to the fragmentation of landscape. According to the hydrological data analysis of Lijin Hydrological Station, from the year 1950 to 2005, the Yellow River runoff decreases prominently in flood season and dry season, and the decline rate is achieved respectively 3.87×10^8 m^3/a and 0.62×10^8 m^3/a. With the rapid socio-economic development of the Yellow River basin, and the extensive use of water resources, the Yellow River water resources are increasingly short. Since 1972, the Yellow River downstream Lijin hydrological station began to appear no flow. Henceforth, the Yellow River no flow occurs frequently. In the 1990s, the no flow time of Yellow River brought forward constantly, the no flow length, the no flow duration increased constantly, and the flow decreased constantly. Especially in 1997, the Yellow River occurred the most serious no flow in the channel, no flow appeared 13 times, and days of no flow added up to 226 d, and no flow channel length reached 704 km.

The coming sediment in the Yellow River presents reducing trend in accord with the change of runoff, the coming sediment decreased from 11.72×10^8 t in 1960s to 6.55×10^8 t in 1980s, in

1990s the annual average sediment further reduced to 3.45×10^8 t. The Yellow River no flow and coming water and sediment decrease continuously, which break the water and salt balance in the wetland soils, and make the mineralization degree of groundwater increase and soil salinity rise, the surface vegetation communities begin to evolve to salinity tolerance, and wetland vegetation coverage decreases, leading to ecological degradation of newborn estuary wetland ecosystems. Since 2002, although united water and sediment regulation in the Yellow River has been carried out, the coming water and sediment still are less, and the ecological characteristic of estuary wetland ecosystems and landscape patterns structure, wetland fragmentation will still maintain the current changing tendency.

4.2 Human driving forces

The population of the Yellow River Delta is increasing more and more, and regional socio – economy increases rapidly, human activities such as wetlands reclamation, oilfield exploration, urbanization construction, coastal aquiculture, and so on, all that can make an important impact on wetland degradation and wetland landscape fragmentation. Therein, human disturbances such as wetland reclamation, coastal aquiculture and oilfield development have a prominent effect on wetlands.

4.2.1 Wetlands reclamation

Since 1950s, the large area of lands in the Yellow River Delta has been reclaimed. At that time, with the one – sided emphasis on "take grain as the key link", human reclaimed a large area of grassland, forest land, floodplains and wetlands, which caused arable land increases substantially. Whereas, at the same time, the original wetland ecosystems were destroyed badly, soils conditions changed, secondary salinization aggravated, abandoned farmland area began to increase ceaselessly and wetland area decreased widely. Since 1980s, paddy planting is one of the important measures for reclamation of saline and alkaline soils and lands explorations, which causes the tremendous increase of paddy fields. But because of the lack of unified planning and effective management, the blind development of paddy fields is extremely apt to cause the regional secondary salinization widely. Finally, due to human's unreasonable reclamations, it forms a vicious circle of reclamation—abandoned farmland—reclamation again – abandoned farmland again. And agricultural reclamations cause wetland area to decrease substantially, thereby wetland landscape fragmentation and discretization characteristics have been very obvious, and so those wetland landscape fragmentation degrees are deepened ceaselessly.

4.2.2 Coastal aquiculture

The Yellow River Delta coastal beaches have been the development and construction of aquiculture ponds, breeding area to be expanded ceaselessly. Since 1980s, coastal aquiculture ponds become more widely distributed. According to remote sensing monitoring data in the paper, the total coastal aquiculture ponds area was $1,633.5$ hm^2 in 1986, $7,421.3$ hm^2 in 2000, and $21,710.5$ hm^2 in 2008, of which from 1986 to 2000 and from 2000 to 2008, the total coastal aquiculture ponds area increased by $5,787.8$ hm^2, $14,289.2$ hm^2 respectively. Furthermore, aquiculture ponds wetland area accounted for the proportion of artificial wetlands is also increasing, particularly in 2008 the percentage reached 56.7%, which indicated that the aquiculture ponds has been the most important part of the artificial wetlands. It is also can be seen from the Modern Yellow River Delta wetland distribution map, the aquiculture ponds are mainly distributed in coastal region. It is thus clear that with the continuous expansion of aquiculture ponds, coastal beach wetlands are more and more occupied, which thereby brings about cutting off of wetland habitats and aggravating estuary wetland landscape fragmentation.

4.2.3 Oilfield development

Shengli Oilfield is China's secondary largest oilfield located in this region. According to the statistic date from Shengli Oilfield administration, by the end of year 2005, there were 36 verified oilfields, involved 7 oil picking factories altogether such as Shengcai, Donginin, Gudao, Hekou and Binnan. That the oilfield development and construction form a compact district interconnected or interwoven with oil wells, stations, roads, power lines, and underground pipelines in this region. Oilfield development drives the development of Dongying City and regional industrialization and urbanization, but at the same time oilfield development has also brought some negative impacts and destruction to the regional wetland ecosystems. For example, oil and gas causes the partition of wetland ecological integrality, wetland bird's habitat occupied, wetland vegetation deforestation, and so on.

5 Conclusions

Study on wetland landscape fragmentation has an important significance to recognize the wetland degradation mechanism and wetland ecological restoration and protection. In the paper, that the maximum patch area, average patch area, patch density and landscape mean proximity index of the Modern Yellow River Delta wetland has been calculated and analyzed in order to identify the spatial and temporal characteristics of estuary wetland landscape fragmentation. The result shows that wetland landscape fragmentations of artificial wetlands and natural wetlands in the Modern Yellow River Delta have different change characteristics. Of that, artificial wetlands distributions from inland to coastal tend to concentrate, landscape connectivity between patches improves and wetland landscape fragmentation is to mitigate. Meanwhile, mainly due to human's disturbances, natural wetlands area reduce rapidly and landscape connectivity between patches become worse, thereby landscape fragmentation degree are tend to deepen.

Natural factors and human activities are the two main driving factors of all driving forces of wetland landscape fragmentation of the Modern Yellow River Delta. In the long run, that the Yellow River runoff and sediment load reduction plays a decisive role on wetland landscape fragmentation of the Modern Yellow River Delta. While in a short time, that agricultural reclamation, coastal aquiculture and oilfield development causes wetland landscape fragmentation directly. In short, the wetland landscape fragmentation of the Modern Yellow River Delta is a very complex process affected by both natural changes and human's disturbances.

Acknowledgements

The study was financially supported by the National Natural Science Foundation of China (Grant number 50709044 and 51179207), and Research subject of the China Institute of Water Resources and Hydropower Research (Grant number HJ1135). Thanks should be extended to Jian Chen and Liang Wang for their great efforts in processing the remote sensing images.

References

You C, Zhou Y B, Yu L F. An Introduction of Quantitative Methods in Landscape Pattern Fragmentation[J]. Chinese Agricultural Science Bulletin,2006, 22(5): 146-151.

Liu J F, Xiao W F, Jiang Z P, et al. A Study on the Influence of Landscape Fragmentation on Biodiversity[J]. Forest Research,2005, 18(2): 222-226.

Naiman R J, Decamps H, Pollock M. The Role of Riparian Corridors in Maintaining Regional Biodiversity[J]. Ecology,1993, 3: 309-312.

Lian Y, Wang X G, Huang C, et al. Environmental Flows Evaluation Based on Eco-hydrology in Yellow River Delta Wetlands[J]. Acta Geographica Sinica,2008, 63(5): 451-461.

Zhao Y M, Song C S. Scientific Survey of the Yellow River Delta National Nature Reserve[M]. Beijing: China Forestry Publishing House,1995.

Fang Y, Liu Y L. Study on Wetland Vegetation Restoration of the Yellow River Delta[M]. Beijing: China Environmental Science Press, 2010.

Mitch W J, Gosselink J G. Wetland [M]. 3ed. New York: John Wiley & Sons, Inc, 2000.

Beth Middleton. Wetland Restoration Flood Pulsing and Disturbance Dynamics[M]. New York: John Wiley & Sons, Inc, 1999.

Paul A, Keddy. Wetland Ecology-principles and Conservation[M]. Cambridge: Cambridge University Press, 2000.

Allan Crowe. Quebec 2000: Millennium Wetland Event Program With Abstracts [M]. Canada: Elizabeth MacKay, 2000.

Davis S M, Ogden J C. Everglades—the Ecosystem and its Restoration [M]. Delray Beach: St. Lucie Press, 1994.

Chen W F, Zhou W Z, Shi Y X. Crisis of Wetlands in the Yellow River Delta and its Protection [J]. Journal of Agro – Enviornment Science, 2003,22(4):499 – 502.

Cui B S, Liu X T. Ecological Character Changes and Sustainability Management of Wetlands in the Yellow River Delta [J]. Scientia Geographica Sinica, 2001, 21(3): 250 – 256.

Zong X Y, Liu G H, Qiao Y L, et al. Study on Dynamic Changes of Wetland Landscape Pattern in Yellow River Delta [J]. Journal of Geo-Information Science, 2009. 11(1): 91 – 97.

Zhang G S, Wang R Q. Research on Dynamic Monitoring of Ecological Environment in Modern Yellow River Delta [J]. China Environmental Science, 2008, 28(4): 380 – 384.

Wang G Q, Wang Y Z, Shi Z H, et al. Analysis on Water Resources Variation Tendency in the Yellow River [J]. Scientia Geographica Sinica,2001, 21(5): 396 – 400.

Wang X Q, Wang Q M, Li H G, et al. Spatial Pattern of LUCC in Different Micro-geomorphic Types of Huanghe River Delta [J]. Scientia Geographica Sinica,2008, 28(4):513 – 517.

Zhao G X, Zhang W Q, Li Y H, et al. GIS Supporting Study on Current Siltation and Erosion Dynamic Changes of Yellow River Mouth [J]. Scientia Geographica Sinica,1999, 19(5): 442 – 445.

Wang R L, Huang J H, Han L Y, et al. Study on the Yellow River Delta Wetland Landscape Pattern Evolution [J]. Yellow River,2008, 30(10): 14 – 16.

Tian J Y, Wang X C, Cai X J. Protection and Restoration Technology of the Yellow River Delta Wetland Ecosystems [M]. Qingdao: China Ocean University Press,2004.

Han M. Ecological study of the Yellow River Delta Wetland [M]. Jinan: Shandong People's Press, 2008.

Shi W Z. Theory and Methods of Spatial Date error Processing [M]. Beijing:Science Press,1998.

Yang G J, Xiao D N. Forest Landscape Pattern and Fragmentation: a Case Study on Xishui Natural Reserve in Qilian Mountain [J]. Chinese Journal of Ecology,2003, 22(5): 56 – 61.

Li B, Wang P T, Zhao J Y, et al. Human Activities on the Environmental Impact of the Yellow River Delta and the Countermeasures [J]. Yellow River,2010, 32(4): 70 – 71.

Li Shengnan, Wang Genxu, Deng Weng, et al. Influence of Hydrology Process on Wetland Landscape Pattern: A Case Study in the Yellow River Delta[J]. Ecological Engineering, 2009 (35): 1719 – 1726.

Li S N, Wang G X, Deng W. et al. Effects of Runoff and Sediment Variation on Landscape Pattern in the Yellow River Delta of China [J]. Advances in Water Science,2009, 20(3): 325 – 331.

Wang R L, Huang J H, Han Y L, et al. Study on the Evolution of Yellow River Delta Wetland Landscape Pattern [J]. Yellow River,2008, 30(10): 14 – 16.

Li J, Zhao G X, Fan R B. Analysis of Driving Forces of the Land Use and Land Cover Change at the Yellow River Delta [J]. Jour. of Northwest Sci-tech Univ. of Agri. And For. (Nat. Sci. Ed.),2003, 31(3): 41 – 46.

Xu J X. Sediment Transferring Function of the Lower Yellow Riveras Influenced by Discharge and Sediment Load Conditions[J]. Scientia Geographica Sinica,2004, 24(3): 275 – 280.

Liu X D, An Z S, Fang J G, et al. Possible Variations of PreciPitation Over the Yellow River Valley Under the Global—warming Conditions [J]. Scientia Geographica Sinica,2002, 22(5): 513 – 519.

Study on Utilization of Flood Resources in the Sanmenxia Reservoir

Li Xingjin, *Duan Jingwang*, *Wang Haijun* and *Lou Shujian*

Sanmenxia Multipurpose Project Administration, YRCC , Sanmenxia, 472000, China

Abstract: The utilization of flood resources in Sanmenxia Reservoir should comprehensively considered several requirements such as flood control, sediment ejection and hydroelectric power production, for different water and sediment characteristics in flood peak, flood volume, sediment concentration and flood process in each time flood, and it has different effects for reservoir requirements. This paper used the actual flood process during 2011 flowed in Sanmenxia Reservoir, analyzed water and sediment characteristics of 3 types of big flood, small flood with high sediment concentration and small flood with low sediment concentration of flood process, and reservoir real operation effects, associated with comprehensive benefits exertion of Sanmenxia Reservoir, put forward the optimization operating suggestions in different types of flood process.

Key words: Sanmenxia Reservoir, flood resources, utilization

1 Foreword

"Generating electricity in non flood period and ejecting sediment in flood period" is an effective mode that were searched after for suited the characteristics of water and sediment in the Yellow River and got multipurpose benefits. Sanmenxia Reservoir now operates with the mode that ejecting thoroughly in flood period, it will open all equipments discharging when inflow more than 1,500 m³/s in practical adjustment. With shortage of water resources and high demand of flood water resources using, needs more study about flood water and sediment characteristics of the reservoir inflow (in this paper, the inflow means the discharge at Tongguan station if without special indication), in different flood events, collect corresponding operation with predominance of flood characteristics to meet Sanmenxia Reservoir's function such as flood prevention, sediment adjustment, and electric power production further, and to improve flood water usages.

2 Flood water resources usages requirements analysis in Sanmenxia Reservoir

Without regard the operation effect on projects in reservoir area and sedimentation of downriver Xiaolangdi Reservoir, the grade which flood water resources using requirements in Sanmenxia Reservoir from high to low as the order of: flood prevention, sediment adjustment, and electric power production.

2.1 Flood prevention requirement

In recent flood regulation scheme of the Yellow River middle and its lower reaches, the operation rule of Sanmenxia Reservoir is that "Sanmenxia Reservoir discharges thoroughly when flood occurring, and operates as other needed mode when Xiaolangdi Reservoir can not meet the needs of flood prevention." The flood prevention requirements of Sanmenxia Reservoir on the one hand is to discharging water when big flood occurring in upper river reaches to sure flood water can be discharged in time; on the other hand, regulated with Xiaolangdi Reservoir and other flood prevention project to impound flood water for lighten the losses of the flood when flooding in lower reaches or upper and lower reaches together.

2.2 Sediment ejection requirement

Based on sediment sources area and area precipitation character, the sediment volume distribution in each month is obvious odds in Sanmenxia Reservoir, most sand volume comes in flood season especially form July to October, even focuses in several flood courses in flood season. During 1952 to 2006, there is 82.9% sediment volume flow into and 87.8% flow out of the reservoir in flood season (Tu Xinwu et al., 2010). So flood season is main period for sediment ejecting. Sediment ejection requirement of Sanmenxia Reservoir, on the one hand is to discharge out sediment taken by flood; on the other hand, use flood water to eject out sediment that deposited in the reservoir area to realize the sedimentation balance in a year or several years and maintain effective reservoir storage.

2.3 Hydro electric power production

Hydro electric power production is a clean production mode for a built hydro project, after met the needs of flood prevention and sediment ejection, get more electric power is one important object of flood water utilization in Sanmenxia Reservoir.

3 Assorting of the flood types for different requirements

The flood flow in Sanmenxia Reservoir can be assorted by its sources as mainstream or tributaries flood, by its course as pyknics or spiky flood, by its sediment concentration as low or high sediment concentration flood and so on. Here, by reservoir's requirements of flood prevention, sediment ejection, power production, and by operating experience, assorts flood as three types: big flood, small flood with high sediment concentration and small flood with low sediment concentration.

3.1 Big flood

Big flood is the type that should operate for flood preventing. When upper or down or both upper and down river reaches having big flood courses (such as serial number flood events that flood peak volume over 5,000 m^3/s in Longmen or Tongguan station in middle reaches; or flood peak volume over 4,000 m^3/s in Huayuankou station in middle reaches), the dam will operate to discharge or maintain flood water to insure the upper and down reaches flood prevention safety firstly.

From resent year inflow fact of the reservoir, both flood time and flood peak from mainstream has dropped off gradually with some large projects construction in upper reach of the Yellow River (Tu Xinwu, et al., 2010). With flood forecast and information transfer level improved, there are more time to prepare for preventing flood operation when big flood occurring.

3.2 Small flood with high sediment concentration

Small flood with high sediment concentration is the flood that peak between 1,500 m^3/s to 3,500 m^3/s, and highest sediment concentration over 100 kg/m^3. This type flood has higher current dynamical power, can get stronger sediment ejecting power with reservoir operation, and take sand out which caught by flood or deposited in reservoir, this type flood is the important flood type to reach reservoir sediment ejection requirement. On the other hand, in many years operated practices of the reservoir, the first natural flood during flood season have excellent brushing effect in sediment ejection, and the sediment concentration out of the reservoir during ejecting operation is nearly insensitive with inflow volume, and dam discharge constructions combination used in ejection period, even insensitive with inflow sediment concentration (Yellow River Conservancy Commission of the Ministry of Water Resources, 2008), this type flood should be used for sediment ejection without thinking about its sediment concentration.

In recently years, the flood water resources can be used in sediment ejection tended to decrease with inflow flood time-events reduced, some years even no one flood occurred (such as 2008), if there are no water can be used to brush reservoir, or have no chance to eject sand lastly, the reservoir will get continued deposit. From 2002, water and sediment regulation was put into practice, artificial flood events made by regulated several reservoirs totally were used for reservoir brushing which very benefit for reservoir's deposition, but limited by the water volume and so on, and the brushing result can not entirely meet reservoir sediment ejection requirements. So it is the key aspect in flood water resources utilization in Sanmenxia Reservoir that discharging sediment in time reasonably when inflows have suitable water and sediment condition

3.3 Small flood with low sediment concentration

Small flood with low sediment concentration is the flood that peak between 1,500 m^3/s to 3,500 m^3/s, and highest sediment concentration lower than 30 kg/m^3. In this type flood courses, the reservoir is safety in respect of flood prevention, the effect of water using be low for its low sediment ratio when used for sediment ejection, and can not meet brushing function. This type flood can be used for power production to improve water resources benefits. The suitable courses include a time-event flood or some stages of a time-event flood all can use for power production.

4 Case analysis of the 2011 flood and reservoir operation

During flood season in 2011, expect water and sediment regulation in beginning, there are 3 times flood flow in the reservoir, each flood has typical represent characters that can primely meet one of requirements of the reservoir, now analyze reservoir flood water using effect link with reservoir operation by the time order of flood occurred. (Three events of flood eigenvalues see Tab.1, three events of flood hydrograph see Fig.1.) (The data used below in this paper are flood reporting value.)

4.1 Situation of three times flood and reservoir operations

4.1.1 Introduction of three times flood

The first flood belongs to the type of small flood with middle sediment concentration, and importantly it is the first natural flood in flood season. It caused by storm water of mainstream and tributaries. The flood last 5 d, its flood and sediment volumes are 9.608 $\times 10^8$ m^3 and 0.176,1 $\times 10^8$ t respectively; the highest daily average flow and average sediment concentration are 2,820 m^3/s (Sep.9) and 25.2 kg/m^3 (Sep.8) respectively.

The second flood belongs to the type of small flood with low sediment concentration. It caused by storm water of mainstream and tributaries. The flood last 5 d, its flood and sediment volumes are 11.91 $\times 10^8$ m^3 and 0.142,6 $\times 10^8$ t respectively; the highest daily average flow and average sediment concentration are 3,290 m^3/s (Sep.14) and 17.2 kg/m^3 (Sep.14) respectively.

The third flood belongs to the type of big flood that happening probability low in resent years. It caused by strong lasting rain water of mainstream and tributaries, and the main water come from Weihe River tributaries. Huaxian station in Weihe River flood peak gets 5,260 m^3/s which is the highest flow since 1981; Tongguan station flood peak reaches 5,720 m^3/s, numbered as "the first flood peak in middle reaches of the Yellow River in 2011", it also the first numbered flood in middle reaches since 1998. The flood last 8 d, its flood and sediment volumes are 27.21 $\times 10^8$ m^3 and 0.290,6 $\times 10^8$ t respectively; the highest daily average flow and average sediment concentration are 5,480 m^3/s (Sep.21) and 13.4 kg/m^3 (Sep.21) respectively.

Tab. 1 Three events of flood eigenvalues

Events of Flood	Item	Station		
		Longmen	Huaxian	Tongguan
First (Sep. 6 ~ 10)	Peak flow(m^3/s)	1,080	2,290	2,830
	Occurring time(d, h:min)	7,14:54	9,6:00	9,14:54
	S_{max}(kg/m^3)	—	—	33.4
	Flood volume($\times 10^8 m^3$)	3.787	5.642	9.608
	Sediment volume($\times 10^8$ t)			0.1761
	$\bar{s}$(kg/m^3)			18.3
Second(Sep. 12 ~ 16)	Peak flow(m^3/s)	1,350	2,190	3,350
	Occurring time(d, h:min)	14,6:45	14,5:00	15,0:00
	S_{max}(kg/m^3)	—	—	—
	Flood volume($\times 10^8 m^3$)	4.455	6.998	11.91
	Sediment volume($\times 10^8$t)			0.1426
	$\bar{s}$(kg/m^3)			12.0
Third(Sep. 18 ~ 25)	Peak flow(m^3/s)	2,100	5,260	5,720
	Occurring time(d, h:min)	18,20:06	20,20:18	21,15:30
	S_{max}(kg/m^3)	—	—	—
	Flood volume($\times 10^8 m^3$)	9.769	14.88	27.21
	Sediment volume($\times 10^8$ t)			0.290,6
	$\bar{s}$(kg/m^3)			10.7

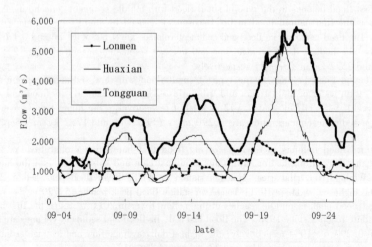

Fig. 1 Three events of flood hydrograph

4.1.2 Operation of the reservoir

During three times flood courses, reservoir keeps thoroughly discharging operation when flood peak flowed in. In the first and second flood courses, there are 12 bottom holes (bottom elevation is 280 m), 2 tunnels (bottom elevation is 290 m) opened to discharge; in the third flood courses, the peak flow is bigger than before, so open 12 bottom holes, 2 tunnels and 12 deep holes (bottom elevation is 300 m) to discharge entirely. Operation period of time see Tab. 2.

4.2 Flood specialties and reservoir operation analysis

In three times floods, the peak flow and the peak sediment of Tongguan station in one flood courses occur at the same time nearly, it indicates the rule that sand transferred mainly in flood period in middle reaches of the Yellow River (See Fig. 2), the rule is objective basis in utilization of flood water resources in Sanmenxia Reservoir.

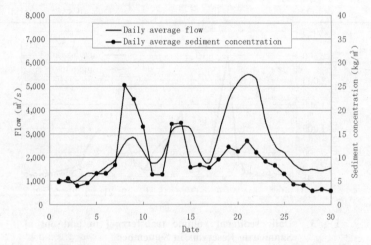

Fig. 2 Daily average flow and sediment concentration of Tongguan Station in September

Reservoir operations during three times flood flowing in are nearly same, but the effects of operation are different for different flood characters and reservoir real situation.

From the characters of floods, the highest sediment concentration of the first flood is large than other two times flood, the first time flood can be assorted to high flood courses under the tendency that sand volume coming in the reservoir reduced in resent years, and at same time, this time flood is the first natural flood in flood season, its water and sediment characters are fitness for sediment ejection. From the results, both peak flow and water volume of the first time flood are lower than other two times flood courses, and the first operation lasting time is the shortest too, but the highest sediment concentration out of the reservoir in this operation is the largest one, and this time operation average sediment concentration reaches 2.4 and 3.4 times compared with the second and the third time operations, totally, in first operation the water volume used in sand discharging is lower and the sediment ejection efficiency is higher than other two times operations. (Reservoir operation characteristic values in each flood see Tab. 2)

The second time flood with low sediment concentration, and which happened short interval after the first flood, for reservoir just ended sediment ejecting operation, discharge sediment again within little time, the result and brushing efficiency of this time is lower more than the first time flood.

On the third flood, the reservoir is operated mainly for flood prevention. The reservoir discharges flood water at full capacity in that water level during flood peak flows in, the highest water lever gets 305.52 m, over than 305 m restricted water level during flood season. For large total water volume, the sediment volume out of the reservoir is the largest one in three time's operation. But influenced by high flood detention water level and after twice discharging operation, this time sediment ejection efficiency is lower than other two time's, the average sediment concentration out of the reservoir during operation period just gets 18.8 kg/m³.

Tab. 2 Reservoir operation characteristic values in each flood

Flood event	Using period (day, hour)	Using hours	$Q_{max,out}$ (m³/s)	Occurring time (d, h:min)	$S_{max,out}$ (kg/m³)	Occurring time (d, h:min)	Wout ($\times 10^8$ m³)	Sand volume ($\times 10^8$ t³)	$\bar{d}$ (kg/m³)
1	7,20 ~ 10,12	64	3,850	8,2:00	120	8,8:00	6.50	0.413,9	63.7
2	12,12 ~ 16,14	98	4,730	13,0:06	97.9	13,8:00	10.93	0.293,8	26.9
3	18,18 ~ 24,8	134	5,960	22,12:00	29.8	19,14:00	23.14	0.435,7	18.8
Total		296					40.57	1.143,4	28.2

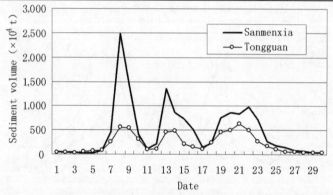

Fig. 3   Daily sediment volume transferring in and out of Sanmenxia Reservoir in September

In 2011 flood season, Sanmenxia Reservoir operations basically according with flood characteristics. Discharge flood water with full capacity to prevent flood disasters when big flood flow in. Sediment ejecting operation courses correspond with high sediment concentration courses, monthly sediment volume out of the reservoir gets $1.423,9 \times 10^8$ t, monthly sediment brushing volume in reservoir area is $0.745,7 \times 10^8$ t, the benefit of sediment ejection is better (daily sediment volume transferring in and out of Sanmenxia Reservoir in September, see Fig. 3). At the same time, try to optimize flood resources utilization, use flood courses that can be foresaw tendency, on the promise that insure contenting requirements of flood prevention and sediment ejection, improve flow grade which begin to discharge thoroughly, and change to power production with sediment ejecting operation in the end of flood courses to use low sediment concentration and peak flood water resources which had low sediment ejecting efficiency; the average beginning flow value of thoroughly discharge in Tongguan Station of three times operations is 2,600 m³/s (higher than 1,500 m³/s that usual thoroughly discharge flow grade in Tongguan Station), the average flow value of impounding in three times operations gets 2,300 m³/s, the courses of flood head and tail that with low sediment concentration are used to product power and add electric quantity 40×10^6 kWh. From operations results, there still have space to optimize operation further, for example in the second time flood, the flood sediment concentration is low, and the reservoir may operate as power production with sediment ejection under flood season control water level instead of entire discharge, by this way, can eject most of sediment as well as can increase water volume for power production, that improve flood water resources utilization comprehensive benefits.

5 Suggestions

Under the rule of discharge during flood period, choose suitable operation according with flood characteristic, can meet different requirements of Sanmenxia Reservoir and improve its comprehensive benefits. To realize flood resources utilization optimizing, need further study combine with deposition and erosion in reservoir area, there just base on actual water and sediment flowing in and out of the reservoir give some suggestions as below.

(1) Confirm big flood flow grade that reservoir operating as flood prevention base on flood prevention requirements. For example, prescribe reservoir putting into flood prevention operation when has numbered flood peak over than 5,000 m^3/s in Longmen or Tongguan stations; flood courses under the grade of flood prevention operation, can be used as sediment ejection or power production with sediment ejecting.

(2) For small floods can not confirm thoroughly discharge operation or not by a single flow score, it must consult flood peak, flood water volume, sediment concentration and reservoir fore condition together. For example, for low sediment concentration flood courses and the reservoir just had sediment discharge operation, then may choose operation mode as power production with sediment ejecting instead of thoroughly discharge operation.

(3) In one time flood courses, its different stage can be used as different modes and should switch operation among flood prevention, sediment ejection and power production timely. For example, change to power production mode in time at the end of a flood courses.

References

Tu Xinwu, et al. Hydrological Rules Study in Sanmenxia Reservoir Area of the Yellow River[M]. Xi'an:Shanxi Technology Press,2010:92 – 93,122 – 122.

Yellow River Conservancy Commission of the Ministry of Water Resources. Water and Sediment Regulation Test[M]. Zhenzhou:Yellow River Conservancy Press , 2008:109 – 113.

The Strategic Status and Role of the Northern Mainstream of the Yellow River in the System of "Keeping Healthy Life of the Yellow River"

Tan Shengbing, *Zhang Xinghong* and *Guo Guili*

Yellow River Shanxi Bureau of Yellow River Conservancy Commission,
Yuncheng, 044000, China

Abstract: "The difficulty of taming the Yellow River is great due to the issues of sediment and lack of coordination between water and sediment". As a result, the sediment problem is the core issue of the management of the Yellow River. In the system of "keeping healthy life of the Yellow River", the core of the nine methods is to solve problems such as "less water", "much sand" and "imbalance between water and sediment". Based on the warping pilot project in Lianbo Beach, the article analyzes the strategic status of Lianbo Beach in the small northern mainstream, a natual flood- and sand-detention place, on the warping pilot project. Besides, considering the connecting role of the small northern mainstream in the peach flood scour test, the article focuses on the feasibility and scientificity of the transformation from the peach flood scour test into the production run during the period after Guxian Reservoir of the Great Northern Mainstream put into use. During this period, the joint regulation of water and sediment between reservoirs is more optimized and the way to manage the sediment is more complete and scientific. To conclude, this paper holds the view that the northern mainstream of the Yellow River plays a strategic and irreplaceable role in the system of "keeping healthy life of the Yellow River".

Key words: sediment control, north mainstream, warping, the Yellow River

1 River reaches' overview

The North Mainstream of the Yellow River refers to the Yellow River running from Laoniuwan, Pianguan County to Fenglingdu, Ruicheng County in Shanxi Province. With a length of 752.5 km, it belongs to the middle reaches of the Yellow River. The reaches from Laoniuwan to Yumenkou is known as the great northern mainstream and the reaches from Yumenkou to Fenglingdu is commonly known as the small northern mainstream.

1.1 Great northern mainstream overview

The reaches from Laoniuwan to Yumenkou is known as the great northern mainstream. The total length of the reaches is 630 km and it flows through the longest stretch of canyon on the Yellow River - Shanxi-Shaanxi Canyon. The water level drop is 611 m and the slope is 0.84 ‰. The left bank of the reaches flows through Xinzhou, Luliang, Linfen and Yuncheng, four districts and 12 counties in Shanxi Province. Wanjiazhai Water Control Project is constructed in Pianguan County, Xinzhou City. The Tianqiao Hydropower Station in Baode County. Hequ County and Baode County are located in the shore of the Yellow River and dikes were built to protect the counties. The counties in Luliang region like Xingxian, Linxian and Liulin County built sporadic dikes, mainly to protect the village and farm land.

The counties along the river are poverty-stricken areas of the Loess Plateau with poor production conditions due to high mountains and steep slopes. Overall, the river is a canyon river without great regional floods, but every local bends suffer different flood disaster almost every year. Since the 1970s, the masses along the river began to build dams to protect the banks and the beach and have achieved a certain effect. After impoundment of Wanjiazhai Water Control Project, the unique water and sediment conditions of the Yellow River changes and water in the downstream river bends become clear and the flow is undercutting the river bed. With erosion increasing, the basis of the existing embankment and unprotected high sloping bank is affected by the obvious erosion hazard.

1.2 Small northern mainstream overview

The small northern main stream refers to the reaches from Yumenkou, Hejin City to Fenglingdu, Ruicheng County in Shanxi Province, the natural border of Shanxi and Shaanxi provinces. The left bank of the river flows through the cities of Hejin, Wanrong, Linyi, Yongji and Ruicheng of Yuncheng District, Shanxi Province. This reach is a kind of valley river cut into the loess mesa and a typical accumulation of wandering river. The stream gradient is 0.3% to 0.6 ‰ with an average of 0.4 ‰ from the top to the bottom. In history, due to serious flooding, frequent swing of the mainstream and erosion of the riverbed, the high slope of the bank frequently collapsed and the villages were destroyed and thus caused untold disasters to the people on both sides. So, People say that a site is located for "thirty Years in the east bank of the river and thirty years in the west bank of the river", which refers to the frequent change of the Yellow River route. The total beach area along the left bank is 420,000 mu. Six large and medium-sized electric pumping stations have been built.

The project management of The small northern mainstream Shanxi Province began in 1968 and as of January 2012, a total of 27 river training construction works have been constructed with a total length of 91.65 km and 520 stacks. To be specific, eight places of 22.74 km and 53 stacks are under local management and 68.908 km and 467 stacks are under management of the Yellow River Conservancy Commission (YRCC). The construction of the project has preliminary straightened the river regime and ensured the life safety of the people along the river.

2 The strategic status of the small northern mainstream in taming the Yellow River

2.1 The status of warping in the small northern mainstream of the Yellow River on the sediment treatment

Water and soil conservation of the Loess Plateau is the fundamental measure to reduce the sediment into the Yellow River. However, due to wide areas of treatment and heavy task, in short term there is still a lot of sediment to enter watercourses. Therefore, a variety of measures such as sediment retention by the main stream key projects and warping in the area of the small north mainstream are necessary to trap and reduce sediment into the downstream.

The small north main stream is the typical accumulation Channel. The river terrain conditions determine the place to be the natural flood and sediment detention area of the Yellow River. In history, flood and sediment was in a natural detention state and the mainstream again and again wandered and swung. Sediment was paving on a wide riverbed, and beaches on both sides was a slow uplifting state under the constraints between the high wall platform shaped land. In recent decades, with the rapid development of economy and society and the development and utilization of water resources, average frequent flooding is getting smaller and smaller, and this situation will exist for a long period. Under this situation, the natural water and sediment detention effect of the small north river beach is greatly reduced. Compared with natural water and sediment detention with artificial warping, the artificial warping can maximize the use of beach space and play a greater role of flood control and sedimentation reduction. The small north main stream lies between the lower end of the coarse sediment source region of the Yellow River and Xiaolangdi, covering a total area of about 1,100 km^2, including the beach area of 682 km^2 and is likely to be deposited 10×10^9 m^3 of incoming sediment. Therefore, the implementation of the artificial warping of the small north river plays an important strategic position and role in the overall layout of the control of coarse sediment of the Yellow River.

Since 2004, YRCC began to conduct warping test in Lianbo Beach, Hejin City and offered scientific parameters and technical support for large warping. Up to now, the test total run 576.5 h, to fill up 5.708,5 × 10^6 t of coarse sand, coarse sand deposition in the proportion which the particle size of greater than 0.025 mm reach more than 40% and accomplished the initial realization the key goal, detaining the coarse sediment and discharging the fine.

2.2 The effect of the small northern mainstream on riverbed scouring during the peach floods

Tongguan section is located in the downstream of the Yellow River and Weihe River confluence. Riverbed rise in Tongguan has an important impact on flood control of the lower Weihe River and small north mainstream. Since 1986, because the large reservoirs of Longyangxia and Liujiaxia put into use and water use for agricultural and industrial purposes is increased, the Yellow River water, especially during flood period decreased substantially and the water level at Tongguan had a tendency of going upward. To lower the riverbed elevation of Tongguan has become one of the important goals of the work of the Yellow River management. Each year in late March to early April, floods are happened in the reaches of the Ningxia and the Inner Mongolia due to ice melting, when it is the peach blossom season in that area, so called the floods as "peach flood". The peach flood has a large peak and has some effect on decreasing bed elevation. Since 2006, YRCC has carried out continuous tests to use and optimize the erosion process of the peach flood to reduce the bed elevation and has achieved some success.

The small Nnorthern mainstream of the Yellow River just lies in joints of Shanxi- Shaanxi Canyon and Sanmenxia Canyon. Its special geographical position determines its important role in Tongguan sedimentation reduction test. The small Nnorth main stream, after years of treatment, has initially formed the complete system of flood control. However, due to the disapproval of the "recent research report", 132.5 km of the river is still far under control.

By analyzing the test indicators in flood peak of Tongguan station in peach flood season from 2006 to 2010, we can seen, except in 2009, the actual peak values reach the design value. During the 2009 trial, due to the occurrence of floodplain in Hanjiazhuan, Yongji City of the small northern mainstream river reach, floods ran out of groove resulting in reduced peak discharge and flood volumes of Tongguan station to a certain extent. How to ensure the peak discharge and flood volume indicator of Tongguan station mainly depends on the flood situation of small north mainstream. From the above analysis we can see the important role of small north mainstream on Tongguan scour test.

Tab. 1 Comparison of actual characteristic values and design values of the peach floods at the Tongguan station

	items	2006	2007	2008	2009	2010
Design value	Peak flow (m^3/s)	2,500	2,800	2,600	2,500	2,500
	Flood flow ($\times 10^8 \ m^3$)	13(10 d)	9.0(5 d)	13.4(10 d)	13. (10 d)	13.0(10 d)
Actual value	Peak flow(m^3/s)	2,620	2,850	2,790	2,340	2,750
	Flood ($\times 10^8 \ m^3$)	12. 7	8.93	12.9	11.01	13.88

3 The strategic position and role of great northern mainstream of the Yellow River in the Yellow River water and sediment control system

Owing to excellent regional location of the great northern mainstream reaches, in the seven main stream backbone control projects of the Yellow River water and sediment control system, two are on this reaches. One is Qikou Water Control Project, the other being Guxian conservancy hub. According to the arrangement of the Yellow River watershed planning, Guxian conservancy hub will be first built.

Guxian Reservoir is proposed to be built at Guxian village, 10 km upstream of Hukou Waterfall, in downstream section of the northern main stream of the Yellow River. It will control the drainage area of 490,000 km^2 including sandy and coarse sandy area of about 60,000 km^2. Guxian Reservoir is mainly for flood control and sedimentation reduction and takes into account the power generation, water supply and irrigation, regulation of water and sediment, comprehensive utilization.

3.1 The strategic position of Guxian Reservoir in the Yellow River sand control system

"The difficulty of taming the Yellow River is greatly due to the issues of sediment and lack of coordination between water and sediment". In recent years, by joint operation of the Yellow River Wanjiazhai, Sanmenxia, Xiaolangdi Reservoirs and its tributaries Guxian, Luhun Reservoirs, YRCC explore the practice of the process of shaping the coordination of water and sediment relations, the initial realization of the "1 + 1 > 2" regulation operating results. Wanjiazhai Reservoir storage capacity is limited, and the Wanjiazhai Reservoir is far from the Xiaolangdi Reservoir, water and sediment regulation are subject to greater restrictions prevent full realization of the regulatory capacity of reservoirs in Yellow River water and sediment

After the completion and use of Guxian Reservoir, it has a great storage capacity, and can greatly reduce the distance between the upstream reservoir and Xiaolangdi Reservoir to optimize the water and sediment regulation reservoir mode of the reservoir. Together with its sediment trapping period, it can greatly reduce the amount of sediment into the downstream. sediment retention capacity designed for Guxian Reservoir is 11.8 × 10^9 m^3, and over the same period of Guxian Reservoir construction, warping construction projects in the small northern mainstream can trap sediment about 10 × 10^9 m^3. With these two projects, about 25 × 10^9 of sediment can be reduced for Sanmenxia and Xiaolangdi reservoirs and the lower Yellow River.

As of 2011, YRCC has organized six times of Tongguan scouring test, and achieved certain results. But by the objective conditions of Ningxia and Inner Mongolia river, i. e. ice thawing time, channel opening manner and channel water storage capacity, et al., combined with that the Wanjiazhai Reservior is small in water storage and far away from the Sanmenxia Reservoir, lack of water supplement ability, it has not been converted to production run. After the completion and use of Guxian Reservoir, peach flood scouring will turn into the production run and play a crucial role for lowering the Tongguan bed elevation.

Tab. 2　Comparison of capacity of Guxian Reservoir and Wanjiazhai Reservoir with the distance to the Sanmenxia Reservoir and Xiaolangdi Reservoir

items	Guxian Reservoir	Wanjiazhai Reservoir
Design capacity(×10^8 m^3)	165.6	8.96
Distance to Sanmenxia Reservoir (km)	438	1,000
Distance to Xiaolangdi Reservoir(km)	568	1,130

3.2 The important role of Guxian Reservoir in the Yellow River flood control system

The Yellow River "big flood" refers to the floods in the Sanmenxia area, the measured maximum) flood, such as in 1843 (Qing dynasty, twenty-three years) and 1933 (the Shanxian Station since 1919). The 1843 flood is the flood once in thousand years, 12 days flood volume of 11.9 × 10^9 m^3, most of them produced in the area above Longmen. The 1933- flood had 51.43% of the runoff from the areas above Longmen. Combined in recent years, the Yellow River ice flood disasters are frequently exacerbated, and Hukou Waterfall Scenic Area and the section of the north small northern mainstream of the Yellow River have caused the disaster.

It is evident that Guxian Reservoir can greatly defend the "big flood" of the Yellow River, reduce the flood control burden of Sanmenxia and Xiaolangdi Reservoirs and decrease the probability of the lower Yellow River flood plain being affected. During ice flood Guxian Reservoir can intercept the ice and release the ice hazards from Hukou scenic spot and the small northern mainstream and reduce ice control task of Sanmenxia and Xiaolangdi Reservoirs.

As a strategic water project in great nothern ainstream, Guxian Reservoir can fundamentally alleviate water shortages in Shanxi Province. It is of great significance to the ecological restoration of Linfen, Lvliang and Yuncheng City in Shanxi Province. With an annual water supply of about 1.5×10^9 m³, more than 4.0×10^8 mu of irrigation region can realize the gravity irrigation and thus water supply costs are significantly reduced. Annually 3.0×10^8 m³ of water can be injected to downstream of the Fen River and Su River. Its backwater length is 233 km and the end of backwater reaches Liulin County. In this way, a huge artificial lake comes into being, which will be extremely important to change the extreme water shortage conditions and ecological environment in Lvliang District, Shanxi Province.

4 Conclusion

The Small Northern Main Stream not only contribute to the soil and water conservation of the Loess Plateau and the containment of the remaining sediment left by the managemeant of the check dam, but also plays a key role in reducing sediment depositionsiltation in Xiaolangdi Reservoir and the downstream channel by changing the spatial distribution of sediment. In addition, the warping of the small northern main stream can greatly improve relatively poor farming conditions on both sides of the river and promote local economic and social development.

The small northern mainstream is key flood control sections in Shanxi Province. Measures such as strengthening the remediation of the river, improving flood control system, and gradually adjusting and changing the unfavorable river regime to protect banks and villages must be taken. At the same time, we must work out measures to implement the flood control responsibility system with flood control chief executive responsibility system as its core, strengthen the construction of non-engineering flood control measures, intensify to remove the river obstacles to protect flood space, and provide life and property security protection for people along the river and offer protection for sediment regulation, the bed elevation, and other major Yellow River strategies.

As key water control project of great northern mainstream of the Yellow River, Guxian Reservoir is proposed to build during "12th Five-Year" period. After the completion of Guxian Reservoir, it will control the main source area of the Yellow River floods and sediment, especially coarse sediment source area. It will enrich the modes of the joint regulation of water and sediment between reservoirs. While reducing the influence of the upper reaches of the river and boundary conditions to the Tongguan scouring test, it will provide strong protection for the bed elevation and thus its strategic position is extremely important.

After the completion and use of Guxian Reservoir, it can reduce the flood control burden of Sanmenxia and Xiaolangdi Reservoirs and decrease the probability of the lower Yellow River peach flood area being affected. During ice flood, Guxian Reservoir can intercept the ice and release the ice hazards from Hukou scenic spot and the small northern mainstream and reduce ice control task of Sanmenxia and Xiaolangdi Reservoirs. At the same time, Guxian Reservoir can fundamentally alleviate water shortages in Shanxi Province and it is of great significance to irrigation, water supply and ecological restoration of Linfen, Lvliang and Yuncheng City in Shanxi Province.

To conclude, the paper holds the view that the northern mainstream of the Yellow River plays a strategic and irreplaceable role in the system of "keeping healthy life of the Yellow River"

References

Yellow River Shanxi Bureau. Records of the Small Northern Mainstream of Shanxi of the Yellow River[M]. Zhengzhou: Yellow River Conservancy Press, 2002.

Jiao Enze, Jiang Enhui, Chang Ching. Guxian Reservoir's Benefit Assessment and Related Issues [J]. Yellow River, 2011.

Hu Jianhua, Song Haixia, Wu Hailiang. Warping Potential and Related Issues in the Small Northern Mainstream of the Yellow River [J]. Yellow River, 2007.

Hou Suzhen, Lin Xiuzhi, Tian Yong, et al. Key Technology Studies on Lowering the Elevation of Tongguan by Peach Flood Washing [M]. Zhengzhou: Yellow River Conservancy Press, 2010.

C. Strict and Efficient Management of Water Resources in the River Basin

Legal, Institutional and Technical Instruments of Rational Management of Water Resources in the Niger Basin

Drissa Naman KEITA[1] and *Robert DESSOUASSI*[2]

1. Niger Basin Authority, Box 729, Niamey, Niger
2. Niger Basin Observatory. PO Box 729, Niamey, Niger

Abstract: The Niger River with a length of 4,200 km, drains an area of 2,100,000 km^2 divided among nine states which have created the Niger Basin Authority (NBA): Benin, Burkina Faso, Cameroon, Cote d'Ivoire, Guinea, Mali, Niger, Nigeria, Chad. In the context of sustainable and agreed development of the Niger Basin, the Member States have strengthened their cooperation on the solidarity and reciprocity for a sustainable, reasonable and equitable use of water resources. To this end, legal, institutional and technical instruments have been implemented:

(1) The Water Charter of the Niger Basin Water: it comes to effect on July 19, 2010 and determines the rules relating to: ① the fair and reasonable allocation of water resources between member states on the one hand and the various other uses, in the other hand; ②the consultation between Member States in the planning of projects likely to have significant adverse effects on water resources; ③the protection and preservation of water resources, ④the prevention and resolution of conflicts over the use of water resources. It obliges the concerned States to manage water resources so as to maintain the quantity and quality of these resources to the highest possible level (control of land degradation and these waters, silting and pollution). To achieve these aims, the NBA has set up water management institutions which are: ① the Permanent Technical Committee, responsible for monitoring the rational use of water in the basin; ②the Regional Consultative Group, in charge to establish consensus in the design and implementation of structural works; ③the Panel of Experts for advising on the specific technical issues related to structural developments in the basin and④the Sub-basin Commissions, responsible for proposing the terms of use of water resources in each sub-watershed basin and to help resolve issues related to water use. (2) Allocation and forecasting models of water resources that allow, among other things: ①to optimize the management of the resource; ②analyze the hydraulic impacts of the planned measures;③coordinate the management of dams;④alerting water actors in case of emergencies and ⑤manage the input and output reservoirs of dams.

Key words: Water Charter, fair and reasonable distribution of water resources, Consultation, protection and preservation

The Niger River with a length of 4,200 km, covers an area of 2,100,000 km^2 divided among nine states that have created the Niger Basin Authority (NBA): Benin, Burkina Faso, Cameroon, Cote d'Ivoire, Guinea, Mali, Niger, Nigeria, Chad. In the framework of sustainable and concerted development of the Niger Basin, member states have strengthened their cooperation based on solidarity and reciprocity for sustainable, reasonable and equitable use of water resources. To this end, legal, institutional and techniques tools have been implemented:

The Water Charter of the Niger Basin is the main legal instrument for implementing the principles of solidarity for a sustainable and equitable use of the Niger Basin waters.

Came into force on July 19, 2010, the Water Charter of the Niger basin determines the rules relating to:

(1) the fair and reasonable distribution of water resources between member states on the one hand and the various uses on the other.

For this purpose, following factors should be considered especially: relevant geographical, hydrological and climate data of each of the states of the basin, previous, current and future use of the water resources of the basin, the economic and social needs of states and populations, the need to avoid wasteful use of water in the basin, the principle of compensation to a State that is obliged to waive an activity in order to reconcile divergent uses, the right to water of populations, the effects of water use by a State on other States, the existence of a minimum environmental flow to maintain wetlands services in the basin etc..

(2) Consultation between Member States in the planning of projects likely to have significant negative impacts on water resources.

This rule cleared the conditions for the planning and operation of projects and programs initiated in the basin.

To this end, States undertake to inform the Executive Secretariat of all projects and works they would propose to undertake in the basin.

Furthermore they undertake to abstain from carrying out on the portion of the river, its tributaries and sub-tributaries within their territorial jurisdiction, all works likely to pollute water or negatively alter the biological characteristics of the fauna and flora.

Indeed, before a State implements in its territory any projects likely to have significant negative effects for the other states of the basin, it must provide them, and in a good time, a notification through the Executive Secretariat. If the State which has received the notification or the Executive Secretariat believes that the proposed measures are likely to cause significant harm, consultations and negotiations are undertaken to achieve a fair and consensual solution.

(3) the protection and preservation of water resources. According to this rule, States Parties undertake to prevent any deterioration and enhance the state of terrestrial aquatic ecosystems and meet their water needs, and preserve wetlands that depend from the basin, strengthen the protection of the aquatic environment, ensure the progressive reduction of transboundary pollution and prevent further pollution, systematic environmental assessment procedures, conservation of biological diversity, determine the potability and release standards in the Niger basin, prevent and manage emergency situations, help to mitigate the effects of harmful situations as floods, droughts, siltation and climate change, etc..

(4) The prevention and resolution of conflicts related to the use of the water resources. This rule requires States Parties to favor conciliation and mediation as means of settlement of any dispute which might arise between states over the use of water resources of the Niger Basin.

To achieve these objectives, the NBA has set up institutions of water management which are:

(a) The Permanent Technical Committee, responsible for monitoring the rational use of water in the basin: Advisory body to the Executive Secretariat, the Permanent Technical Committee (PTC) is charged with, among other things, to:

Ensure the rational and equitable use of waters of the Niger Basin;

Develop information tools to organize consultations on any planned project within the basin;

Give informed advice on technical aspects of projects, economic and social studies and their coherence with the action plan for sustainable development of the basin;

Facilitate dialogue, consultation, negotiation and mediation of disputes or conflicts related to the use of the water of the Niger Basin.

(b) The Regional Advisory Group, responsible to establish consensus in the planning and construction of structural works.

(c) The Panel of Experts is responsible for giving technical advice on specific issues relating to developments works in the basin, especially on large dams.

The mission of the panel includes among other things, following activities:

Issue an opinion on the quality of studies (environmental and social assessments, the conception, management and safety of works and their social and economic dimensions);

Follow up of ongoing studies, identify any additional studies necessary to drive according to international standards in this area;

Support efforts aimed to place all existing or planned works in a regional perspective to avoid potential conflicts and minimize costs in order to achieve a maximization of the resource use and benefits sharing among member countries of the NBA;

Support the Permanent Technical Committee and the Regional Advisory Group in building consensus on appropriate standards of planning and sustainable development of water resources and management of works;

Issue a relevant opinion on the formulation and implementation of legal, regulatory, technical, economic and financial instruments for the optimal management of hydraulic works.

(d) Sub-basin commissions, responsible for proposing the terms of use of water resources in each sub-watershed and to help resolve issues related to the water use.

Real instrument of concerted and sustainable management of water resources, the Water Charter guarantees every user the right to information on water resources in the basin. It also sets an imperative for the conservation of the water resources of the basin, in quantity and quality, and ecosystems, especially wetlands. By defining the status of joint works or works of common interest, the Water charter makes possible the co-management of dams by at least two member States of the Niger Basin Authority.

The Charter is the most important legal support, complementary to the legal arsenal of the Niger Basin Authority. It is a binding instrument but agreed to a mutual agreement by the States Parties.

It is accompanied by a series of implementing regulations which are the appendices. These relate to:

The protection of the environment in the Niger basin;

The rules of coordinated management of large hydraulic existing or planned works in the basin;

The sharing of costs and benefits associated with these works;

The general status and legal regime of common and common interest works.

Appendix No. 1 on the protection of the environment was adopted by the 30th Ordinary Session of the Council of Ministers held on 30 September and 1 October 2011 in N'Djamena, Chad. It aims to ensure better protection of the environment of the basin to promote its sustainable development. It applies to all sectors of the environment and all natural resources of the basin, excluding land issues and marine environment which are governed by national laws and international conventions binding the States Parties. The Environmental Annex sets out general obligations of States Parties, including a requirement to have: ① a national policy document on the environment; ② a framework law on environmental matters; ③ a national action plans for the environment; and ④ to proclaim the environmental protection of national interest. The final beneficiaries of the Water Charter are basin populations whose rights have been recognized both for access to water and sanitation and for the exercise of their economic activities. The environment and the large existing or planned works in the basin are also the final beneficiaries because annexes on the environment and management of dams will empower users of water resources of the basin on the principles of polluter-pays and user-pays, to reconcile the economic and social development requirements and the need for environmental protection on the one hand, and to determine the rules and procedures for coordinated management of dams, taking into account other uses of the water on the other.

Digital River Basins: Trends and Critical Techniques

Yu Xin[1] , *Kou Huaizhong*[2] , *Wang Wanzhan*[1] and *Zhai Jiarui*[2]

1. Yellow River Institute of Hydraulic Research, Zhengzhou, 450003, China
2. Yellow River Conservancy Commission, Zhengzhou, 450003, China

Abstract: The paper introduces the recent advances of the four river modeling systems in China, including Digital Heihe River Basin, Digital River Basin Model, Nature – Human Dual Evolution Modeling System and Yellow River Numerical Modeling System. Following the introduction , new trends are given in developing river basin modeling systems in terms of the philosophy, approach, quality control, models integration and support sub – system setup. The key point of the tendencies is that a river basin modeling system, which is expected to do fully coupled modeling of the changing processes of water, ecology, and economy, must be formed as a comprehensive system integrated with multiple modules on a public developing platform, be tested and evaluated in a systematic and scientific way. Besides, data assimilation and model parameterization techniques are helpful in the set-up, test, evaluation, and application of the river basin modeling system. At the end, the authors suggest the research activities should be orientated towards: ① uncertainty analysis; ② mega complicated data management and visualization; ③ large scale multi – discipline river basin modeling system integration with cloud computation; ④ application of Karman Filtering and 3D/4D variation assimilation techniques; ⑤ mixed river basin virtualization; ⑥ water cycling processes and transport/changing processes of multiple matters and their modeling.

Key words: coupled modeling of water, ecology and economy, modular integration, quality control, comprehensive models integration environment

1   River basin modeling systems

1.1   Digital Heihe River Basin (DHRB)

Digital Heihe River Basin (DHRB) , developed by the Institute of Environment and Engineering in Cold and Arid Regions, is an integrated modeling system consisting of data management sub – system, models platform and observation sub – system. The most fundamental work in the setup of DHRB is to setup the information infrastructures. DHRB includes all applications of the core models.

1.1.1   DHRB development philosophy and methodology

Considering the fact that the system consisting of water, soil, atmosphere, ecology and human is very complicated, it is necessary to simulate water, ecology and economy as a whole by using integrated river basin modeling model and management model, together with the mega spatial data, so as to not only have a good understanding of the processes and mechanisms in the river basin, but also answer the questions arisen in making strategies about issues in macro – scales. So it is requested to make two kinds of integrated models to realize the scientific research goals and the management goals.

One kind of integrated models, called scientific research – orientated integrated model, is able to simulate various processes in the basin especially hydrological and ecological processes so as to have more in – depth understandings of water, biological, chemical cyclings in the basin.

The important efforts are directed to set up an integrated river basin model which is hoped to be able to perform comprehensive modeling. Firstly, by coupling a distributing hydrological model and an overland model (which serve as the skeleton) and an underground water model, a water resources model, and an ecological models together, an integrated model is formed, which is able to represent the hydrological and ecological processes in the basin, driven by a regional atmospheric model. Secondly, by coupling the integrated model with the social – economy model, the required integrated river basin model is made.

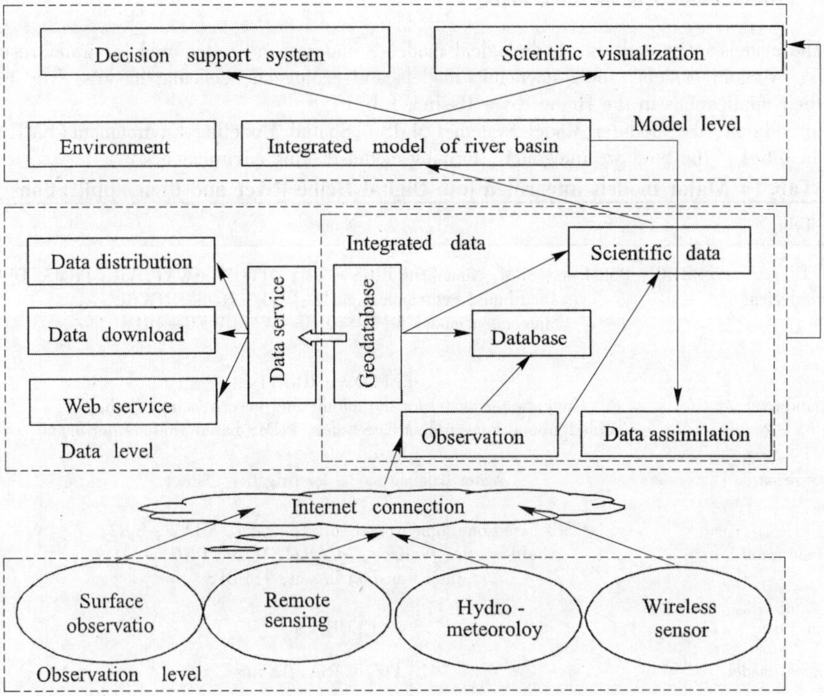

Fig. 1　Framework of Digital Heihe River Basin

The other kinds of integrated model, called management – orientated integrated model, which is based on the integrated river basin model, is expected to be a spatial decision support system about how to utilize water resources, other natural resources, and socio – economic resources in a sustainable way.

Besides the two kinds of integrated models, it is necessary to set up a model – making environment which is based on advanced information techniques. The model – making environment will serve as a powerful tool in making integrated river basin models.

1.1.2　Tasks in the recent years

（1）To establish runoff models which are suitable for the upper, middle and lower Heihe River Basins.

Of the upper basin models to be made, the core models are distributing hydrological models. The models, with the output from climate modeling and regional climate modeling as the input, are used to forecast runoff processes at the outlet. The hard nut to crack is how to do a rather satisfactory coupled modeling in the complex system of climate, plant, soil, frozen soil and snow.

The models to be established for the middle Heihe River Basin are the models which can do coupled modeling of water, ecology and economy, i. e., the eco-hydrological models and the eco-economic models. The integrated models are hoped to be used to find a suitable strategy for the sustainable development in the basin, with the sustainable development of the water resources as the core.

The models for the lower basin to be made are coupled models of water, and ecology and economy, but with the coupled models of surface runoff, groundwater, and ecology as the core.

（2）To set up model-making environments.

1.1.3 Major models to be integrated

The models integrated are: hydrological models, underground water models, water resources models, overland models, social-economic models and ecological economic models. The models have their applications in the Heihe River Basin (Tab. 1).

In addition, the Modular Model System (MMS), Spatial Modeling Environment (SME) and other facilities of the kind are integrated, forming model-making environments.

Tab. 1   Major models integrated into Digital Heihe River and their applications

Type	Name
Hydrological	DiHBV, TOPModel, SRM, Xinanjiang River model, DTVGM, SWAT, VIC, PRMS, DLBRM Distributed hydrological model: WEP – Heihe, DWHC Integrated model: MM5 与 DHS – VM, WEB – DHM
Groundwater	FEFLOW、MIKE11 Groundwater model for the middle and lower reach of Heihe, Coupled groundwater and surface water, PGMS based groundwater model
Water resource	Water balance model for irrigation district
Surface water	SiB based one dimensional surface model, SHAW, SiB2, Shuttleworth – Wallace, NCAR/LSM, HYDRUS – 1D water transport in soil, TSEBPS
Land use	CLUE – S
Ecological model	C – Fix, CASA, TESim
Social, ecology and economic model	Coupled water with eco – economic model of the Heihe River, Environmental and economic model, water resources optimal deployment

1.1.4 Critical issues to settle

About the Heihe River integrated models of water, ecology, economy and managment, critical issues to settle are about the model paremeters and modeling uncertainty.

In the upper basin of the mountainous area, the research activities will be focused on more in-depth understanding of both the physical processes in the water frozing/ thawing and snow accumu-lating and the effects of the human activities on hydrological processes on the mountains so as to im-prove the perfomances of both the frozen land hydrological model and the snow accumulating model.

In the middle basin, the reserch activities will be focused on: ①the observing of the small – scale morphology both in the seasonally frozen land and in the area where groundwater emerges onto the plains so as to have complete sets of data; ②water resources transfer in the basin between the low lands; ③the applying of SME and PLM (Patuxent Landscape Model) in the whole middle ba-sin.

In the lower basin, efforts will be made to address the coupled modeling of underground water and overland runoff.

1.2 Digital River Basin Modeling System (DRBMS)

Since 2001, Tsinghua University, taking the Yellow River Basin as the research object, has shaped basically a DRBMS, which is to simulate hydrological and sediment transport processes in large river basins. Its basic structure, major achievements, and next tasks are given below.

1.2.1 Framework

DRBMS, the framework of which is shown in Fig. 2, is an integrated model characterized by multiple spatial resolutions, easily – determined model parameters, and capacity of parallel computation. DRBMS is able to simulate flow, and matter transport processes in as large areas as river basins.

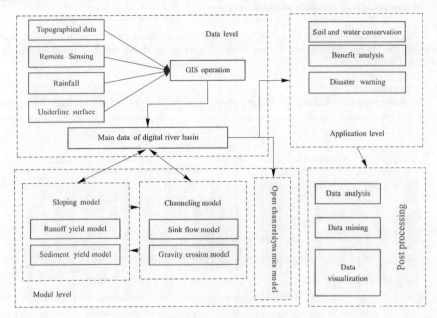

Fig. 2   Framework of Digital River Basin

1.2.2   Major achievements

The framework of DRBMS has been completed.

Based on DEM's data and access system, the framework divides a river basin into four kinds of hierarchical morphological areas: slope faces, small basins, regional (branch) river basins and the whole river basin.

The slope face is the place for which the soil erosion and runoff model are made while the small basin, the branch river basin and the whole river basin are the places for which the sediment and runoff assembling models are models. In the hierarchical way the models are assembled into a complete DRBMS.

DRBMS includes five key techniques: ①saving and access to the DEM data in a large river basin, ;②acquisition of parameters about the gullies in the river basin;③model parameters determination using remote sensing data;④management of distributing data (i. e., save and access);⑤parallel computation.

DRBMS is able to simulate rainfall – runoff, calculate the water resources, flooding processes, continuous runoff assembling process in a large river basin.

Combined with Rainfall forcast models, DRBMS is able to forecast flooding processes and sediment – yielding and transport processes.

1.2.3   Critical works to do in the near future

It is planned to focus on the development of:①the models for the regions (branch rivers) in the Yellow River Basin, for example, the models for grass lands in the high mountains in the Yellow River Head area (including the snow melting model), integrated flow – sediment model for the Yellow loess plateau; ②a integrated model of rainfall forecast models, runoff models, groundwater models, non-point-source pollutants transport/changing models; ③a model of a network of main rivers in the whole basin, large reservoir operation models, irrigation district water moist movement models, and check-dam effect models, et al.

1.3 Water resources dualistic evolution system

Nature – Human Dualistic Evolution system has been developed by China Institute of Water Resources and Hydropower Research since 1999. The following shows the system framework, core models and critical works.

1.3.1 Framework
The water resources dualistic evolution system includes basic platform, database management and application system (Fig. 3).

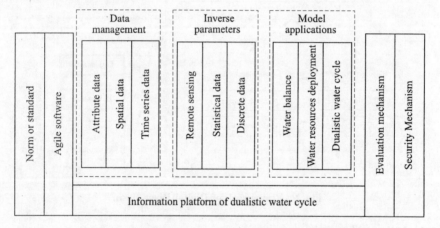

Fig. 3   Framework of Dualistic Evolution System

1.3.2 Core models
Two core models are coupled into the system. One is the model of water and energy transfer process in a large river basin (WEP – L). The other is the model of Water Allocation and Regulation Model (WARM).

The primary feature of WEP – L is the integration of the distributing hydrological models with SVATS. WEP – L is able to simulate not only the physical processes of various factors involved in water cycling, such as plant interception, various kinds of evaporation from land and plants, subsurface flow, underground flow, flow converging, snow accumulating and meting, but also the energy cycling processes such as radiation.

WARM includes both the model for rational allocation of water resources and the water resources regulation model.

The water allocation model includes water resources supply-demand balancing sub-model, quantitative economic sub-model, population forecast sub-model, national economic water demand forecast sub-model, multiple water sources regulation sub-model, ecological water demand forecast sub-model, et al.

The water allocation model has a large time scale such as months or ten-day and has a large spatial scale as the allocation units while the water cycling modeling uses a day as time scale and smaller spatial scales such as various irrigation areas, land uses and plant structures. The large differences in time and space scales between the models necessitate dual-directional coupling ways to connect the models: one way to decompose large scales into small scales and the other way round to assemble small scales into large scales (Fig. 4).

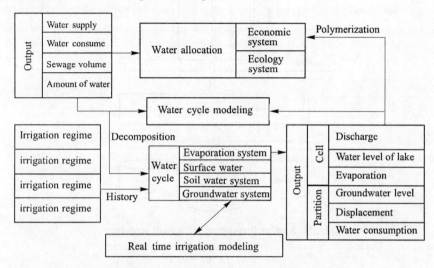

Fig. 4　Water resource allocation information coupling

1.3.3　Critical works

Critical works include: ①data collection and assimilation; ②comprehensive modeling of the water cycling from atomspere to soils, and underground; ③comprehensive modeling of water cycling processes in a river basin and concurring water chemical processes, sediment transport processes, and ecological processes, and; ④techniques for the setup of digital river basin platforms based on river basin environment virtualization and integration.

1.4　The Yellow River numerical modeling system (YRNMS)

In 2003, Yellow River Conservancy Commission (YRCC) completed the coding of Plan for the Setup of the Yellow River Numerical Modeling System of the Digital Yellow River Program, which specifies clearly the goals and tasks. The team has completed the integration platform, the models and other tasks so far.

1.4.1　The goals

The goals are to develop a virtual Yellow River Basin, by establishing water – related professional models, macro – economic models and ecological models and using advanced water – related and information techniques. The virtual Yellow River Basin is expected to be able to perform coupled modeling of the natural, economic and ecological fields so as to give early warnings, forecast, and scenario simulation results in the course of the Yellow River training, utilizing, protecting, management and other important decision makings.

1.4.2　Framework

The system, YRNMS, using three – layered .NET framework, integrates the Enterprise Service Bus (ESB) with technique of J2EE and other components. In the years to come, by applying the design idea of SOA, and using web service and other advanced techniques, the team will have the models(which are planted in other platforms at the other different locations) connected, integrated and running interactively. Fig. 5 and Fig. 6 illustrate the logical structure and models base respectively.

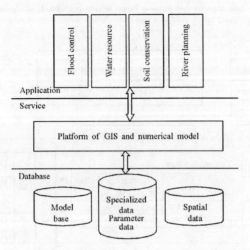

Fig. 5　System framework of logical structure

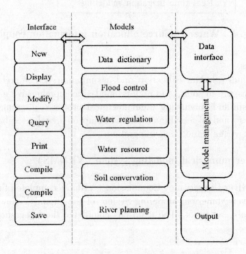

Fig. 6　System framework of model base

1.4.3　Major achievements

What have been completed are given as follows.

(1) The setup of YRNMS integration platform able to basically perform the integration and running of various professional models.

(2) Six types of models, including the model for the soil and water loss in loess plateau, two dimensional model of hydrodynamics and sediment transport, one dimensional model of water, sediment transport and water quality, one dimensional model of hydrodynamics and sediment in reservoirs, three dimensional model of turbulent flow in reservoirs, two dimensional model of current and sediment transport in the Yellow River estuary. The models have been integrated in the system.

(3) A super computing center has been established for weather forecast, and numerical modeling of hydrodynamics and sediment transport in large spatial scales.

(4) Guides for Coding the Yellow River Models, and Method for evaluating the Yellow River

models, which have standardized the development of the models.

1.4.4　Critical issues to settle

(1) To improve index for evaluating the Yellow River models, and to setup an service environment for evaluating the shared models.

(2) To setup a river channel – river basin modeling platform so as to integrate YRNMS with the digital river basin of TU – BASIN.

(3) To develop the second-phase model of soil erosion in the Yellow loess plateau.

(4) To setup a fully coupled model of hydrodynamics, sediment transport and morphological changes in the Yellow River, a MPI – based 2D model of hydrodynamics and sediment transport along a longitudinal vertical profile in a reservoir, a hybrid model with 1D main channel and 2D floodplains of hydrodynamics and sediment transport with real-time feedback and correcting.

(5) To develop a model of dynamic ice in the Ningxia – Innner Mogolia reach of the Yellow River.

(6) To develop a water quality model able to give early warning and prediction of pollutants changing along the stem Yellow River downstream of Longmen in case of a severe pollution happening.

2　Trends in digital river basin developing

From the developing of the digital river basins, it is seen that 3S techniques, especially GIS, and computional steering by both using pre/post-processing visualization and intervening simulation in an interactive way, have gained wide applications in the setup of the digital river basins. Besides, new trends have emerged in terms of the philosophy, approach, quality control, models integration and support sub-system setup in developing river basin modeling systems.

2.1　Developing philosophy: from uncoupled modeling of few matters to coupled modeling of multiple factors

Considering the fact that many factors, human and natural, in a river basin are mutually connected and affected in complex ways, it is necessary for the uncoupled modeling of a single medium or process to change into the comprehensive coupled modeling of multiple matters or multi – processes, i. e. , water cycling and concurring processes of other matters. Scientific discoveries have proven that the way in which multiple concurring physical fields are able to be represented by solving the PDFs, is able to provide a powerful technical support for the comprehensive river basin modeling.

In addition, multiple-goal management in a river basin makes it necessary to do comphrehensive coupled modeling of dynamic processes of water, ecology and economy by using an integrated model.

2.2　Developing approaches: from source coding to modules/components assembling

Previous practice in water-related model developing is whoever needs a model codes it in person, and whoever codes a model uses it. Because of this, the model developed would be always accessed and used by one user or limited number of users, and would be always less modularized. Therefore it would be difficult to expand and use by the people other than the code writers.

A digital river basin require that one modular/component be able to be accessed and used by numerous users on various levels in modeling and that multiple modules/components be able to be accessed and used an individual user.

It means that a necessary change from the previous practice of non-expandable source codes to modules/components should be able to be assembled. The modularized /or components assembled software is easy to be updated in parts, wide applicable, flexible, and easy to be integrated with users' codes.

In addition, the software is always equipped with rich documents and examples, which help users to get familar quickly, and with good measures for mistakes preventing and checking, which help to have the software maintained easily, function well, have a long life period, and build up knowlege.

2.3 Quality control: from focusing on instantaneous or isolated result only to process and study on uncertainty of model and parameters

Because of the lacks of in-depth understanding of uncertainities of the model and the parameters, previous users would highlight the instantaneous modeling results at the time specified, ignoring the whole process.

In facts, a numerical model contains all sorts of errors resulting from computer, numerical algorithms, and errors in formulating equations, boundary conditions and others. In the modeling, it is found that the sources of resultant errors are rather a lot, and even shocking. Modeling participants should recogize the uncertainties, try to identify the errors and find out the effects of the errors on policy making.

2.4 Transforming to public developed environment

Fig. 7 illustrates the evolving stages of the methods for river basin numerical model integration: ① by using public interface; ② by building the management model on top of scientific models; ③ by transforming scientific models to modules of the management model.

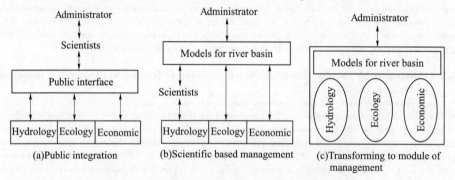

Fig. 7 Methods for integrating river basin models

The method①, which puts the professional models under the public interface, leaving users to select and run the models needed one by one, is, in fact, only suitable for scientists or engineers.

The method② makes it possible for the professional models to be selected indirectly, or for scientists who run the professional models to parameterize the models. The method has its limitations due to the model data defining, the temporal and spatial segmenting methods, sequential running of the models, et al. Therefore, it is apparent that the method ② will not work if one model needs constant modifications to the other models as the condition to its starting.

The method ③ abandons the professional model interfaces, transforms the professional models into the modules of the management model, and adopts the public executing environment where the modules are running simultaneously (like MMS made by USGS). The method ③ are still limited due to the data defining and data formats of the modules, low efficiency in communicating, et al.

From the perspective of computer science, the most effective way of software developing is to start from the scratch (source coding) so as to avoid the complicated connections between interfaces of the softwares to be integrated due to the fact that different softwares are developed at different times and places, using different languages and methods.

One of the challenges at present is to setup a public developing environment where modules can be made for integration into softwares.

Xie Jianchang proposed a kind of knowledge visualization integration platform, which provides a new approach and a platform for river basin model integrating and application. Fig. 8 shows the developing method and procedures in using the platform, given as follows.

①Form a theme by experts or decision makers; ②form concepts; ③identify the relationship between concepts; ④set the models to represent the concepts; ⑤connect the data and the models; ⑥apply for the theme.

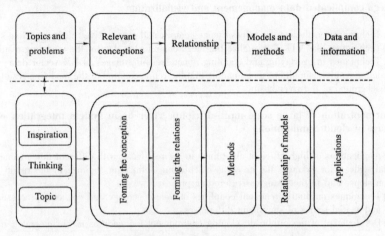

Fig. 8　Applications carried out on a platform

2.5　Data assimilation and fundamental research intensification

Data assimilation, which, as a method, was used in data analysis in the early meteorology, harmonize both data measured at different places, at different times, in different tools with the results of numerical models in an optimal standard method. Data assimilation makes a harmonic and optimal relationship between data and models. Several algorithms, including polynomial, optimal interpolation, blending, Nubging, Kalman filtering and variation method, have been widely used in the application of meteorology and hydrology.

Data assimilation can help to provide data for initial conditions, and boundary conditions in numerical modeling, and can optimize some parameters. So data assimilation helps to extract the effective data from the measured data and improve performance of forecast systems.

On one hand, applications in practice are the goal of scientific research and model setup, and provide requirements and the platforms where the results of scientific research and models are tested. One the other hand, research on natural laws offers fundamental support for models and their applications. Models serves as the middleware between applications and research on natural laws, and can find out reliability of the research results about the natural laws in applications.

Considering numerical models are based on the results of scientific research on natural laws, research on fundamental laws governing flow and concurring processes of the other matters, especially the key model parameters and formulations of the laws, should be intensified.

3　Hot and difficult issues

The hot and difficult issues in the setup of numerical modeling systems can be listed as follows.

3.1　Uncertainty evaluation

(1) Applications of data assimilation such as Mont Carol simulation.

(2) Identification of the sources, probability distributions, and quantitative expressions of modeling errors.

(3) Codes for the evaluation of the performances of the numerical models, setup of standard cases to be used for the evaluation.

(4) Others.

3.2　Mega complicated data management and visualization

(1) Techniques on the management of mega complicated data in river basin modeling systems.

(2) Techniques like VTK to visualize the mega complicated data of the modeling results.

(3) Techniques in overlaying and coupling remote sensing images, DEM vector data, complicated modeling results, simultaneous animating graphic representation of the modeling results on the graphic background of the river basin.

3.3　Harmonization of large scale multi-discipline river basin system integration with the usage of cloud computation.

(1) Applications of high efficient algorithms in the real cases with different flow regimes, various spatial scales (for example the cases as calculation of the flow regimes in the runoff forming process on slopes and in the retrogressive eroding process).

(2) Techniques on dual-directional couplings of water, ecology and economy processes in the temporal and spatial scales of significant differences.

(3) Development of the public developing environment for modules to be integrated into softwares.

(4) Techniques for cloud computation in the digital river basin platforms/softwares.

3.4　Application of Karman Filtering and 3D/4D variational assimilation techniques

(1) Find out a right way of using data measured at points and in areas to finalize the best model parameter, and setup suitable river basin models with the minimum modeling errors.

(2) Find out a right way of balancing complicated model selection and mega data to enhance accuracy in hydrological process predictions given no data or incomplete information.

3.5　Mixed river basin virtualization

About mixed river basin virtualization, which means synchronizing numerical modeling, physical modeling and the real river dynamics. Hot and difficult issues are:

(1) Establishment of the theoretical framework of theories for virtual river similarities by referring to the methods and techniques for physical modeling.

(2) Exploring the correspondence, connection, and mutual complement between the real river, physical model and numerical model.

(3) Design and realization of both the network of the sensors to link physical rivers and virtual rivers and the techniques needed for real-time information transmission.

(4) Setup of such an environment able to provide calculating support, integrating, coordinating, and discussing.

3.6　Water cycling processes and transport/changing processes of multiple matters and their modeling

(1) Relationships between hydrological processes and morphological characters in river basins and their quantitative formulation.

(2) Processes of gully erosion coupled with slope erosion and their modeling.

(3) Modeling techniques for modeling of both slope eroding processes and retroeroding proces-

ses in the yellow loess plateau.

(4) Theories and techniques for fully integrated modeling of flow, sediment transport, bed deformation.

(5) Mechanisms for changes in plane-forms and their modeling.

(6) Fully coupled modeling of waves, current and sediment transport in estuaries.

(7) Settling, re-suspending, absorbing, de-absorbing, and decaying processes of different categories of pollutants in hyper-concentrated environmernt and their modeling.

(8) Ice blocks heat-absorbing and relasing processes and their modeling.

(9) Processes of changes in resistence of ice-jammed river and their modeling.

(10) Techniques for optimalizating of the regulation of linked reservoirs.

(11) Other relevant important issues.

References

Chen Guodong. Integrated Management of Water, Ecology and Economics of the Heihe River [M]. Beijing: Science Press, 2009.

Li Xin, Chen Guodong. On the Watershed Observing and Modeling Systems [J]. Advance in Earth Science, 2008,23(7):756 – 764.

Li Xin, Chen Guodong. Digital Heihe River Basin. 3: Model Integration [J]. Advances in Earth Science, 2010(8):851 – 865.

Wang Guangqian, Liu Jiahong. Digital River Basin Model [M]. Beijing: Science Press, 2006.

Wang Guangqian, Li Tiejian. Digital Yellow River Model [J]. Sciencepaper Online, 2007(2): 492 – 499.

Wang Hao. Water Resource of the Yellow River and Its Evolution [M]. Beijing: Science Press, 2010.

Wang Hao, Yan Denghua. Subject System of Modern Hydrology and Water Resources and Research Frontiers and Hot Issues [J]. Advances in Water Science, 2010, 21(4): 479 – 489.

You Jinjun, Gan Hong, Wang Hao. Advance in Water Allocation Model and Prospect [J]. Journal of Water Resources & Water Engineering, 2005, 16(3):1 – 5.

Yellow River Conservancy Commission. Planning for Digital Yellow River Numerical Modeling System [R]. Zhengzhou: Yellow River Conservancy Commission, 2009.

Development Tendency of Numerical Modeling Software [EB/OL]. http://hi. baidu. com/sunwzhp/blog/item/2e94b8ff1e5e7d385c600883. html.

Ning Jianguo. Development Tendency of Mechanics of Explosion [EB/OL]. http://www. d3dweb. com/Documents/201010/31 – 21373324214. html.

Hu Zhendong. Development Tendency of Multi-phase Physics. [EB/OL]. http://www. cntech. com. cn/news/media/2009 – 11/Multiphysics – Modeling – Trend. html

Huang Liuqing, Wang Manhong. Component China [M]. Beijing: Tsinghua University Press, 2006.

Huang Liuqing. Software's Nirvana [M]. Beijing: World Book Press, 2005.

Gerald M Weinberg, Daniela Weinberg. General Principles of System Design [M]. Beijing: Tsinghua University Press, 2005.

James West. Modeling for River Basin Management [M]. Zhengzhou: Yellow River Conservancy Press, 2004.

Li Jiren, Pan Shibin. Digital River Basin of China [M]. Beijing: Publishing House of Electronic Industry, 2009.

Xie Jiancang, Luo Jungang. Integrated Service Platform for the Information Explosion Process in Water Resources Industry and its Application Pattern [J]. Water Resources Informatization, 2010,10(4):18 – 23.

Han Guijun, Ma Jirui. Nonlinear Tide Model Using Four Dimensional Variational Assimilation [M]. Beijing: Ocean Press, 2002.

Wang Hui. Review on the Data Assimilation into Marine Ecosystem Model [J]. Advances in Earth

Science, 2007, 22(10): 989 –996.

Fei Jianfeng, Han Yueqi, Wang Yunfeng, et al. Data Assimilation Test on Theextended Kalman Filter —The Study Based on Lorenz(1960) Model [J]. Scientia Meteorologica Sinica, 2004, 24(4): 413 –423.

Yu Yunli, Lai Xijun. 2D Horizontal Unsteady Flow Model for Assimilating Remote Sensing Water Levels [J]. 2008, 19(2):224 –231.

Huang Yong, Wang Ying. Application of Shallow Water Model by Use of Ensemble Kalman Filter Data Assimilation [J]. Journal of PLA University of Science and Technology, 2008, 9(1): 86 –90.

Gong Jianhua, Li Wenhang. Exploring Conceptual Framework and Application of Virtual Geographic Experiments [J]. Geography and Geo – Information Science,2009, 25(1):18 –21.

Gong Jianhua, Zhou Jieping, Zhang Lihui. Study Progress and Theorectical Framework of Virtual Geographic Environments [J]. Advances in Earth Science, 2010, 25(9):915 –926.

How to Fund Basin Management Plans?
"Invest in IWRM – it pays back!"

Jean – François DONZIER

The International Network of Basin Organizations (INBO),
The International Office for Water, 21 rue de Madrid – 75008 Paris(France)

Abstract:Water and energy shortages, pollution, waste, ecosystem destruction: the situation in many countries is serious and is likely to get worse with climate change. Immediate action is urgent and requires the introduction of Integrated Water Resources Management (IWRM), organized at basin level with application of its six fundamental principles.

The question of its financing is essential and additional specific means of leverage must be envisaged to create incentives for limiting wastage and for cleaning up pollution: modern financing systems must be tailored to the specific situation of each country.

But in general everyone must contribute: the tax-payers, the offenders and the users

Key words:Integrated Water Resources Management (IWRM), investments, financing, "user – polluter – pays" taxes, river basin organizations

1 Introduction

Water and energy shortages, pollution, waste, ecosystem destruction: the situation in many countries is serious and is likely to get worse with climate change. Immediate action is urgent and a global, integrated and coherent water resources policy must be implemented, taking account the legitimate needs of the populations, while protecting aquatic and land ecosystems, to preserve the future and the heritage of mankind.

The Millennium Development Goals for drinking water and sanitation cannot be achieved without significant progress being made towards simultaneous introduction of Integrated Water Resources Management (IWRM), organized at the relevant river basin, lake and aquifer levels, whether local, national or transboundary.

Around the world, there are 263 large rivers and lakes, including 69 in Europe and 63 in Africa, plus several hundred aquifers (for example 69 on the American continent), which share their basins between at least two and sometimes many more neighbouring countries (up to 18 in the case of the Danube).

Basin-level management has developed rapidly in many countries, which have used it as the basis for their national water legislation, or are experimenting with the system on national or transboundary pilot basins. The International Network of Basin Organizations for example has 185 members or observers in 68 countries.

The experience acquired in this way means that it is now clear that integrated water resource management at basin level offers a very real advantage for governance, that is necessary for providing concrete solutions to the population's demand for various uses of water: agriculture, drinking water, energy, tourism, aquaculture, navigation, and so on.

If it is to be efficient and benefit all users, IWRM must be built around 6 main principles:

(1)Water resource management must be organized and discussed at the geographical level at which the problems occur, in other words at local, national or transboundary river basin, river, lake or aquifer level.

(2)It must be based on integrated information systems identifying the resources and their uses, pressure from pollutants, ecosystems and how they work, risk identification and monitoring of trends. These information systems must constitute an objective basis for discussion, negotiation, decision – making and assessment of the action taken, as well as for coordinating the financing from the various funding sources.

(3) It must be built around management plans, or master plans, setting the medium and long term objectives and giving a common vision of the future.

(4) It must involve the implementation of successive, multi-year, priority action and investment programmes, according to the financial resources available.

(5) It must mobilize specific financing, in particular based on application of the "polluter – pays" principle and "user-pays" systems.

(6) It should allow participation in the decision – making process by the local authorities concerned, representatives of the various user categories and environmental protection associations or those working in the general interest, alongside the competent Government departments. Through a process of discussion and consensus, it is this participation that will guarantee the social and economic acceptability of the decisions reached, taking account of the real needs, the level of acceptance and the ability to contribute by the social and economic stakeholders. Decentralization is the key to water policy effectiveness.

2　A legal and institutional frameworks must enable these six principles to be applied

A clear legal framework must in each country specify the rights and duties, the possible levels of decentralization, the institutional competence of the various parties, and the procedures and means essential for good water governance.

Special account must also be taken of the particular situation worldwide of transboundary rivers, lakes and aquifers.

3　The question of IWRM financing is essential

In nearly all countries, water as natural resources – is considered to be a "common heritage", which cannot be appropriated. Even if nearly everywhere, the water resource, which is considered to be a "natural raw material", is free, its governance, mobilization, transport, storage and treatment, and the protection of aquatic ecosystems do however have a management, investment and operating cost, and this cost has to be paid.

Meeting the diversity of demands and organizing global and integrated management of resources and environments implies that a whole series of functions must be carried out in a complementary and coherent manner across all territories in the river basins.

All of these functions are generally not performed by a single organization and the most frequent situation is one in which numerous bodies and initiatives, both individual and collective, both public and private, exist alongside each other in the same territory.

All of these functions must be organized in a lasting way and their management, investment and operational financing must be mobilized and guaranteed for the long term, whatever methods are used.

4　Mobilization of funds

4.1　Needs twice as great as the available funding for water

The essential investment needed to overturn current trends and meet the numerous water requirements (drinking water and sanitation, navigation, hydro – power, irrigation, leisure and tourism, environmental management, etc.) in the emerging and developing countries, is estimated at about $ 180 million per year for at least 25 years. At present, water funding stands at $ 80 billion per year, so another $ 100 billion per year must be found over the next 25 years at least.

In most cases, these investments and the collective service management and equipment, operating and maintenance costs cannot be covered, or at least not completely, by the traditional national and local public budgets alone.

Even if public development aid were to double, it would at best represent 10% to 15% of investments.

4.2 How to mobilize the additional funding necessary to enable needs to be met

With about 1% of global GDP, or nearly US $ 300 billion per year, the amounts paid every year by the various users of water for consumption and treatment, are already significant. Although they only represent about 10% to 15% of the volumes consumed, as opposed to about 70% for agricultural users, urban consumers provide most of this financing. The consumers in rural and underprivileged areas, industries and irrigated agriculture, usually only pay low prices or fees which to date only represent a very small part of the real cost of the water services.

In many countries, revenues from users of the services are not therefore sufficient to meet all the costs, in particular those linked to paying off investments and, even if the operating costs are covered at least in part, this is still rarely the case with the financial depreciation and provisions for equipment renewal, which raise enormous problems in terms of the durability of the investments made.

This then leads to the application of the "user – polluter – pays" principles, with costs being recovered by making the contribution of each party proportional to its uses or to the damage it causes. This is a credible approach towards mobilizing the enormous financial resources necessary, while creating economic incentives among the users to reduce wastage and pollution.

4.3 Whatever the situation, governments cannot bear all the costs and conventional public financing has now reached its limits

A balance needs to be found between the ability to contribute by each category of consumer, the economic cost of water and the public participation options, which are up to each State, according to how it defines its scope of action

4.4 Investments in the water sector are particularly capital – intensive

The creation of large infrastructures on the scale of a river basin, or for inter – basin transfers, large water mains, treatment and sewerage plants, as well as the distribution, drainage and wastewater collection networks, represent major initial unit costs which need to be staggered over time, and which can realistically only be paid off over a very long period of several decades.

This policy indeed needs to be scheduled for the medium and long term, taking account of the time needed to mobilize the partners and design and build the projects, not forgetting the general straitjacketing of available funding, which means that it is impossible to do everything, everywhere, immediately.

The objectives to be reached and the necessary means of all types must be specified in water planning and management master plans, covering a time frame of 15 ~ 20 years. The drafting of successive Priority Intervention Programmes, for which a realistic duration can be four, five, or six years, is the instrument for implementing these master plans.

Additional specific means of leverage must be envisaged to create incentives for limiting wastage and for cleaning up pollution: modern financing systems must be tailored to the specific situation of each country, but can in general be built around the following three notions

4.4.1 Administrative taxes

There are at present two main categories:

General administrative taxes for issue of the necessary licenses (registration and other fees) or for use of the public domain (granulate quarrying taxes, hydroelectric concession, land used for infrastructures or reservoirs, river transport, etc.) as well as fines for non – compliance with regulations and standards or for intentional or accidental violations leading to damage.

A new form of "ecological taxation". A number of industrialized nations, particularly in Europe, are examining or experimenting with systems of "general taxation on polluting activities" aimed at guiding polluters of all types towards more appropriate practices, by means of "internalization of external costs". The polluting products are then taxed directly on production at the company manufacturing them, with the effect of increasing the cost to the end customer, who will therefore reduce the quantities purchased or will use alternative products that are cheaper and less environmentally harmful. The European Water Framework Directive proposes that the member States set up systems to remunerate "environmental services" in the river basins, asking for an estimate of the cost of damage to the environment caused by a given activity, along with the cost of alternative uses of the resource…

There is also the case of CICOS, in which the countries concerned levy a duty of 1% on imports from third party countries and allocate it to the working of the Commission.

4.4.2　Pre-allocated fees and taxes

(1) Pre-allocated parafiscal taxes, based on the principle that "water must pay for water" allowing financing actions or equipment of common interest, which cannot be directly covered by the individual users or collective services. The pre-allocated fees are collected specifically from water uses (intake of raw water for hydroelectric production, thermal power station or industrial cooling, irrigation or supply of drinking water) and/or from discharge of waste water. The income passes through specific, individualized financial circuits (which are not therefore included in the centralized general public budgets) and is totally or partially devoted to improving the water sector.

(2) National systems, passing through "Special Appropriation Accounts" with the credits being reallocated either directly to major projects or programmes decided on at a central level or, more generally, distributed in the form of regional or decentralized budgets made available to regional administrations, local authorities and local public development agencies.

(3) Territorial systems, in particular organized at the level of the river basin communities; in this case, all the funds collected from water uses or pollution in the basin are reallocated to projects to improve the resource or uses in the basin itself. They pass through the budget of a specialized river basin organization.

These are economic incentive and solidarity instruments.

4.4.3　Industrial and commercial pricing of collective services linked to water uses

This consists in having the consumers and users of collective services pay all the direct and, if possible, all the indirect investment and operating costs of the services they receive, through various types of pricing (flat rate, proportional, quantitative, geographical or social equalization, etc.), with or without external compensation mechanisms (subsidies, direct coverage by the public authority of structural works, of administrative costs, and so on).

These services, whether organized by public or private organizations, must then offset their expenditure with revenue from the prices charged to the users, calculated pro-rata to the services received or to consumption (drinking water, waste water, industrial raw water, irrigation, etc.), which implies the development of metering and measurement systems.

The implementation of subsidy systems to limit exceptionally high costs and/or equalization systems between user categories can, if transparent, constitute a means of adapting to the diversity of situations encountered.

Hydroelectric power frequently finances not only the investment in and operation of the hydraulic works directly or indirectly linked to electricity production, but also the general costs related to improving the water resources.

An industrial and commercial approach to water services is compatible with equalization systems allowing the financial effort required of the various user categories to be balanced and equalized whenever necessary, provided that it is done ensuring full transparency of costs and pricing. There are a variety of possible systems:

(1) "Territorial" equalization between the services of an individual administrative area in order to balance between the users the cost of access to resources or of pollution control and clean-up.

(2) Equalization between sectors, for example between the water and electricity sectors.

(3) Equalization between users, to encourage access to water by the most underprivileged.

(4) Equalization between functions, to ensure upstream-downstream solidarity and to finance

general administration and data acquisition functions, or the building of infrastructure, or performance of planning and development work in the general interest.

In many developing countries, the poorest are often those who pay the most for water (albeit in small quantities) owing to the speculation surrounding the shortage of this essential asset. These population categories either already resort to individual or semi-collective alternatives which have a significant cost to them, or buy water, sometimes at a proportionally very high price, from distributors/transporters who deliver directly to the districts, in conditions of hygiene that often leave much to be desired···

5 The creation of a basin organization could be a "profitable" project

The creation of new basin organizations is a major project which, depending on the institutional system of the country concerned, will take somewhere between 5 and 10 years or more and will represent a significant initial investment. The local situation and the size of the river basin naturally vary widely, but it can be roughly estimated that the unit costs for a new river basin organization are several million US dollars over a 5 – year period .

Even if detailed and extensive economic studies are required in each case, it is generally shown that with a small annual participation on the part of the users, based on taxes levied on water intake and pollutant discharges, a basin financial mechanism can mobilize considerable overall sums, although this is a gradual process, which would allow significant funding of structural and priority investments and the correct working of the equipment, as well of course as paying off the initial unit costs.

6 Conclusions: everyone must contribute

In any case, if there is insufficient financing, plans will never get off the drawing board and somewhere, in one form or another, someone will have to pay:

(1) The tax-payers, who pay their taxes into the central or local general budget.

(2) The offender, who is required to pay a fine in the event of negligence or a breach of laws, standards and regulations.

(3) The user, who pays for the services received either directly (supply of drinking water to the tap, of raw water to the factory or irrigated plot of land, connection to the collective sanitation network, and so on), or indirectly (reforestation of upper catchment basins, flood protection, restoration of ecosystems, combating upstream pollution, or creation of a dam-reservoir, but also data systems, research and training).

If things are to improve significantly, what is important is to ensure that payment is levied at the "right place": in other words, on the one hand all those who by their actions have a negative impact on the water cycle must be given incentives to reduce this negative impact so that they pay less and contribute to sustainable development and, on the other hand, all users of the resource must pay a price for the services they receive.

The organization of IWRM at basin level with application of the six fundamental principles recalled in the introduction offers a pertinent framework for planning and for coherent mobilization of funding, whatever the country's level of development. It provides an overview of the problems to be resolved and mobilizes all stakeholders at the most appropriate scale to find the best solutions, define priorities and time-frames, achieve economies of scale and more profitable investments, ensure solidarity and an upstream/downstream balance that encourages all parties to contribute to reaching commonly defined objectives.

Problems on Water Resources Management of the Xiaobeiganliu Reach of the Yellow River and its Countermeasures

Zhang Yali, *Wu Yunyun*, *Guo Quanming* and **Hu Jieqiong**

Yellow River Shaanxi Bureau, YRCC, Weinan, 714000, China

Abstract: With rapid social and economic development along the Xiaobeiganliu Reach of the Yellow River (XRYR), demand for the Yellow River water resources has been increasing, and contradiction between supply and demand with respect to water resources has become more serious. Based on survey of the current situation of water resources, application and demand for water resources by industrial and agricultural development plans in the basin under the jurisdiction of the Weinan City, Shaanxi Province on the right bank of the XRYR, this paper analyzes the main problems in utilization and management of water resources, and, in the light of the situation of the basin and current situation of water resources of the XRYR, presents several countermeasures for future water resources management.

Key words: the Xiaobeiganliu Reach of the Yellow River (XRYR), water resources management, existing problems, countermeasures

In recent years, with rapid social and economic development in the Xiaobeiganliu Basin of the Yellow River, demand for the Yellow River water resources has been increasing, and the contradiction between supply and demand with respect to water resources has become more serious. Therefore, being aware of the current situation of water resources utilization in the Xiaobeiganliu basin, finding out the main causes of water resources utilization, strengthening water resources management, and carrying out effective development, utilization and protection of the Yellow River water resources by legal, administrative, economic, technical and educational means are the guarantee of achieving sustainable utilization of the Yellow River water resources and keeping healthy life of the Yellow River.

1 Current situation of water resources in XRYR

The XRYR refers to the reach from Yumenkou to Tongguan. It is 132.5 m long, and flows through four counties (cities) of Hancheng, Heyang, Dali and Tongguan of Shaanxi Province, with basin area of 1.38×10^5 km^2. On the way, it admits tributaries of the Weihe River, the Jushui River, the Fenhe River and the Sushui River, et al. The water and sediment of the reach mainly come from the area upstream of the Longmen Hydrological Station. In recent years, due to decreased rainfall in the Yellow River Basin, increased industrial and agricultural water consumption, regulation by reservoirs and water and sediment reduction by water and soil conservation, great change has taken place in the incoming water of XRYR. According to statistics, from 1919 to 2005, the mean annual incoming water at Longmen Hydrological Station was 2.896×10^{10} m^3. Since 1995, the Yellow River has entered low water period. The incoming water from 1996 to 2005 was 48.9% less than that from 1919 to 1986. The water has not only to meet industrial and agricultural needs, but has also to meet the need of the Yellow River ecology. Now there are four water diversion works on the right bank of XRYR, of which three are for agriculture and one is for industry. As approved by YRCC, the four works can draw 2.38×10^8 m^3 of water. The effective irrigation area of the three agricultural diversion works is 2.390×10^6 mu (1 mu = 666.67 m^2), and the annual average water intake quantity is 3.45×10^8 m^3. The existing water resources are far from satisfying the needs of life and production.

According to statistics of the data in Weinan's special plan for "the 11th Five – Year" of water resources utilization and protection, the total water supply of the city in 2010 was 1.453×10^9 m^3, while the annual water need of Weinan was 2.27×10^9 m^3.

2　Main problems in water resources utilization and management

In recent years, with development and utilization of water resources of the XYYR, the basin administrative organization and the local government have strengthened water resources management. Yet there still exist some problems, which are as follows.

2. 1　Runoff and sediment are distributed unevenly within a year, and flood and drought disasters are serious

The annual runoff and sediment distribution of the XRYR is uneven, with runoff and sediment concentrated in the flood period. According to statistics, at Longmen Hydrological Station, the mean annual flood – period runoff is 1.754×10^{10} m^3, accounting for 57. 5% of the yearly runoff. The mean annual flood – period sediment is 8.14×10^8 t, making up 87. 7% of the yearly sediment. The maximum annual runoff of the Station was 5.521×10^{10} m^3 (in 1967), 4 times of the minimum annual runoff of 1.394×10^{10} m^3. The maximum annual sediment was 2.424×10^9 t in 1967 (hydrological year, namely from July of the year to June the next year, the same below), being 12. 4 times of the minimum annual sediment of 1.96×10^6 t (in 2000).

According to statistics, the mean annual runoff each decade from 1950 to 2010 at Longmen Hydrological Station were 3.215×10^{10} m^3, 3.366×10^{10} m^3, 2.846×10^{10} m^3, 2.762×10^{10} m^3, 1.895×10^{10} m^3 and 1.757×10^{10} m^3 respectively. The 1950s and 1960s were the high water period, when the runoff was larger than mean annual runoff by 19. 7% and 25. 3% respectively; the 1970s and 1980s were the medium water period, when the runoff was less than the annual runoff of the 1960s by 15. 4% and 17. 9% respectively; the 1990s and recent years (2001 ~ 2010) were low water period, in which the runoff was less than the mean annual runoff of the 1960s by 41. 1% and 51. 5% respectively, less than the mean annual runoff of the 1980s by 28. 3% and 40. 9% respectively, and less than mean annual runoff by 29. 4% and 34. 6% respectively. The runoff changes of past years showed a decreasing trend.

Since there are no regulating reservoirs on the reach, there are no means to control and regulate flows. In high water years, abundant rainfall and large runoff result in flood disasters. The "Aug. 1967", "Aug. 1994" and "Aug. 1996" floods and many ice runs have caused river to rise sharply, destroyed roads and villages, submerged farmlands, caused house collapse, people and livestock casualties and traffic disruption, and caused great economic losses. While in low water years, less rainfall and small runoff result in drought. According to historical record of natural disasters in recent few hundred years of Shaanxi Province, in the 110 years from 1840 to 1949, there have occurred 12 great droughts, about once in ten years. In the 20 years from 1963 to 1982, there have occurred 51 different kinds of droughts, roughly once in 2 to 3 years on average, and most of them are spring or summer droughts. According to the records, there are no effective rainfall in the period from late October 2008 to March 2009 in most part of Weinan, Shaanxi Province, and the temperature in the same period is 1 ~ 2 ℃ higher than that of normal years, thus the most severe drought since 1997 occurred. 3.800×10^6 mu of crops is suffered from drought, of which 1.170×10^6 mu is severely stricken by drought. And when there is need to draw water from irrigation, there is no water at some intakes, The droughts not only directly threaten the normal development of industry and agriculture, but also seriously affect the social and economic development.

2. 2　Shortage of water resources to restrict local economic development

With the development of industry and agriculture and steady growth of urban population, the water resources and water supply capacity can not meet the needs of industrial and agricultural production and people life, resulting in water crisis. According to statistics of Weinan, the city's water supply capacity in a normal year is 2.015×10^9 m^3, less than the current annual water demand of 2.27×10^9 m^3 and the forecast water demand of 2.51×10^9 m^3 in 2015 by 2.60×10^8 m^3 and

4.90×10^{8} m^{3} respectively. Weinan City has a population of 5.31×10^{8} and 1.30×10^{8} km^{3} of farmland. The per capita water resources consumption is only 380 m^{3}, and the arable land per capita water share is 230 m^{3}, ranking ninth among the ten cities (prefectures) of Shaanxi Province. And the figures are the 1/3 and 1/4 of those of Shaanxi Province in a normal year, showing serious imbalance distribution of water and soil resources. The shortage of water resources has seriously restricted local economic development. Besides, the potential of water resources is limited. The available water resources of Weinan are 1.273×10^{9} m^{3}, 1×10^{9} m^{3} has been exploited, and the remaining is less than 0.3×10^{9} m^{3}. The shortage of water resources has seriously restricted social and economic development.

2.3 Irrational water consumption mechanism to make the serious waste of water resources

No institutional constraint on water consumption, weak awareness of water saving, and irrational charge for industrial and agricultural water consumption combined have resulted in serious waste of water resources. According to the existing mechanism of water consumption, no award will be given for water saving, and no punishment will be imposed for waste of water resources, i. e. there is no different treatment between economy and waste. Such a water consumption mechanism imposes little economic constraint on water users, resulting in weak awareness of water saving. Besides, according to the relevant provisions of the State on the management of water resources in the XRYR, no charge shall be made for agricultural water consumption, and industrial water consumption rate is only around 0.1 Yuan/m^{3} and is collected by the local water administrative organizations. Low water price has led to serious waste. For example, with flood irrigation widely used, the effective water usage is only 40% ~ 50%. At the same time, due to aging of irrigation infrastructure, the efficiency of water supply can not be brought into full play. All of this, together with insufficient awareness of water scarcity and weak consciousness of water resources protection, has led to widespread waste.

2.4 There is an upward trend in pollution

The XRYR is located in the golden triangle area between Shaanxi, Shanxi and Henan Provinces. With population growth along the Yellow River and national economic development, domestic and industrial waste water discharge has been increasing sharply, and each year domestic waste water discharge has accounted for about half of the total waste water discharge. Since the urban environmental infrastructure in counties and cities along the XRYR is very weak, waste water treatment is not up to the standard, and some places even have no waste water treatment equipment. More than 90% of industrial and domestic waste water is directly discharged into the Yellow River and its tributaries without any treatment.

According to the survey in 2010, the sum of the annual runoff of the 19 tributaries of the XRYR was 7.402×10^{9} m^{3}, and the annual pollutant transport was 4.944×10^{7} t, of which the Weihe River had the largest annual runoff, 6.011×10^{9} m^{3}, accounting for 81.2% of the total of 19 tributaries, and its pollutant amount was 4.831×10^{7} t/a, making up 97.7% of the total. Main tributaries on the right bank of the XRYR are the Weihe River, the Xushui River, the Tonghe River, et al. Along these tributaries there are five sewage outlets, which are mainly outlets of domestic waste water from Hancheng City and waste water from the Longmen Steel Plant of Hancheng. The total annual waste water discharge is 5.5×10^{7} m^{3}, and the total contaminant discharge is 4.1×10^{4} t /a, of which, the annual waste water discharge of the outlets of Hancheng domestic waste water is the largest, 2.0×10^{7} m^{3}, and the contaminant discharge is 9.3×10^{3} t/a, accounting for 36.4% and 22.7% respectively. These industrial and domestic pollutants entering the river has made the Yellow River water quality badly deteriorate and caused huge economic losses.

2.5 The legal status and right limits of the basin organization have not been clearly defined and the organization lacks direct management right

The basin organization is subject to intervention of local authorities while exercising water management within the region, and it can not carry out effective management of the water resources within the region. According to YRCC's "Three Determine's" program with respect to institutional reform, the Yellow River Basin organization, as the administrative body of the XRYR, is only the agency of YRCC, rather than a level of administrative unit, and it can only exercise water administration within the region according to authorization of YRCC. Since it is authorized, it has not direct, complete "right", and its administrative function differ from that of local water administrative departments defined by law, or it is much smaller than the latter, therefore, it can not perform effective management of water resources within the basin. As a most remarkable example, when the incoming water of the Yellow River was too small to meet the need of ecological water consumption, and we tried to close the industrial and agricultural intakes along the bank, the action would be interfered in by local authorities. Though the objective was finally achieved at much work, much manpower and material resources had been wasted, bringing passivity and difficulty to water resources management.

2.6 The function of supervision is difficult to fulfill thoroughly

The "Rules for Implementation of Yellow River Water License Management" provides that the basin administrative organization shall exercise supervision over water diversion from the main stem of the Yellow River downstream of Yumenkou within Shaanxi Province, and it also provides that units or individuals applying for water diversion shall firstly submit application to the water administrative departments of the government of the province within which the water intakes are located. Under this system, since the low water period of the Yellow River is just the time when the drought is serious and there is a need for water diversion in great quantity, the supervision and management of basin administrative organization would be interfered in by local governments and obstructed by water users, the contradiction between demand and supply was very sharp, so the function of management is difficult to fulfill completely.

2.7 There are not perfect the supporting laws and regulations on the legal safeguard

Since basin management is adopted for the Yellow River, YRCC is an agency of the Ministry of Water Resources, it often lags and has much difficulty for formulating relevant regulations and policies, so that there is no sound legal basis for the development, use, management and protection of the Yellow River water resources. The typical example is the legal basis for levy of the Yellow River water resources fee. The new Water Law stipulates that "The State Council shall provide the specific measures for implementation of the water collection license system and the levying of water resource fees." YRCC has issued "Rules for Implementation of Yellow River Water Collection License System", but has not issued regulations on levying of the Yellow River water resources fees.

3 Suggestions on the water resources management of XRYR

In view of the above-mentioned problems in the development and use of water resources, and according to the actual situation of the basin, we put forward the following suggestions on future water resources management.

3.1 To establish and improve a clearly defined integrated basin water resources management system

To establish and improve an integrated basin water resources management system is an effective way to achieve sustainable use of water resources, is the inevitable requirement for sustainable social and economic development, and is also the institutional guarantee for optimal allocation of water resources. The basin administrative organization should manage affaires relating to water resources as the representative of the State. To ensure reasonable development and use of the Yellow River water resources, we must strengthen unified water resources management, establish and improve an integrated basin water resources management system, and clearly defined division of right limits between the basin administrative organization and local authorities. We mush exercise unified water regulation, implement planned water use system, water collection license system, et al. , so as to play the maximum benefit of the limited water resources.

3.2 To strengthen legal construction with respect to the Yellow River water resources management

We must strengthen legal construction with respect to the Yellow River water resources. Firstly, on the basis of the existing laws and regulations such as "Law on Prevention and Control of Water Pollution", "Provisions on Management of Urban Water Economy", "Water Law", "Environmental Protection Law", "Law on Water and Soil Conservation" and "Regulations on Technical Policies on Prevention and Control of Water Pollution", et al. , formulate highly feasible supporting regulations suitable for the characteristics of the Yellow River, so as to form a system of laws and regulations with respect to the Yellow River water resources protection. Secondly, to strengthen law enforcement, carry out various types of supervision on law enforcement such as economic sanctions, accountability, et al. , openly and severely crack down on illegal cases.

3.3 Construction of regulating reservoirs to improve the Yellow River water quality

According to statistics, at Tongguan and Longmen, the flood volume of each flood event is usually larger than 6.0×10^8 m^3, and the annual flood runoff is generally larger than 3.0×10^9 m^3. Since there are no regulating reservoirs or necessary means for water regulation, the valuable flood resources flow away without producing any benefits, which not only brings threat to flood control on the lower reaches, but also results in waste of valuable water resources. During the flood period, most water is wasted, while after the flood period there is no water to use, resulting in floods and droughts. It is suggested speeding up construction of regulating reservoirs such as the Guxian Reservoir and the Qikou Reservoir on the middle reaches of the Yellow River, so as to perform effective flood management with engineering measures and to keep the river in good ecological condition. It is also suggested accelerating infrastructure construction in the Yellow River irrigation area to improve utilization ratio of water resources.

3.4 To establish water saving system

The shortage of the Yellow River water resources has been becoming more and more serious, and this was mainly caused by the non-standard economic behavior of mankind. Water saving is a kind of social work, and to do the work well requires understanding and support of the whole society. It is suggested carrying out publicity and education by various means so that the masses may be aware of the hardship; perfecting water consumption mechanism and exercising a management system of fixed quotas with respect to various kinks of water consumption, charging progressive fees for over – quota water consumption, and charging punitive price of water for the behavior of wasting water resources, and at the same time commending and giving rewards to those who are outstanding in water saving; improving usage ratio of water resources by vigorously spreading water – saving facilities and technology and improve water supply facilities so as to achieve sustainable use of water resources.

3.5 Strengthening management and strictly carrying out YRCC's water regulation schemes

Total quantity control and fixed quota management are the core of water resources management, and are also the important means to achieve reasonable development and sustainable use of water resources. The development of the Yellow River water resources has reached a very high level and has initiated a series of environmental problems. Therefore, it is necessary to gradually decrease water consumption and to restrict the total development quantity to a reasonable limit by means of total quantity control and fixed quota management. For this reason, the YRCC has been formulating and issuing the Yellow River water regulation program each month. For various areas and water using units, the total quantity is a rigid objective, is a principle that can not be broken through, and is also the situation about the whole water resources that can be provided for the local economic development. As a basin administrative organization, we must strengthen the function of supervision and management and earnestly fulfill water regulation programs.

3.6 To reduce pollution to water resources

In the process of economic development, it is very difficult to eliminate pollution completely. Therefore, we need to determine an acceptable pollution level. After determining this acceptable level, we should do the following to reduce pollution to water resources. Firstly, to establish and improve a license system with respect to pollution discharge, namely clearly stipulate in laws and regulations the determination of control index of total pollution discharge, legal liability that should be borne for violation of the license system of pollution discharge, and licensing procedures, and give full play to the system in pollution reduction and combination of prevention and control. Secondly, to take water quantity, water quality and water pollution prevention and control into consideration and make overall plans from the point of view of the whole Yellow River Basin.

4 Conclusions

To achieve harmonious development of the Yellow River water resources development, utilization and protection, it is far from enough just to relying on the above – mentioned measures. In the future supervision and management, we should continuously strengthen the consciousness of water resources protection, and effectively develop and use the Yellow River water resources by legal, administrative, economic, technical and educational means, so as to achieve harmonious development between human being and water resources use and to sustain the healthy life of the Yellow River.

Strict and Efficient Management of Water Resources in River Basin—A Case of the Cross Border Basin of Niger

Kone Soungalo[1] and *Robert Dessouassi*[2]

1. Niger Basin Authority(NBA), PO Box 729 Niamey, Niger
2. Niger Basin Observatory, PO Box 729 Niamey, Niger

Abstract: The nine member states (Benin, Burkina Faso, Cameroon, Ivory Coast, Guinea, Mali, Niger, Nigeria and Chad) sharing the Niger River Basin have early recognized the concerted management of water resources by creating the Niger River Commission in 1963 and the Niger Basin Authority (NBA) in 1980.

The NBA, in the aim to lead a concerted, sustainable, equitable management of the basin water resources, has developed a shared vision with the support of the 9 Member States and signed a Declaration called "Declaration of Paris". This statement sets out the management principles and good governance for sustainable and shared development of the Niger Basin. The implementation of the Shared Vision has developed legal, institutional and technical instruments for the planning and good governance of water resources.

– Legal tools: The agreement creating NBA signed in 1980 by nine member states and the Water Charter adopted in 2008 and ratified by member States.

– Institutional Bodies tools of good governance of the basin are: The Summit of Heads of State and Government, The Council of Ministers, the Technical Committee of Experts, the Executive Secretariat with the Niger Basin Observatory, the National Focal Structures, the Permanent Technical Committee, the Panel of Experts and the Sub-basins Commissions.

– Technical tools for planning and management for the development of the basin resources: A network of 105 hydrometric stations and a database which serves to support the models of hydrological forecasts and allocation of water resources, an Action Plan for Sustainable Development (APSD) by 2027, an Investment Program by 2027, a Hydraulic model for allocation and water resources management and a model for hydrological forecasting.

These models are used to analyze the impacts of new development on the basin and alert in case of imbalance. As a tool for decision support, they can anticipate emergencies and deficits, plan achievements and manage hydraulics structures.

Key words: the Niger Basin, IWRM model, Water resources allocation and management, flow forecast, irrigation, hydropower

1 Introduction

The nine member states (Benin, Burkina Faso, Cameroon, Ivory Coast, Guinea, Mali, Niger, Nigeria and Chad) who shared the Niger River Basin was early aware of the joint management of water resources by creating the Niger River Commission in 1963 and the Niger Basin Authority (NBA) in 1980.

To conduct a concerted, sustainable and equitable management of water resources of the basin, the NBA has developed a shared vision with the support of the nine member states and sign a statement "called Paris Declaration." This statement sets out the principles of management and good governance for sustainable and shared development of the Niger Basin.

The implementation of this Shared Vision has developed legal, institutional and technical tools for the planning and good governance of water resources.

2　Legal tool: the water charter of the niger basin

Adopted by the 8th Summit of Heads of State and Government of the NBA held in Niamey, 30 April 2008, the Water Charter of the Niger Basin is the legal instrument which provides a framework for cooperation among States members of the Niger Basin, according to management principles and good governance for a sustainable and shared development of the basin.

2.1　The objectives of the charter

The objectives of the charter are:

(1) Promote cooperation based on solidarity and reciprocity for a sustainable, equitable and coordinated use of the water resources in the catchment of Niger.

(2) Promote and simplify dialogue and cooperation between states in the concept and implementation of programs and projects affecting or likely to affect the water resources of the basin.

(3) Define procedures for review and approval of new projects users of water or likely to affect the water quality.

(4) Regulate principles and conditions for allocation of water resources between different sectors of use and associated benefits.

(5) Establish principles and rules to prevent and resolve conflicts related to the use of water resources in the basin.

(6) Define modalities of participation of water users in the management decision-making of water resources.

2.2　General principles

The Charter provides that, the use of the basin water takes into account a number of principles including:

(1) Participation and fair and reasonable use of the water resources of the basin.

(2) The non – harmful use of water resources.

(3) Precaution and prevention and.

finally, the principles of "'Polluter Pays and User Pays".

2.3　Uses and reports between uses

In the use of water resources, no use is priority over the others. However in case of conflict between uses, special attention is given to vital human needs.

2.4　Projected measures

The Charter required that, before a state should be unable to implement on its territory, measures likely to have significant negative effects for other states, it should have, through the Executive Secretariat, provided in due time to the latest, a notification and all relevant information. He must also abstain from implementing the proposed measures during the period between the notification and that given for the answer.

However, a special treatment is given in cases of extreme urgency to public health, public safety and other interests of equal importance to immediately proceed in the implementation, notwithstanding the notification requirement.

2.5 Public participation

The Charter guarantees every user the right to be informed about the status of water resources and to participate in the development and implementation of decisions regarding the valuation of the basin resources.

2.6 Common works and common interest infrastructures

A common works is a work for which Member States decided by a legal act that is their common property and invisible.

A common interest works is a work of interest to two or more Member States of the Niger Basin Authority and for which they have decided by mutual agreement of the coordinated management.

Specific agreements will determine the status as well as procedures to participate in their financing, management and sharing of benefits arising from their construction and management.

3 Institutional tools

Bodies of good governance of the basin are:

(1) The Summit of Heads of State and Government, supreme guidelines and decision making body.

(2) The Council of Ministers, supervisory body of the Authority.

(3) The Technical Committee of Experts, responsible for preparing sessions of Council of Ministers.

(4) The Executive Secretariat with a Niger Basin Observatory, implementing body of the Authority.

(5) The National Focal Structures.

(6) The Permanent Technical Committee.

(7) The Panel of Experts.

(8) The sub-basins Commissions.

4 Planning and management technical tools for the development of the basin resources

The international character of the Niger River and its sharing between nine riparian countries enabled the development of planning and simulation tools, negotiation, coordination and management to allow an optimal management and development of infrastructure (in terms of timing, sizing, strategic management rules) for a rational allocation of satisfactory and sustainable water resources. The development of these tools and their application started with the implementation of the Shared Vision.

In application of the Niger Basin Water Charter approved in 2008, member countries are expected to notify the NBA of any project that may influence or affect water resources in transboundary levels. However, such notification is coming on stream only gradually. To examine the impact of some of these projects, the NBA dedicated executives are required to carry out simulations on the IWRM model and to interpret them, using the configurable dashboard related to the model.

4.1 Observatory network

To allow the collection of necessary information for different models, the NBA through the Niger HYCOS Project has assisted the national hydrological services through the creation and rehabilitation of 105 gauging stations within the nine member countries of the NBA.

This Niger HYCOS Project funded by the French Development Agency (FDA) and the African Water Facility, for the first phase lasted from 2006 to 2010.

The phase 2 of this Project being called IWRM 2 focuses on data collection.

The network consists mainly of classical water level stations and stations containing data collection platform (DCP) with remote.

4.2　Modeling tools

The allocation and management of water resources model or IWRM model:

This hydraulic model was developed with Mike Basin in ArcGIS environment with the assistance of DHI and BRL engineering.

It allows to hydraulic consistency, the various projects identified throughout the basin and is the basis for technical studies of programming actions.

The characteristics of this model are:

(1) The basin was divided into 60 sub – basins.

(2) 445 nodes (calculations points) were introduced.

(3) 66 gauging stations were used.

(4) 93 sampling points for different uses (water supply, irrigation···).

(5 (Data from the 23 existing dams have been included in the model (storage dams and hydro-electric dams over the water).

Data relating to projects, existing structures, samples collected in the basin have been introduced to the model.

There has been the reconstitution of missing data.

Model calibration was performed year by year in generating series of runoff and that, over the period of 1966 ~ 1989. Fig. 1 is sampling points in the basin

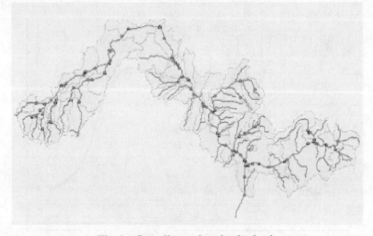

Fig. 1　Sampling points in the basin

In the inner Delta which is a particular area (wetland), the timing has been made to ensure the continuity of the hydrological system to crossing the delta and examine the impact of development on local flooded areas.

IWRM model allows to:

(1) Analyze impacts of carried out Projects or to carry on water resources.

(2) Respond to requests from countries on the feasibility of their projects.

(3) Assist the political decision-making.

Hydro-economic module is coupling of IWRM model + an economic module, allows to:

422

Analyze the scenarios of basin development in economic terms: IRR calculation, sensitivity calculation, calculate the number of days of navigation by tailbay, etc.

Management Module of dam reservoir it is used by dams specialists to optimize management of reservoirs to meet the relevant sectoral needs based on specific hydrologic conditions.

The forecast model Computer Prediction System (CPS) has been developed with the assistance of ISL in order to alert involved communities and managers of dams on the occurrence of extreme events in order to take the necessary steps to minimize impacts.

It allows to forecast flows at various set of time horizons (in 2, 5, 10, 15 days, 1 month, trend of high and low flow rates), to predict the date and the maximum flood.

Fig. 2 is IWRM model interface. Fig. 3 is Forcast and observed data.

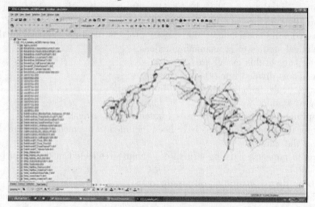

Fig. 2   IWRM model interface

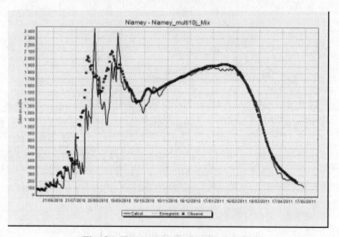

Fig. 3   Forcast and observed data

5   Conclusions

5.1   Products and impacts

(1) The choice in 2007 of a Watershed Management consensually approved by countries at political level, through simulations;

(2) The illustration of the Fomi dam impact (Guinea) on the Inner Niger Delta (Mali) for taking into account in its management procedures;

(3) Initiation of funding agreements for dams;

(4) Rise of technical functions of the NBA for developing quality information leading to increased confidence of countries;

(5) This rise of technical functions, and of hydrological monitoring, is likely to encourage users to participate in the financing of services.

5.2 Achievement indicators

(1) Effective use of tools by experts in the form of analysis notes, forecast bulletins, capitalization sheets on the project cycle of dams;

(2) Adjustment of management methods and / or characteristics of works and practices based on analyzes produced by the NBA (eg. restriction of the irrigated area into campaign);

(3) Improved process of validation of steps of the dams Project cycle (ESIA, feasibility, etc.).

To make these tools more effective and sustainable, hydrometric data collection must be improved and strengthened.

Study Conclusions and Suggestions of Compensation Policies for Application of Floodplain Area in the Lower Reaches of the Yellow River

Teng Xiang[1], *Zhou Li*[2], *Huang Shuge*[1] and *Li Lixiang*[3]

1. Flood Control Office of Yellow River Conservancy Commission (YRCC),
Zhengzhou, 450003, China
2. Construction and Management Department of YRCC, Zhengzhou, 450003, China
3. Xiaolangdi Multipurpose Dam Project Management Center, Zhengzhou, 450000, China

Abstract: A big contradiction has been existed for a long time between production and living of the masses in floodplain area, and flood discharge, and flood and sediment detention. The more developing the basin economic and social and higher requirements flood control safety, the more prominent the contradiction. Because flood and sediment detention can not be compensated, masses in floodplain area constructed "production dike" illegally for needs of living and development constantly. Through study by YRCC, compensation was regarded as the effective way to solve the contradiction. "Instructions of using the compensation policy at the floodplain area after flood inundation in lower reaches of the Yellow River" was reported to The State Flood Control and Drought Resistance Headquarters and The Ministry of Water Resources (MWR) by YRCC in 2004 and 2005. The problem attracted attention by relevant departments of the state. "Flood Inundation Compensation Policy Study Working Group for Floodplain Area of the Lower Reaches of the Yellow River" was formed by MWR, The Ministry of Finance (MOF), the National Development and Reform Commission (NDRC), Henan and Shandong provinces and YRCC. From 2006 to 2010, the working group had organized 19 units, more than 200 people to participate in study work, and hold expert advisory council five times. Flood inundation compensation policies are not only the requests of 1.895×10^6 masses of in the lower reaches of the Yellow River, but also the requests of implementing lower Yellow River harnessing strategy, maintaining compensation policy continuity of floodplain area of the Yellow River, and building new socialist countryside. Statistical calculation shows that national finance can undertake flood inundation economic compensation. Under present condition, the implementation of the policy compensation after flood inundation in floodplain area of the lower Yellow River is necessary, also be feasible. With the national flood retention and storage compensation approach as reference, based on the actual situation of floodplain area, the compensation content, standard and capital share proportion are defined. Flood inundation loss of autumn crops and permanent residents housing are compensated only. Crop compensation ratio is 60% ~ 80% of the previous three years average annual output value; housing compensation ratio is 70% of the main part loss value. Which 80% of the total compensation funds is paid by central national finance, and 20% is paid by local.

Key words: floodplain, compensation policy, the lower reaches of the Yellow River, study

1 Background

The river course in the lower reaches of the Yellow River from Mengjin County of Henan Province to estuary in Kenli County of Shandong Province with total length of 878 km involve $3,154$ km^2 area of floodplain within the $4,860$ km^2 of total area of river channel, which affects 15 cities, 43 counties (districts), $1,928$ villages, 1.895×10^6 people, 3.401×10^6 mu farmland and 0.481×10^6 mu forest land in Henan and Shandong Provinces. The channel morphology of lower reaches of the Yellow River is wide in upstream and narrow in downstream. The flood discharge capacity is big in upstream and small in downstream. Upper broad floodplain assumed for flood peak reduction, flood storage and sediment retention, is important component for the Yellow River flood control engineering system. It has an important role for ensuring flood control safety of the lower reaches of the Yellow River. A big contradiction has been existed for a long time between production and life of the masses in floodplain area, and flood discharge, and flood and sediment detention. The more developing the basin economy and society and higher requirements of flood control safety, the more prominent the contradiction. Because flood and sediment detention can not be compensated, masses in floodplain area constructed "production dike" illegally for needs of living and development constantly, which impedes the exchange of water and sediments between the channel and the floodplain, accelerates sediment retention in main channel, reduces flood discharge capacity, and aggravates the situation of "secondary hanging river". The dikes will be lashed by transversal channel, diagonal channel and rolling channel during medium and normal floods. It is current key problem of the lower Yellow River harnessing to solve the contradiction between long period stability of the river and economic and social development of the floodplain area, which causes social concern closely. Because many experts have considered the floodplain of lower reached of the Yellow River as special zone for flood storage and sediment retention, and the floodplain has displayed as river morphology, the harnessing strategy of the river channel in the lower reaches of the Yellow River at present and in the future is put forward by the YRCC through study, that is stabilizing the main channel, regulating water and sediment, widening river channel and reinforcing dykes and compensating through policies. Compensating through policies was regarded as the effective way for solving the contradiction between flood control and masses production in floodplain. Therefore, the YRCC reported to the State Flood Control and Drought Resistance Headquarters and the Ministry of Water Resources in 2004 "Ask for instructions about the floodplain in lower reaches of the Yellow River is treated as same compensation policy as flood storage and detention area". In 2005, Henan and Shandong Provinces government were also submitted to the State Council for national policy compensation for floodplain in the lower reaches of the Yellow River after flood inundation.

2 Study process

Instruction was given by Vice Premier Hui Liangyu in "Domestic Dynamic Proof" (1410th) for "the Yellow River downstream floodplain people rebuild production dykes to increase the danger situation for flood control" on May 21, 2005. The instruction said "It must be paid more attention, and handled properly. It should be managed not only to ensure flood control safety of the Yellow River according to "Flood Control Law" and "River Channel Management Regulations", but also safeguard the interests of the masses in floodplain by establishing reasonable compensation mechanism. MWR should provide an opinion through discussing with related localities and departments.

MWR paid attention deeply to it and began to work actively in accordance with spirit instructed by the leadership of the State Council. "Flood Inundation Compensation Policy Study Working Group for Floodplain Area of the Lower Reaches of the Yellow River" was formed by MWR, MOF, NDRC, Henan and Shandong Provinces and YRCC. The group leader is vice minister EJingping of MWR.

The group conference was hosted by vice minister E Jingping in Beijing on Sep tember 2006. The State Flood Control and Drought Resistance Headquarters Office is responsible for overall coordination. The necessity and feasibility of policy compensation will be studied in stages by YRCC mainly.

The study of policy compensation necessity for floodplain of the lower Yellow River was finished from Oct ober 2006 to March 2008. The study of policy compensation feasibility for floodplain of the lower Yellow River was finished from March 2008 to April 2010. The general report was finished on October 2010. The executive plan of floodplain operation of the lower Yellow River (draft) was finished in 2011.

From 2006 to 2010, the working group had organized 19 units, more than 200 people to participate in study work, and hold expert advisory council five times. Detailed and reliable data has been collected through organizing 300 people totally to investigate and study from more than 4,000 people, 900 doors in more than 120 villages, 50 townships, 10 counties. The working group had ever discussed with Henan and Shandong Provinces' government 5 times during study process, reported, communicated and asked for opinions 4 times with MOF, and 8 times with NDRC.

3　Main study contents and objectives

Main study contents for lower reaches of the Yellow River include: flood risk analysis; production dike problem; safety management mode; basic properties, function and correlation; floodplain analogy analysis between lower reaches of the Yellow River and China's other main rivers; status of floodplain in the Yellow River flood control system; range determination of flood discharge and storage, sediment retention in floodplain; scheme analysis of flood discharge and storage; and related policy study et al.

The overall goal is to get conclusion whether compensation is needed after flood discharge and storage, sediment retention in floodplain through the study of necessity and feasibility of implementing policy compensation after inundation in lower reaches of the Yellow River to provide macro decision-making basis for country.

4　Conclusion

Through joint study, conclusion had been reached that under the current conditions, the implementation of the policy compensation for floodplains in lower reaches of the Yellow River after flood inundation is necessary, also be feasible.

4.1　The implementation of the policy compensation in floodplain in lower reaches of the Yellow River after flood inundation is necessary

4.1.1　The implementation of the policy compensation in floodplain after flood inundation is the urgent requirements of 1.8 million masses in floodplain

According to Henan and Shandong Provinces survey analysis in 2007, net income per capita in floodplain of Henan Province is 2,835 Yuan, which is equivalent to the province's rural average level 10 years ago; net income per capita in Heze floodplain of Shandong Province is 2,637 Yuan, which is lower 2,348 Yuan than 4,985 Yuan of the province's rural per capita net income. The 4 national level and 5 provincial level impoverished county distribute in floodplain above Taochengpu, the impoverished population is amounted to 0.8 million. The status can not meet the harmonious society development and the new rural construction. Once floodplain inundation happens, if there is no clear compensation policy, the production and life of floodplain masses will face a very difficult situation, they may become the social unstable factors, the consequence is very serious. Therefore, policy compensation for floodplain in lower reaches of the Yellow River after inundation is very necessary, which enable floodplain masses to share economic development benefits with whole country masses.

4.1.2　The implementation of the policy compensation in floodplain after flood inundation is the urgent requirements of realizing long period stability in lower reaches of the Yellow River

Because of the flood characteristics of coming quickly and subsiding steeply, and extremely serious sediment problem, harnessing way of "slow flow and subside silt" was summarized by predecessors. From "wide river course for flood passage" advocated by Jia Rang during Western Han Dynasty to "wide channel and solid dykes" river harnessing strategy after the founding of People's Republic of China, wide river course are used for decreasing flood peak and subsiding silt to allow the Yellow River to return to the status of "wide channel and slow flow but no longer pressing". flood disaster by levee breach has been decreased effectively. The flood peak of Huayuankou hydrologic Station in 1958 was 22,300 m^3/s. Because of floodplain water and sediment retention, flood peak was decreased to 12,600 m^3/s in Aishan Hydrologic Station, 10,400 m^3/s in Lijin Hydrologic Station. Sedimentation amount in floodplain account for 70% of the whole lower reaches of the

Yellow River since 1946. Because of no inundation compensation, there was not safeguard for masses production and life. On one hand, in order to decrease the inundation losses, Xiaolangdi Reservoir was required to store medium and small flood, which will accelerate reservoir sedimentation. On the other hand, production dyke was constructed to avoid floodplain inundation, which cause no exchange of water and silt between main channel and floodplain, the situation of "secondary hanging river" will further deteriorate. At present, 600 km of production dyke have been constructed in lower river course of the Yellow River, which is the largest scale, closest to main channel and most solid production dyke since 1946. Now Xiaolangdi Reservoir has entered the late stage of intercepting sediment, large amount of sediment will be forced to enter the downstream channel. Due to production dyke barrier, water and silt between main channel and floodplain can not exchange freely, which lead to sediment in main channel, flow capacity of main channel will be severely weakened, flood control safety of lower reaches of the Yellow River will be threatened. The basic premise to solve the problem is to get rid of worry of floodplain masses, that is to implement policy compensation in floodplain after flood inundation.

4.1.3 The implementation of the policy compensation in floodplain after flood inundation is the urgent requirements of maintaining policy continuity in lower reaches of the Yellow River

As early as in 1974, the State Council issued document number 27(1974) with provisions about the production and living of the people in floodplain that production dyke should be abandoned and tableland should be constructed to avoid flood, and one wheat crop a year should provide the grain for the whole year. This policy exempt agriculture tax of the masses, it is compensation actually on backward economic in floodplain. In 2006, exempting agriculture tax policy was implemented in China, the policy in floodplain of lower reaches of the Yellow River was no longer have compensation character, which result in interruption of floodplain compensation policy. Therefore, the policy compensation implementation in floodplain after flood inundation is actually a continuation of state compensation in new situation.

4.2 The implementation of the policy compensation in floodplain in lower reaches of the Yellow River after flood inundation is feasible

4.2.1 National financial can bear floodplain economic compensation after flood inundation

Studies show that from 1949 to 1999, before Xiaolangdi Dam reservoir was put into use, floodplain in lower reaches of the Yellow River was inundated in different degree in 29 years among 51 years, the flood frequency occurs every 1.8 years, with 0.5×10^6 mu annual average inundated farmland. Longyangxia Reservoir was put into operation in 1986. Because the reservoir storage effect, the river base flow conditions were changed. Xiaolangdi Reservoir was put into operation in 2000, flood boundary conditions of downstream were changed, the floodplain natural inundation probability was reduced greatly. Therefore, using actual flood series of lower reaches of the Yellow River since Longyangxia Reservoir was put into operation, combined with the current main channel discharge capacity, through flood process calculation of Xiaolangdi Reservoir regulation, the flood inundation frequency in different floodplain of downstream of the Yellow River is about once every 5 years. The current main channel bankfull discharge in downstream of the Yellow River is about $4,000 \text{ m}^3/\text{s}$. Submerged floodplain farmland by flood once every 5 years is of about 1.1×10^6 mu, occupying 32.3% of total floodplain farmland area. According to above analysis, floodplain inundation can only happen 5 times among 24 years from 1986 to 2010 in downstream of the Yellow River, which is about once every 5 years. The total submerged farmland in these 5 times is about 5.11×10^6 mu, about 1.01×10^6 mu every time averagely, which occupies 30% of total floodplain farmland. It can be borne by the central and local financial resources.

4.2.2 The effect of the policy compensation implementation will be remarkable

According to the typical investigation, the policy compensation after flood inundation is urgent desire of masses in floodplain. The floodplain masses said, as long as the safety and security can be ensured, and loss can be compensated, from the overall flood control situation, production dyke can be removed gradually.

The contradiction between production and living of the masses in floodplain and flood and sediment detention can be solved through the policy compensation after flood inundation. The harnessing strategy of the river channel in the lower reaches of the Yellow River, stabilizing the main channel, regulating water and sediment, widening river channel and reinforcing dykes and compensating through policies, will be implemented frequently. The situation of "secondary hanging river" and flood control will be improved greatly. Therefore, the policy compensation implementation after inundation is not only the urgent desire of masses in floodplain, but also the requirements of harnessing of lower reaches of the Yellow River.

5 Compensation Suggestion

The floodplain of lower reaches of the Yellow River is a special zone for flood and sediment retention, distinct from other flood storage and detention area and other rivers. The formulation of compensation policy should fully consider its particularity. The formulation principle should be advantageous to long period stability of the Yellow River, to floodplain economic and social development, disaster prevention and reduction, and population migration from floodplain and decrease naturally. In order to avoid the policy orientation of encouraging blind development, the compensation content, standard and capital share proportion were determined based on comprehensive consideration of the agricultural production resumption and the national fiscal capacity, according to the actual situation of inundated floodplain in lower reaches of the Yellow River, referring to the spirit of national " Interim Measures of Compensation after Flood Detention Area Inundation " (No. 286th decree of the State Council).

5.1 Compensation content

According to sampling survey of flood inundation in August 1996, water and sediment regulation in 2002, and autumn flood in 2003, the family agricultural production machinery, draft animals and major family durable consumer goods per hundred households have less or easy to transfer, which do not have submerged loss basically. Submerged loss is mainly arable land and houses. Considering longer of flood forecast period in floodplain, it is recommended that inundation loss of crops (not including high crops impacting flood control) and housing are compensated only.

5.2 Compensation standard

Considering large proportion of national and provincial level impoverished county in floodplain of the lower reaches of the Yellow River, compensation proportion which is slightly higher than the national flood storage and detention basin is recommended. That is, crops is recommended to compensate according to 60% ~ 80% of average mu production value of autumn main crop during 3 years before flood inundation; housing is recommended to compensate according to 70% of inundation loss value (main part of residents housing is compensated only, other ancillary buildings do not belong to the scope of compensation).

5.3 Compensation capital and proportion

Submerged loss compensation fund is shared by central finance and provincial public finance. Considering floodplain population of Henan Province accounts for 2/3 of the total population of the lower reaches of the Yellow River, the finance income level of Henan Province is low. Therefore, 80% of the total amount of compensation funds is recommended to be borne of by central finance, and 20% by provincial public finance.

6 Conclusions

Through many years of common efforts by many departments, the State Council agreed to implement policy compensation in floodplain of lower reaches of the Yellow River after inundation in 2011. This is the Yellow River's expectation, but also the floodplain people's expectation. It embodies water harnessing concept of harmony between human and water of the Chinese nation. Henan and Shandong Provinces' government and YRCC are suggested to formulate matching institutions and measures as soon as possible for implementing this benefiting masses policy cogently.

River Water Purification, Riverbed Digging
and Water Resource Acquisition Using Gate Control
and Streamwise Pebble Levees

K. Akai[1], *K. Ashida*[2], *K. Sawai*[3], *T. Kusuda*[4], *Feng J T.*[1], *H. Isshiki*[5],
Shen J H.[6], *S. Takada*[7], *Li Z G.*[1], *Wang G.*[1], *Sheng G M.*[8] and *Bao S L.*[1]

1. "UTSURO" Research Group
2. Kyoto Univ. (Prof. Emerius), Japan
3. Setsunan Univ., Japan
4. Univ. of Kitakyushu, Japan
5. Saga Univ., Japan
6. CTI Engineering Co., Ltd., Japan
7. Ministry of Land, Infrastructure, Transport and Tourism, Japan
8. Shanghai, Investigation, Design & Research Institute, China

Abstract: Yellow River discharges a vast amount of sediment every year. Since the sediment accumulate at the river mouth, the river length becomes longer and the bed slope reduces gradually. As the result, the tractive force decreases, sediment accumulate on the riverbed, the level of the riverbed becomes higher, flow capacity reduces, and it causes the breaking of the embankment. If the tractive force is increased by narrowing the water passage through the operation of the movable weirs and by deepening the hydraulic radius through the adjustment of the section shape of the passage, the large quantity of sediment would be transported to downstream safely, and the riverbed would be stabilized. Furthermore, if the river water is reserved in a part of river course, the running-dry of Yellow River would be prevented. The purification of the river water would also be realized. Since the water channel would have a function of fishway, the ascent of fishes would increase, and the fishery resource of Yellow River would become rich. For the purpose, we propose "Continuous river water Purification and riverbed Digging system (CPD)".

Key words: Yellow River, riverbed digging, water purification, water resource acquisition, CPD, UTSURO

1 Outline of a Continuous river water Purification and riverbed Digging system (CPD)

The first author of the present paper has invented a continuous water purification system in 1987 (International patent: WO 2008/068872 A1). In order to utilize for the lowering of the riverbed level, he extended the function and propose Continuous river water Purification and riverbed Digging system (CPD) this time. As shown in Fig. 1, movable weirs are set up at the upstream and downstream positions of the river, and penetrative levees made of pebbles are installed streamwise or longitudinally. If we control the movable weirs at upstream and downstream positions, we may expect the following effects: ① riverbed digging and sediment transport; ② contact oxidation; ③ purification by sedimentation; ④ oxidation pond; ⑤ fishway; ⑥ garbage collection; ⑦ storage of water and prevention of running dry; ⑧ channel for water level preservation.

2 Concept of a CPD

A concept of a CPD is shown in Fig. 2. As an example of a movable weir, a rubber dam is shown in Fig. 3. We can store water for preventing running-dry. If we open the weirs, it would not prevent flow capacity.

In order to achieve the three major themes on Yellow river and the four targets of the flood control, we show the related and/or required items on the CPD below.

(1) The bed slope of Yellow River is about 1/10,000. The distance between the adjoining

weirs G is about 20 ~ 50 km depending on the dam upheight (See Fig. 3).

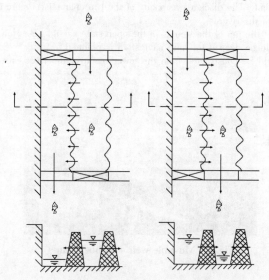

**Fig. 1   A Continuous river Water Purification and
Riverbed Digging system（CPD）**

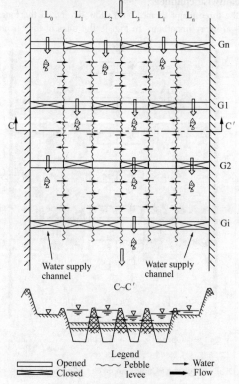

Fig. 2   Concept of CPD

(2) The breadth L of a channel is 50 ~ 500 m. The two channels in the center are used for the discharge of sediment. The discharge capacity of the two channels is more than 3,000 t/s that is the usual flow rate of the river.

(3) The height of the top of the weir G in the open state should be as low as possible. Especially, the depth should be deep if the ship navigation is planned.

(4) A CPD should be constructed from the lower reach of the river except the water supply channels.

Fig. 3   Movable weir (Rubber dam)

3   Details of a CPD

3.1   Utilization of streamwise channels

In order to achieve the three major themes on Yellow River and the four targets of the flood control, the roles of channels are very important. In Fig. 4, the roles of the channels in a CPD are shown.

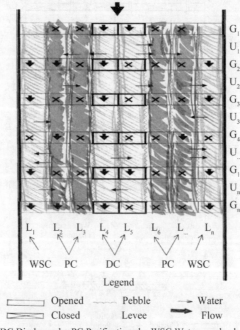

DC:Dischage ch.; PC:Purification ch.; WSC:Water supply ch.

Fig. 4   Roles of a CPD

The two central channels are used for the discharge of the sediment. They transport a vast a-mount of sediment discharged upstream. The two channels on both sides of the central channels are used for water purification and storage. They prevent for the muddy water in the discharge channels to enter into the purification channels by utilizing the water level difference between the purification and discharge channels. In the purification channels, the water is purified by the gravitational sedi-mentation and by filtering through the purification levees. The channels at both sides of the river are used for the water level preservation channels.

3.1.1 Channels for discharge or transport of sediment

Yellow River discharges a vast amount of sediment upstream, and the sediment accumulates on the riverbed. It has raised a big problem. How to discharge the sediment to the river mouth safely? It is an important problem. In the proposed system, we strengthen the tractive force by narrowing the width of the water passage, deepening the hydraulic radius R and increase the transportation a-bility of the sediment. We operate the control of the movable weirs repeatedly and transport the sed-iment to the river mouth.

3.1.2 Channels for purification and storage of water

Two channels on both sides of the discharge channels are used for water purification and stor-age. The water level of the clean water is higher than or equal to that of the muddy water. They store clean water and prepare for the running-dry of Yellow River.

3.1.3 Channels for preserving old water level in water use along the river

There are many cities, factories and farmland along Yellow River. The water of Yellow River is used by them. The channels at both sides of the river are used to preserve the water level for the convenience of users. They allow people to use old facilities. They supply water of higher quality to users abundantly.

3.2 Effects of a CPD

By combining the opening and closing operations of the movable weirs, we can expect the fol-lowing effects: ① digging and sediment transport; ② contact purification; ③ sedimentation purifi-cation; ④ purification by oxidation pond; ⑤ ascent of fishes, ⑥ garbage collection; ⑦ water stor-age and use.

3.2.1 Transport of sediment produced by riverbed digging

The tractive force τ is given by

$$\tau = \gamma_w RI \tag{1}$$

where γ_w, R and I are the specific weight of water, hydraulic radius and bed slope . Hence, if the hydraulic radius R is increased by narrowing the water passage, tractive force τ is increased and the riverbed is lowered.

We can narrow the passage of water by controlling the gates of the weirs. The flow capacity is increased by lowering the riverbed.

3.2.2 Riverbed digging by spring water and overflow

If we close and open the gates of the upstream and downstream weirs as shown in Fig. 5, the water is stored in the two side channels. It causes a big difference of the water levels between the center and side channels. As the result, the water in the side channels spring out into the center channel and digs the riverbed of the center channel. When the water levels in the side channels are increased further, the water flows into the center channel over the levees as shown in Fig. 6 and digs the riverbed of the center channel. If we control the gates of the weirs appropriately and repeat the gate operation, we would be able to lower the riverbed of the whole river.

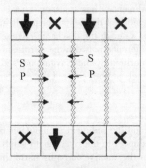

SP:Settling Purification

Fig. 5   Opening and closing of weirs (Plan)

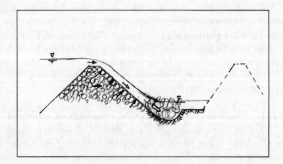

Fig. 6   Digging by flow over a streamwise levee

3.2.3   Water purification

(1) Water purification through gravel.

Polluted water becomes clean after penetrating through gravel. The phenomena is called gravel pore purification. In case of fresh water, some experimental results are obtained as shown in Fig. 7, Tab. 1. The first author of the present paper has confirmed the similar effect in case of water penetration through a pebble levee as shown in Fig. 8.

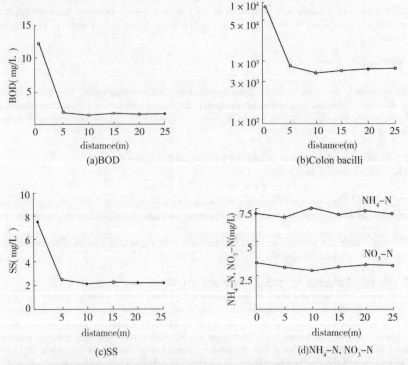

Fig. 7   Penetration length and removal of polluted substance in gravel pore purification

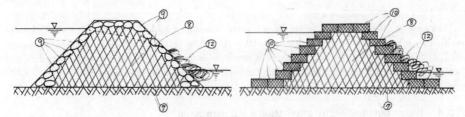

Fig. 8　Purifying levee

If water passes about 5 m through pebbles and blocks having many cracks, SS, BOD, COD and co-lon bacilli are improved or reduced substantially. Nitrogen and phosphorus are absorbed by plank-ton and removed by purifying levees.

Tab. 1　Experimental results of gravel pore purification measured at Kii river, Wakay-ama, Japan

Item	Penetrated water	Main stream	Removal rate
SS (mg / L)	0.2	4.6	96%
BOD (mg /L)	$\geqslant 1$	4.4	Very high
Muddiness (deg)	$\geqslant 1$	3	Very high

(2) Verification of water purifying levees.

When Mr. Feng Jinting of Yellow River Conservancy has visited Japan, we made some verifi-cation on water purification by gravel pore purification at Kii river, Wakayama, Japan as shown in Fig. 9 . The breadth of the levee, water level difference and width of outflow are 45 m, 1 m and 10 m. The flow rate is 200 L/s.

(a)Water quality of main stream　　　(b)Quality of penetrated water

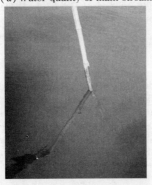

(c)　　　　　　　　　　　　(d)

Fig. 9　Verification of effect of purifying levees

(3) Amount of purified water.

In general, if the purification ability of a levee is assumed 20 L/(s · m), the amount of water purified by a levee with length 50 km is 0.10×10^8 m^3/s. If we use 5 ~ 7 parallel levees are used, $0.5 \times 10^8 \sim 0.7 \times 10^8$ m^3/s of water is purified. If we install a CPD between Zhengzhou and Dongying, the total length is 500 km and the amount of the purified water reaches $5 \times 10^8 \sim 7 \times 10^8$ m^3/s.

3.2.4 Water purification by gravitational sedimentation

If muddy water is stored in calm water region, the water becomes clear because of gravitational sedimentation, since small particles such as mud sink due to gravity. The closed water region is a huge sedimentation pond. The accumulated sediment is processed in the CPD.

3.2.5 Purification by sunlight

If water becomes clean and sunlight penetrates into water, nitrogen and phosphorus change to the algae due to photosynthesis. Then, the carbon dioxide assimilation becomes active, the oxygen rich water regions are created and the ecological circulation is promoted.

3.2.6 Purification by living bodies

Living creatures eat nutritious foods for maintaining their lives. Since fishes and shellfishes eat nutritious foods in water, the water becomes non-nutritious and clean. Furthermore, when number of fishes increases in a river, the water in the river becomes cleaner.

3.2.7 Ascent of fishes

Most of fishes and shellfishes living in a river repeat to go up and down the river according to the season and end their lives. Since a CPD makes the water cleaner and the ascent becomes easier, it makes the fishery active. A CPD is a huge fishway.

3.2.8 Collection of garbage

If water penetrates a levee, garbage floating on water surface is filtered. Especially, a big amount of floating garbage remains around HHWL (top of the levee) of the levee. Garbage springing out from the opposite channel and remaining below the infiltrating surface is washed away. However, most of garbage is attached to the levee and becomes harmless by being exposed to wind and rain and sunlight.

3.2.9 Effect of water storage

When a running-dry takes place in Yellow River, we close gates of movable weirs of CPDs and store water. If the average dam upheight, channel width and distance between the adjoining movable weirs are 2.5 m, 300 m and 50 km, we can store water of 37.5×10^6 t. Hence, if we plan to store water between Zhengzhou and Dongying, the storage capacity reaches to 25×10^8 m^3, since the distance between the two cities, the breadth of the river and the average dam upheight are 500 km, 2 km and 2.5 m. It corresponds to the capacity of Sanmenxia dam.

3.2.10 Advanced water purification

As mentioned above, if the mud water is made clean by the gravitational sedimentation and then penetrated through purifying levees, the pollution of water is reduced substantially. If we repeat the garbage collection, contact oxidation by purifying levees, gravitational sedimentation in reservoir, oxidation pond by sunlight, ascent of fishes and ecological circulation, an advanced purification of the water in Yellow River becomes possible. Many useful effects can be brought to us through the combination of the opening and closing operations of weir gates in CPD(in Fig. 10 and Tab. 2).

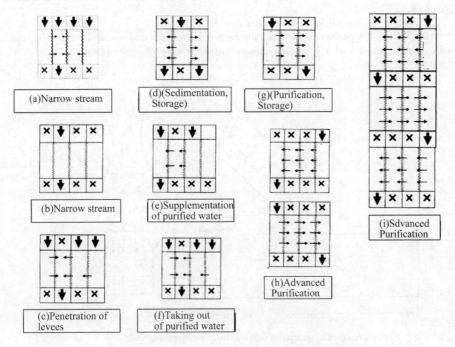

Fig. 10   Roles of streamwise channels of CPD

Tab. 2   Effects of opening and closing of weir gates

Effects	Pattern								
	(a)	(b)	(c)	(d)	(e)	(f)	(g)	(h)	(i)
① Digging and transport of sediment	○	○	○			○			
② Purification by contact	△		○	○	○	△	○	○	○
③ Purification by sedimentation	○		○	○	○	○	○	○	○
④ Oxidation pond (Purification by sunlight)					○		○	○	○
⑤ Ascent of fishes	○	○	○	○	○	○	○	○	○
⑥ Collection of Garbage	△		○	○		△	○	○	○
⑦ Water storage (Prevention of running – dry)	○	△	△	○	○	○	○	○	○
⑧ Channel for Water level preservation		○	△	○			○	○	○

4   Steps of building a system

A CPD discharges a vast amount of dug sediment. Since the discharged sediment accumulates downstream, it is inevitable to carry out in advance a construction to countermeasure floods and sediment at the lower reaches of the river. For this purpose, we must do first the construction of "a system to control and utilize river water by a tidal current generator using UTSURO" (Fig. 11) , (International patent: WO2004/090235AI; Chinese patent: 608805).

An UTSURO A or tidal current generator digs the riverbed and increase the flow capacity at the river mouth.

② Reuse of discharged mud and sand at river mouth

An UTSURO B separates mud and sad discharged downstream into mud and water. The mud is used to reclaim a vast land and tideland. Clean water is discharged in the process.

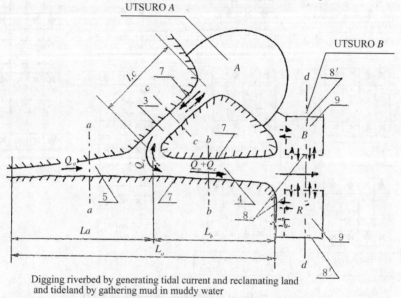

Digging riverbed by generating tidal current and reclamating land and tideland by gathering mud in muddy water

Fig. 11　Water control system by a tidal current generator using UTSUROs

5　Rough estimation of budget for constructing a CPD

Rough estimation of budget for constructing a CPD is in Tab. 3.

Tab. 3　Rough estimation of budget for constructing a CPD

Items	Quantity	Unit price	Amount of money
Movable weir (Rubber dam)	2,000 m	750,000C ¥/m (10,000,000J ¥/m)	7.5B. C ¥ (10B. J ¥)
Streamwise levee	350 km (50 km × 7)	15,000C ¥/m (200,000J ¥)	5.25B. C ¥ (70B. J ¥)
The others			1.5B. C ¥ (20B. J ¥)
Subtotal			7.5B. C ¥ (100B. J ¥)
Total	Total length = 500 km	Abt. 75B. C ¥ (Abt. 1.0T. J ¥)	

Note: C ¥ : Chinese Yuan; J ¥ : Japanese Yen; M. : Million, B: Billion; T. : Trillion.

6 Conclusions

In order to achieve the three major themes on Yellow River and the four targets of the flood control, we proposed a system called Continuous water Purification and riverbed Digging system (CPD)". In the system, we narrow the water passage of Yellow River to make the hydraulic radius R deeper and increase tractive force. It makes a big amount of sediment to be transported downstream safely. It adjusts the shape of the stream and stabilize the riverbed. Furthermore, for preventing the running-dry of Yellow River, the CPD stores a vast amount of water in the river course. The water is purified by the force of nature. The water is used to meet the water demand in the region adjacent Yellow River. The water activate the ascent of fishes and enrich the fishery resource of Yellow River.

References

Li Guoying. Maintaining Healthy Life of Yellow River [M]. Zheng zhou: Yellow River Conservancy Press, 2005.

Experiment Report of Sewage Purification [R]. Ministry of Construction, Arakawa Jyoryu River Work Office (1979).

K. Akai. A Proposal to Control of Huang River Using UTSURO [M]. Y. Feng, J. Feng, Translated. Zhengzhou: Yellow River Conservancy Press, 2010.

Implementation of Strict Water Resource Management Measures for the Yellow River in Henan

Liu Tian[1] , *Niu Min*[1] and *Li Fan*[2]

1. Yellow River Henan Bureau, YRCC, Zhengzhou, 450003, China
2. Xiaolangdi Multipurpose Dam Project Management Center, Ministry of Water Resources, Zhengzhou 450000, China

Abstract: Water is short in Henan Province along the Yellow River, The water resource along the Yellow River in Henan Province has great importance to Henan Province and prop up sustained development of economy and society, It is necessary to implementation of strict water resource management measures for the Yellow River in Henan Province for the sustainable utilization of water resources. We put forward overall requirements and basic principles for establish strict water resource management. It is essential to introduce measures in various areas, including discharge calculation, preplan for drought – relief emergency response and measured respectively agricultural and non – agricultural water supply, in this way, to establish management system which adapts to strict water resource management and establish index system of strict water resource management. The total water quantity of the Henan Province along the Yellow River was subdivided according to national plan for water allocation in 1987 and delimit total amount control line. We practice the water permit system, monthly regulation scheme, ten days water order and five days rolling water order for control the diverted water volume of the Yellow River. We should perfect the enforcement systems from the following respects: to implement the water regulations act and carry out its detailed rules of the Yellow River; to carry out the responsibility system for water regulation; to take dual controls for total amount control and discharge at trans-provincial border sections control; to accelerate talents team construction; to earnestly implement the supervising and examination of water regulation, online superintend and check by remote monitoring system; to tighten on the spot inspections and to establish technical supporting system of strict water resource management. The system includes the research on enduring effect mechanism of water-withdrawing licenses superintend, functional no flow interruption indicator, the plan of water regulation in Qinhe River, water resource managements system and mechanism of the Xiaolangdi Reservoir area of the Yellow River, water regulation from passive to active regulation, following and supervising whole process of water supply, and the inter-regional water supply. By adapting strict water resource management, It is guaranteed the continuous utilization of the Yellow River water resource and sustained development of economy and society along the Yellow River in Henan Province.

Key words: water resource management, strict, measure, Henan Yellow River

Henan Province is one of the provinces famous for its population and farming production. As one of the most important food production areas, Henan Province has a severe scarcity in water resources. The quantity of water per capita is 440 m^3, only about 1/6 of the national average. The quantity of water per capita in the area along the Yellow River in Henan Province is 275 m^3, which far below the internationally acknowledged minimum warning line of 1,000 m^3, only about 1/10 of the national average. The economic district construction of the Central Plains officially becomes the national strategy. The kernel of build economic district of the Central Plains is explore the new way for the new industrialization, new urbanization and new agricultural modernization scientifically and harmoniously development, which must not be carried out at the cost of agriculture. The strategic arrangements of build economic district of the Central Plains are constructed the group of city of central plains, grain producing core area and ecology function zone. The water resource along the Yellow River in Henan Province has great importance to economic district of the Central Plains and prop up sustained development of economy and society. It is necessary to implementation of strict management measures for the Yellow River in Henan Province for the sustainable utilization of water resources.

1 General requirements and basic principles

Establishing the strict management system for water resources in Henan Yellow River, Deeply implementing the scientific concept of development and actively practicing river training philosophy: "keep the healthy life of the Yellow River", the focus will be placed on the total amount control, quota management, optimization distribution, protecting and saving of water resources, with water resources argumentation and water-withdrawing permit supervision and management as the tool, taking reform and innovation as the motivator, and ability construction as the precondition, it is set up and improved the management, index, enforcement and technical supporting system for practice a strict water resource management system, working hard to improve the efficiency of utilization of water resource of the Yellow River and realizing the sustainable development of economic society along the Henan Yellow River by sustainable utilization of Water Resources in the Yellow River.

The establishment of strict water resource system for Henan Yellow River is based on several principles as follows: to take realization of the harmony of human beings and nature and keeping the healthy life of the Yellow River as core idea, sustainable utilization of water resources in the Yellow River to prop up the sustainable development of economic society along the Henan Yellow River as starting point and destination of water resource management, carrying out and implementation of all relevant national laws, regulations, rules and technical specification as basic guarantee, water resources optimize-allocation and improvement on the efficiency of water utilization as key work, drawing up and perfecting various emergency preplan and guaranteeing measurers as important measures, and strengthening the development of the water resource management administrators as important assurance.

2 Establishment of the strict water resource management system

2.1 Management system

2.1.1 Establishing water regulation management system

We had established many rules and regulations for the purposes of strengthening water regulation management, to improve the standardization of management, and fully derive the economic benefits and social benefits of water resources in Henan Yellow River. For example, it has established the Henan Yellow River water regulation management methods, water regulation responsibility system of the Henan Yellow River, emergency disposal regulations of the water regulation, reporting preplan for major water pollution accident in Henan Yellow River, and so on. Which define the establishment principle of the water regulation and responsible institutions responsibility, take dual controls for total amount control and discharge at trans-provincial border sections control, set measures and procedures of emergency disposal, establish water regulation consultation system, watch system, water supervision and management system and major water pollution accident report system.

2.1.2 Establishing water regulation order management system

In order to strengthen water planned allocation management and improve on the efficiency of water utilization, the YRCC carried out water regulation order management, the Yellow River Henan Bureau (YRHB) formulated some regulations concerning the order for diversion from the Yellow River and measured respectively agricultural and non-agricultural water supply which set the order principle of total amount control, dynamic management, formalities of reporting and responsibility divided at different levels, executed order management of declare, approving, examining and adjusting, measured respectively agricultural and non-agricultural water supply real time metering, and automatic metering management system of non-agricultural water supply.

2.1.3 Establishing discharge management system

YRHB had formulated the measurement methods of the diversion water measurement for strengthening discharge measurement and exact measurements of discharge, carrying out the planning in the use of water and regulation water according to instructions, and for that making total amount control and discharge at Gaocun section control meet the requirements of the YRCC control indexes in order to achieve the objectives of no flow drying up. the management systems of "discharge measurements, daily/monthly/yearly report , supervision and inspection", executing, exercise, job responsibility,card mount guard and signature system are set-up.

2.1.4 Establishing water-withdrawing licenses management system

YRHB formulated water-withdrawing licenses management for strengthening supervision and inspection, for safeguard in accordance with the law the legitimate rights and interests of the parties involved, for require strict economy in the use of water and reasonable development and utilization of regional water resources, and for establish the long effect mechanism for water-withdrawing licenses management. Due attention is to be paid to the daily management and specific supervision of the water-withdrawing licenses of total amount control, water abstraction permits, water-withdrawing quantity,water use efficiency, water pollution, yearly/monthly/ten-days water allocation planning, measuring device,water use statistics. If water-withdrawing quantities above the quota will adoption of progressive price markup, take a statistical report and special statistical report system for water use statistics.

2.1.5 Establishing preplan for drought-relief emergency response

In order to do a better job of drought-relief emergency response in Henan Yellow River, we had formulated preplan for drought-relief emergency response, to clear the duties and division of responsibilities, set up emergency response program and measures, establish command system, safeguard mechanism, information monitoring, and handling and reporting mechanism for drought-relief emergency response.

2.2 Index system

2.2.1 Total amount control line

The total water consumption from the Yellow River of the Henan is 5.54×10^9 m^3 in accordance with State regulations,the main stream of the Yellow River water consumption is 3.567×10^9 m^3, and its tributaries is 1.973×10^9 m^3. Henan Province subdivide the total water consumption to the city along the Yellow River, the main stream of the Yellow River water consumption is included twelve cities such as Zhengzhou, Kaifeng, Luoyang, Anyang, Xinxiang, Jiaozuo, Puyang, Xuchang, Sanmenxia, Shangqiu,Zhoukou and Jiyuan. Tributaries of the Yellow River water consumption is included eight cities such as Zhengzhou, Luoyang, Pingdingshan, Sanmenxia, Jiaozuo, Jiyuan, Anyang,Xinxiang and Puyang.

2.2.2 Water efficiency control line

Analyses and checking water consumption of customer are carried out according to the water use quota of the Henan province, to supervise and check construction of facilities for water-saving, set up water efficiency control line in order to against waste water in the allotted time, outdated technologies and equipment prohibited by the State. To carry out water right transfer in time cities that have not spare water-withdrawing permit should be get a license by water right transfer.

2.2.3 Cross-section discharge control index

Water regulation ordinance of the Yellow River practice hydrologic section discharge control according to the "water regulations of the Yellow River", Gaocun Hydrological Station is trans-provincial border hydrologic section. Its cross-section discharge must meet the water regulation plans and real-time regulation instruction requirements for cross – section discharge control. Monthly and ten – day average discharge in the Gaocun Hydrological Section can not be lower than 95% of control discharge, and daily average discharge can not be lower than 90%. When the discharge upper Gaocun Hydrological Station has bias or the section actual flow has bias with the control index, the Gaocun Hydrological Section control discharge may increase or decrease but may not be below early warning discharge.

Rules for the implementation of the water regulation regulations of the Yellow River set minimal discharge and guarantee rate of main tributaries section control. The minimal discharge of Heishiguan station in the Yiluo River is 4 m^3/s, and its guarantee rate is 95%. Minimal discharge of Yuncheng, Wulongkou and Wuzhi in the Qinhe River are 1 m^3/s, 3 m^3/s, 1 m^3/s respectively, and the guarantee rate are 95%, 80% and 50% respectively.

Rules for the implementation of the water regulation regulations of the Yellow River set early warning discharge of trans-provincial border hydrological section and main control section in mainstream of the Yellow River. When the discharge of control section reached early warning discharge, YRCC took measures to reduce the amount of water diverted, until close water intakes, and relevant departments and units must comply. Early warning discharge of main control section in Henana are Huayuankou, Gaocun and Sunkou, and the early warning discharge are 150 m^3/s, 120 m^3/s and 100 m^3/s respectively.

2.2.4 Diversion water error control index

Measures of diversion water in the lower Yellow River practice water order management for monthly, ten-days and five-days rolling water-withdrawing plan. The diversion water order error control is: the relative errors of the average daily discharge of diversion water between actual or approved discharge are no more than ±10%, in case of water and sediment regulation and emergency, the relative errors of the average daily and ten-days average discharge of diversion water between actual or approved discharge are no more than ±5%.

2.3 Enforcement system

2.3.1 Strict carrying out the responsibility system

The water regulation ordinance of the Yellow River provides that all water regulation plan and executing instruction carry out by responsibility of administrative leadership of the local people government and main leader of YRCC and subordinate units and departments. Since the implementation of the ordinance in 2006, every year the Ministry of Water Resources government announces person-in-charge for water regulation of the Henan Province and YRHB by bulletin. Under ordinance, the YRHB is responsible for mainstream of the Yellow River in Henan Province, the departments of water administration of the Henan people's governments is responsible for mainstream of the Yellow River. Up till now, responsibility person for tributaries has not announced, it has been further clearly defined.

Under rules for the implementation of the water regulation ordinance of the Yellow River, the departments of water administration of the local people 's governments at or above the county level and YRCC its subordinate units and departments shall be clearly the management organizations and person-in-charge for water regulation, and instituted responsibility system for water regulations. YRHB has established water regulation responsibility system of the Henan Yellow River, but the departments of water administration of the local people 's governments at or above the county level have not instituted responsibility system and not yet to be practicable. The responsibility of administrative leadership of the local people government should be further strengthened, person-in-charge for water regulation of the departments of water administration of the local people 's governments at or above the county level along the Yellow River is announced every year to the society, and improve the training of administrative leadership in this field so as to enhance their ability of dealing with water regulations and has played an important role in decision-making and supervision to improve executive ability of water regulation plans.

2.3.2 To carry out the performance appraisal for water resource management in the Yellow River

To strictly enforce the water regulation ordinance of the Yellow River and water permit and water resources fee levy management regulations, water regulations, water permit approval, cross-section discharge control and total discharge amount control implementation shall be announce periodically in various forms using mass media and other means. To implement the punishment and compensation system for water consumption beyond the plan or reduce water using of economic and ecosystem, the cross-section discharge failing to meet the limits or exceed regulation instruction requirements will be announced until limit approved water-withdrawing license for a new construction project. The research and practical exploration to mechanism of performance appraisal for total discharge amount and water use efficiency, government performance appraisal at different levels should include water resource management, water use efficiency as a restricted index shall be incorporated into integrated evaluation index system for the local economic and social development.

2.3.3 To strengthen the supervision and management of water resource and water regulation

Based on the characteristics of water resource management, the supervision and inspection can be classifies into three types of regular, water use peak and emergencies. Internet and on-the – spot inspections measures are taken to strengthen the water resource management and water regulations inspections. Internet inspections is use the remote monitor and control system to inspect the execute instruction of water regulations, diversion discharge, total amount of diversion water and section discharge. On-the-spot inspections mainly take patrolling or garrison supervision and examination.

Regularly supervision and inspection are mainly based on-the-spot inspections to performs water-withdrawing permit, The main contents of the regularly supervision and inspection are as follows: planned allocation, total amount control, monthly report of water-withdrawing statistics, water quality, installation and usage for measuring device and water-saving facilities, water-withdrawing without a license, exceed authority to give the certificate, unauthorized change object of water-withdrawing water-withdrawing purposes, measuring device, water-saving facilities and fulfillment circumstance of wastewater treatment.

Water use peak is water use in the period from March to June, The main contents of the regularly supervision and inspection are as follows: trans-provincial border and key hydrologic section releasing discharge, the Yellow River mainstream reservoirs releasing discharge, formulating and implementing water – withdrawing plan, execution of real-time regulation and implementation of the responsibility system for water regulations. In March to June, the trans-provincial border sections or the key hydrologic sections are inspected daily by internet. Comprehensively organizing and implementing on – the – spot inspections, if necessary, garrison inspection to the main water intakes will be implemented, when the daily average discharge in trans – provincial border and key hydrologic section lower than 80% of control discharge for three consecutive days and bring a negative effect of water use in the downstream。

Emergencies supervision and inspection are used for emergency of water regulation or emergency response to combat drought。 It mainly includes that: formulating and implementing of emergency plans, execution of real – time regulation, implementation of the contingency dispose measure and drought-relief measures, Establishing emergency drought – relief organizations, and report the drought – relief Information. Emergencies supervision and inspection are mainly based on – the – spot inspections. When the emergency belongs to the Category I , Category II or Category III, the patrolling supervision and inspection, should be organized and implemented if necessary, to implement the garrison inspection at the main Water Intakes.

2.3.4 To strengthen the construction of management teams

YRHB has a specific water resource management and regulations department, but the specialized organs have not set up for water resource management and regulations from the city down to the county levels, which is incompatible with the practice a strict management system of water resource and measures. For providing a guarantee for practice a strict management system of water resource, the municipal and county bureau should set up specialized management organizations and have professional knowledge comprehensive and greatly strengthened administrative personnel, to improve the training of specialists in this field so as to enhance social management and public managing competence of the officers at different managerial levels.

2.4 Technical supporting system

2.4.1 Accelerating water regulation management system construction

Construction of the second phase of water regulation management system will be initiated, YRHB emphatically constructs the monitor ability for water intakes of the branches of the Yellow River with construction and management by the localities. Combining with the existing remote monitor control system, remote monitor control for all water gates in the mainstream of the Yellow River and water intake of main tributaries of the Yellow River are practiced, to realize water regulation on main streams and tributaries of the Yellow River by construction of water regulation management system, and to facilitate the Water consumption of total amount control and strengthen the implementation of the water regulation and increase management level of water regulations.

2.4.2 Accelerating establishment of tributary regulation management mechanism

The tributary regulation must be included into water regulation of the Yellow River according to the provisions of "water regulation ordinance of the Yellow River" and "Rules for the implementation of the water regulation ordinance of the Yellow River". But the tributary regulation is relatively late; there are some problems such as that the management system is not consummate, management mechanism not perfect, duty not clear, and management not efficient in the tributary regulation. According to the provisions of "water regulation ordinance of the Yellow River", management order of the tributary regulation which is featured by normal and orderly should be established and improved as soon as possible, to formulate the water regulation management of Qinhe River and the Yiluo River, and define administrative department at all levels and implement responsibility of tributary regulation management. It must be carried out the study on the establishment of water allocation plan in dry season and mechanism of emergency water regulation for the Yiluo River and Qinhe River base on minimal discharge of main tributaries section control, water-withdrawing permit and water resources utilization. setting up tributary regulation coordination mechanism to reconcile different provinces water use, settling the contradictions of the water resources competing uses in the upper and middle and lower Yellow River. strengthening the water management and measurement, strict water use plan management, strengthening supervision, and preventing and seasonable eliminating low discharge emergency to ensure tributary water consumption be less than water quota and main section control discharge meeting the requirements.

2.4.3 To strengthen the basic research

The main basic researches are as follows: study and practice of implementing the strict management system for water resources in Henan Yellow River, Research on establish enduring effect mechanism of water-withdrawing licenses superintend, research on functional no flow interruption indicator, research on water resource management system and mechanism of the Xiaolangdi Reservoir area of the Yellow River, research on tracking supervision whole process of water supply, study on sustainable development and utilization of water resources in Qinhe River, study and practice of the inter-regional water supply, and study on water regulation from passive to active regulation. All of these provide technological support for implement strict water resource management.

3 Epilogue

In 2009, The Ministry of Water Resources puts forward "Advancing strict water resource management system". In 2010, YRCC has raised a very clear request in implement strict water resource management system. In 2012, the State Council promulgated the "Proposals on implement strict water resource management system" to comprehensive arrangements for implement strict water resource management system. On the basis of summarizing experience, considered the practical situation, YRHB brings forward establish strict water resource management system. The basic guarantee is to establish and improve a system compatible with the strict water resource management, the key is to establish index system conforming with the strict water resource management, the chief means are, to establish enforcement system coincident with the implement strict water resource management, and the Important foundation is, to establish technical supporting system fitting in with the needs of strict water resource management. By means of establishing strict water resource management system and implementing the requirements of strict water resource management, it will be realized that to effectively protect, rationally exploit, omprehensively manage, optimize allocation and overall save the water resource of Henan Yellow River, and provided the support of , the social and economic sustainable development along the Yellow River in Henan Province.

References

Liu Tian, et al. Analysis on Real Time Water Regulation of Henan Yellow River[J]. Yellow River, 2006.

Xin Zhong, Liu Tian, et al. Research on Measures of Sedimentation Reduction of Diversion Canals [J]. Journal of Henan Water Resources,2008.

Liu Tian. Existing Problems of Diversion Management and Its Countermeasures for the Yellow River Diversion Project in Henan [R]. 2008 YRCC Technical Workshop—the Summer Session, 2008.

Liu Tian, et al. Preliminary Discussion on Content of Water Resources Meticulous Management in Henan Yellow River[C] //. Proceedings of the 4th Internation Yellow River Forum. Zhengzhou: Yellow River Conservancy Press, 2009.

Decision Support System for River Basin Water Quality Management

Chen Wen, *Zhang Haiping* and *Lu Qianming*

DHI China, Shanghai, 200000, China

Abstract: Water quality management is the planning for the protection of water quality for various beneficial uses and regulating and enforcing programs to accomplish the planning goals, and laws and regulations dealing with water pollution control. Therefore, water quality management is a very complicated task in relation to hydrology, hydrodynamics, pollution load from various sources and water quality, which requires the consideration of a wide scope of social, economic and environment aspects. An integrated and systematic approach is required to manage water quality over a river basin. Over the past decade, rapidly advancing computational ability and the development of user – friendly software/operation systems have made the use of numerical models commonplace in water resources management. In many large projects related to water quality, a large number of studies have relied on the use of numerical models developed by DHI, which couple hydrological, hydrodynamic and water quality simulations, in concert with a GIS based pollution load evaluation model, not only for early warning and forecasting of emergency pollution accidents, but also for water quality assessment and environmental strategy evaluation. With the rapid development of computer and information technology, mathematical modelling combined with database, GIS and network technology have become the most relevant and efficient tool to manage water quality in an integrated way. DHI China has developed a well – structured decision support system platform for basin – wide water quality management, which combines all relevant models, GIS, database and network technology into a whole. The system has been successfully applied in a number of projects and proven to be a powerful tool for water resources management authorities.

Key words: water quality management, river basin, decision support system, integrated

1 Introduction

For sustainability of the human survival and better quality of life, it is necessary to manage the environmental resources at scales ranging from local to global. This requires an assessment of the extent of these resources, including an understanding of their variation. And then a model can be used to simulate the response to future scenarios of changing environmental conditions.

Generally, the models can be divided into physical model and mathematics model. With the development of the computer capacity, the mathematics model is more convenient, more versatility, and competitive economically. The mathematical model can predict an expected future state, and thus is an important part of informed decision making in management. While single models can (and do) directly support this decision making process, often the complexity of environmental systems, and the multi-faceted nature of many environmental problems, means that decision makers commonly require access to a range of models, data and other information. Software systems that integrate models, or databases or other decision aids, and package them in a way that decision makers can use are commonly referred to as Decision Support Systems (DSS). With the quick development of computer and information technology, it is agreed that an environmental DSS has become the most relevant and efficient tools to manage environments in an integrated way.

Water, atmosphere and soil resources are three basic elements for environment management. Water environment is decisive on survival and development of human society, especially river basin is most closely watched for human life. The mathematization of water environment in river basin has been studied for a long time from hydrologic system to water quality system. The river basin management authorities over the world have traditionally mainly focused their work on the hydraulic construction aspects related to hydropower and flood control. Then with the global economic growth, water pollution has been very common and serious. Water bodies are polluted through continuous e-missions as well as spills during emergencies. Continuous emissions are from industrial and munici-pal point sources, as well as from non − point sources such as livestock and fertilizers. Thus human are also becoming more demanding on water quality, and water quality management has become more important. Compared with the developed countries water management in developing ones star-ted relatively late, but allows of no delay.

Over the past decade, rapidly advancing computational ability and the development of user −friendly software/operation systems have made the use of numerical models commonplace in water resources management, and there are many more currently under development. DHI China has been involved in more than 15 large projects related to water resources and water quality all over China (Shanghai, Sichuan, Chongqing, the Yangtze River Basin, the Yellow River Basin, the Huai River Basin, the Pearl River Basin, the Tai Lake Basin and the Songhua River Basin). Based on the studies, we have accumulated extensive expertise in water quality management, as well as identified the issues to be addressed on its modelling improvement. On that basis, DHI Chi-na has developed and implemented a DSS for water quality management, which combines all rele-vant models, geographic information system, database systems and network technology into a whole. According to actual effects, it is considered that the system fits the management require-ments in China. The paper seeks both to communicate some aspects of water quality modelling to modellers, and to communicate the DSS framework design to system developers.

The paper begins by outlining our modelling approach for river basins. It then states our con-siderations of DSS, and describes the framework and functions of our system. Finally, some pros-pects and comments are made on water quality DSS development.

2 Water quality management decision support system

2.1 Two-level approach

Numerical model of water quality is now common, both as a method for scientists to test hypot-heses and so better understand essences, and also as a predictive tool for those who manage water quality system. Rizzoil and Young pointed out that three categories of users (scientist, manager and stakeholder) can be identified. Obviously, the scientist needs to have a tremendous amount of knowledge, and be able to organise, develop and modify the model, and then the manager just needs the prepared models which are usually integrated in a DSS, and finally the stakeholders just want to see intuitive and dynamic results to understand the difference of different management deci-sions.

As above, Soncini-sessa et al. put forward a "two-level DSS" concept. A DSS which can be really used in management should align modelling developer and end user. The modelling developer can develop some new models and integrate them in the DSS, while the end users (managers and stakeholders) can obtain the modelling results through some easy operations. The water quality DSS developed by DHI China follows this concept and develops our own way as follows.

(1) The DSS focuses on manager's needs (specific domain, easy to use, and easy to under-stand). Under the assistance provided by DSS, the manager's only tasks are data collection, mod-elling setup selection and so on. So managers can easily use the integrated/calibrated models to ob-tain the results in an easy way, once they have been trained.

(2) The DSS can be used in similar area merely by developing a new model, without having to rewrite the code.

(3) The DSS is convenient for secondary development. This means that the system allows for the transfer of knowledge from science to management.

2.2 Primary considerations of the system

The DSS includes water quality modelling for water quality management, geographic information system (GIS), WebGIS technology, database system, and network technology.

The basic concept is quick and easily to set up a scenario simulation (including the boundary set up). Once a scenario has been defined through the interface system, only a few steps are required to get results. Model relevant information, such as boundary conditions, point sources, gate operations etc. could be automatically extracted from the database. This means that the domain base allows the separation of data from models. It is very important for model re – usability.

The DSS offers a visualization decision – making support assistant for the water quality management. The simulation results are automatically generated in pre – defined tables or maps. Depending on whether the results are to be used in planning (e. g. ranking of different treatment investments), design, surveillance of river and lake water quality or the model results shall be used to issue warnings in case of accidental spills, different options for dissemination of results are implemented. Automatic generation of html pages for the Web is also a possibility.

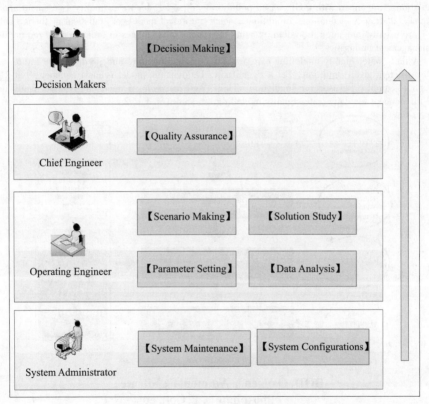

Fig. 1   Roles and workflow

Based on the management mechanisms in China, we further separate "manager" into four categories: system administrator, operating engineer, chief engineer and decision maker. Thus, the system adopts integration of BS (Browser Server) and CS (Client Server) structures. The CS application is provided for the system administrator (responsible for system configurations) and operating engineers (responsible for scenario making and parameter setting). The user could use the CS application to finish all the data flow and work flow for an emergency simulation and scenario study. Some simulations or scenarios may need to be published for the decision makers and this could be achieved by a BS application. As a webpage application, it can be accessed widely by intranet users or remote devices according to the desired authority arrangements. Anyway, a manager can do all things through CS application which implements all the functionalities including what BS application does. So the DSS works as: the system administrator maintains and configures the system for proper functioning; the operating engineer works for the data collection, scenario making and analyzing, then submit the scenarios to the chief engineer for verifying; the chief engineer checks the simulations and publishes the verified scenarios; the decision makers then analyze the submitted data and simulations for decision making. The detailed workflow is shown in Fig. 1.

2.3 Water quality modelling approach for river basins

Water quality modelling is the core of DSS. With the rapid development of computer technology, the use of modelling and simulation software tools is generally appropriate and efficient when time and human resources are limited. So in our experiences, all the studies relied on the use of numerical models developed by DHI. The DHI models can couple hydrodynamic and water quality simulations, in concert with a GIS based pollution load model, for water quality assessment and environmental strategy evaluation. In addition, some traditional models or statistical methods (still in use) are usually integrated in system by using programming languages in order to meet requirements of management authorities.

A daily water quality modelling can simulate current situation, and predict some future states under different given condition. For a river basin, 1D dynamic model is most common, but 2D or 3D model can also be used for simulation of lakes and reservoirs if necessary. Model components and the concept for daily water quality modelling are shown in Fig. 2.

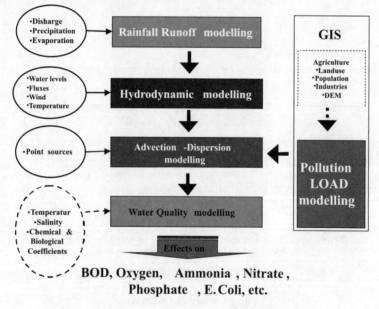

Fig. 2 Structure of daily water quality modelling system

In water quality management of river basin, early warning and emergency response forecasting for sudden water pollution accident is a specific issue, other than daily water quality modelling for predicting an expected future state. It is usually believed that 1D dynamic model in principle provides a good balance between modelling accuracy and efficiency and are suitable for emergency response modelling. In addition, in order to improve forecast accuracy, it is considered to be necessary that the real – time measurement information of flow and concentration during a pollution incident must be reflected in the forecast modelling. That is, if emergency monitoring data (concentration and/or discharge) at downstream of the pollution accident location are available, the real – time data will be used for model updating by using the data assimilation facility of MIKE software.

Model components and the concept for emergency water quality modelling are shown in Fig. 3.

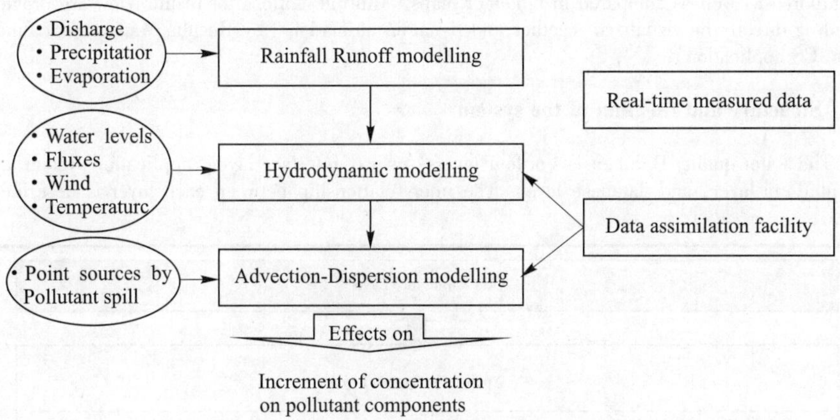

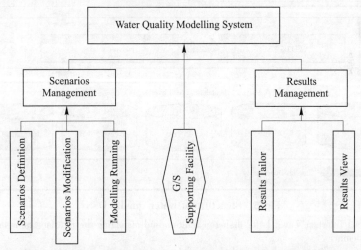

Fig. 3   **Structure of emergency water quality modelling system**

All water quality modelling are embedded in the CS application. The CS application provides a friendly and easy – to – use interface for modelling operation and results view to avoid complex works on modelling and results pre – / post – processing. This is developed based on the components of MIKE model software, database and GIS technology. Through a Show Map button on the CS interface, a GIS map pops up and the spatial locations of specific objects (e. g. branches, water quality monitors, and wastewater outlets) can be identified. And the operations on the active interface are interactive with the GIS interface layer. The functionality structure of water quality modelling system in CS application is illustrated in Fig. 4.

Fig. 4   **Functionality structure of water quality modelling system**

Two baseline scenarios are integrated in the CS application respectively for the daily water quality management and emergency management, with relevant modules previously set up. Based on the baseline scenarios, various scenarios for daily water quality forecast on consideration of, e. g. boundary condition changes or pollutant load changes as well as scenarios for spill accident simulation can easily be defined and executed (by Scenario Management function in the CS application). The integrated computational modules can automatically extract all relevant data from the da-

452

tabase for scenario simulation. Simulation results from different scenarios can be flexibly extracted and tailored as well as compared in tables or maps. Multiple options for results view are provided, including directly be visualised together and dynamic animation (by Results Management function in the CS application).

2.4   Structure and function of the system

The water quality DSS consists of four layers: user & interface layer, application system layer, GIS platform layer, and database layer. The inter-relationship between each layer is described in Fig. 5.

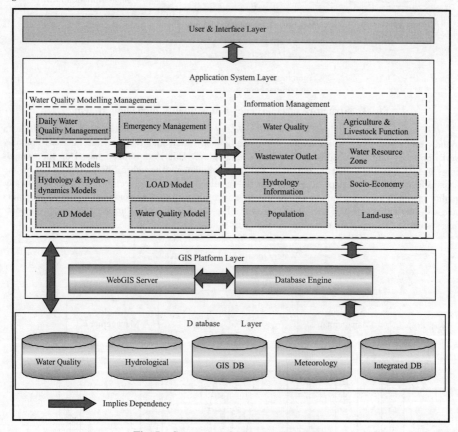

Fig. 5   Structure of water quality DSS

The User Interface Layer is for dialog generation and managing the interface between the user and the system.

The Application System Layer includes the information management and the water quality modelling management.

Information management includes the information query, water quality information assessment, public report management, and system management etc, which provide the input to the water quality modelling management. This system enables the users to retrieve all relevant information about the monitoring stations and the results, wastewater outlets etc. Based on this information the assessment report can be prepared and visualized by GIS maps.

As mentioned above, Water quality modelling management includes two parts, i. e. , daily water quality management and emergency management. The former implies daily water quality forecast simulation combined with various modelling scenarios ("what if"), while the latter is essential for contingency management planning in case of spill accident. Through the appropriate application of the relevant modules, numerous factors (including different hydrology, operation and pollution load) can be assessed.

The integration of above functions will form the platform for sewage control, water environment security monitoring and water quality management.

The GIS Platform Layer is the GIS software system supporting the system development and operation. The ArcIMS, ArcGIS Engine and ArcSDE of ESRI will be used to implement the development. ArcSDE will be used to save the map data, while ArcIMS and ArcGIS Engine will set up the map service (including different scales and different zones in real world) based on the map data retrieved from ArcSDE.

The Database Layer provides the basic data support for the system and MIKE models and comprises data of water quality, hydrology, meteorology, operation, socio – economy, as well as model results.

3　Application experience

As mentioned in the introduction, this water quality DSS has demonstrated the strength in several large river basins of China. Two examples of recent applications are briefed below.

(1) EU – China River Basin Management Programe (RBMP) – Development of Early Warning Emergency Response (EWER) forecasting system for pollution incidents in the middle reaches of the Yellow River Basin.

Development of the EWER system was one of the RBMP research programs, and the overall aim was to build the EWER system enabling it to forecast pollution incidents and to make the technical tools available to the YRCC Water Resources Protection Bureau to assist in pollution incident management in the middle reaches of the Yellow River. It is based on the mathematical models on a mature DHI software platform which provide a series of user – friendly interfaces, functions of data input management and visual output etc. Based on 1D dynamic models, some improvements have been done in order to overcome the inherent deficiency of 1D model and to address the unique conditions of the Yellow River.

Development of Aqua – Environment Risk Assessment and Precaution System in Tai Lake Basin.

This project was a part of national Major Science and Technology Program for Water Pollution Control and Treatment. A series of mathematical models (involving hydrology, pollution load, 1D and 2D hydrodynamic, water quality, lake eutrophication modelling) and an aquatic environment risk assessment and early warning platform have been built up for Tai Lake Basin. The system is composed of forecast, database, data exchange, GIS, emergency management and information system modules. And that all modules have been integrated into one standard interface.

4　Conclusions

The main objective of water quality management is to achieve sustainable use of water resources by protecting and enhancing their quality while maintaining social and economic development. Given the increasing complexity and disciplinary breadth of water problems, DSS become a necessary tool to make traditional water models more useful and thus give support to the water quality management. A water quality DSS is an integrated and interactive computer system, consisting of analytical tools and information management capabilities, designed to aid decision makers in solving relatively large, unstructured water quality related problems.

The water quality DSS developed by DHI China follows the "two-level DSS" concept, and offers a different level of service to a different target user. It has been implemented by adopting the integration of Browser Server and Client Server structures. As the technological bases, the two most efficient ones are the mathematical models and the model integration techniques. In our DSS, the water quality modelling management includes two parts, i. e. , daily water quality management and emergency management. In either case, the numerical models developed by DHI are adopted primarily as the core components. Meanwhile, new advances in software engineering are providing stronger operability for the model integration. This means that models with the different origin will be able to communicate through some standard interfaces. This DSS consists of four layers: user & interface layer, application system layer, GIS platform layer, and database layer, and is characterized for its flexibility, generality, compatibility and serviceability with the use of this multi - level structure, each specializing in one particular function.

Our water quality DSS has clearly scored an initial success in a lot of projects in several large river basins of China, but the development and application is far from a mature field. It is considered that the DSS will further focus on the new knowledge acquisition and representation and integration with more efficient expert help systems for better planning/management/ optimization.

References

Rizzoli A E, Young W J. Delivering Environmental Decision Support System: Software Tools and Techniques[J]. Environmental Modelling & Software, 1997,12(2 - 3):237 - 249.

Soncini - Sessa R, Nardini A, Gandolfi C, et al. Computer Aided Water Reservoir Management: A Prototype Two Level DSS. Invited Paper: NATO ARW on Computer Aided Support Systems in Water Resources Research and Management, Ericeira, Portugal.

Martin Volk, Sven Lautenbach, Hedwig van Delden, et al. How Can We Make Progress with Decision Support Systems in Landscape and River Basin Management? Lessons Learned from s Comparative Analysis of Four Different Decision Support Systems[J]. Environmental Management, 2010(46):834 - 849.

Zeng Yong, Cai Yanpeng, Jia Peng, et al. Development of a Web - based Decision Support System for Supporting Integrated Water Resources Management in Daegu City, South Korea[J]. Expert Systems with Applications, 2012(39):10091 - 10102.

Cao Yu, Yan Jing. Descion Support System for Watershed Management[J]. Chinese Journal of Applied Ecology, 2012(23):2007 - 2014.

Ngirane. Integrated Water Resources Planning as a Factor in Environmental Pollution Control[J]. Water Science & Technology, 1991(24):25 - 34.

Solutions to the Problems in the Management of River Basin Water Resources

Chen Hongwei

Yancheng Water Conservation Office, Yancheng, 224001, China

Abstract: China's vast territory includes seven river basins, such as the Yangtze River basin, the Yellow River basin and the Huaihe basin. The relevant water resources management agencies have made great contributions to promote a sound and rapid economic and social development of the watersheds. The water resources management involves a wide area of river basin and needs the overall planning of its upper, middle and lower reaches, banks, tributaries, surface and underground, water quantity and water quality, and exploitation and protection and so on. In 2011, No. 1 Central Document and the Central Conference of Irrigation Work clearly put forward the most strict water resources management system. In February 2012, the State Council issued the "Views on Implementation of the Most Strict Water Resources Management System", which brought new and more serious challenges to river basin water resources management. With the background, based on a comparison between the river basin water resources management systems at home and abroad and aiming at the possible solutions to the weakness of China's basin-region combined management system, this study provides some countermeasures for further improvement of China's water resources management from the following aspects: legal system construction, management operating system, power and authority division, monitoring ability, and safeguard mechanism.

Key words: river basin, water resources management, solutions

1 Evolution of the basin water resources management system in China

1.1 The 1988 "Water Law"

On January 21, 1988, the 24th Session of the Sixth National People's Congress Standing Committee passed the "Water Law" and made an important step to promote the unified management of water resources. The Article 9 defines: "the state implements the water resources system combining the unified management and classification with the departmental management. The water administrative department of the State Council is responsible for the unified management of water resources. The other relevant departments of the State Council, in accordance with the division of their responsibilities, are responsible for the administration of water resources together with the water administrative department. Water administrative departments and other departments of local people's governments above the county level, in accordance with the division of their responsibilities of the people's government at the corresponding levels, are responsible for the administration of water resources. " It is obvious that the established water resources management system "combining the unified management and classification with the departmental management" is essentially based on the region unit. This management system has not clearly mentioned the legal status, the responsibility and the management authorities of the river basin management agencies.

1.2 The 2002 "Water Law"

On August 29, 2002, the 29th Session of the Ninth National People's Congress Standing Committee revised and passed the "Water Law". The Article 12 defines: "the state implements the water resources system combining the basin management and regional management. The water administrative department of the State Council is responsible for the unified management and supervision of water resources. The river basin management agencies established by the water administrative department of the State Council for the important rivers and lakes can implement the relevant laws and administrative regulations of water resources management within their jurisdiction, and they are also authorized by the water administrative department of the State Council to be responsible for the management and supervision of water resources. The water administrative departments of local people's governments above the county level, in accordance with their authorities, are responsible for the unified management and supervision of the regional water resources. " Therefore, the "Water Law" in 2002 strengthens the unified management of water resources, deletes the Article 9 of the "Water Law" in 1988, and makes clear the responsibilities of the basin water resources management and the local people's government water administrative departments on the unified management of water resources.

1.3 No. 1 Central Document

The "Decision of the Communist Party of China (CPC) Central Committee and State Council on accelerating the development of water conservancy reform" (hereinafter referred to as "Decision") was announced formally on January 29, 2011, focusing on water conservancy reform development. This is the first time for the CPC Central Committee to put forward a reform decision on the overall development of water resources since the People's Republic of China was founded. The Article 23 of the "Decision" clearly points out that the water resources management system should be improved through combining the basin management with the regional management, and the mechanism for water resources management should be established and be characterized with clear responsibility, clear division of labor, code of conduct, and coordinated operation. The Article 28 requires establishing and perfecting the water law system, to perfect the laws and regulations for water resources allocation, conservation and protection, flood control and drought relief, water conservancy in rural area, soil and water conservation, and river basin management. The "Decision" puts forward new requirements to perfect the management system of water resources in China and to manage water resources according to laws.

1.4 The State Council No. 3 Document

In order to implement the "Decision" issued on the 2011 CPC Central Committee conference on water conservancy, in February 2012, the State Council issued the third policy document for 2012 for a strict water resources management system. The document requires further improvement of the water resources management system combining the basin management with the regional management to strengthen the unified planning, management and scheduling of basin water resources. The exploitation and utilization of water resources should be in accordance with the unified planning of the basins and regions, thus giving full play to the multiple functions of and the integrated benefit from water resources. The document also requires to work out the allocation scheme for the basin water volume of the main rivers, to establish the controlling index system of the total amount of water use, covering the river basin and the three levels of regional administration (i. e. province, city and county), and to implement the total amount control of the basin and regional water abstraction. River basin management agencies and the water administrative departments of the local people's governments at and above the county level should formulate and improve the scheduling schemes, emergency dispatching plans and schedules of water resources, and to implement the unified scheduling of water resources, and the regional water resources regulations should be subject to the unified management of the river basin water resources. River basin management organizations should strengthen the provincial boundary water quality and quantity monitoring of the important rivers and lakes.

2 Reference from foreign river basin water resources management system

Taking basin as a unit for water resources management is the common experience of the current world water resources management.

2.1　Japan as a representative of Asian countries

After the Meiji Reform, the integrated management of river basin water resources has started to be implemented in Japan. In 1896, Japan enacted the "River Law", which abolished the private ownership of the rivers and rivers were administrated by the local governments and their administration was supervised by the Minister of Construction. With the rapid economic development, water consumption greatly increases, which led to the further need for the improvement of the river basin water resources management. In the 1960s and 1970s, a series of laws were enacted for the basin water resources management. For example, the "Water Resources Development Corporation Law" and the "Water Resources Promotion Law" were enacted in 1961, the "Water Pollution Prevention Law" in 1970 and the "Special Measures Act of Water Source Region" in 1973. Especially, the "River Law" was amended in 1964, which becomes the basic law of the basin management and which defines that rivers are public resources. Rivers (including their management facilities) are divided into the first class and the second class, managed by the Minister of Construction and the magistrates respectively. It also clearly defines the river management, development, utilization, protection, water resources regulation and some special cases of dam.

2.2　France as a representative of European countries

In France, the "Water Law" enacted in 1964 makes a clear division of six river basin regions and established six basin water authorities to implement the integrated management of water quality and water volume in terms of different river basins. These water authorities have no administrative power and they are only the agencies of the "Basin Committee" (i. e., "Water Parliament") formed by representatives from basin water users, local governments, the sector of central government responsible for water resources. Their main responsibilities are to make river basin plans, levy, provide financial support, and collect and release water information. The basin committee is not a standing body, which is different from the River Water Conservancy Committee in China, and it is only equivalent to the water conservancy "parliament" at the basin level. It is a "parliament" form of the democratic management of water resources. It can let all kinds of water users participate in the decision-making process of water resources exploitation and utilization. It can, therefore, enhance the democracy and legality of decision-making and provide authoritative suggestions for the water authorities when making river basin plans, the water resources exploitation and utilization policies, water levies and other water regulations as well. Due to this "Water Law", the ecological state of French rivers has been significantly improved, and even in Paris, a region with a dense population, the quality of drinking water can meet the relevant requirements.

2.3　South Africa as a representative of African countries

South Africa is a country with a serious shortage of water resources. To strengthen the water sources management and protection, the first "Water Law" was issued in 1956 and revised later many times, the "Water Research Law" was enacted in 1971, the "State Water Resources Policy" in 1977, and the "White Paper of Water Supply and Health" and the "White Paper of Water Policy" in 1994. In 1997, the "Water Service Law" was enacted and the amendment to the "Water Law" and other regulations and policies of water resources were also issued. The amended "Water Law" clearly defines that water resources belong to all the people and are completely managed by the central government. It also defines the basin management strategies through the particular articles: to propose the strategies, goals, plans, criteria and procedures of the water resources exploitation, use, protection, maintenance, management and control; to consider the different aspects of the water resources management area such as geology, population, land use, climate, plants and water conservancy project; and to consider the needs and expectations of the current and the potential water users. The water abstraction must be licensed by the Ministry of Water Conservancy and Forestry, Basin Management Agency (Water Committee) and local governments.

2.4 The United States as a representative of North American countries

In United States, the federal government and state governments can enact their own laws for river basin water resources in terms of their own needs. The legislation is characterized by the integration of water resources exploitation and use by both the federal government and river basin management agencies. For example, in 1933, the "Tennessee Valley Authority Act" was enacted and the Tennessee Valley Authority was set up as the authoritative basin management agency. The Tennessee Valley Authority was defined as a governmental agency, responsible for comprehensive exploitation and management of Tennessee River Basin, including such affairs as flood control, waterway transport, and irrigation. At the same time, the Tennessee Valley Authority was authorized with administrative power and its relationship with other agencies was also made clear. Therefore, the Tennessee Valley Authority, according to its basin water resources, can fully consider the requirements of the long-term development of water resources exploitation and utilization and work out the plan for the comprehensive and long-term development, including flood control, power generation, shipping, irrigation, agricultural production, environmental protection and so on, thus appropriately coordinating the industrial economy and environmental protection, and finally arriving at a great achievement (Song Haiou, Pu Guangzhu and Ma Pinyi, 2007). In addition, to implement the integrated river basin management, the United States Congress passed the "Water Resource Planning Bill" in 1965 and the water resources council was set up. Considering the overall interests the river basin, the different Basin Commissions were set up to be responsible for the comprehensive exploitation and utilization of the basin water resources and land resources.

2.5 Brazil as a representative of South American countries

Brazil is one of the richest countries in their freshwater resources in the world. However, its population is concentrated in cities and other economically developed areas, so water shortage is also a problem for Brazil. To solve the problem, Brazil implements an "integral" approach to treating river basins as land unit in its water resources planning and management and this approach not only comprehensively considers all kinds of needs, problems and goals, but also take into consideration the adjacent basins, aquifer and state and municipal administrative regions. In water resources management, power is decentralized because different river basins have different physical and biological characteristics. If the stakeholders within the basin are very clear about their needs and problems of water resource management, then the decisions will not need to be made by the central government. However, to solve some complex problems of water resources, the central government will be involved. In Brazil, the water resources management system includes National Water Resources Committee, State and Federal District Water Resources Committee, River Basin Commission, Federal, State and Municipal Water Resources Organizations and water management agencies. The National Water Resources Committee is composed of the representatives from the departments of federal government responsible for the water resources management, state water resources committee, water users and non-governmental organizations. The national and state secretariat as the specific management agencies are also the actuators of the National Water Resources Committee. River Basin Commission, known "water parliament", is a special organization. Water management agencies are the executive secretariats of River Basin Commissions, which are like private or state-owned enterprises, but the establishment of water management agencies must be proposed by River Basin Commissions and authorized by State or Federal Water Resources Committees.

2.6 Australia as a representative of Oceania countries

In Australian Constitution, states are authorized to exercise the legislative power over basin water resources. Therefore, the different states enact their own water laws to exploit, use, protect and manage water resources in terms of their own characteristics. For example, the state of Vitoria enacted the "Vitoria Soil and Water Conservation Law" in 1881; the "Vitoria Integrated Irrigation

Law" was enacted in 1886; and in 1905, the "Vitoria Water Law" was enacted and Water Supply Committee was set up as an agency to comprehensively manage the river basin. In 1976, South Australia enacted a new "Water Law" to conduct a comprehensive management of water resources. The integrated river basin management practice in Australia has formed a unique total catchment management. For example, to promote and coordinate the effective planning and management of Murray - Darling Basin, the "Murray - Darling Basin Agreement" was formulated in 1985, and according to the agreement, the basin management agency was established. In 1993, the federal government and state governments enacted specific laws to define its legal status (Don Blackmore, 2003). For the protection of the river basin, its ecological system and watershed health, the "2001 ~ 2010 Murray - Darling River Basin Comprehensive Management Strategy to Ensure a Sustainable Future" was formulated in 2001 as a new strategy of natural resources management.

3 Problems in China's basin-region combined water resources management system

According to the "Water Law", China has implemented the basin-region combined water resources management system, but there are still some problems, which can not satisfy the needs of the sustainable development of river basins and the most strict water resources management.

3.1 Imperfection of laws and regulations

China has seven large river basins. Due to the geographical locations, the natural conditions of their water resources are greatly different and as a result, their water resources management should also have different focuses. So far there have not been the laws and regulations of water resources management and allocation schemes which can indicate the intrinsic features of river basins except the "Taihu Basin Management Regulations" and the "Yellow River Water Scheduling Regulations". In addition, many other cases have not been covered by the present water law and regulation system such as the exploitation of water resources in the air, the paid transfer of water resources, the restoration and compensation of water ecology, the construction teams' illegally digging wells, the water resources waste, the destruction of water supply facilities, the users' refusal to pay water rate, the polluted water utilization, the supervision on the total pollutant discharge after proposing the limit amount, and the management of sand mining at trans-provincial regions (Chen Hongwei, 2008).

3.2 Unclear division of responsibilities

According to the "Water Law", the authority of the river basin management agencies is endowed by laws and regulations and the Ministry of Water Resources as well, and the water administrative departments in local people's governments at and above the county level exercise their corresponding administrative powers. Although the "Water Law" enacted in 2002 has clearly defined the specific work of the basin management agencies, it has not defined the specific responsibilities and authorities in practice, for example, the licensing system of water abstraction, the jurisdiction of water resources paid use system, the authority scope of water administration over the water supervision and examination and the implementation of water administrative punishment and so on. These result in some overlaps of the responsibilities and authority exercise of the basin management agencies and the water administrative departments in local people's governments. Especially, in the process of water administrative law enforcement, it is likely to result in the dislocation, omission and overreaching of exercising authorities and performing responsibilities because of the unclear division of responsibilities and authorities in the existing water laws, regulations or departmental rules (Chen Hongwei, 2008).

3.3 Inappropriate control system

As the administrative division is related to the water resources management, rivers and river systems are artificially divided, which is contrary to the nature of water resources. In addition, the local departments of water administration are under the administration of the local governments and they are required to serve the local economic development. There also exist the overlap of the responsibilities and authorities in different departments. All these may result in disasters (Wang Tianlan, 2003). Take the Yellow River as an example. The Yellow River goes through nine provinces and municipalities, so its water resources are managed by different authorities in its upper, middle and lower reaches. The provinces in its upper and middle reaches take their advantages of terrain to use the water resources, which to some degree results in the water drying up of the river in the lower reach. The water drying up of the Yellow River has mainly caused by natural factors, but the lack of the unified management and scheduling is also a factor that can not be ignored. It is obvious that the present control and management system can not satisfy the needs of new situations.

3.4　Unreasonable artificial segmentation

The current system of water resources management artificially divides the whole process into water transfer, water supply, water use, water saving, drainage, sewage treatment and water reuse and it also results in the urban-rural segmentation, thus destroying the natural circulation of the exploitation and utilization of water resources and the conservation and protection of water resources as well. It also results in the following phenomena: the department responsible for water transfer does not care about water supply, water saving and drainage; the department responsible for waste treatment does not care about water reuse. In other words, there exists an unreasonable separation between different departments such as the department for pollution control, the department for exploitation and the department for drainage. Even more seriously, some departments charge for pollution control but do nothing; the developers burden more and they not only exploit the water resources, but also need to control water pollution; the dischargers just pay for their drainage but do nothing for pollution control. As a result, water resources have not been effectively saved and protected (Yuan Hongren and Wu Guoping, 2002).

3.5　Local interest maximization

At present, although the basin-region combined management system has been implemented, the local governments at all levels, to achieve the maximization of their economic interest, excessively use their authorities to vigorously exploit water resources in their administrative regions instead of initially and actively saving and protecting water resources. They have also failed to exploit and save water resources in their administrative regions from the perspective of the whole basin interest, the overall development of river basins and the sustainable development, thus to some extent resulting in the conflict of the river basin management and the regional management.

4　Measures to improve China water resources management

4.1　Perfecting the relevant legal system

The successful river basin management needs the legal guarantee. China has enacted various laws and regulations for water resources such as the "Water Law", the "Flood Control Act", the "Law of Soil and Water Conservation", the "Law of Water Pollution Prevention", the "Management Regulations of Water Use Licensing and Charge for Water Resources Use", the "Taihu River Basin Management Regulations" and the "Yellow River Water Scheduling Regulations". However, the laws and regulations for the basin management can not satisfy the needs of the most strict water resources management system. In addition, the existing legal system can satisfy the basic needs of the current national conditions and water conditions, and at the same time the water resources environment management has also been enriched by some successful experience from the western countries and is gradually moving towards the legalization and standardization, but the cases of the non-enforcement and lax enforcement of laws continue to exist. Therefore, it is necessary to summarize and draw on the successful experience both at home and abroad and to accelerate the revision of the present "Water Law". Aiming at different rivers, it is necessary to formulate the laws for the basin management, for example, the "Yangtze Law", the "Yellow River Law" and the "Huaihe Law", and the corresponding regulations as well, for example, the "Yangtze River Water Allocation Scheme" and the "Huaihe Water Allocation Scheme", to make clear the macro-responsibilities and authorities of the basin management agencies to implement the unified planning, make the overall arrangement, provide macro-guidance and supervision and examination, to make specific the annual trans-provincial schedule and the water scheduling plans for drought emergency, and finally to conduct the real implementation of the unified basin water resources management and to perfect the legal system of water resources.

4.2 Reforming the management system

The basin water resources management system has been implemented by many countries and it is a common way to manage water resources. Different countries have adopted different models in terms of their own situations. As far as China is concerned, it is necessary to set up river basin management committee or the general office of state for basin management to be mainly responsible for the overall water resources management, to supervise the performance of the responsibilities performed by the basin management agencies, and to coordinate the relationship between the basin management and the regional administrative management. It is necessary to establish and perfect the exchange mechanism in allocation work, the information sharing mechanism, the all parties participation and negotiation mechanism, the pollution control and corresponding punishment and restoration mechanism, the ecological compensation mechanism and the joint law enforcement mechanism in the exploitation, saving and protection of the basin water resources (Lu Guihua, 2011). It is also necessary to set up the special institutions for supervision and inspection of some important areas to avoid the segregated basin management, the confusion of responsibilities, and the regional protection in the process of water resources exploitation, saving, protection and allocation and finally to improve the efficiency of river basin management.

4.3 Dividing the responsibilities and authorities correctly and clearly

A river should be treated as a whole by considering the mutually dependent relationship of its upper, middle and lower reaches, its banks and its branches. River basin and its passing regions have different features and they are a loosely structured community, but they are closely related to form a unity. Therefore, it is necessary to make clear the responsibilities and authorities of the basin management and the regional management, to lay emphasis on the overall planning and coordination development. It is necessary to make clear that basin agencies and local governments should work in terms of their own responsibilities and authorities when they implement water use permission, examine the river construction projects, work out the pre-arranged scheduling plans, establish the evaluation system of quantity control of pollutants discharge into the water function area, strengthen the management of sand excavation at borders of provinces, and enhance the supervision and inspection system (Chen Xian, Wang Guizuo, Liu Dingxiang, et al., 2011). Regional administrative management must obey the unified management of the river basin and the regional departments should accept the macro-guidance and actively perform their responsibilities in water resources management (Lu Guihua, 2011). On the other hand, the river basin management must be based on the regional administrative management and should fully consider the need of regional development, optimize the developing layout, coordinate the relationship of the upper, middle and lower reaches of the river, its banks and its branches to keep the overall planning of the basin exploitation and protection in line with the regional social and economic development. The river basin management should also serve the regions, promote a sound interaction of basin and regional development to realize multi-win, thus improving the efficiency of combining basin management and regional administrative management (Guan Yexiang, 2011).

4.4 Improving the monitoring ability

To implement the most strict water resources management system is to establish the system of the water resources exploitation and control, the system of the water use efficiency control and the system of quantity control of pollutants discharge into the water function area. These systems cover the whole process of exploiting, saving and protecting water resources. Among them, the most important one is to strengthen the water resources monitoring. It is necessary to accelerate the establishment of the monitoring and management platform and the information management system at the national, basin and regional levels, to the monitoring system covering the important water users, the water function area and the main trans-provincial sections. It is also necessary to further optimize the existing network of hydrologic stations, to improve hydrological monitoring, pre-warning and forecast, to promote the groundwater monitoring construction, to improve the groundwater monitoring network and to strengthen the ability of monitoring, pre-warning, management and rapid response, thus ensuring the appropriate functioning of the control index of monitoring, evaluation, assessment, rewards and punishment.

4.5 Perfecting safeguard mechanism

In order to effectively implement the most strict water resources management system, it is necessary to perfect "five safeguard mechanisms". Firstly, the target responsibility system of water resource management needs to be implemented to make clear that the principals of the river basin agencies and the local governments at all levels are responsible for the basin and the regional water resources management and protection. The binding targets of the water resources exploitation, saving and protection can be considered as one of the important aspects of the comprehensive evaluation system of the economic and social development; the routine assessment and the annual assessment can be conducted by the higher water administrative departments, together with the relevant departments and the assessment results can be used as an important aspect to evaluate the management achievement (Chen Hongwei, 2011). Secondly, it is necessary to establish the scientific accountability index system of water resources management, including the total quota control index of water use, the control index of water use efficiency, the water quality index of trans-provincial sections or important water function areas within the administrative regions (COD, ammonia nitrogen, total phosphorus) and so on. It is also necessary to standardize the accountability procedure and perfect the accountability system to achieve the accountability normalization, institutionalization and routinization in water resource management as soon as possible, and especially to strengthen the accountability of serious water pollution accidents. Thirdly, it is necessary to improve the water resources management by promoting the standardization of the basin and the regional water resources management and by increasing the equipments and facilities for water resources management and the related technicians as well. Fourthly, it is necessary to highlight the leading role of local governments at all levels in water resources management, to expand the channels for investment and financing, and to establish a long-term and stable financial support mechanism for water resources management. It is also necessary to perfect the finance transfer and payment system, to explore and promote the water rights transaction between the upper, middle and lower reaches and the compensation of the downstream regions to the upstream regions for their protecting water resources and water environment, and at the same time, to establish the accountability system to deal with the excessive pollution discharge of the upstream regions or the water environmental accidents caused by the upstream regions. Finally, it is necessary to enhance the public's awareness of cherishing water, saving water and protecting water resources in various ways, to establish the management monitoring system with the involvement of the public by following various ways such as to inform the public before project approval, to convene meetings to listen to the public, to supply online information and to assess the projects after they are finished. It is necessary to let the public know the relevant information, participate in water resources management and express their opinions, thus realizing the fairness and transparency in decision-making (Chen Hongwei, 2006).

5 Conclusions

The essence for improvement of river basin water resources management is to establish a set of multi-functional management system which adapts to the characteristics of natural water resources basins, to optimize the allocation of the limited water resources in different river basins, to maximize the comprehensive benefits and to safeguard and promote the sustainable economic and social development. Only by thoroughly implementing the spirit of the 2011 No. 1 Central Document, the Central Conference of Irrigation Work, the State Council No. 3 Document for 2012 and timely revising the "Water Law", can we enhance the legal status of the basin water resources management, coordinate the relationship between the basin management and the administrative regional management, strengthen the comprehensive basin management and water affairs management, effectively improve the basin water resources carrying capacity, enhance the coordination of water resources and economic development, and support the sustainable development of the basin and regional economy by the sustainable exploitation and utilization of water resources.

References

Song Haiou, Pu Guangzhu, Ma Pinyi. On the Improvement of River Basin Management System in China [Original title in CNKI: On the Completed of Our River Basin Management System] [A]. China Law Society - Environment and Resources Law Society. Law-governing Environmental Protection and the Harmonious Society Construction—2007 China Law Society - Environment and Resources Law Society (Annual Conference) [C]. 2007, 88-90.

Don Blackmore. The Key Points in Murray - Darling River Basin Management—The Confluence Regional Integration Management [J]. China Water Resources, 2003 (11A): 55 - 56.

Chen Hongwei. A Theoretical Discussion on the Revised "Water law" and its Supporting Policies [J]. Water Resources Development Research, 2008 (3): 37 - 39.

Wang Tianlan. Discussion on Water Rights and the New "Water Law" [J]. Journal of Lanzhou Institute of Education, 2003 (2): 54 - 64.

Yuan Hongren, Wu Guoping. Water Resources Protection and Legislation [M]. Beijing: China Water and Power Press, 2002: 161.

Lu Guihua. A Philosophical View on the Strategies for Water Resources Management [J]. Jiangsu Water Resources, 2011 (7): 7 - 8,11.

Chen Xian, Wang Guizuo, Liu Dingxiang, et al. Some Suggestions on the Division of Responsibilities in Basin Management and Regional Management [J]. Water Resources Development Research, 2011 (7): 88 - 92.

Guan Yexiang. On the Relationship between the River Basin and Regions [N]. China Water Conservancy Newspaper, 2011 - 07 - 28 (01).

Chen Hongwei. To Implement the Most Strict Water Resources Management System in Yancheng City: Practice and Thinking [J]. China Water Resources, 2011 (7): 42 - 44.

Chen Hongwei. On the "Water Abstraction Licensing and Water Resources Levy Management Regulations" and Suggestions on its Supporting Regulation System [J]. China Water Resources, 2006 (14): 48 - 49,51.

Research on Management of System Coupled Mode and Institutional Innovation of Water Resources in the Yellow River Basin

Qin Zhaohui[1] and *Tan Qijia*[2]

1. School of Economics and Management in China Three Gorges University, Yichang, 443002, China
2. School of Foreign Language in Xihua University, Chengdu, 610039, China

Abstract: The management of water resources in the Yellow River Basin has been pay highly attention and discussed at home and abroad. Analyzing the management of water resources in the Yellow River basin includes cross – section water resources, generalized water resources, water supply in the earth's surface, supplied amount of water resources, usable amount of water resources, water conservancy project in the earth's surface, well engineering, sewage utilization project and rainwater collection project. On this basis, System Coupled Mode of Water Resources has been established. Finally, putting forward a sound system is a set of effective safeguard to promote the management of water resources. Implementing the regulation of tradable permit of water using right and strictly implementing water allocation are to promote rational development and utilization of water resources in the Yellow River Basin and its sustainable development.

Key words: the Yellow River Basin, management of water resources, System Coupled Mode, System Innovation

1 Research on Management of the Water Resources in the Yellow River

151 academicians of the Chinese Academy of Sciences and Engineering in 1998 co – authored the appeal for action to save the Yellow River , and made report to the State Council on countermeasures and suggestions to ease the Yellow River cutoff based on inspection of the Yellow River Basin. As a result, management of the water resources in the Yellow River has aroused great concern at home and abroad. Hu Angang (2000) adopt new thinking in The Public Policy of the Allocation of Water Resources in transitive period to analyze the allocation of water resources under the new situation in China, conducted in-depth study of water resources in the Yellow River Basin as an example and made a set of institutional arrangements around water market of the Yellow River Basin. In the book of *Management of the water resources in river basin* written by Ruan Benqing (2000), he made the Management of the water resources in the Yellow River as an example to put forward a set of system management around water market of the Yellow River Basin. ChangYunkun (2001) in his book *Study on Yellow River* cross-section *and system of water rights the Yellow River* writes rationalizing water price in the Yellow River Basin, reforming of the current management system of water resources in the Yellow River Basin , implementing separate management of water governance and water services. Conservancy Commission(2002) in *The Recent Focus on Management and Planning of Development of Yellow River* proposed overall layout, the basic train of thought, key management measures, development target, and safeguard measures to solve major issues such as the contradiction between supply and demand of water resources, threatening flood of the Yellow River , and the deterioration of ecological environment. Wang Sucheng in Resource Water – People in Harmony with Nature (2003) put forward new management for the sustainable use of water resources to support economic and social environment and sustainable development of water resources. Tang Weiqun in the article Management of the Water Resources in the Yellow River (2004) says new ideas for the allocation of water resources and also attention of the ecosystem within the river basin during the management of the Yellow River Basin. Luo Qing (2006) in the text of Study on Carrying Capacity of Water Resources in the Yellow River Basin, put out standard

structure of characterization between comfortable and rich class, then reasonably allocate water resources in the Yellow River. On the basis of socio – economic development in the river and forecast of water utilization. In summary, the experts study the management of water resources in the Yellow River Basin from different angles and put forward the problem of water resources management in the new era status which has made an important contribution to theory and practice of water resources in the Yellow River Basin.

2 Evaluation of management of water resources in the Yellow River Basin

On the basis of analysis of the status of Management of Water Resources in the Yellow River, evaluating Management of Water Resources in the Yellow River is as follows: cross – section water resources, generalized water resources, water supply in the earth's surface, supplied amount of water resources, usable amount of water resources, water conservancy project in the earth's surface, well engineering, sewage utilization project and rainwater collection project.

2.1 Water resources of cutoff in the Yellow River Basin

The hydrological cross – section of water resources also includes two parts: water resources of the surface and ground. Surface water refers to the real water consumption non – human use flows through the hydrology section of the river runoff; ground water means the total amount of water and non – reusable groundwater in the sub – basin of water, as shown in Tab. 1.

Tab. 1 Water resources of cutoff in the Yellow River Basin　　　(Unit: $\times 10^8$ m^3)

Main cross – section	Water resources of the surface	Water resources of the ground.		Total water resources
		Total water resources	The amount of non – reusable water resources	
Guide	201.7	65.3	1.9	203.6
Lanzhou	308.0	102.2	5.3	313.3
TouDaoguai	313.6	160.9	40.5	354.1
Longmen	351.5	200.9	47.4	398.9
SanMenxia	448.9	325.9	86.5	535.4
HuaYuankou	485.4	361.0	97.5	582.9

Note:: Luo Qing, *Carrying Capacity of Water Resources in the Yellow River Water Basin*. Research Institute of Water Resources and Hydropower in China, 2006.

Annually natural runoff of section of Guide in the Yellow River Basin is 2.017 $\times 10^{10}$ m^3, non – reusable groundwater resources is 1.90 $\times 10^8$ m^3, Total water resources is 2.036 $\times 10^{10}$ m^3; Annually natural runoff of HuaYuankou section in the Yellow River Basin is 4.854 $\times 10^{10}$ m^3, the amount of non – reusable groundwater resources is 9.75 $\times 10^9$ m^3, the total water resources is 5.829 $\times 10^{10}$ m^3. Visibly, water distribution of the west and east resources is uneven, a distributive pattern of rich water in the east and little in the west.

2.2 The general water resources of the Yellow River Basin

General water resources including total water resources of the cross – section, the effective earth's surface and soil evapotranspiration, which is divided into farmland, home and work site, woodland and grassland, as shown in Tab. 2.

Tab. 2　Evaluation of the general water resources of the Yellow River Basin(Unit: $\times 10^8$ m^3)

| Partition of Water resource | Precipitation | The general water resources | | | | |
| | | | Effective evapotranspiration | | | |
		Runoff of water resources	Farmland	Home and work site	Woodland	Grassland
The Yellow River District	3,563.0	676.4	890.9	15.9	439.5	733.8
Above Longyangxia	632.3	212.1	5.5	0.1	29.9	197.3
Longyangxia to Lanzhou	433.0	116.1	48.5	1.0	61.8	121.0
Lanzhou to Hekou Town	427.6	53.7	119.6	2.8	22.3	94.4
Hekou town to Longmen	480.2	49.2	144.6	0.6	61.0	82.0
Longmen to Sanmenxia	1,038.9	143.5	386.8	6.7	165.8	164.3
Sanmenxia to Huayuankou	274.7	50.3	89.6	1.6	86.1	32.4
Below Huayuankou	157.8	32.0	85.8	2.9	10.8	10.0
Internal flow area	118.6	19.5	10.5	0.1	1.7	32.5

Note: Luo Qing, *Carrying Capacity of Water Resources in the Yellow River Water Basin*. Research Institute of Water Resources and Hydropower in China, 2006: P43.

Distribution of precipitation in the Yellow River Basin is uneven, and relatively small and scattered. Generalized water resources are 2.7566 $\times 10^{11}$ m^3. Its runoff water resources accounts for 24.5% , and effective evapotranspiration accounts for 75.5%, which is 3.1 times more. The effective precipitation is 3.563 $\times 10^{11}$ m^3, which show that the effective precipitation of soil water and other forms of storage play an important supporting role to the economic and ecological systems, which, precipitation of Longmen to Sanmenxia is relatively high and water resources is relatively a-bundant, while precipitation of other areas is less and has the shortage of water resources.

2.3　Supply quantity of surface water in the Yellow River Basin

Supply quantity of surface water is directly related to water supply of the Yellow River Basin, water supply also has a profound impact on the production and living conditions of the Yellow River Basin. Following is to analyze the supply quantity of all kinds of the Yellow River and supply a-mount of surface water from 1980 to 2000.

Tab. 3 Supply quantity of all kinds water in the Yellow River   (Unit: $\times 10^8$ m^3)

Water sources	1980	1985	1990	1995	2000	Variation from 1980 to 2000	Variation from 1980 to 2000
The total water supply	342.9	333.1	381.1	404.6	418.8	75.8	22.1
Surface water	249.2	245.2	271.7	266.2	272.2	23.1	9.3
Shallow groundwater	80.9	73.6	92.5	116.8	122.9	42.0	51.9
Deep artesian water	12.4	13.6	16.3	20.8	22.6	10.2	82.7
Other water sources	0.5	0.7	0.7	0.8	1.1	0.6	106.3

Note: Luo Qing, *Carrying Capacity of Water Resources in the Yellow River Water Basin*. Research Institute of Water Resources and Hydropower in China, 2006.

From 1980 to 2000, the total water supply of the Yellow River Basin was overall upward trend, but the ascending speed of water supply of the different sources are different. From Tab. 3, water supply of groundwater is much higher than that of the surface. Among them, water supply of the deep artesian grows relatively fast. The ascending rate of Water Supply of other sources in the table is relatively high, but the absolute amount is still small, the proportion is small too.

Tab. 4 Supply quantity of surface water in the Yellow River Basin   (Unit: $\times 10^8$ m^3)

Partition	1980	1985	1990	1995	2000	Variation from 1980 to 2000	Variation from 1980 to 2000
The whole valley of the Yellow River	249.2	245.2	271.1	266.2	272.2	23.1	9.3
Above Longyangxia	1.7	2.0	2.2	2.3	2.5	0.8	47.0
Longyangxia to Lanzhou	20.8	23.3	23.6	26.9	28.6	7.8	37.5
Lanzhou to Hekou Town	132.8	141.9	159.2	159.2	160.9	28.1	21.1
Hekou Town to Longmen	6.3	5.8	6.4	7.0	7.3	1.0	16.2
Longmen to Sanmenxia	50.3	45.1	47.3	45.1	47.0	−3.2	−6.4
Sanmenxia to Huayuankou	13.7	8.6	12.3	9.5	9.0	−4.7	−34.5
Below Huayuankou	23.5	18.4	20.7	16.2	16.7	−6.8	−28.9
Internal flow area	0.1	0.1	0	0	0.2	0.1	91.6

Note: Luo Qing, *Carrying Capacity of Water Resources in the Yellow River Water Basin*. Research Institute of Water Resources and Hydropower in China . 2006.

From Tab. 4, the supply quantity of the Yellow River Basin is unevenly distributed, and the water supply is relatively small. Generally speaking, water supply of the Yellow River Basin changes little during 20 years, but the speed of demand is developing in the fast trend. The contradiction of supply and demand highlights. What's more, the surface water supply above Longmenxia is in a growing trend, while the surface water supply below Longmenxia is in a downward trend.

2.4 Utilization of water resources in the Yellow River Basin

Utilization of surface water resources in the Yellow River Basin is equal to the natural surface water resources minus the demand of ecologically environmental water of river way and the abandoned quantity of discharged flood which is outside river way where is an area difficult to control.

Tab. 5　Utilization of water resources in the Yellow River Basin　(Unit: ×10⁸ m³)

Regional stations	River runoff	Minimum volume of ecological water	Floods difficult to control	Available volume
Tangnaihai	202.0	129.0	18.5	54.5
Lanzhou	321.8	112.3	10.3	199.2
Hekou Town	325.8	139.1	0	186.7
Longmen	367.7	98.5	0	269.2
Sanmenxia	468.4	120.1	2.0	346.4
Huayuankou	522.1	204.5	0	317.6
Lijin	523.7	162.4	0	361.2

Note: Luo Qing, *Carrying Capacity of Water Resources in the Yellow River Water Basin*. Research Institute of Water Resources and Hydropower in China, 2006.

Tab. 5 shows that utilization of water resources in the upper reaches of the Yellow River is relatively small, and faced with flood losses which is difficult to control, especially above Longmen station. Middle and lower reaches can use a relatively large, but the minimum demand of ecological water is increasing. In general, the river runoff is increasing, but increase of the amount of available volume is smaller than that of river runoff.

2.5　Project evaluation water conservancy in the Yellow River Basin

Since the reform and opening up, the water conservancy in the Yellow River Basin has made great achievements. The following is an evaluation and analysis according to water conservancy in the surface of the Yellow River Basin (Tab. 6) and motor – pumped well engineering in the Yellow River Basin District (Tab. 7), Sewage utilization, rainwater collection project and water conservancy projects in the Yellow River Basin District (Tab. 8).

Tab. 6　Water conservancy in the surface Yellow River Basin in 2000　(Unit: ×10⁶ m³)

Size of the project	Projects of water storage			Project of water diversion			Project of lift irrigation		
	Quantity	Total storage capacity	Status of water supply	Quantity	Pilot scale	Capacity of water supply	Quantity	Scale of water lift	Status of water supply
Large	21	62,753	1,324	26	1,904	9,967	5	166	534
Medium	157	5,153	1,290	93	432	7,735	74	217	1,972
Small	2,456	3,196	12,771	12,731	1,420	4,666	22,257	637	3,787
Micro	16,391	496	232	2	2	220	2	10	2

Note: Luo Qing, *Carrying Capacity of Water Resources in the Yellow River Water Basin*. Research Institute of Water Resources and Hydropower in China, 2006.

According to Tab. 6, the number of water conservancy of large and medium – sized water storage, diversion of water and pumping is relatively small, but capacity of water supply is relatively strong, especially the capacity of project of the water diversion. Projects of small – scale water is more, capacity of water supply of water storage is stronger than that of medium – sized projects of water pumping, which is the main water supply forces of the Yellow River Basin. The number of micro – engineering is numerous, but capacity of water supply is limited. Thus, the small water conservancy project is a major component of water supply of the Yellow River Basin. Capacity of water supply of the medium and large water diversion project is strong. The potential for development is good.

Tab. 7, the number of shallow producing wells of the Yellow River Basin is in a large number as well as mechanical electric. Shallow groundwater is the main source of water supply, while deep groundwater of producing wells is small. But from vertical view, it is in a deeper and deeper trend with years.

Tab. 7　Motor – pumped well engineering in the Yellow River Basin District in 2000

Region	Shallow groundwater			Deep groundwater		
	The number of producing wells	The number of supporting mechanical electric	Status of capacity of water supply ($\times 10^4$ m^3)	The number of producing wells	The number of supporting mechanical electric	Status of capacity of water supply ($\times 10^4$ m^3)
The Yellow River Basin	581,002	529,604	1,213,619	22,203	20,635	268,656
Above Longyangxia	435	107	1,315	0	0	0
Longyangxia to Lanzhou	2,860	2,824	52,996	7	0	363
Lanzhou to Hekou Town	69,647	62,810	160,369	12,703	11,483	86,165
Hekou Town to Longmen	37,969	32,446	44,060	2,133	2,004	11,561
Longmen to Sanmenxia	193,953	181,450	550,662	3,082	3,064	106,117
Sanmenxia to Huayuankou	59,134	53,063	116,229	1,141	1,141	38,119
Below Huayuankou	176,745	161,925	268,492	644	631	8,977
Internal flow area	40,258	34,978	19,496	2,493	2,312	17,354

Note: Luo Qing, *Carrying Capacity of Water Resources in the Yellow River Water Basin*. Research Institute of Water Resources and Hydropower in China, 2006.

Tab. 8　Sewage treatment utilization and rainwater collection project in the Yellow River in 2000

Region	Sewage treatment and use			Project of rain collection	
	Number (Unit)	Processing capacity ($\times 10^4$ t/d)	Annual use ($\times 10^4$ m^3)	Quantity (at)	Annual use of ($\times 10^4$ m^3)
The Yellow River Basin	36	211	13,595	2,248,984	9,069
Above Longyangxia	0	0	41	23,018	93
Longyangxia to Lanzhou	4	6	0	341,058	1,048
Lanzhou to Hekou Town	3	27	663	533,903	1,287
Hekou Town to Longmen	1	2	650	98,597	1,289
Longmen to Sanmenxia	25	151	10,438	1,109,110	4,197
Sanmenxia to Huayuankou	1	20	0	116,299	1,060
Below Huayuankou	2	6	1,803	891	9
Internal flow area	0	0	0	26,108	87

Note: Luo Qing, *Carrying Capacity of Water Resources in the Yellow River Water Basin*. Research Institute of Water Resources and Hydropower in China, 2006.

From Tab. 8, the number of sewage treatment works and the rainwater project in the Yellow River Basin is few water supply which is limited and mainly concentrated in the Longmen to Sanmenxia.

In summary, although water conservancy construction in the Yellow River Basin has made great achievements, but there is still a lot of regional and structural contradictions, the contradiction between supply and demand of water resources has become increasingly prominent and more serious, not only affect the level of economic development of the Yellow River Basin, and has profound impact on the quality of people's lives, but also affect the ecological environment, so we need to come up with a rational management of water resources to promote the rational use of water resources.

3　Management of system coupled Mode of Water Resources in the Yellow River Basin

Managing system of coupled model of water resources is mainly to improve on the total quantity of water supply and control water demand in the areas of production, reducing the production and consumption processes in water consumption, making use of advanced technologies for waste water treatment so that water resources can be cycled many times.

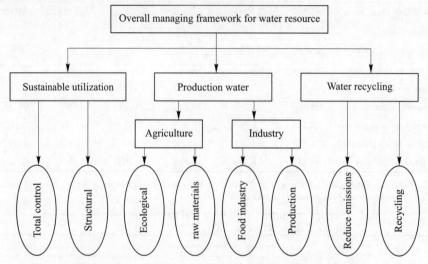

Fig. 1　Overall managing framework for water resource

Management of system coupled model of water resources is to achieve efficient use of water resources, improve the output efficiency of water resources and minimize discharge of water waste, and achieve water conservation and water recycling. It is need to increase the investment of wastewater treatment, establish the recovery and use systems of wastewater to ensure that wastewater emission are below the carrying capacity of the environment, and promote the reasonable adjustment of the internally economic structure of the Yellow River Basin, promote industrial upgrading and sustainable development.

4　Management strategies and institutional innovation of water resources in the Yellow River

4.1　A sound system is the effective protection management of water resources

Firstly, it should build a sound system, establishing an effective organization at the government level, responsible for coordinating the longitudinal management of the relevantly governmental departments, organization, coordination and supervision of managing planning of water resources in the Yellow River Basin, and also make effective solutions to problems appeared in the process of management of water resources. Throughout the watershed level it should formulate and improve relevant system of laws and regulation, strengthen macro – control functions, to encourage enterprises to implement cleaner production by policy levers and market tools, create a favorably incentive environment, vigorously promote technological innovation, and related technical support and incentive system, in order to support the recycling of water resources.

4.2 Systemic innovation of water rights

Building mechanism of water compensation and trading system of water rights. It should develop distributing system of water use, managing mechanism and the developmental mechanism in basin – wide based on socio – economic and ecological problems of the Yellow River Basin, and rationally determine the value of water resources. And also establish Long-term water resource compensation mechanism, implement the establishment and perfection of trading system of water rights to create needed marketing mechanism of water in the Yellow River Basin on basis of strengthening management of water resources in basin – wide.

References

Easter K W, Rosegrant M W, Dinar Ariel. Formal and Informal Markets for Water: Institutions, Performance, and Constraints[J]. World Bank Research Observer, 1999, 14(1).

Easter K W, Rosegrant M W, Dinar Ariel, et al. Market for Water: Potential and Performance [M]. Boston: Kluwer Academic Publishers, 1998.

Thobani, Mateen. Formal water Markets: Why, When, and How to Introduce Tradable Water Rights [J]. World Bank Research Observer, 1997, 12(2).

Ruan Benqing, Liang Ruiju, Wang Jie. Management of Water Resources in River Basin[M]. Beijing: Science Press, 2001.

Chang Yunkun. Research on Cutout and System of Water Rights of the Yellow River[M]. Beijing: China Social Sciences Press, 2001.

Yellow River Conservancy Commission. The Recent Focus on Management and Development Planning of Yellow River[R]. 2002:1 – 5.

Wang Sucheng. Resource, Water, Man and Nature Live in Harmony[M]. Beijing: China Water and Power Press, 2003.

Tang Weiqun. Research on Management of Water in Yellow River[D]. Wuhan: Wuhan University, 2004.

Luo Qing. Carrying Capacity of Water Resources in Yellow River Basin[D]. Beijing: China Academe of Water Conservancy and Hydropower, 2006.

Li Xu. Social System Dynamics[M]. Shanghai: Fudan University Press, 2009.

Thoughts on Improving the Management System of the Heihe River Basin

Qiu Jie[1], *La Chengfang*[2] and *Wang Daoxi*[1]

1. The Heihe River Basin Authority, Lanzhou, 730030, China
2. Upper Hydrological and Water Resources Bureau, YRCC, Lanzhou, 730030, China

Abstract: Heihe River implements the unified management of water resources continued 11 years, initially established the combination of watershed management and regional management of the River basin management system, ecological environment is worsening the situation got preliminary keep within limits. At the same time, also need to see, "The Heihe River preliminary regulation planning", which is official replyed by the State Council, requirements to recovering the downstream ecology to nineteen eighties level's target has not became a reality. Watershed management and regional management combining the management mode has not truly established.

Heihe River is a Inland River which is a tans – provincial River, for sustaining the healthy life of Heihe River, on the lower reaches of the realization of harmonious development, must implement the unified management of River basin. Heihe's current water resources management system has many drawbacks, such as single relying on administrative coordination, lacking rigid constraints and legal authority, and so on. These drawbacks restrict Heihe River Basin management institution of integrated the watershed management control ability, also directly affect on water dispatching and watershed management effect. In order to solve the problems, implementation of watershed management and regional management combining unified basin management mode, should be from strengthen catchments organization function and control ability, constructing by the River basin agencies responsible for the backbone of regulating reservoir at the technical level; setting of subordinate units for the River basin management institution to control the water conservancy projects in the management level; to promulgate and put into practice the "Heihe River Basin water resources management regulations" in the perspective of law. Multi level security, Multiple range of comprehensive watershed management system, in order to realize the unified management of River basin, to lay the foundation for the development of watershed health.

Key words: Heihe River Basin, management system for inland river, ecological

Heihe River is located in northwest inland of China, downstream is of extreme drought with little rain. The last century since the 1960's, a large number of reclamation of the middle reaches of the Heihe River, agriculture is a significant increase in water, crowding out ecological water, resulting in downstream vegetation decline and fall, the desert is expanding rapidly, make Ejina which In the lower reaches of Heihe to be the principal source of sandstorm in China. The State Council approved that the Yellow River Conservancy Commission set up the Heihe River Basin Authority in 1999, to responsible for carrying out the unified management and scheduling of the Heihe. Clearly the State Council proposed the lower reaches of Heihe River ecological restoration around 2010 to 1980s levels in "The Heihe River preliminary regulation planning", and demanded the implementation of watershed management and administrative regional management combined with the management model in the Heihe River Basin, to establish authority, efficient, coordinated watershed management system.

The year ended the spring of 2012, Heihe implementation of integrated water resources management has been 11 years. By the positive efforts of watershed organizations and local governments at all levels, the trend of ecological deterioration of the Heihe River Basin has been initially contained, watershed management achieved initial success. But at the same time, due to the lack of effective management system to protect, and has been no construction of the backbone of storage project, the single management model which relies on administrative means to coordinate has been came to the bottleneck, the target that make the lower reaches of Heihe River ecology to the 1980s levels has not realized.

1 The problems in Heihe River Basin management

1.1 The principal contradiction in Heihe River Basin

Heihe River Basin is located on the Hexi Corridor which is located in the Inland Northwest of China, strategic position is very prominent. Limited carrying capacity of water resources here, historically, the water on the contradiction was very outstanding. During a certain age of Yongzheng of Qing Dynasty, Shaanxi and Gansu Governor Nian used military force to mandatory implementation the average allocation of water resources system; after the founding of new China, In Hexi People's Liberation Army had been invited to the establishment of the Heihe River Basin Water Management Committee by Gansu Province for the coordination of regional water contradictions. Now, the State Council approved the establishment of the Heihe River Basin Authority. But as China's largest inter – provincial inland River, Heihe crosses the three provinces and two military units, coordination of the work carried out is not easy, to implement unified management is difficult. Mainly there are several conflicts: ① Regional economy is not water – saving high – speed development and valley water resources carrying capacity is limited incompatible; ② Single administration can not meet the complexity of the watershed management requirements; ③ Uniform basin – wide laws and regulations vacancies can not be done according to the law of water.

1.2 Difficult to combine watershed management and regional management, authoritative, efficient and coordinated River basin management system has not been established

Heihe River Basin Authority is responsible for the Heihe water resources management and scheduling, responsible for organizing the preparation and reporting of water resources development and conservation planning, responsible for formulating the water allocation scheme and annual water use plan; supervision in charge of important hydrological control stations within the valley and River basin water resources monitoring; responsible for organizing the water permit system, inspection, supervision of basin water allocation implementation of the plan; responsible for organizational basin water project construction, operation scheduling and management; responsible for coordinating the water dispute between the treatment basin introspection (autonomous regions) by the State Council approved.

Local government is responsible for the implementation of related to integrated watershed management and water resource management policies, regulations, and implementation of River basin water allocation plan, the yearly water allocation program, the annual water use planning and water conservation measures, and implement the responsibility of Chief Executives, will be the implementation of the assessment of government performance of main content.

No significant water projects, especially the main stream of the backbone of the project scheduling management, there is no basin – wide laws and regulations for administering water, the Heihe River Basin unified management rely on administrative means: Basin agencies to develop plans, conduct supervision and inspection, coordination of water disputes, and local governments in their respective regional management level responsible for the concrete implementation. Which, as a simple central basin agencies sent in the Heihe River Basin agencies, with no direct administrative leadership all levels of government within the basin, can not be directly involved in the implementation of policies and programs; at the same time, local Government is to implement the unified management of the Heihe River Basin water resources, but also share the interests of the Heihe River Basin Water Resources. River Basin Integrated Water Resources managers can not restrict the regional managers. Lead to basin agencies to control the ability to weaken, watershed management and regional management are difficult to combine, "authoritative, efficient and coordinated River Basin management system" and did not set up the situation.

2 Thoughts on improving the management system of the Heihe River Basin

2.1 Improve the basic principles of the Heihe River Basin management system

According to the State Council on the approval of the Heihe River Basin management planning as well as views on the most stringent water management system(Guo Fa [2012] No. 3, hereinafter referred to as "opinions"). To strengthen and establish the position of authority in a unified management structure of the Heihe River Basin of the Heihe River Basin Authority, the establishment of an efficient and coordinated River basin management system; implement the most stringent water management system in the Heihe River Basin, promote coordinated economic and social development of the Heihe River water resources, water environment carrying capacity in Heihe River Basin. Of water given the need, the amount of water lines, water system should be to coordinate good Heihe downstream along the rivers, surface water and groundwater, need strict planning and management and the water resources assessment, and strengthen water resources in a unified manner.

2.2 Approach of implementation

2.2.1 Establish the authority of the watershed management organizations, combined with watershed management and local administration

As a severe water shortage in the inter – provincial inland River, Heihe River Basin, the regional economic and social development must be coordinated with the water resources carrying capacity. Watershed management organizations should develop practical water allocation plan and take the total amount of water to control the index system, levels of government should be responsible for the implement. According to the requirements of the "opinions", the yearly water allocation task is completed into the performance evaluation.

Watershed management agencies need to establish the lower – level management units, in conjunction with local governments at all levels, joint participation in local Heihe water administration, and performance evaluation in local government; River Basin Authority and local governments to jointly develop water scheduling task appraisal system, to verify the local government's annual water scheduling the task is completed by the basin agency; yearly water allocation task is satisfactorily completed, the government performance assessment of the appropriate points, the yearly water allocation task is not completed, the government performance assessment shall not be qualified.

2.2.2 Strengthen the capacity of basin agencies on integrated water resources management and scheduling control

Should authorize the Heihe River Basin Authority is responsible for the construction of the Heihe River Basin water project planning consent approval. The Heihe River Basin Authority to formulate a basin – wide planning consent censorship system of the water resources assessment and water license management approach, and promulgated as a basin – wide policy documents. Acts of violation of the above systems and methods, the river basin organization should be mandatory specifications and local governments jointly.

Watershed management organizations to be directly responsible for water projects (cited main stream of the backbone of regulating reservoir and the main stream, carrying water engineering), construction, operation and management, achieve reasonable regulation of the inflow, the strict management of water, from the engineering level to provide protection for the unified management of basin water resources and scheduling. For river basin management institutions established short, weak institutional capacity – building objective reality from the framework for regulatory agencies to carry out the construction of infrastructure to comprehensively strengthen the management capacity of the river basin organization.

2.2.3 Strengthen the construction of the basin – wide laws and regulations, establish and improve the Heihe water administration and law enforcement teams

Formulates of the Heihe River Basin Management Ordinance and other regulations as soon as possible, improve surface water and groundwater resources development, utilization, management and protection policies, in order to achieve according to the law. Strengthen law enforcement status and functions of the Heihe River Basin Authority to implement the scheduling and monitoring tools. The establishment of the Heihe River Basin water administration brigade directly under the system of water governance at all levels of law enforcement teams, and strengthen the construction and implementation of water administration and enforcement systems.

2.3 Security measures

2.3.1 Improve watershed management procedure of consultation mechanisms to protect the basin's major democratic decision – making on matters

Heihe River Basin authorities and bodies to establish the procedure of consultation platform, Basin provinces (autonomous regions), the armed forces to participate in, negotiate the Heihe River water allocation plan, the annual water use plan and related system approach involving watershed unified management of the major issues and important matters, coordination between regions, to the distribution of benefits between the department, basin major decisions on the basis of democratic consultation and watershed and regional needs to be fully reflected in the decision – making process.

2.3.2 Establish a reasonable system of supervision and protection of watershed management is conducted in an orderly

The relevant departments of the State Council and Central Military Commission, need to set up the Heihe River Basin management and supervision of institutions, the establishment of the Heihe River basin management and supervision system. The Heihe River Basin management oversight body should contain the watershed parties, monitoring should cover the whole basin of the Heihe River, the key sections, including both banks.

Supervisory bodies the right to supervise the implementation of the Heihe River Basin in the unified management laws and regulations, institutional policies and program plan, water scheduling tasks, the local government performance evaluation and administrative affairs, the preparation of watershed management planning and implementation, preliminary work and construction of water project operation and management and capital usage, soil conservation and water resources, protection of water administrative law enforcement, the important hydrological and watershed groundwater monitoring information, flood control and drought, water security, ecological protection and construction, and economic regulation. The Heihe River Basin Authority and the local levels of government need to implement the monitoring results.

2.3.3 Increase the unified management of the Heihe River Basin and watershed management propaganda to build a harmonious atmosphere of public opinion of the basin

Publicity efforts to increase the unified management and watershed management in Heihe River Basin and improve the understanding of the necessity of the basin the people of the Heihe River water resources to implement a unified management, and strengthening national awareness of water conservation, water conservation awareness. The Heihe included in the national – level public service announcement category, for the realization of the governance of the whole people concerned about the dust storms, concerned about the Heihe to create a good atmosphere of public opinion. The Heihe River Basin Authority and the local governments at all levels in accordance with the laws and regulations and relevant provisions of watershed management and governance information to the community and accept the supervision of the whole society.

3 Expectations and outlook

The system smoothly is the premise of the Black River Integrated Water Resources Management. Only the State Council and Central Military Commission, the relevant departments under the coordination of deployment, the Yellow River Conservancy Commission in Heihe River Basin Authority and local governments to work together, combination of administrative, engineering, regulation, legal protection, harmonious, organic combination of River basin management and regional management can really achieve, in order to truly realize the combination of unified management of water resources and watershed administration the large Heihe management model, can we truly expected to implement the State Council on the Heihe River Basin governance.

Unified management of the Heihe River Basin long way to go. Smooth system to realize the unified management of water resources is only a first step. Heihe River Basin at all levels of government should be based on integrated water resources management as an opportunity to co-ordinate the arrangements for the coordinated development of regional industrial, agricultural, animal husbandry, forestry, environmental protection, tourism and other. To be fixed in the water required, the amount of water line, due to the water system should be thinking, actively carry out protection and restoration of ecological, engineering, agriculture, livestock, industrial restructuring and transformation of water-saving society, the rational allocation of basin water resources and the region's rapid economic development win-win.

Research on Spatial Data Sharing Service Application Platform of the Yellow River Basin

Xu Zhihui[1] , *Wang Yimin*[1] , *Wei Yongqiang*[1] , *Hu Man*[2] and *Zhang Yanli*[1]

1. Information Center of YRCC, Zhengzhou, 450004, China
2. Xi'an ARSC Information Industry Limited, Xi'an, 710054, China

Abstract: The natural environment and socio-economic conditions of the Yellow River Basin can be shown by the spatial data, which provide important supporting role of spatial information for the Yellow River Basin management and development. By analyzing Yellow River Basin spatial data management and application, and on the basis of the research about service GIS technology, the platform's overall framework, key technologies and construction ideas about the sharing service application platform based on the first national water census spatial data of the Yellow River Basin is discussed. The result shows that, based on the first national water census spatial data of the Yellow River Basin, the establishment of spatial data sharing service platform technology advanced and feasible for flood control and drought, water resources regulation and management, soil conservation supervision and control, engineering construction management, etc. to provide unified spatial data sharing services to meet the needs of the Yellow River Basin management and development for spatial information, will have a huge social and economic benefits.

Key words: spatial data sharing services, spatial data sharing service application platform, spatial data management of the basin

1 Introduction

1.1 The Yellow River spatial data management and application

The natural environment and socio-economic conditions of the Yellow River Basin can be shown by the spatial data, which provide important supporting role of spatial information for the Yellow River Basin management and development. At present, large scale digital topographic maps and high-resolution aerial remote sensing images and satellite remote sensing images in the local area of the Yellow River Basin provide an important spatial information support for flood control and drought, water resources regulation and management, soil conservation supervision and control, and engineering construction management.

However, for the spatial data application in the Yellow River Basin, there are the following key issues: ① the lack of large-scale spatial data, that watershed conservancy commission have the largest scale of the basin-wide spatial data is 1: 250,000 digital line graph, can not meet the need of the watershed development and management; ② backward mode of spatial data sharing, which mainly relies on the offline copy, can not provide unified map services online to be called by the application systems; ③ the repeated purchases of the same type of basic GIS system in different units and in different project, those are resulted in a lot of money to be wasted. The lack of spatial data resources and the backward mode of sharing can not meet the need for spatial information of the modernization about the Yellow River Basin management and development.

1.2 Service GIS technology development

Service GIS is a GIS technology system based on service-oriented software engineering approach, which regards GIS functionality in accordance with certain specifications as services that are cross-platform, cross-network, and cross-language to be called by a variety of client. Service GIS have service aggregation capabilities to integrate GIS services from other servers.

Service GIS supported by a variety of SOA (Service-Oriented Architecture) standards of practice and spatial information service standards, it can be used in a variety of SOA-based system, seamless heterogeneous integration with other IT business systems, which allows application system developers to rapidly build business applications. Compared to component GIS based component-oriented software engineering approach, Service GIS inherits the technical advantages of the former, but at the same time there is a qualitative leap. Service GIS has gradually become the main GIS technology system and play an increasingly important role in promoting spatial data sharing.

1.3 Water census spatial data resources

River basin management agencies through the first national water census can get the following basin-wide spatial data: thematic spatial data about the protection of rivers and lakes, water conservancy projects, economic and social water supply, river and lake development and control, soil and water conservation, water sector capacity building, irrigation area and wells taking groundwater for irrigation; 1:50,000 basic geographic information data; and 2.5 m high-resolution remote sensing images. The authoritative data is very precious, which is very necessary in-depth development and used of the spatial data to service the modernization of the Yellow River development and management.

2 The overall framework of sharing service application platform

2.1 The platform goal of building

The main purpose of the Yellow River Basin water census spatial data sharing is to meet the needs of users and application systems. Water census spatial data and the thematic spatial data from business sector can be integrated and standardized to be a variety of data sets for sharing service application platform. Based on Service GIS technology, to build sharing service application platform of the Yellow River Basin spatial data with secure and reliable, will meet the needs of spatial data sharing online in a heterogeneous environment.

2.2 The platform task of building

To develop management practices of watershed spatial data sharing use and construction standards and specifications of sharing service platform combining with the existing management approaches and standards at home and abroad; to build a distributed spatial database system by integrating water census spatial data resources and thematic spatial data resources; to design and develop sharing service application platform of the Yellow River Basin spatial data, which provides technical support for the realization of spatial data sharing, and provides unified map services for the business application systems. The main building task can be summarized as follows: ① to develop the specifications of the management practices and standards, ② to establish a spatial database system, and ③ to develop sharing service application platform software system.

2.3 The overall framework of platform

The overall framework of the sharing service application platform of the Yellow River Basin spatial data in Fig. 1 includes support layer, data resource layer, service layer and application layer. The support layer is component of networks, servers, storage and backup hardware and software environment. The data resource layer mainly includes various types of the water census spatial data. The service layer is the core of the platform, which includes spatial information display subsystem, the sharing and exchanging subsystem, and database management subsystem, operation and maintenance management subsystem. Application layer is composed of all kinds of application systems based on the platform spatial data services.

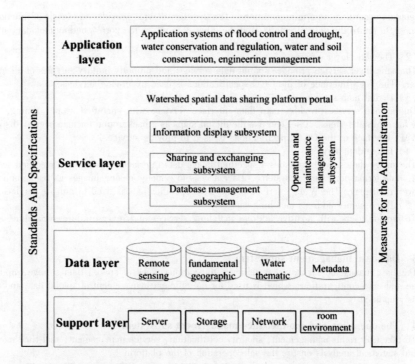

Fig. 1   The overall framework of sharing service application platform of the Yellow River Basin spatial data

2.4   Spatial database design

Spatial database provide functions including spatial data storage, management, and maintenance. The first national water census spatial data of the Yellow River Basin should be stored and managed by the spatial database. Based on unified technical standards, the first national water census spatial data of the Yellow River Basin will be processed into remote sensing image database, foundation geographic database, water thematic spatial database. Business units and sectors may refer to the relevant standards to establish thematic spatial database.

2.5   Platform subsystem design

2.5.1   Information display subsystem

Information display subsystem service adopts the mode of resources catalog to provide complete and detailed spatial information display for users, which includes digital maps, image maps, and other forms of exhibition. Information display subsystem achieves seamless overlay of water census spatial data and thematic spatial data from business units, and also provide the Yellow River Basin spatial data browsing, query, statistics, spatial analysis, maps annotation and correction online.

2.5.2   The sharing and exchanging subsystem

The sharing and exchanging subsystem provides data sharing service interface, which follows the OGC standards, achieves the exchanging and sharing of spatial data services in a heterogeneous environment.

(1) Directory service.

User can get spatial data classification and data description by the directory service, and query the metadata according to certain conditions. The directory service provides two levels of functionality: one is to return the data directory, and the other is to return the corresponding metadata attributes.

(2) Thematic layer service.

Thematic layer service provides spatial data services and property inquiry services of all layers for user. The WFS interface of the OGC specifications is used to provide services.

(3) Digital map service.

Digital map service provides map data interface services with specified scope, timing, data format for external systems. User can use the digital maps as a background for map overlay display. The WMS interface of the OGC specifications is used to provide services.

(4) Image data service.

The image data service provides a specified image data interface services with specified scope, timing, data format for external systems. User access to remote sensing images as background for map overlay display. The WMS interface of the OGC specifications is used to provide services.

(5) Place names address matching service.

To provide users with positive address matching services and reverse address matching services.

2.5.3 Database management subsystem

Database management subsystem achieves the various types of spatial data storage, update, manage and map configuration, which is used for the efficient management of spatial data and continuously updated.

2.5.4 The operation and maintenance management subsystem

To take user rights management, security certification, service management, real-time monitoring and statistical analysis ensure the safe operation of the platform.

2.6 Standards and specifications

Standards and specifications are important parts of the sharing service application platform of the Yellow River Basin spatial data, which provide guidelines for the construction, management, application of spatial data and ensure that the platform is established under unified technical specifications.

The development of spatial data sharing management approach will provide the basic rules of spatial data sharing and an important guarantee for all aspects of spatial data sharing. The development of standards will provide standardization guidance for the construction and application about sharing services platform to ensure the smooth implementation of sharing services.

3 Key technologies of platform

3.1 The OGC and ISO interoperability specifications

In order to achieve openness and interoperability of spatial data, the OGC and ISO develop the implementation of norms. The sharing service application platform of the Yellow River Basin spatial data must support OGC standards: Web Feature Services (WFS), Web Coverage Service (WCS), and Web Map Service (WMS). WMS provides GetMap interface, which will have the various layers with the same spatial reference together, when given the spatial coordinates and boundaries, you can get the corresponding map, and query feature elements on the map to the server. WFS provides the returned GML encoding of the elements and elements added, modified and deleted. WCS for the image data will contain the location of spatial data as "coverage" in the online exchange. In addition to the OGC interoperability specifications, the platform also supports the ISO service specifications: GeoRss can be used for geographic analysis, geographic search and geographic polymerization; and Keyhole Markup Language (KML) can be used describe the surface features. The interoperability specifications from OGC and ISO are mainly supporting technologies of sharing service application platform of the Yellow River Basin spatial data.

3.2 Spatial data engine technology

The spatial data engine provides functions about spatial data storing, reading, indexing and changing. Spatial data engine integrates the GIS spatial geometry object and attribute into large relational database, which is able to efficiently index, append, update, delete and query. The level of functional integrity and performance of spatial data engine will directly affect the level of functional integrity and operational efficiency of GIS applications. Sharing service application platform of the Yellow River Basin spatial data should use mature spatial data engine technology.

3.3 Image pyramid technology

Image compression and efficient image pyramid technique can greatly compress image data, which will provide a good and fast browsing solution of very large image data display speed problem. Remote sensing image data of the Yellow River Basin water census need to be effectively stored and managed by image pyramid technique.

4 Research on platform construction ideas and operating mechanism

4.1 The construction ideas

The sharing service application platform of the Yellow River Basin spatial data is an intensive technology, heavy workload, complexity of the business, and long periodic technical work. The construction ideas that the overall plan for phased construction, typical design, and gradually promotion are recommended.

The overall plan is to focus on information technology development goals of the river basin management agencies to construct the platform based on uniform data standards and unified technical policy. Phased construction is first of all rely on water census spatial data to carry out the initial construction of the platform, to solve the difficulty of the platform, and be accumulated experience in building, and then gradually to construct in the business sectors, and ultimately the formation of a physically distributed, logically centralized service-oriented sharing service application platform of the Yellow River Basin spatial data. Typical design is a comprehensive summary of the experience in building similar projects for the clear purpose of the platform, design specifications, technical conditions, organization and management forms. In accordance with the typical design program, the construction can be gradually extended to the conditional business units and sectors.

4.2 Platform service model and operational mechanism

On the one hand the sharing service application platform of the Yellow River Basin spatial data directly provides authoritative and reliable water census spatial data to be browsed by users, on the other hand it provides a variety of development interfaces to be called by application systems.

The sharing service application platform of the Yellow River Basin spatial data adopts distributed sharing services application mode. Water census spatial data is managed and daily maintained by the river basin management agencies' information center, while thematic spatial data is managed and daily maintained by the business units and sectors. Water census spatial data services are released into platform by the river basin management agencies' information center, at the same time thematic spatial data services from the business units and sectors are aggregated by Service GIS of sharing service application platform of the Yellow River Basin spatial data. Users can get spatial data classification and data description by the directory service. These spatial data sharing services can be used by the data source unit on the application for review. The river basin management agencies' information center will monitor this process. So the sharing service application platform of the Yellow River Basin spatial data operating mechanism is established. (Fig. 2)

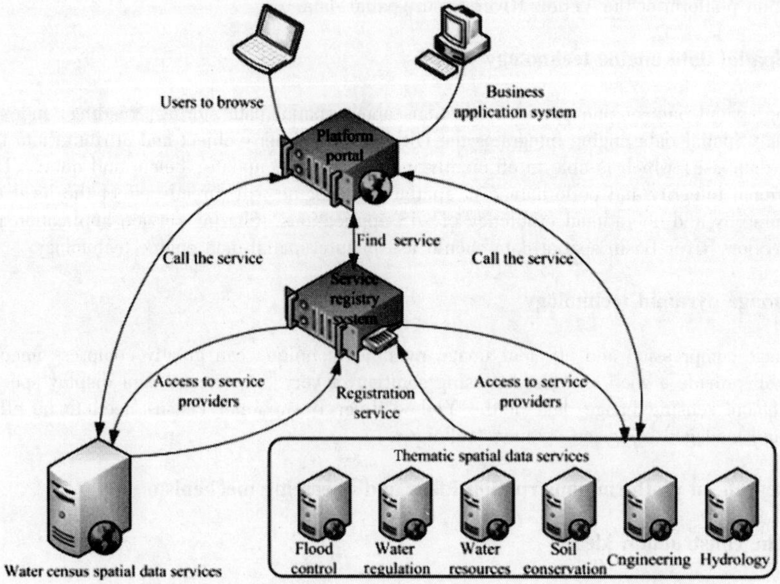

Fig. 2　Platform service model and operational mechanism

5　Conclusions

Based on uniform standards and norms, water census spatial data resources and thematic spatial data resources integration, the river basin management agencies' computer network environment, data storage facilities, and water census hardware and software equipment, the establishment of sharing service application platform of the Yellow River Basin spatial data will gradually provide unified watershed-wide large-scale spatial data sharing services for business application systems and support for decision-making and business management, which will have a huge social and economic benefits.

References

Li Deren, Huang Junhua, Shao Zhenfeng. Sharing Platform Framework for Digital City Service-oriented Design and Realization [J]. Wuhan University of Information Science, 2008, 33(9).

Yu Hu, Ye Zhixuan, Wang Qun. Research on sharing Service Application Platform of Hangzhou Fundamental Spatial Database [J]. Surveying and Mapping, 2010, 35(4): 181 – 183.

Hangzhou Government Information Resources Sharing Research Group. Share Research and Practice of Hangzhou Government Information Resources [J]. Information Systems Engineering, 2009(6) 117 – 120.

Chen Deqing, Ma Hongyi, Chen Zidan. The Digital Map Database of 1:50,000 Water Infrastructure [J]. Water Resources Information, 2010(4): 30 – 32.

The Future of the Internet of Things Technology Application in the Yellow River Management and Development

Ming Hengyi[1] , *Guo Li*[2] and *Du Kegui*[1]

1. Yellow River Jinan Bureau, Jinan, 250032, China
2. Yellow River Shandong Material Reserves Center, Jinan, 250032, China

Abstract: The paper tracks the evolutionary process of the Internet of Things, introduces the Internet of Things application in river management and development at home and a-broad, analyses the progress and problem of the Digital Yellow River, proposes the solutions and ideas that can be applied in the Yellow River management and development. The Internet of Things was considered as the third wave in the information industry after the Computer and the Internet. It is becoming more and more important in river management and development. The Digital Yellow River was proposed based on the Digital Earth and technical background at that time. But today the Information Technology has made great changes. The Digital Earth has evolved into the Smarter Planet. And the Digital Yellow River will be upgraded to the Smarter Yellow River. The Internet of Things has a bright future in Yellow River management and development. All the automated and intelligent basic information collection will depend on the Internet of Things technology. To integrate flood control projects and information technology facilities based on the Internet of Things technology will make it smarter, which will be the way to achieve the modernization of the Yellow River management and development. The Yellow River management and development will be pushed to a new level by the Internet of Things technology. The smarter Yellow River will keep healthy life.

Key words: the Internet of Things technology, the Digital Yellow River, the Smarter Yellow River

1 The concept of the Internet of Things and its evolutionary process

The concept of the Internet of Things was first used by Kevin Ashton in 1999. He proposed a great idea of creating the Internet of Things to achieve global goods information sharing in real time by RFID based Internet.

In 2005, the International Telecommunication Union released the "ITU Internet Report 2005: Internet of Things". The definition of the Internet of Things has already changed in this report. It was no longer only the Internet of Things based on RFID technology. This report said the active data exchange between things and things via the Internet are no longer out of reach. All objects in the world from tires to toothbrushes, from housing to the tissue can exchange information via the Internet actively. Radio Frequency Identification (RFID), sensor technology, nanotechnology, intelligent embedded technology will be more widely used. In 2009, IBM Corporation proposed the concept of the Smarter Planet. IBM raised the idea, the task of IT industry for the next phase was to apply the next – generation IT technology into every field of industry, embedded sensors into the power grid, railways, bridges, tunnels, highways, buildings, water systems, dams, oil and gas pipelines and other objects, connected each other to form the so – called "Internet of Things". Then the Internet of Things integrates with the existing Internet to achieve the integration of social and physical world. On this basis, human beings can manage production and life more refined and dynamic, so as to achieve "smart" state.

2 The Internet of Things application in river management and development

The Internet of Things was considered as the third wave in the information industry after the Computer and the Internet. The Internet of Things was listed into the national development strategy in USA, European, Japan and Korea. "Twelve Five" plan of the Internet of Things has been launched in my country. Its application is becoming more and more widespread in many industries. It is becoming more and more important in river management and development.

2.1　The IBM Smart Water Management

The Hudson River is a 520 km watercourse that flows from north to south through eastern New York State in the United States. It is blessed with a rich natural, historical and cultural heritage. After World War II, some big companies, for example General Electric, General Motors, IBM, etc set up many factories along the Hudson River. Due to industrial development, the river had been polluted. IBM established a Centre of excellence water management in Hudson River with New York's Beacon Institute for New York State's Hudson River protection plan. It is the first monitoring/forecasting network design for river estuary in the world. To understand the river will help state government to formulate a long term plan for ecological protection and economies development model in this area. IBM used a large number of the Internet of Things technology to achieve real – time river environmental monitoring in this project. IBM mounted a variety of sensors on the buoys, then scattered them throughout the river. They formed a distributed sensor network via the wireless network for transferring various indicators of the river includes chemical, physical, biological. These indicators are computed by some mathematical models to form a visual river for policymakers and researchers. It will be better for environmental protection policymakers, regional economic planners, and aqua – organism managers to manage the river based the visual model.

2.2　Blue – green algae prevention in Taihu Lake

Wuxi is a water city with dense network of rivers and lakes in south of the Changjiang River. Water area is 30% of city area. Water is highly eutrophic due to rapid industrial development in recent years. The large bloom of blue – green algae in Taihu Lake has seriously affected production, living of residents and Wuxi city's image. From 2009, Wuxi city used the Internet of Things technology to monitor the water quality in Taihu Lake. It replaced the old method that water sample was analyzed in laboratory after manual getting. They scattered sensor buoys to monitor the 24 h water changes in the lake. The monitor device mounted on buoys can rapidly detect seven water – quality data of the PH, dissolved oxygen, turbidity and blue – green algae etc, and then they were transferred to water – quality monitor application platform. Wuxi set 21 inspection points, 13 distributed blue – green algae video – monitor system along Taihu Lake, a water environment emergency monitor ship and an emergency monitor car. Now they are studying the method that they can use flying robots to monitor blue – green algae, use remote sensing satellites to enhance monitor. Wuxi had built 86 water – quality monitor stations. A system of blue – green algae alarm, salvage, and resource utilization was formed, which limited the large – scale bloom of blue – green algae more effectively.

3　The progress and problem of the Digital Yellow River Project

The information infrastructure, application services platform and application system have been completed initially after ten years building from YRCC implemented the Digital Yellow River Project. It supported the primary work of the Yellow River management and development effectively, promoted the informatization and modernization level of the Yellow River management and development.

YRCC had completed the infrastructure including telecommunication, data transport, computer network, data collection, data storage and management and high performance computing platform etc., and built the distributed application service platform consists of several nodes implemented at the YRCC Information Center, the Hydrology Bureau and the Water Resources Protection Bureau etc. on this basis. The application system principally contains ones such as flood control and disaster mitigation, management and allocation of water resources, protection of water resources, ecosystem environment monitoring for soil conservancy, construction and management of projects, E – Government, etc. It was built with using in build time and urgent project first strategy. It had played an important role in the Yellow River management and development.

The information Technology has developed rapidly during the building process of the Digital Yellow River. Some technology used in the Digital Yellow River project has already lagged behind the times. In the Digital Yellow River project, some information collection methods are semi – automatic and manual as limited technology condition at that time. These methods are not only inefficient, but also can make error easily. So it can effect the environmental simulation and decision. It can't achieve high frequency and real – time information collection under all weather conditions.

Before the Internet of Things and Smart Earth theory was proposed, the physical infrastructure and IT infrastructure are two different things. One is the airport way and buildings, and the other is the data center PC and network. But in the Internet of Things times, reinforced concrete, cable, chip and network will be integrated into a unified infrastructure. In the sense, the infrastructure is like a construction site, the world with economic management production social management and personal life is running on it. The concept of the Internet of Things has completely subverted traditional thinking. Due to the traditional thinking influence, the Digital Yellow River was positioned as non – engineering measures, which affects the effect of the IT technology applied in the Yellow River management and development. Flood control projects and information infrastructure are always designed and constructed alone, they are considered as two absolutely different facilities, so it is difficult to have a good effect like $1 + 1 > 2$ to achieve the long term goal of the Digital Yellow River.

4 The future of the Internet of Things technology application in the Yellow River management and development

The Information Technology has made great changes today. The Digital Earth has evolved into the Smarter Planet. The Digital Yellow River will upgrade to the Smarter Yellow River. The Internet of Things has a bright future in Yellow River management and development. All the automated and intelligent basic information collection will depend on the Internet of Things technology. If we integrate flood control projects and information technology facilities based the Internet of Things technology, make it smarter. It will be the way to achieve the modernization of the Yellow River management and development.

4.1 The automated and intelligent basic information collection will depend on the Internet of Things technology

Semi – automatic and manual information collection method has some obvious drawbacks. The Internet of Things technology can be used to solve some problem that information can't be collected automatically in the past. We can use RFID to identify materials, staff and projects, make it connected by the Internet of Things. The information of them will be collected automatically, and we can trace their position and transfer process. We can solve the problems that make mistakes and corrupt behaves easily by manual input and report level by level. It can achieve the informationization and automation of assets management, personnel administration and project management. The information of hydrogeology, water quality, rainfall, and soil moisture can be collected automatically by a variety of sensors with buoy and buried method. Because the Yellow River is a heavily sediment – laden river, many sensors can't work in it. Cooperative development can solve some problems, and make sensors working in the Yellow River environment.

4.2 Integrating flood control projects and information technology facilities based the Internet of Things technology, making it smarter

Dam, critical levee section etc. flood control projects and information infrastructure are always designed and constructed alone. All kinds of informationization methods can't play a role. If we integrate flood control projects and information technology facilities based the Internet of Things technology, make it smarter. It can solve many difficult problems in the past easily. First we can integrate a variety of sensors into dam, critical levee section, make it as an entity. Second we can use the middle and lower reaches of the Yellow River optical cable in plan to transfer all kinds of basic data that sensors collect. Next we can extend the existing application platform to process the data. Finally we can achieve the smart monitor of flood control project.

Now we deal with dangerous situation of flood control project by the way of manual inspection and then reporting it level by level. Using these methods not only consume a lot of time, money, manpower and material resources but also make mistake and omission easily. If we bury strain sensor into flood control project, identify the parts of project with RFID and set sensor deformation threshold level, we can monitor the positions and transformation of project dynamically. All kinds of project inspection will complete automatically and intelligently. For example, we can implement the root – stone detection with HF RFID. After embedding the RFID into every root – stone, the detector in critical levee section will trace the position of every root – stone. This method can be completed automatically, it is better than manual work and ultrasound.

Under the present condition, we reported emergency rescue plan of flood control project and material usage level by level. If we use the Internet of Things technology, the material usage can be computed automatically, lost root – stone can be traced, and the collapsed parts of project can be computed automatically. So we can save a lot of rescue time, limit the damages, and overcome all kinds of disadvantage by manual work.

5 Conclusions

The Information Technology has made great changes today. The Digital Earth has evolved into the Smarter Planet. The Digital Yellow River will upgrade to the Smarter Yellow River. The Internet of Things has a bright future in Yellow River management and development. All the automated and intelligent basic information collection will depend on the Internet of Things technology. Integrating flood control projects and information technology facilities based the Internet of Things technology, making it smarter. It will be the way to achieve the modernization of the Yellow River management and development. The Yellow River management and development will be pushed to a new level by the Internet of Things technology. The smarter Yellow River will keep more healthy life.

References

Li Guoying. Digitalization of the Yellow River [J]. Yellow River, 2001, 23(11):1 – 4.

Li Jingzong, Kou Huaizhong. Achievement and Future Development of the "Digital Yellow River" Project [J]. Yellow River, 2011, 33(11):1 – 6.

International Telecommunication Union. Internet Reports 2005: The Internet of things[R]. Ceneva: ITU, 2005.

China Environment Network. Environment IOT technology Taihu Lake case [EB]. http://www.cenews.com.cn/ztbd1/hjwlwjsthal/.

IBM Smarter Planet [EB]. http://www.ibm.com/smarterplanet.

Study on the Index System of Water-Draw and Utilization Assessment of a Water Supply Project Planning in the Yellow River Basin

Wei Hao, *Sun Zhaodong*, *Sun Xiaoyi* and *Shi Ruilan*

Yellow River Water Resources Protection Institute, Zhengzhou, 450004, China

Abstract: Lacking water and frequent occurrence of drought in the central of Gansu and Ningxia Province leads to the low living standards of farmers in the area. On the contrary, there are well lands and sufficient sunlight. In order to improve the farmers' living level, to promote the sustainable development of society and economy and to realize the regional rational development and effective protection of water resources, a water supply project, to continuously develop land resources and expand the irrigation area, is planned on the basis of irrigation water saving in their existing irrigation districts. Absorbing experts' suggestions and considering the index usability. The index system is of multi-level, gives the meanings of indexes, puts forward the evaluation method, and discusses the threshold values for the indexes. In the index system, six aspects such as water drawing, water supply, water use, water saving, water return and influence on society are involved and several indices. This index system would be applied to evaluate the planning area water resources conditions, development and utilization, carrying capacity, etc. and to screen the planning water use schemes. The evaluation results will provide basis for recommending reasonable planning water use scheme, improving water resources management level, increasing water resources utilization efficiency, and realizing optimization allocation of regional water resources.

Key words: Yellow River, water supply project, planning, water-draw and utilization assessment, index system

The shortage of water resources has become one of the main factors restricting China's sustainable economic and social development, especially under the influence of global climate change and large-scale development and utilization of water resources, water resources shortage in our country is becoming more and more serious. At present, water resources in our country is per capita 2,220 m^3, accounting for a quarter of international level, and the spatiotemporal distributing of water resource is extremely inhomogeneous. Precipitation is concentrated, and there are large differences of annual runoff in different regions. Imbalance of water resources natural conditions cause fragile ecological environment in northern region of China, the severe contradiction between water resources supply and demand, and frequent droughts and floods.

In recent years, drought continues in China's northern regions and some areas are of serious water shortage. Deterioration of ecological environment and water pollution further exacerbated the shortage of water resources with serious impact on sustainable social and economic development. The core mission of planning water-draw and utilization assessment is to promote the coordination between economic and social development and water resources carrying capacity at the macro level. By the establishment of the scientific, perfect, and reasonable index system, water resources situation in planning regions will be scientific reappraised, and near future and long-term trends of water demand will be predicted. On the basis of water resources conservation and protection, optimize water resources allocation and make overall arrangement of water resources optimized allocation, effective protection, rational development, efficiently utilization, and scientific management.

Planning water-draw and utilization assessment is a systematization and comprehensiveness job and is still at the exploratory stage. There has not yet been a perfect index system applied to the engineering planning water-draw and utilization assessment. Therefore, this paper aims to establish a set of reasonable design and operational evaluation index system.

1 Overview of project planning

Lacking water and frequent occurrence of drought in the central of Gansu and Ningxia Province leads to the low living standards of farmers in the area. On the contrary, there are well lands and sufficient sunlight. In order to improve the farmers' living level, to promote the sustainable development of society and economy and to realize the regional rational development and effective protection of water resources, a water supply project, to continuously develop land resources and expand the irrigation area, is planned on the basis of irrigation water saving in their existing irrigation districts.

According to the requirement of a relevant document issued by the MWR of China in 2010, it should be done to carry out a water-draw and utilization assessment for the planning of the water supply project. Planning water-draw and utilization assessment technical requirements (for Trial Implementation) and the upper and lower levels planning, it is necessary to carry out the project planning water-draw and utilization assessment. In order to make sure that planning water-draw and utilization assessment will go on smoothly, it is needed to establish a set of index system to evaluate conditions, utilization and carrying capacity of water resources in planning area scientifically and reasonably. So it will provide reference for recommending rational planning water supply scheme, promoting water resources management level, improving the regional water resource utilization efficiency, and realizing the optimal allocation of regional water resources.

2 Methods

The establishment of index system is mainly using literature analysis and expert consultation method. According to domestic relevant professional planning, laws, rules and management system, extract indexes and establish planning water-draw and utilization assessment index system. On the basis of the actual situation of this project, screen the indexes and set up planning water-draw and utilization assessment index system suitable for the water supply project planning.

3 Establishment of index system

The premise of index system establishment is related water resources management system, and the core is water-saving society construction, and the base is the consistency of economic structure system with regional water resources carrying capacity, and the goal is optimal allocation of water resources. By comprehensive analysis and evaluation, index system is established systematically, comprehensively, and scientifically.

At present, the multi index comprehensive evaluation become the important method of comprehensively knowing about things, and the research scope related to society, economy, technology and other aspects. Similarly, to comprehensively evaluate a regional planning, the indexes will also involve many aspects, and it is not enough to analyze and evaluate a regional planning issue using one or several indexes. With this premise and background, we put forward the need to establish a planning water-draw and utilization assessment index system for comprehensive evaluation.

In multiple indexes comprehensive evaluation, index system construction is a key problem, which is the premise of scientific evaluation. Analysis and evaluation of a regional planning, in addition to the qualitative description and analysis, it is more important to qualitatively describe and quantitatively analyzes. The so-called quantitative analysis is to find or to build a metric ruler with which to measure an area planning, and the conformity, consistency and coordination of planning will be answered. Consideration of many factors and guided with the related policies and regulations, the evaluation index system involves water resources development, utilization, conservation, protection and so on.

3.1 Establishment principle of index system

In this paper the establishment of planning water-draw and utilization assessment index system mainly guided by the following principles:

(1)Select indexes comprehensively. Evaluating index system is guided with the scientific theory. Seize the essence of evaluation object, which is targeted. The selected indexes should cover all aspects of water-draw and utilization assessment as fully as possible.

(2)Reflect the levels. The selected indexes should reflect the overall situation and each classification and each individual case.

(3)Relative independence of indexes. The selected each index will reflect one side situation, and each index should be independent of other index.

(4)Combination of qualitative indexes and quantitative indexes. The selected indexes are not only qualitatively described but also quantitative described.

(5)Combination of comprehensive indexes and single indexes. The selected index system should reflect both comprehensive and individual situation.

3.2 Indexes selection

Based on methods and framework of domestic universal index system, selection of planning water-draw and utilization assessment indexes is completed by following steps .

(1)Determine the planning water-draw and utilization assessment development goals. On the basis of extensive reference to the development target and strategy put forward by domestic relevant agencies in planning water-draw and utilization assessment and domestic institutions and scholars' research results, sum up and determine the connotation and development target of planning water-draw and utilization assessment.

(2)Determine the indexes classification framework. According to the plan water-draw and utilization assessment target and domestic general evaluation index system classification framework, determine the classification framework of index system through expert consultation.

(3)Determine the indexes selection standard. Based on the planning water-draw and utilization assessment development requirements and domestic the authority of the index system of selecting standard, put forward the indexes screening criteria of planning water-draw and utilization assessment index system in light of China's actual situation of water resources development.

(4)Determine the index library. Taking indexes classification framework as a guide, determine primary index library of the index system with comprehensive selection through consulting a great deal of literature.

(5)Screen indexes. According to the indexes selection standard and relevant literature, determines the final index system by comprehensively consulting experts and case investigation method.

3.3 Planning water-draw and utilization assessment index system

Tab. 1 Planning water-draw and utilization assessment index system

Attribute layer	Factors layer	Index meaning
Water drawing	total amount control of water consumption	According to a basin or regional water resources and water environment capacity and the regional water environmental quality index and standard, control the total water amount in the natural environment carrying capacity range
	Quota management	A management method of using the quota to rationalize the water use
	Effect on water resources allocation	In water resources planning and management process, fully coordinate the relationship among water and society, economy, ecology, environment and other elements, improve the matching degree of the water resource with the others, realize the reasonable utilization of water resources, and promote sustainable economic and social development
	Hydrological effect	Calculate influence on river flow with different design flow of project water
	Effect on river pollutant capacity	On the basis of water quality meet the water function area requirements, water could accommodate the maximum number of pollutants
	Effect on other water user	Analysis on effect on water user and compensation in the area implementing water saving transformation

Continued Tab. 1

Attribute layer	Factors layer	Index meaning
Water supply	Water supply pipe leakage rate	In the process of water supply, the inevitable loss in pipeline caused by itself, certain along and partial loss and the pipeline aging brings other losses accounted for all the water supply volume ratio
	Water loss rate	Percentage of difference between total water quantity and charge quantity of water plant accounts for the total water quantity
	Owned water supply metering rate	Percentage of water supply metric owned by all enterprises and institutions accounts for total water supply
	Other water sources of percentage	Percentage of brackish water, rain water, reclaimed water and other water sources amount converted into alternating conventional water resources accounts for total
Water use	The various sectors of water use ratio	Town life, rural people and livestock, agriculture, industry, construction industry and the three industry, the ecological environment water demand, mainly used for the analysis of different industry water use and water level
	Ten thousand yuan GDP water	Water use per ten thousand yuan GDP, reflecting the macro water level indicator of city with certain economic strength. Water use per ten thousand yuan GDP is essential for water resources quota index, is a measure of a country or region, water efficiency, water saving potential of water resources carrying capacity and the important index of sustainable economic and social development.
	Water use added Ten thousand yuan of industrial value quantity	Evaluation quantity of water use added ten thousand yuan value, statistics industrial total water use, all industrial added value, calculated water added ten thousand yuan
	The main industry product water use	Main industry product actual water consumption quota, depending on the region of determining the actual high water industry and its main product
	Comprehensive water use per capita	Planning of regional water use average total population
	Urban residents living water	A daily average water consumption of urban residents
	Rural residents living water	Refers to the rural residents per capita average daily consumption of water
	Large livestock water	A large livestock per head average daily consumption of water
	Small livestock water	A small livestock each average daily consumption of water
	Poultry water	A poultry each average daily consumption of water
	Irrigation quota	A unit of irrigation area on an irrigation water
	Irrigation quota (net irrigation quota, gross irrigation quota)	Sum of each irrigation quota in crop before sowing, inserted before and during growth periods
	Farmland irrigation water use per mu	Actual agricultural irrigation water use per mu on average
	Field water utilization coefficient	A storage in the field of primary root layer of water and poured into the field of the ratio of water
	Utilization coefficient of irrigation water	Crop growth needs of water for irrigation water ratio
	Water efficiency of canal system (stem, branch, bucket, agricultural drainage)	The canal water and main canal headwork water diversion volume ratio
	Water productivity	Concentrated reflection of water use efficiency of crops in a comprehensive index
	Plan water rate	Percentage of the actual water consumption included in the plans accounts for the total quantity of water
	Main crop water use	Area (normal year) each kind of main crop actual quantity of irrigation water per mu, according to the actual situation determine the main crop for evaluation

Continued Tab. 1

Attribute layer	Factors layer	Index meaning
Water saving	Management system and management mechanism	Integrated management of water-related, county and government above the county level have water-saving management agencies, county government and below have the people responsible for unit, enterprise management
	laws and regulations	Water rights allocation, transfer and management system; license system for water drawing and pay for water natural resources use system; water-draw and utilization assessment system; pollutants discharge permit system and polluter-pays system; water-saving product certification and market admittance system; water measurement and statistical system; systematic water resources management regulations and rules, especially the water plan, save water laws and regulations
	Water saving society construction	County level and above county level governments' planning for developing water saving society construction
	the water price mechanism for promoting water saving and pollution prevention	Establishment the water price mechanism fully reflecting the shortage of water resources and serious water pollution situation to promote water saving and pollution prevention
	Water saving investment security	The government should support the construction of water-saving society in a stable investment, broaden the financing channels, and actively encourage the private capital investment
	Water saving publicity and public participation	Take water resource saving and protection into education and training system, carry out publicity in a variety of forms; win support for water saving consciousness among the people to form the water saving consciousness in the whole society
	Water saving irrigation ratio	Percentage of implemented water-saving irrigation area for water-saving potential evaluation
	Water saving irrigation project area	Percentage of water saving irrigation project area account for the effective irrigation area
	Coverage of water-saving devices (including public domestic water)	Ratio of The tertiary industry and residential water use water-saving appliances and total water appliances
	The residents living water door device rate	Percentage of household water meter
	Channel lining rate	Channel lining intact rate
	Percentage of other water sources	Percentage of brackish water, rain water, reclaimed water and other water sources amount converted into alternating conventional water resources accounts for total
	Channel efficiency (stem, branch, bucket, agricultural drainage)	The effective water utilization degree of without splitting channel in the conveying process
	Repeated use rate of industrial water	Percentage of industrial water recycling accounted for the total industrial water
	The main products of large industrial water quota (power, coal)	Water consumption of unit production, mainly used to analyze the water consumption level of regional high consumption aquaculture
	Living water recycling rate	Percentage of reuse living water
	Sewage treatment and reuse rate	Percentage of reuse sewage treatment quantity

Attribute layer	Factors layer	Index meaning
Water return	Annual output of industrial waste water	Wastewater generated by all industries in the particular year
	Annual output of domestic sewage	Sewage quantity generated by all residents in the particular year
	Industrial wastewater up to discharge standards rate	Percentage of industrial wastewater up to emission standard accounted for the total volume of industrial waste water discharged
	Sewage standard up to discharge standards rate	Percentage of domestic sewage discharge up to emission standard
	The influence on surface water	Industrial wastewater, domestic sewage, irrigation return water and other pollutants discharge into rivers, lakes, reservoirs and other forms of surface water, which result in the decline of water quality
	On the influence on groundwater	Industrial wastewater, domestic sewage, irrigation return water and other contaminants are into the ground, which result in the deterioration of groundwater quality.
Influence on society	Political influence	Analyze on the problems of poverty population and regional population drinking water solved after the implementation of the project
	Policy toward nationalities	Relevant measures and regulations for the regulation of ethnic relations and ethnic problem
	Poverty alleviation policy	Relevant policies to help poor households or poor regions for developing production and changing poverty

As the index system established principles and steps mentioned above, the index system has been established on the basis of existing research results and planning water-draw and utilization assessment target and task. In the index system, six aspects such as water drawing, water supply, water use, water saving, water return and influence on society are involved. There are 55 indexes such as total amount control, water supply pipe leakage rate, the various sectors of water proportion, management system and management mechanism, with an annual output of industrial waste water, political influence and so on. Each index threshold value is determined with local water quota standard and the surrounding area and countrywide level in order to provide basis for comprehensive analysis of regional planning and water level. Primary index system is shown in Tab. 1 with each index meaning.

3.4 Planning water-draw and utilization assessment index screening

In the primary index system, science, comparability, decision-making relevance, availability, conciseness, universality, sensitivity characteristics and other characteristics of each index are analyzed separately. Based on the index screening principles and the project characteristics and taking existing relevant research results for reference, indexes in the primary index system established above have been screened by absorbing experts' suggestions and field research. Screened index system involves the same six layers as the primary index system but the total number of indexes is 41, such as total amount control, water supply pipe leakage rate, comprehensive water use per capita, management system and management mechanism, sewage standard discharge rate, political influence, etc.. Screened index system is shown in Tab. 2.

Tab. 2 Index System of Water-Draw and Utilization Assessment of a Water Supply Project Planning in the Yellow River Basin

Attribute layer	Factors layer
water drawing	total amount control of water consumption
	Quota management
	Effect on water resources allocation
	Hydrological effect
	Effect on river pollutant capacity
	Effect on other water user
water supply	Water supply pipe leakage rate
	Water loss rate
	Owned water supply metering rate
	Percentage of other water sources
water use	Comprehensive water use per capita
	Urban residents living water
	Rural residents living water
	Large livestock water
	Small livestock water
	Poultry water
	Irrigation quota
	Irrigation quota (net irrigation quota, gross irrigation quota)
	Farmland irrigation water use per mu
	Field water utilization coefficient
	Utilization coefficient of irrigation water
	Water efficiency of canal system (stem, branch, bucket, agricultural drainage)
	Water productivity
	Main crop water use
water saving	Water saving publicity and public participation
	Water saving irrigation ratio
	Water saving irrigation area
	Coverage of water-saving devices (including public domestic water)
	The residents living water door device rate
	Channel lining rate
	Percentage of other water sources
	Channel efficiency (stem, branch, bucket, agricultural drainage)
	Water saving publicity and public participation
	Water saving irrigation ratio
	Living water recycling rate
water return	Sewage standard met discharge rate
	The influence on surface water
	On the influence on groundwater
influence on society	Political influence
	Policy toward nationalities
	Poverty alleviation policy

4　Conclusions

　　Comprehensively considering several factors and guided with the relevant policies and regulations, a reasonable design and operational evaluation index system has been established. In this index system, the water resources development, utilization, conservation and protection are involved. The index system is of multi-level, gives the meanings of indexes, puts forward the evaluation method, and discusses the threshold values for the indexes. In the index system, six aspects such as water drawing, water supply, water use, water saving, water return and influence on society are involved and several indices. This index system would be applied to evaluate the planning area water resources conditions, development and utilization, carrying capacity, etc. and to screen the planning water use schemes. The evaluation results will provide basis for recommending reasonable planning water use scheme, improving water resources management level, increasing water resources utilization efficiency, and realizing optimization allocation of regional water resources.

　　Being still in the pilot phase, there has not yet been a comprehensive evaluation index system for planning water-draw and utilization assessment, the water-draw and utilization assessment index system for planning water-draw and utilization assessment and the planning of supply project in the central of Gansu and Ningxia province put forward in this paper is expected to be further improved and perfected in the future.

References

Li Guibao, Zhang Wenli. The Introduction of Evaluation Index System of Water-saving Society Construction[J]. China Standardization,2007(6):6-8.

Operation Management and Maintenance of Remote and Distributed Control System of Sluices in the Lower Reaches of the Yellow River

Shi Jing, *Song Tao* and *Han Song*

Information Center of Yellow River Conservancy Commission , Zhengzhou, 450004, China

Abstract: Remote and distributed control system of sluices in the Yellow River has already been built and put into operation, which brought significant economic and social benefits. However, the operation and maintenance of the system faces some problems and difficulties. Basing on the problems occurred in daily operation, the author expounds the importance of system maintenance and management. According to the present status of the sluice monitoring system, this paper discusses respectively from the management specification of sluices monitoring system, the talent team construction of the system operation and maintenance to the maintenance of system hardware and software maintenance. The article analyzes the problem of the sluices monitoring system and corresponding solutions are put forward. In order to ensure the system in good running, the system maintenance ideas were summarized. All of these are of great value to improve water gate automatic operation management level and good condition rate of equipment.

Key words: sluices of the Yellow River, monitoring system, operation management, maintenance

1 Introduction

The Yellow River Basin is short of water resources. With the rapid development of social economy, the demand of water resources in the Yellow River is increasing. The question of water – resource supply – demand is becoming more and more serious. In particular, because of the backward of downstream sluices monitoring methods, illegal diversion, unauthorized diversion occurred from time to time. Because of the lack of effective control method for the sluices of the Yellow River, it is difficult to ensure the river continue to flow in case of emergency. In 2001, the Yellow River Conservancy Commission (YRCC) launched the "Digital Yellow River" project, making out comprehensive planning and putting up the construction of informatization. We built remote and distributed control system of the Yellow River for sluices by using modern advanced computer technology, automatic control technology, sensor technology and digital and video transmission technology. The system is made up of five grades: Yellow River Water Regulation Center, Henan and Shandong province Bureau, 13 city bureau, 32 county bureau, 84 gate. The system diagram was shown in Fig. 1.

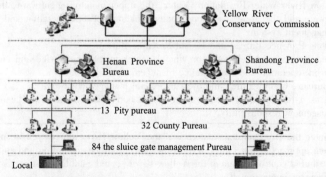

Fig. 1   The diagram of Remote and Distributed Control System of Sluices in the Yellow River

The construction and operation of the monitoring system realized the timely information collection of the river diversion, to ensure the continuous flow of the lower Yellow River, maintain the base flow of the river life, and provide modern remote monitoring method.

2 The operation and maintenance system and method of the Sluices Monitoring System

After monitoring system had put into operation, research and development work was announced finished. And there are a large number of tedious works in the following phase of monitoring system operation, maintenance and management. In order to extend its life cycle and make the investment benefit play to the better level, the monitoring system should constantly updated and improved. Based on nearly 30 years of statistical data at home and abroad, no matter the system maintenance costs or the time spend account for more than 70% of the entire lifetime of the workload (see Tab. 1).

Tab. 1 The percentage of costs of development and maintenance in each period accounted for the entire lifetime cost of the sluices monitoring system

Decade	Development cost	Maintenance cost
1980 ~ 1990	60% ~ 70%	30% ~ 40%
1990 ~ 2000	40% ~ 60%	40% ~ 60%
2000 ~	30%	70%

Above table shows the importance of system maintenance clearly. So the study of sluice monitoring system maintenance of the basic law, summarize the basic method of monitoring system maintenance, have realistic significance on monitoring system maintaining practice and practical guide providing. To do a good job on the sluice monitoring system operation and maintenance work, guarantee the safe operation of monitoring system, we should consider from the following aspects.

2. 1 System construction and standardized management of the monitoring system

Monitoring system's reliability, safety and stable operation, not only requires reliable hardware and software facilities and sophisticated technology, but also need a set of more important effective management system .

(1) Gradually straighten out the system operation management system and mechanism, and establish two main responsibility body for system operation and maintenance management system in the Commission level and under the Commission level. The tasks of each unit are shown in Tab. 2.

Tab. 2 Operation and maintenance management system

- ● Information Center of the Yellow River Conservancy Commission
  - ☞Yellow River Water Regulation Center, the upper monitoring software, the monitoring software and the upper backbone communication network operation and maintenance management system
- ● Shandong Province Bureau and Henan Province Bureau
  - ☞Local stations and below the provincial center monitoring device and communication network operation and maintenance management
- ● City Bureau, County Burea and the sluice Gate management Burea
  - ☞Not as a responsibility for maintaining the body, only to assume the operation and management responsibility

(2) Guarantee the strict implementation of "the Yellow River sluices remote monitoring systems management approach "and "the Interim Provision for sluices operation and application of the remote and distributed control system of the Yellow River", and establish a supervisory system and make the management implemented.

(3) The Yellow River Conservancy Commission (YRCC) and subordinate departments will take the annual target assessment methods and implement system of rewards and punishment to improve water scheduling management capabilities.

(4) To set down a Quality Supervision Regulation of sluice gate operation maintenance costs of budget management and operation maintenance to strengthen the implementation of funding. Using and maintenance of the quality supervision and inspection further strengthen the operation and maintenance funding for the unified management.

(5) To establish the file system of maintenance. The personnel movement is serious in the department of research and development. To keep ready for technical documentation in the store development process will provide important technical support for the staffs who will take over the work later.

(6) To establish the cooperation operating mechanism of the sluice water diversion monitoring between hydrology bureau, Shandong Bureau and Henan Bureau, and carry out regular inspections of monitoring and flow rate calibration, which will give effective protection of the monitoring accuracy.

2.2 The construction of operation and maintenance personnel of monitoring system

To ensure the monitoring system features to play with the normal operation, it is necessary to build a well organized, skilled and business master personnel monitoring system team. Each unit should be combined with the actual situation and needs, to formulate feasible training plan. The operators and management personnel at all levels especially the sluice gate by operators should master the operation and maintenance of the system on basic theory and skill. They can skilled use of the system, and to analyze and get rid of the general failure. Professional maintenance personnel for the system should deeply understand the system of software and hardware technology, timely clearing system faults. We should organize special training and exchanges from time to time for the changing of the guard personnel. Strengthening the system of certificates, provide long – term protection of monitoring system for sluice gate management development.

2.3 System maintenance and management

When the monitoring system failure or the local problems occurred, we should maintenance and improve immediately. The monitoring system maintenance is composed by system hardware maintenance of equipment and system software maintenance.

2.3.1 Monitoring system hardware maintenance

The normal operation of the hardware is the premise of ensuring the good operation of the monitoring system. Hardware equipment covers all the physical equipments of the monitoring system. Hardware maintenance is composed by daily repair and fault treatment. Classification of the hardware equipment in the monitoring system, maintenance personnel and maintenance methods are shown in Tab. 3.

2.3.2 Maintenance classification of the monitoring system software

The reasons for the software product's maintenance can be divided into four types:

(1) Corrective maintenance.

We find out the system errors in the early operation of the Monitoring system, including the problems of diagnosis and correction of errors. Generally accounts for 20% of the total maintenance.

(2) Adaptive maintenance.

With the continued use of the monitoring system, the computer hardware configuration, software configuration replace, or external data environment changes, we also need to modify the software program to adapt to a change. For example, the calculation flow equation replacement of 16 Sluices, such as Pan Zhuang, improved the measurement accuracy of the Yellow River water. Generally the workload of adaptive maintenance accounts for 25% of the total maintenance.

Tab. 3　The summary table of monitoring system hardware maintenance

The monitoring system hardware			Maintenance staff	Method
The local control system of intake sluices	Gate room device	PLC control cabinet, hoist, the video encoder, Switch, Gate level meter, Load sensor, Limit switch etc.	The gate of operation and management staff	Daily repair: ● Be institutionalized ● Notice of hoist room power supply is normal or not, changes in temperature and humidity. Whether the control room equipment is operated or not ● Gate level meter, the water level of gate and other outdoor equipment safety ● On equipment for routine examination, when the system take place non－normal conditions, the person should report to departments immediately
	Control room equipment	Sluice monitoring machine, UPS, power supply, Broadband wireless access equipment etc		
	Outdoor equipment	The water level of gate after and gate before, Video Camera of after the gate and before the gate		
Superior monitoring system		Sluice monitoring machine, GR server, domain server, web server, the total database server, cities database server and video server	Routine inspection/ Equipment suppliers	Fault treatment: ● Equipped with a certain amount of standby devices, such as switches, video decoder, when the machine fails, we first replace the failure to restore emergency equipment then repair the faulty equipment ● When the machine fails, we first use a storage medium on the database and monitoring system software and other information backup, and a certain number of standby servers is need ● Checked and adjusted on a regular basis to the hardware monitor system, we should take protective measures in advance to minimize the incidence of failure

(3) Perfect maintenance.

During the use of the software process, the users tend to put forward new functionality and performance requirements to software. We need to modify or develop software to expand the software capabilities, enhancing the software performance. For example, New function of the artificial input of measured stream flow. Generally the workload of perfect maintenance accounts for 50% of the total maintenance.

(4) Preventive maintenance.

This maintenance is to improve the software maintainability, reliability, etc., and for the future lay a good foundation for further improvement of the software. Generally the workload of preventive maintenance accounts for 5% of the total maintenance. As shown in Fig. 2.

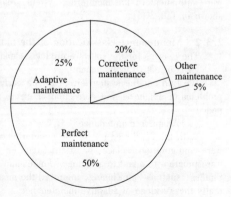

Fig. 2　Maintenance ratio

3   The system problems and solutions

3.1   Daily inspection of sluices' operating situation

Every thursday, operation and maintenance team of the Information Center of the Yellow River Conservancy Commission will send full – time staff to inspect on sluices of the Yellow River at the Yellow River Water Regulation Center. When there are something wrong with the sluices, the problems will be reported to the superiors, and after examination and approval by the higher authorities the detail will notify the maintenance staffs.

3.2   Application system maintenance

Monitoring System is composed by the operating system, monitoring system software, database software and application software, etc. The Information Center of Yellow River Conservancy Commission was appointed to undertake the software maintenance. The system is set remote maintenance software "pcanywhere", remote maintenance staff do not have to go to the scene to modify the parameters. They can easily control the remote computer on the local computer and deal with the system trouble.

3.2.1   Software maintenance workflows

When the system need to maintenance, province, city and county maintenance personnel apply to their superiors first. The YRCC's Information Center maintenance personnel determine the change request. Specific software maintenance workflow process is shown in fig. 3.

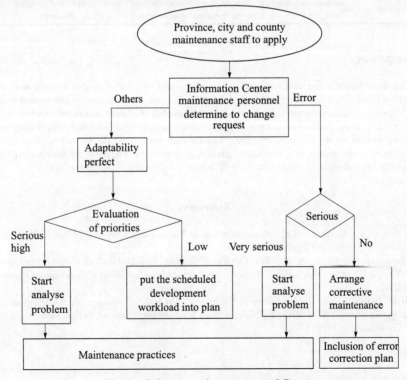

Fig. 3   Software maintenance workflows

3. 2. 2 Software maintenance content

Maintenance of software system needs daily operate monitoring, monitoring function modification and improvement, software upgrades, system maintenance of antivirus on monitor machine and server (see Tab. 4).

Tab. 4 The contents of Monitoring System software maintenance

Server/monitoring machine	Maintain content
GR server	The daily operation of the server monitoring, monitor the operation, monitoring function modification and improvement, software upgrades, monitoring object of temporary deployment of adjustment and release
The total database server	The daily operation of the total database server, data dictionary maintenance, maintenance of stored procedures, triggers maintenance, operational maintenance, make changes to the original structure of the database for the changing demands
Cities database server	Data dictionary maintenance, maintenance of stored procedures, triggers maintenance, operational maintenance etc, make changes to the original structure of the database for the changing demands
Web server	The daily operation of the server monitoring, the optimal adjustment of the interface and the kernel, data storage and display etc
Domain server	The daily operation of the server monitoring, the definition of domain, management of the member of a domain etc
Sluice monitoring machine	Operating system updates, virus library upgrades, system anti – virus, monitoring system software maintenance, data backup, system recovery, work in the local company remote debugging, answer units on the monitoring system of all kinds of problems

4 Conclusions

Operation maintenance and management of remote and distributed control system of sluices in the Yellow River is a long – term and formidable job. It is necessary to ensure the normal operation of the system, to further improve the comprehensive benefits of sluices monitoring system by cultivating and improving the maintenance staff maintenance management level and establishing a sluice monitoring system operation and maintenance management system. To do the maintenance and management work of the remote and distributed control system of the Yellow River for sluices well will benefits the improvement of the similar sluices automatic operation management level.

References

Zheng Renjie, Yin Renkun, et al. Practical Software Engineering (Second Edition)[M]. Beijing: Tsinghua University Press, 2005.

Zhu Chenhua, Xie Ming, et al. The Design of Remote and Distributed Control System of the Yellow River for Sluices[R]. Information Center of Yellow River Conservancy Commission, 2007.

Xie Ming, Song Tao, et al. Rectification Technology Program of Remote and Distributed Control System of the Yellow River for Sluices[R]. Information Center of Yellow River Conservancy Commission, 2011.